THE GENERAL ASSEMBLY OF GALAXY HALOS:
STRUCTURE, ORIGIN AND EVOLUTION

IAU SYMPOSIUM 317

COVER ILLUSTRATION:

This is a montage of three images illustrating hot topics of our Symposium, clockwise from the bottom:

1) the Milky Way galaxy arching above the platform of ESO's Very Large Telescope (VLT) on Cerro Paranal, Chile. The extent of our galaxy's cloudy and dusty structure can be seen in remarkable detail as a dim glowing band across the observation deck. Our galaxy is surrounded by several smaller satellite galaxies. Prominent here, to the left, are the Small Magellanic Cloud and Large Magellanic Cloud dwarf galaxies which are members of our Local Group of galaxies. Credit: John Colosimo (colosimophotography.com)/ESO.

2) a contrasting pair of galaxies: NGC 1316, and its smaller companion NGC 1317 (right). Although NGC 1317 seems to have had a peaceful existence, its larger neighbour bears the scars of earlier mergers with other galaxies. Credit: ESO.

3) the surface brightness of a simulated halo (standard halo number 8 in Johnston et al. 2008, ApJ, 689, 936), as viewed from an external perspective. The panel is 300 kpc on a side. Adapted from Fig. 14 in Johnston et al. (2008).

IAU SYMPOSIUM PROCEEDINGS SERIES

INTERNATIONAL ASTRONOMICAL UNION

UNION ASTRONOMIQUE INTERNATIONALE

THE GENERAL ASSEMBLY OF GALAXY HALOS: STRUCTURE, ORIGIN AND EVOLUTION

PROCEEDINGS OF THE 317th SYMPOSIUM OF THE INTERNATIONAL ASTRONOMICAL UNION HELD IN HONOLULU, USA AUGUST 3–7, 2015

Edited by

ANGELA BRAGAGLIA

INAF Osservatorio Astronomico di Bologna, Italy

MAGDA ARNABOLDI

European Southern Observatory, Germany and INAF Osservatorio Astronomico di Torino, Italy

MARINA REJKUBA

European Southern Observatory, Germany

and

DONATELLA ROMANO

INAF Osservatorio Astronomico di Bologna, Italy

Shaftesbury Road, Cambridge CB2 8EA, United Kingdom

One Liberty Plaza, 20th Floor, New York, NY 10006, USA

477 Williamstown Road, Port Melbourne, VIC 3207, Australia

314–321, 3rd Floor, Plot 3, Splendor Forum, Jasola District Centre, New Delhi – 110025, India

103 Penang Road, #05–06/07, Visioncrest Commercial, Singapore 238467

Cambridge University Press is part of Cambridge University Press & Assessment, a department of the University of Cambridge.

We share the University's mission to contribute to society through the pursuit of education, learning and research at the highest international levels of excellence.

www.cambridge.org
Information on this title: www.cambridge.org/9781107138193

First published 2016

A catalogue record for this publication is available from the British Library

ISBN 978-1-107-13819-3 Hardback

Table of Contents

Preface

The IAU Symposium 317 titled "The General Assembly of Galaxy Halos: Structure, Origin and Evolution" was held during the IAU General Assembly XXIX in Honolulu, Hawaii, from August 3 to 7, 2015.

The IAU Symposium 317 aimed at bringing the studies of the Milky Way halo together with the perspective on external galaxies' stellar halos and their evolution. The motivation for understanding the physics of galaxy halos begins with the statement from Eggen, Lynden-Bell and Sandage (1962, ApJ 136, 748): "The time required for stars in the (Milky Way) halo to exchange their energies and angular momenta is very long compared with the age of the Galaxy. Hence, knowledge of their present energy and angular momenta tells us something of the initial conditions under which they formed." This statement, which is more than half a century old, illustrates the wealth of information contained in the dynamics and chemical composition of stars in the halos of galaxies and their implication for the models of galaxy formation. This quest starts right at our doorsteps with the study of the Milky Way galaxy. Because we live in it and its light dominates our skies, we can study its stars with state-of-the art instrumentation out to its farthest outskirts.

Measuring the physical parameters of the structural components of the MW thin and thick disk, bulge and halo with imaging and spectroscopic surveys from the ground tells us when these components formed and whether they were dominated by accretion or by dissipative collapse of the MW own gas. These surveys also tell us about the frequency and the generations of stars that emit the light that we see today in our sky, and about the stars progenitors that are responsible for their chemical content. And the future looks even brighter: astronomers are now preparing for a big revolution brought about by the Gaia satellite. They are assessing all the observational implications of the most accredited models for the formation of the Milky Way to be tested against the measurements provided by this space mission. The Gaia satellite will measure the parallaxes and proper motions for the MW stars with exquisite precision such that we shall obtain a tridimensional map distribution for nearly one billion stars around our Sun and a kinematical and chemical census of all Galactic components!

What about the MW halo then? Our Galaxy halo turned out to be a livelier environment than was previously believed. Chemical tagging of globular clusters permitted to discover distinct multiple stellar generations. The accreted stars found in the halo are associated with several different substructures like for example the disrupted Sagittarius dwarf. This substructure provides 20% of all the debris of the stellar halo in the Milky Way, including multiple extended stellar streams. The modelling of these streams constrains the Galaxy potential and its associated mass to $\sim 5-8\times10^{11}\,M_{\odot}$ within 200 kpc, in agreement with the kinematics of halo stars and satellites in the Local Group. Similar complex network of structures is found in the halo of the Andromeda galaxy (M31), the giant spiral galaxy closest to us, as well. Numerous dwarf galaxies and globular clusters, but also streams without clear progenitors contribute to this network as shown by the map of the red giant stars in the M31 halo from the PAndAS survey.

Are halos and streams found around disk galaxies only? No, galaxy halos are ubiquitous in luminous galaxies and we now know that they also extend out to hundred kiloparsecs, that they have complex morphologies with a maze-like web of tails, plumes and spurs, and that they harbor multiple stellar components, with different chemical content and ages. Vivid testimony of these intricate luminous substructures are shown by the very deep images, reaching surface brightness levels to 1% of the night sky. Also the two dimensional maps of discrete tracers like planetary nebulae and globular clusters allow to see substructures both in space and velocity. The recent beautiful example for the giant elliptical galaxy M87 shows the debris of a satellite disrupted in its halo forming a crown and provides evidence that the outer halo is still growing.

To understand how the galaxy assembly took place we need to combine observations with theoretical modelling. This is because we deal with a long sequence of events, where mass accretion, i.e. stars and gas brought in by smaller satellites, and dissipative collapse of the galaxies own gas play important roles, leading to the formation of different structural components that we see today in our own Milky Way and in external galaxies.

The IAU Symposium 317 provided a vibrant forum where experts discussed many different aspects of the global assembly, formation and evolution of galaxy halos and their constituents. Hundred and sixty-eight participants were officially registered, but many more attended the Symposium joining us from the parallel sessions and shared our enthusiasm for the research on stellar halos. By bringing together observers and theorists, we made progress toward a coherent picture of the formation and evolution of the halos in the MW, Andromeda and more distant galaxies.

The Scientific Organizing Committee (SOC) made a terrific job in putting together a stimulating scientific program, and supporting the participation of many young astronomers. The SOC members took an active role during the Symposium also, by chairing the sessions and fostering lively discussions with timely posed and insightful questions. We wish to thank all of them here: Wako Aoki, Kenneth Freeman, Doug Geisler, Ortwin E. Gerhard, Oleg Gnedin, Laura Greggio, Rodrigo Ibata, Alan McConnachie, Poul Erik Nissen, Eric Peng, Chis Sneden, Else Starkenburg and Enrico Vesperini.

The organisation for the Symposium was provided by the IAU personnel. Everything went smoothly and we wish to thank in particular Cathy Cox and Lisa Idem for their help.

The program listed 13 sessions that included 17 invited talks, 39 oral contributions, a summary talk and a plenary talk for the whole Assembly. There were many poster contributions and the scientific program had two poster sessions dedicated to 21 short contributions of 2 minutes each, where the scientific highlights of the poster contributions were presented. We wish to warmly thank all participants for their valuable contributions to the success of our Symposium, and in particular Raffaele Gratton for the summary talk and Ortwin Gerhard for the plenary talk.

During the conference, we received the sad news of the passing away of Dr Nigel Douglas, senior researcher at the Kapteyn Institute in Groningen, Netherlands. He was one of the builders of the Planetary Nebulae Spectrograph (PN.S) and the first Principal Investigator; his contribution to the study of stellar halos was acknowledged by the PN.S team members attending the Symposium and shared with the participants of the IAU Symposium 317.

We include in the following the scientific program of the Symposium, and are proud to share the proceedings of these exciting contributions with the entire IAU community.

Magda Arnaboldi, Angela Bragaglia, Marina Rejuba, Donatella Romano, Editors

Table 1. Program of IAUS 317

Speaker	Title	Type
1. GLOBAL PROPERTIES OF STELLAR HALOS FROM THE MILKY WAY TO EXTERNAL GALAXIES		
I - Monday, 3 August 2015 10:30-12:30 am		
SOC Chairs	Welcome address	
Kathryn Johnston	Origins of Stellar Halos	invited
Marina Rejkuba	Tracing the stellar halo of an early type galaxy out to 25 effective radii	contributed
Alan McConnachie	Stellar halos around Local Group galaxies	invited
Matthias Steinmetz	Investigating the earliest epochs of the Milky Way halo	contributed
Denija Crnojevic	Resolving the extended stellar haloes of nearby galaxies: the wide-field PISCeS survey	contributed
Chair: Else Starkenburg		
II - Monday, 3 August 02:00-03:30 pm		
Chris Mihos	Intragroup and Intracluster Light	invited
Justin Read	Stellar halos: a rosetta stone for galaxy formation and cosmology	invited
Paul Schechter	New axes for the stellar mass fundamental plane	contributed
Johan Knapen	Direct imaging of haloes and truncations in face-on nearby galaxies	contributed
Chair: Oleg Gnedin		
III - Tuesday, 4 August 08:30-10:30 am		
Ortwin Gerhard	The Milky Way, the Galactic halo, and the halos of galaxies	plenary
IV - Tuesday, 4 August 10:30-12:30 pm		
2. HALO STARS AND CHEMICAL EVOLUTION		
Anna Frebel	Chemical abundances of the most metal-poor stars in the Milky Way	invited
Kevin Schlaufman	The Most Ancient Stars in the Milky Way's Halo	invited
Wako Aoki	Very Low Mass Stars with Extremely Low Metallicity in the Milky Way's Halo	contributed
David Yong	Neutron-capture element and Sc abundances in low- and high-alpha Galactic halo stars	contributed
Haining Li	Searching for chemical relics of first stars with LAMOST and Subaru	contributed
Chiaki Kobayashi	Inhomogeneous chemical enrichment in the Galactic Halo	contributed
Chair: Chris Sneden		
V - Tuesday, 4 August 02:00-03:30 pm		
3. DISCRETE CONSTITUENTS OF STELLAR HALOS IN THE MILKY WAY AND IN EXTERNAL GALAXIES : PLANETARY NEBULAE and GLOBULAR CLUSTERS		
Terese Hansen	Explor ing the Early Universe with Extremely Metal-Poor Stars	contributed
Magda Arnaboldi	Planetary Nebulae and their parent stellar populations: tracing the mass assembly of the giant elliptical galaxy M87 and the intracluster light in the Virgo cluster core	contributed
Giuliana Fiorentino	RR Lyrae to build up the Galactic Halo	contributed
Warren Reid	PN populations in the Local Group and distant stellar populations	invited
Judith Cohen	Outward Bound with RR Lyrae Stars: Studies of the Outer Halo of the Milky Way	contributed
Chair: Angela Bragaglia		

VI - Wednesday, 5 August 10:30-12:30 pm		
Eugenio Carretta	Globular clusters and their contribution to the formation of the Galactic halo	invited
Giampaolo Piotto	Single & Multiple Stellar Populations in Globular Clusters: Chemical Tagging, Photometric Sequences, and Dynamics	invited
Corinne Charbonnel	Did globular clusters contribute to the stellar population of the Galactic halo?	contributed
Gary Da Costa	Are the globular clusters with significant internal [Fe/H] spreads all former dwarf galaxy nuclei?	contributed
Douglas Geisler	CaTaclysm in the SMC - star clusters vs. field stars	contributed
Pawel Pietrukowicz	RR Lyrae stars as probes of the Milky Way structure and formation	contributed
Chair: Magda Arnaboldi		
VII - Wednesday, 5 August 02:00-03:30 pm		
Soeren Larsen	Globular clusters in M31, LG and external galaxies	invited
Michael Hilker	Globular clusters as tracers of the halo assembly of nearby central cluster galaxies	contributed
Karoline Gilbert	Recent Results from the SPLASH Survey: Chemical Abundances and Kinematics of Andromeda's Stellar Halo	contributed
Andreas Kupper	Globular Cluster Streams as Galactic High-Precision Scales	contributed
Poster presentation		
Chair: Else Starkenburg		

VIII - Wednesday, 5 August 04:00-06:00 pm		
4. DWARF GALAXIES		
Josh Simon	Satellite systems and halos in the Local Group and beyond	invited
Giuseppina Battaglia	Stellar kinematics and dark matter in dwarf galaxies	invited
Sergey Koposov	Discovery of a large number of Ultra Faint satellites in the vicinity of the Magellanic Clouds	contributed
Eric Peng	Globular Clusters, Ultra-Compact Dwarfs, and the Formation of Galaxy Halos	contributed
Michelle Collins	The Andromeda dwarf galaxies as probes of cosmolgy and galaxy evolution	contributed
Kim Venn	Chemical abundances in metal-poor stars in Dwarf Galaxies	contributed
Chair: Wako Aoki		
IX - Thursday, 6 August 08:30-10:00 am		
Vanessa Hill	Dwarf galaxies around the Milky Way: linking ages, kinematics and chemistry	invited
Donatella Romano	Chemical enrichment in Ultra-Faint Dwarf galaxies	contributed
Antonino Milone	Multi-wavelength photometry of the M54+Sagittarius stellar system, of NGC1851, M22, M2, and other building blocks of the Galactic Halo	contributed
Else Starkenburg	Investigating the earliest epochs of the Milky Way halo	contributed
Poster presentation		
Chair: Doug Geisler		

X - Thursday, 6 August 10:30-12:30 pm
5. PROBING THE DYNAMICS OF GALAXIES WITH SMOOTH HALOS STELLAR SUBSTRUCTURES

Jenny Greene	Metallicity gradient in the halos of external galaxies	invited
Jean Brodie	Constraints from discrete tracers on the assembly histories of ETG halos	invited
Benjamin Cook	Stellar populations of stellar halos: Results from the Illustris Simulation	contributed
Nicola Napolitano	Mass and stellar orbit distribution of Early-Type galaxy haloes	contributed
Bruce Elmegreen	Gas accretion from halos to disks: observations, curiosities, and problems	contributed
Laura Greggio	Studying Stellar Halos with Future Facilities	contributed

Chair: Alan McConnachie

XI - Thursday, 6 August 02:00-03:30 pm

Annette Ferguson	Dissecting Galactic Accretion Events within the Local Group and Beyond	invited
Jorge Penarrubia	The formation of the smooth halo component	invited
Oleg Gnedin	Tracing the assembly of stellar halos with globular clusters	contributed
Antonela Monachesi	Resolving the stellar halos of six massive disk galaxies beyond the Local Group	contributed

Chair: Laura Greggio

XII - Thursday, 6 August 04:00-06:00 pm
6. ORIGIN OF STELLAR HALOS

Amina Helmi	The connection between galaxy formation and the assembly of stellar halos in the LG	invited
Gerhard Hensler	The early gaseous and stellar mass assembly of Milky Way-type galaxy haloes	contributed
Allyson Sheffield	Contributions to the Galactic Halo from In-Situ, Kicked-Out, and Accreted Stars	contributed
Roelof de Jong	GHOSTS: the age and structure of stellar halos around nearby disk galaxies	contributed
Michaela Hirschmann	The stellar accretion origin of stellar population gradients at large radii in massive, early-type galaxies	contributed
Carl Grillmair	Stellar Debris Streams in the Galactic Halo	contributed
Myung Gyoon Lee	Dual Stellar Halos in the Standard Elliptical Galaxy M105 and Formation of Massive Galaxies	contributed

Chair: Eric Peng

XIII - Friday, 7 August 08:30-10:30 pm
7. SUMMARY and DISCUSSION

Raffaele Gratton	Summary talk and Discussion

Chair: Ortwin Gerhard

The General Assembly of Galaxy Halos: Structure, Origin and Evolution
Proceedings IAU Symposium No. 317, 2015
A. Bragaglia, M. Arnaboldi, M. Rejkuba & D. Romano, eds.

doi:10.1017/S1743921315008753

Origins of Stellar Halos

Kathryn V. Johnston

Department of Astronomy, Columbia University
email: kvj@astro.columbia.edu

Abstract. This contribution reviews ideas about the origins of stellar halos. It includes discussion of the theoretical understanding of and observational evidence for stellar populations formed "in situ" (meaning formed in orbits close to their current ones), "kicked-out" (meaning formed in the inner galaxy in orbits unlike their current ones) and "accreted" (meaning formed in a dark matter halo other than the one they currently occupy). At this point there is general agreement that a significant fraction of any stellar halo population is likely "accreted". There is modest evidence for the presence of a "kicked-out" population around both the Milky Way and M31. Our theoretical understanding of and the observational evidence for an "in situ" population are less clear.

Keywords. galaxies: evolution — galaxies: formation — galaxies: halos — galaxies: structure

1. Introduction

Stellar halos contain only of order 1% of the *stars* in a typical disk galaxy, spread out over volumes $\sim 10^6$ larger than the bulk of the other components. While the stellar halo of the Milky Way has been studied for decades using local samples or tracers at larger distances (e.g. Eggen *et al.* (1962), Searle & Zinn (1978)), the low density meant that it was not possible for a long time to detect these components around other galaxies. Interest in these diffuse galactic components really gained momentum with the advent of large-area, all sky surveys (SDSS and 2MASS in particular) which enabled star-count studies to push their sensitivity to sufficiently low surface-brightness (below 30 mag/arcsec^2) that the global structure of and substructure within the Milky Way could be studied (e.g. Newberg *et al.* (2002), Majewski *et al.* (2003)). At the same time, dedicated surveys have mapped Andromeda's halo (e.g. Ibata *et al.* (2007)). These studies have revealed both halos to extend beyond 100kpc from their centers with rich substructure in the form of debris from past accretion events throughout (Newberg *et al.* (2002), Majewski *et al.* (2003), Belokurov *et al.* (2006), Ibata *et al.* (2007)).

It is easy to think of these diffuse components around galaxies as pretty (see surface brightness projections of simulations of accreted halos in left hand panels of Figure 1), but inconsequential to our broader interests of understanding structures in the Universe and how they form since they contain such a tiny fraction of the total mass. However, the existence of coherent debris-structures within halos has several interesting consequences. First, it provides dramatic confirmation of the widely accepted paradigm that galaxies form *hierarchically*. Second, the debris-structures themselves contain signatures (in their phase-space distributions and stellar populations) of the objects that have been destroyed and hence provide window into the nature of smaller galaxies in the past — in some cases to mass scales and epochs that cannot be directly observed at higher redshift. Third, the collective interpretation of all the debris-structures detectable within a halo can give some constraints on the accretion history of the parent galaxy. Fourth, the common origin of stars within an individual debris-structure can be exploited to provide very sensitive constraints on the distribution of mass in the dark matter halo within which

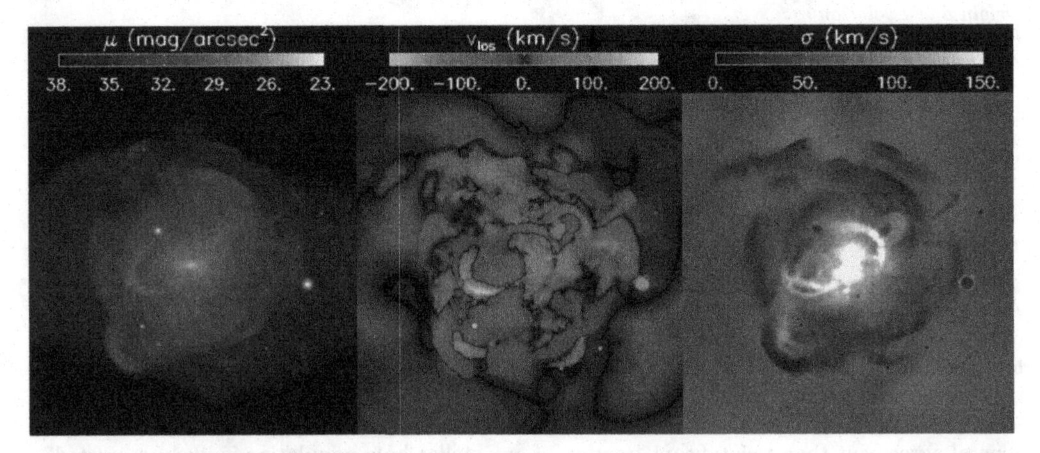

Figure 1. Surface brightness, line-of-sight velocity and line-of-sight velocity dispersion projections of a simulation of a purely accreted stellar halo taken from Bullock & Johnston (2005). Each box is 300kpc on a side. The central galaxy, containing ~99% of the light, is not shown. (Image credit: Sanjib Sharma)

it is orbiting — both the global dark matter distribution (Johnston *et al.* (1999)) and the dark matter subhalos that may also be present (Ibata *et al.* (2002), Johnston *et al.* (2002), Carlberg (2009)).

While the prospects in the previous paragraph are intriguing, the degree to which they can be applied to populations around a galaxy depends on what fraction of the stellar halo is truly accreted. The current observations suggest that accretion is responsible for the formation outskirts of the Milky Way's and Andromeda's stellar halos. However, it is unclear how dominant accretion is more generally in the formation of the inner parts of these halos and for other galaxies. This review will examine our understanding of the origins of stellar halos, concentrating on what is known from theoretical and observational perspectives about Milky Way like spiral galaxies.

2. Stellar Halo Formation Scenarios

For the purposes of this review, stellar halo formation scenarios will be split into three categories.

In situ populations are stars that were born in the stellar halo component and are still on the same (or at least, very similar) orbits that they were born on (as originally postulated by Eggen *et al.* (1962)).

Kicked-out populations are stars that were born in the parent dark matter halo, but within a different stellar component (i.e. the central disk or bulge) than they are now currently: their orbits have evolved significantly (Purcell *et al.* (2010)).

Accreted populations are stars that were born in dark matter halos other than that of the parent galaxy which is now their host (as first put forward by Searle & Zinn (1978)).

Note the above scenarios were chosen in particular because they broadly separate what are likely to be different conditions under which star formation itself occurred, for example between different depths of dark matter potential and, as a consequence, influences on stellar enrichment (importance of feedback and/or accretion of gas). Hence the resulting populations may be currently distinguishable either by their orbits, their abundances or a combination of the two. However, a variety of categories have been used recently in the literature, so the reader is cautioned to be careful in comparing papers. For example: some authors broadly refer to both *in situ* and *kicked-out* stars as *in situ* since both populations are born in the same parent potential (e.g. Zolotov *et al.* (2009)); others have separated *accreted* populations into stars formed in a destroyed dwarf prior to and after its accretion by the parent galaxy (the "endo-debris" stars described in Tissera *et al.* (2013)), which is not a distinction made in this review; and still others have simply referred to stars born in satellites as *ex-situ* rather than accreted (Pillepich *et al.* (2015)).

There is little theoretical guidance as to which of the above formation scenarios is expected to be dominant. Many models in the literature simply explore the observable consequences for a stellar halo that is *assumed* to have formed entirely by accretion using collisionless N-body simulations (Bullock & Johnston (2005), De Lucia & Helmi (2008), Cooper *et al.* (2010)). The advantage of these models is that they are computationally inexpensive and physically simple so it is possible to design them to resolve even small infalling dwarf galaxies. However, no stars in these models are formed in a self-consistent manner, and the contributions of the stellar components of the parent galaxy to the potential are often not considered.

More rigorous explorations of the full range of formation scenarios requires the use of hydrodynamical simulations, which are much more computationally expensive and are only now reaching resolutions necessary to look at the global structure and, in particular, substructures within stellar halos. From the first such studies that attempted to look at these components (Abadi *et al.* (2006)), most have agreed that the simulated stellar halos were formed through multiple mechanisms, in particular both via *kicking-out* populations from the inner galaxy and via *accretion* (Zolotov *et al.* (2009), McCarthy *et al.* (2012), Tissera *et al.* (2012), Pillepich *et al.* (2015)). The simulated halos are dominated by accretion in their outskirts (beyond 20-30kpc) but may have significant (or even dominant) *kicked-out* disk populations within that radius. There is disagreement among the models of whether a true *in situ* population (as defined in this proceedings) is produced, with only a couple of studies finding any evidence for their presence (Samland & Gerhard (2003), Cooper *et al.* (2015)). This disagreement highlights the fundamental limit of these models as they all rely on (differing) prescriptions for star formation and feedback below their respective resolutions. Hence, these results should be considered indicative rather than predictive. They nevertheless provide powerful data sets to understand the distinctive signatures that the different formation mechanisms leave in the simulated stellar distributions, and to suggest how the true contribution of stars formed *in situ*, *kicked-out* or *accreted* might be actually assessed from observations (e.g. separating formation mechanisms, as suggested by Zolotov *et al.* (2010), in the chemical abundance planes).

3. Indicators of Origins

In any discussion of observational signatures of formation mechanism it is important to be clear on the nature of the samples being used, especially when comparing conclusions across different studies.

First, global statements cannot be made without global samples. For example, Solar Neighborhood studies (i.e. within $\sim$ 1 kpc) may not be representative even of the populations in the inner halo due to intrinsic variations around the Galaxy. Moreover, the inner halo (within $\sim$ 10-30 kpc) may have a different dominant formation mechanism than the outer halo.

Second, different stellar populations are expected to be sensitive to different formation mechanisms. For example, studies of M giants (the reddest, most metal-rich giant stars) have found the outer stellar halo to be dominated by a single large accretion event (i.e. the Sagittarius stream, see Majewski *et al.* (2003)) because only the most massive dwarf galaxies are metal rich enough to host this population. The bulk of the stellar halo is actually more metal poor than this, so the contribution of large as compared to small accreted dwarfs should really be assessed in other populations as well before a definitive statement can be made. Indeed, stellar population variations for different substructures have been detected in both the Milky Way (Bell *et al.* (2010)) and Andromeda (Ibata *et al.* (2014)).

Finally, the ability to distinguish the different formation mechanisms in observations depends on the quality, quantity and *dimensions* of information available in the sample.

The importance of these considerations for stellar halo studies is amply illustrated by considering the variety of samples in a few recent studies. Carollo *et al.* (2007) studied the kinematics and metallicities of SDSS calibration stars, probing a wide range of stellar metallicities and ages, and a volume up to 20 kpc from the Sun. Deason *et al.* (2011) analyzed brighter, blue horizontal branch stars in SDSS which could probe the halo to 40 kpc, but with a sample biased towards metal poor populations. Nissen & Schuster (2010) (and subsequently Nissen & Schuster (2011), Schuster *et al.* (2012), Nissen & Schuster (2012)) used a sample of 94, relatively metal rich ($-1.6 <$ [Fe/H] < -0.4) stars and studied their abundance patterns and full orbital elements in great detail, but their sample was limited to within $\sim$335 pc of the Sun. Sheffield *et al.* (2012) took high-resolution spectra of M-giant stars selected from the Two Micron All Sky Survey which had significant velocities relative to the disk (hence likely to members of the stellar halo), a tracer only of the highest metallicity halo population and possibly biased towards populations that may have been recently *kicked-out* from the Galactic disk.

The following subsections briefly outline a few examples of the observational evidence for various formation scenarios, broadly divided by the types and numbers of observed dimensions available, working from the easiest to observe (requiring, at a minimum, projected spatial distributions on the sky) to the most difficult (requiring proper motions and/or high-resolution spectroscopy).

3.1. *Spatial Distributions*

As noted in Section 1 (and apparent in the left-hand panel of Figure 1), abundant spatial substructure in the outskirts (beyond $\sim$ 10-30kpc) of both the Milky Way and Andromeda galaxies indicate the importance of *accretion* in forming these galaxies, even suggesting it is dominant in these regions. Several groups are pushing for analogous star count studies around other nearby galaxies using the Hubble Space Telescope (see discussions in these proceedings and Monachesi *et al.* (2013), Monachesi *et al.* (2015)), but are necessarily limited in their scope by the finite field-of-view.

Another approach could be to look at the global structure of these halos, averaging over (projected) azimuth to produce radial surface density profiles. In the case of the Milky Way, a break in the power law that best-represents the density profile has been reported (Deason *et al.* (2014)), while the stellar halo of M31 seems to be well-represented by a single power law (Gilbert *et al.* (2012), Ibata *et al.* (2014)). It is tempting to attribute breaks as a transition in the physical formation mechanism, e.g. from dominance by a *kicked-out* to an *accreted* population. However, the robustness of this interpretation is not clear. Deason *et al.* (2013) showed that several of the 11 stellar halo models of Bullock & Johnston (2005) — all of which were built by simply superposing accretion events — showed breaks in their radial density profiles.

3.2. *Velocity and Metallicity Distributions*

Low- or medium- resolution spectroscopy can add the additional dimensions of line-of-sight velocity and overall metallicity to maps of stellar populations. The right hand

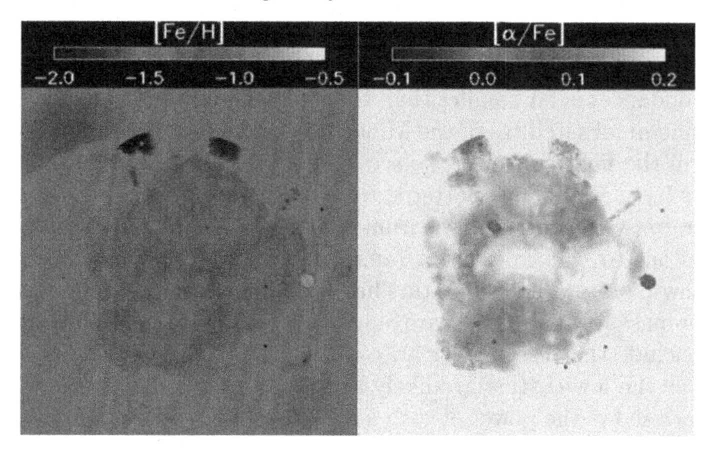

Figure 2. Images of the same simulation as in Figure 1, but this time showing the mean [Fe/H] and [α/Fe] along each line-of-sight (Johnston *et al.* (2008)). The highest surface brightness substructures stand out as higher metallicity, and the substructures in general have lower [α/Fe]. (Image credit: Sanjib Sharma)

panels of Figure 1 show some expectations for a purely accreted halo. For example, around the Milky Way, Schlaufman *et al.* (2009) used SEGUE spectroscopy of a sample of metal-poor main-sequence turnoff stars to find substructures in velocity and use the frequency of their occurrence as an indicator of the importance of accretion in forming the stellar halo. In subsequent work they found these substructures to be chemically distinct from the kinematically-smooth halo component (Schlaufman *et al.* (2011)). They also showed that an increase in spatial autocorrelation of [Fe/H] for stars in their smooth component with increasing distance from the Galactic center (Schlaufman *et al.* (2012)). These observations were consistent with models of purely accreted stellar halos beyond 15 kpc, but not within.

In an analogous spectroscopic survey of giant stars in selected fields around M31 (the SPLASH survey), Gilbert *et al.* (2009) found that stars in kinematically cold structures tended to be more metal rich than those assigned to the kinematically hot components. They also showed that higher surface brightness features are generally more metal rich than lower surface brightness features — a trend that is seen in simulations and can be attributed to more massive (and hence more metal rich) accretion events leading to the most dominant substructures (see illustration in Figure 2).

Dorman *et al.* (2013) extracted HST photometry of M31's disk from the PHAT survey and spectroscopy of 5000 giants from the SPLASH survey and used the combined information on the luminosity function and velocity distribution of stars to show an excess of stars with disk-like luminosity functions moving at high speeds relative to the disk. They concluded this was evidence for a kicked-out population in M31's stellar halo.

3.3. *Full Orbit and Abundance Distributions*

The Milky Way is the one place where the dimensions to be studied can be further augmented with detailed abundances from high-resolution spectroscopy and/or astrometric measurements of proper motions and distances to allow the calculation of orbital elements. Indeed, the original Eggen *et al.* (1962) work that started this field looked at the orbits of stars of different metallicities.

A recent series of papers (Nissen & Schuster (2010), Nissen & Schuster (2011), Schuster *et al.* (2012), Nissen & Schuster (2012)) has used a sample of 94, metal rich ($-1.6 <$ [Fe/H] < -0.4) dwarf stars limited to within $\sim$335 pc of the Sun to investigate halo formation scenarios. The authors found that the quality of their data allowed them to see distinct high- and low-α sequences (separated by less than 0.2 dex in [α/Fe]). (Note that this difference in abundances is far smaller than the almost 1 dex spread in α elements that is seen in dwarf spheroidal satellites of the Milky Way and clearly distinguishes the satellite populations form the field halo stars — see summary in Venn *et al.* (2004).) They find the high-α stars have abundance patterns that suggest an origin in regions with a high star formation rate (with contributions from only SN II and massive stars), are on orbits with apocenters out to $\sim$16 kpc and a range of eccentricities, and have ages 2-3 Gyrs older than the low-α stars. The low-α stars have abundance patterns betraying additional pollution by low-mass AGB stars and are on more eccentric orbits with larger apocenters. The authors conclude that the high-α arise from either *in situ* or *kicked-out* primordial populations while the low-α stars are likely accreted.

The above work shows the power of such a high-dimensional data set. Subsequent work with APOGEE measurements of stellar abundances has allowed these dual populations to be traced over a much larger volume to study the disk/halo interface across a significant fraction of the Galaxy (Hawkins *et al.* (2015), Hayden *et al.* (2015)).

4. Aside: a Case-Study of the Triangulum-Andromeda Stellar Clouds

This section discusses just one group of stellar substructures in our own Milky Way halo as an illustration of how additional dimensions of information can introduce revised interpretations on the contributions of the different possible formation mechanisms. The Triangulum-Andromeda clouds are examples of stellar over-densities in the halo of the Milky Way thought to have formed through the disruption of a satellite and hence contributing to the *accreted* component of the stellar halo. A single over-density was originally identified in M giant stars selected from the Two Micron All Sky Survey (2MASS) covering the region $100° < l < 160°$ and $-35° < b < -15°$ at approximately 20 kpc from the Sun (Rocha-Pinto *et al.* (2004)). Spectroscopic follow-up of these M giants showed them to have a velocity distribution with a small dispersion ($\approx$25 km s^{-1}) and a shallow gradient with mean velocities increasing with decreasing l (in the Galactic-Standard-of-Rest - GSR - frame). A *double* sequence of main-sequence stars in the foreground of the PAndAS survey was later found in the same region (called TriAnd1 and TriAnd2 by Martin *et al.* (2007)). Sheffield *et al.* (2014) found M giant counterparts to both of these sequences in the 2MASS data, at distance of $\approx$18 kpc and $\approx$28 kpc respectively (i.e. Galactocentric $(R, Z) \approx (24, -8)$ kpc and $(33, -12)$ kpc for the central pointing at $(l, b) = (125°, -25°)$. M giants in both TriAnd1 and TriAnd2 followed the same velocity distribution found in the original study. Sheffield *et al.* (2014) assumed that the large Galactocentric radius and distance below the mid-plane combined with the small dispersion was suggestive of neither a disk nor random halo population and used simulations of satellite disruption to show that both TriAnd1 and TriAnd2 could plausibly be due to single accretion event of a satellite on a retrograde orbit.

In subsequent work, Price-Whelan *et al.* (2015) took spectra of RR Lyrae stars in the same region as the TriAnd clouds with the aim of using those stars that had line-of-sight velocities co-incident with the clouds to provide better indicators of their distances. This experiment failed because no RR Lyrae could be found with co-incident velocities. More quantitatively, the number ratio of RR Lyrae to M giant stars associated with the TriAnd clouds was estimated to be $f_{\rm RR:MG} < 0.38$ (at 95% confidence). This ratio is quite unlike any of the known satellites of the Milky Way ($f_{\rm RR:MG} \approx 0.5$ for the very largest and $f_{\rm RR:MG} >> 1$ for the smaller satellites) and more like the population of stars born in the much deeper potential well inhabited by the Galactic disk ($f_{\rm RR:MG} < 0.01$).

These results provide dramatic support for the recent proposal by Xu *et al.* (2015) that, rather than stars *accreted* from other galaxies, the TriAnd clouds could represent stars *kicked-out* from our own disk. N-body simulations of a Milky-Way-like galaxy perturbed by the impact of a dwarf galaxy demonstrate that, in the right circumstances, concentric rings propagating outwards from that Galactic disk might possibly produce similar over-densities (Purcell *et al.* (2011)). If this interpretation is correct, the TriAnd clouds would be the first populations of disk stars to be found in our Galactic halo and a clear signature of the importance of this formation mechanism for stellar halos more generally.

5. Summary of Current Status and Future Prospects

This review has outlined three plausible formation scenarios of stellar halo populations (*in situ* , *kicked out* or *accreted*). While hierarchical structure formation suggests that *accretion* must play a role in the formation of all galaxies, theoretical studies do not yet provide clear predictions for the importance of the other populations: there is still debate about whether a pure *in situ* population, formed in the halo itself and still orbiting there, should exist at all. Observationally, abundant substructure around the Milky Way and Andromeda galaxies confirm the dominance of accretion in forming their outskirts. Within 10's of kpc of the centers of these galaxies the situation is less clear, as orbital times in these regions are much smaller and spatial substructure from accretion has time to phase-mix away. First studies adding velocity and abundance dimensions in these regions have pointed to possible *kicked-out* populations (Dorman *et al.* (2013), Price-Whelan *et al.* (2015)) as well as significant *accreted* populations (Nissen & Schuster (2011)). The clear separation of the local stellar halo into "high-α" and "low-α" sequences at low-metallicity in abundance-space suggests there might also be a true *in situ* component (Nissen & Schuster (2012), Hawkins *et al.* (2015)).

Future data sets promise breakthroughs in this field. Currently, the *GALAH* survey is collecting high-resolution spectra for millions of Milky Way stars with the aim of "chemically tagging" both components (as suggested by Freeman & Bland-Hawthorn (2002)). In the near future, data from ESA's *Gaia* satellite (Perryman (2002)) should allow the calculation of orbital elements for the nearby stellar halo, so substructures from accretion not apparent in space-alone can be detected as associations in orbit-space. Coincidentally, the samples of external galaxies surveyed on large scales to very low surface brightness will become much larger. For example, the wide field-of-view and spatial resolution of the proposed WFIRST satellite will allow it to look at 50-100 galaxies within 10Mpc and create maps of coverage and depth analogous to those available for Andromeda today. On a slightly longer timescale, LSST promises to create catalogues of *millions* of galaxies surveyed to surface brightness of $\sim$ 29 mag/arcsec^2 (Ivezic *et al.* (2008)) — with 1-10% of these expected to host a detectable remnant of an accretion event (Johnston *et al.* (2008)). We will then be able to see if our results for the Milky Way and Andromeda (an outer stellar halo dominated by accretion) are more generally true.

References

Abadi, M. G., Navarro, J. F., & Steinmetz, M. 2006, *MNRAS*, 365, 747

Bell, E. F., Xue, X. X., Rix, H.-W., Ruhland, C., & Hogg, D. W. 2010, *AJ*, 140, 1850

Belokurov, V., Zucker, D. B., Evans, N. W., *et al.* 2006, *ApJ Lett*, 642, L137

Bullock, J. S. & Johnston, K. V. 2005, *ApJ*, 635, 931

Carlberg, R. G. 2009, *ApJ Lett*, 705, L223

Carollo, D., Beers, T. C., Lee, Y. S., *et al.* 2007, *Nature*, 450, 1020
Cooper, A. P., Cole, S., Frenk, C. S., *et al.* 2010, *MNRAS*, 406, 744
Cooper, A. P., Parry, O. H., Lowing, B., Cole, S., & Frenk, C. 2015, arXiv:1501.04630
De Lucia, G. & Helmi, A. 2008, *MNRAS*, 391, 14
Deason, A. J., Belokurov, V., & Evans, N. W. 2011, *MNRAS*, 416, 2903
Deason, A. J., Belokurov, V., Evans, N. W., & Johnston, K. V. 2013, *ApJ*, 763, 113
Deason, A. J., Belokurov, V., Koposov, S. E., & Rockosi, C. M. 2014, *ApJ*, 787, 30
Dorman, C. E., Widrow, L. M., Guhathakurta, P., *et al.* 2013, *ApJ*, 779, 103
Eggen, O. J., Lynden-Bell, D., & Sandage, A. R. 1962, *ApJ*, 136, 748
Freeman, K. & Bland-Hawthorn, J. 2002, *ARAA*, 40, 487
Gilbert, K. M., Font, A. S., Johnston, K. V., & Guhathakurta, P. 2009, *ApJ*, 701, 776
Gilbert, K. M., Guhathakurta, P., Beaton, R. L., *et al.* 2012, *ApJ*, 760, 76
Hawkins, K., Jofré, P., Masseron, T., & Gilmore, G. 2015, *MNRAS*, 453, 758
Hayden, M. R., Bovy, J., Holtzman, J. A., *et al.* 2015, *ApJ*, 808, 132
Ibata, R., Martin, N. F., Irwin, M., *et al.* 2007, *ApJ*, 671, 1591
Ibata, R. A., Lewis, G. F., Irwin, M. J., & Quinn, T. 2002, *MNRAS*, 332, 915
Ibata, R. A., Lewis, G. F., McConnachie, A. W., *et al.* 2014, *ApJ*, 780, 128
Ivezic, Z., Axelrod, T., Brandt, W. N., *et al.* 2008, Serbian Astronomical Journal, 176, 1
Johnston, K. V., Bullock, J. S., Sharma, S., *et al.* 2008, *ApJ*, 689, 936
Johnston, K. V., Spergel, D. N., & Haydn, C. 2002, *ApJ*, 570, 656
Johnston, K. V., Zhao, H., Spergel, D. N., & Hernquist, L. 1999, *ApJ Lett*, 512, L109
Majewski, S. R., Skrutskie, M. F., Weinberg, M. D., & Ostheimer, J. C. 2003, *ApJ*, 599, 1082
Martin, N. F., Ibata, R. A., & Irwin, M. 2007, *ApJ Lett*, 668, L123
McCarthy, I. G., Font, A. S., Crain, R. A., *et al.* 2012, *MNRAS*, 420, 2245
Monachesi, A., Bell, E. F., Radburn-Smith, D., *et al.* 2015, arXiv:1507.06657
Monachesi, A., Bell, E. F., Radburn-Smith, D. J., *et al.* 2013, *ApJ*, 766, 106
Newberg, H. J., Yanny, B., Rockosi, C., *et al.* 2002, *ApJ*, 569, 245
Nissen, P. E. & Schuster, W. J. 2012, *A&Ap*, 543, A28
Nissen, P. E. & Schuster, W. J. 2011, *A&Ap*, 530, A15
Nissen, P. E. & Schuster, W. J. 2010, *A&Ap*, 511, L10
Perryman, M. A. C. 2002, *ApSS*, 280, 1
Pillepich, A., Madau, P., & Mayer, L. 2015, *ApJ*, 799, 184
Price-Whelan, A. M., Johnston, K. V., Sheffield, A. A., Laporte, C. F. P., & Sesar, B. 2015, *MNRAS*, 452, 676
Purcell, C. W., Bullock, J. S., & Kazantzidis, S. 2010, *MNRAS*, 404, 1711
Purcell, C. W., Bullock, J. S., Tollerud, E. J., Rocha, M., & Chakrabarti, S. 2011, *Nature*, 477, 301
Rocha-Pinto, H. J., Majewski, S. R., Skrutskie, M. F., Crane, J. D., & Patterson, R. J. 2004, *ApJ*, 615, 732
Samland, M. & Gerhard, O. E. 2003, *A&Ap*, 399, 961
Schlaufman, K. C., Rockosi, C. M., Allende Prieto, C., *et al.* 2009, *ApJ*, 703, 2177
Schlaufman, K. C., Rockosi, C. M., Lee, Y. S., Beers, T. C., & Allende Prieto, C. 2011, *ApJ*, 734, 49
Schlaufman, K. C., Rockosi, C. M., Lee, Y. S., *et al.* 2012, *ApJ*, 749, 77
Schuster, W. J., Moreno, E., Nissen, P. E., & Pichardo, B. 2012, *A&Ap*, 538, A21
Searle, L. & Zinn, R. 1978, *ApJ*, 225, 357
Sheffield, A. A., Johnston, K. V., Majewski, S. R., *et al.* 2014, *ApJ*, 793, 62
Sheffield, A. A., Majewski, S. R., Johnston, K. V., *et al.* 2012, *ApJ*, 761, 161
Tissera, P. B., Scannapieco, C., Beers, T. C., & Carollo, D. 2013, *MNRAS*, 432, 3391
Tissera, P. B., White, S. D. M., & Scannapieco, C. 2012, *MNRAS*, 420, 255
Venn, K. A., Irwin, M., Shetrone, M. D., *et al.* 2004, *AJ*, 128, 1177
Xu, Y., Newberg, H. J., Carlin, J. L., *et al.* 2015, *ApJ*, 801, 105
Zolotov, A., Willman, B., Brooks, A. M., *et al.* 2009, *ApJ*, 702, 1058
Zolotov, A., Willman, B., Brooks, A. M., *et al.* 2010, *ApJ*, 721, 738

The General Assembly of Galaxy Halos: Structure, Origin and Evolution
Proceedings IAU Symposium No. 317, 2015
A. Bragaglia, M. Arnaboldi, M. Rejkuba & D. Romano, eds.

doi:10.1017/S1743921315007309

Tracing the stellar halo of an early type galaxy out to 25 effective radii

Marina Rejkuba[1]

[1]European Southern Observatory,
Karl-Schwarzschild-Strasse 2, 85748 Garching bei München, Germany
email: mrejkuba@eso.org

Abstract. We have used ACS and WFC3 cameras on board HST to resolve stars in the halo of NGC 5128 out to 140 kpc (25 effective radii, $R_{\rm eff}$) along the major axis and 70 kpc (13 $R_{\rm eff}$) along the minor axis. This dataset provides an unprecedented radial coverage of stellar halo properties in any galaxy. Color-magnitude diagrams clearly reveal the presence of the red giant branch stars belonging to the halo of NGC 5128 even in the most distant fields. The V-I colors of the red giants enable us to measure the metallicity distribution in each field and so map the metallicity gradient over the sampled area. The stellar metallicity follows a shallow gradient and even out at 140 kpc (25 $R_{\rm eff}$) its median value does not go below [M/H]$\sim$-1 dex. We observe significant field-to-field metallicity and stellar density variations. The star counts are higher along the major axis when compared to minor axis field located 90 kpc from the galaxy centre, indicating flattening in the outer halo. These observational results provide new important constraints for the assembly history of the halo and the formation of this gE galaxy.

Keywords. galaxies: elliptical and lenticular; galaxies: halos; galaxies: individual (NGC 5128)

1. Introduction

Extended galactic components contain important clues for the formation and evolution of large galaxies. While the halos contain relatively small fraction of stellar mass, the properties of stars in the remote halo regions can be used to trace the star formation and assembly history of galaxies. This is a rapidly changing field of research that has benefitted from the new wide field cameras on large ground-based telescopes and from the HST. In this contribution I present the results of our study of the extended stellar halo of NGC 5128 (Centaurus A).

Table 1 lists all early type galaxies with resolved stellar population studies from literature. The closest and easiest to observe among them is NGC 5128, located at a distance of only 3.8 Mpc (Harris *et al.* 2010). Thanks to its vicinity this galaxy became a Rosetta stone for early-type galaxy halo studies (Harris *et al.* 1999; Harris & Harris 2000, 2002; Ferrarese *et al.* 2007; Rejkuba *et al.* 2003, 2005, 2011, 2014; Crnojević *et al.* 2014; Bird *et al.* 2015). The global galaxy characteristics are representative of other galaxies of its class with the optical luminosity $M_V = -21.5$, mass 1.2×10^{12} $M_\odot$ (Peng *et al.* 2004), and the presence of an AGN and supermassive black hole in its center (Neumayer 2010). The fact that this galaxy presents clear evidence of a recent merger is consistent with the fact that 73% of nearby luminous ellipticals show tidal disturbance (Tal *et al.* 2009).

2. NGC 5128 stellar halo studies

In a series of HST studies, starting with the WFPC2 camera, following with the ACS and most recently with both ACS and WFC3 cameras in parallel, stellar population content in 9 different locations in the halo of NGC 5128 (Fig. 1) was investigated.

Table 1. Early type galaxies with stellar halo properties studied through resolved stellar population observations.

Name	Type	M_V (mag)	$(m-M)_0$ (mag)	Distance (Mpc)	Environment	Literature References
Maffei 1	E	−21.6	27.7	3.4	Maffei/IC342 group ($A_V \sim 5.1$)	1, 2, 3
NGC 5128 = Cen A	E/S0 pec; Sy2	−21.5	27.91	3.8	Centaurus A group	4 – 14
NGC 3115	S0	−21.1	30.05	10.2	NGC 3115 group	15, 16
NGC 3379 = M105	E1	−20.9	30.06	10.2	Leo I group	17 – 19
NGC 3377	E5	−20.0	30.17	10.8	Leo I group	20
NGC 4486 = M87	E0 pec; Sy; cD	−22.5	31.08	16.4	Virgo cluster core	21

Literature References: (1) Davidge & van den Bergh (2001), (2) Davidge (2002), (3) Wu *et al.* (2014), (4) Soria *et al.* (1996), (5) Harris *et al.* (1999), (6) Harris & Harris (2000), (7) Harris & Harris (2002), (8) Ferrarese *et al.* (2007), (9) Rejkuba *et al.* (2003), (10) Rejkuba *et al.* (2005), (11) Rejkuba *et al.* (2011), (12) Rejkuba *et al.* (2014), (13) Crnojević *et al.* (2013), (14) Bird *et al.* (2015), (15) Elson (1997), (16) Peacock *et al.* (2015), (17) Sakai *et al.* (1997), (18) Gregg *et al.* (2004), (19) Harris *et al.* (2007b), (20) Harris *et al.* (2007a), (21) Bird *et al.* (2010)

The three WFPC2 fields, F1, F2 and F3, probed the inner halo at projected distances of 8, 21 and 31 kpc (equivalent to 1.5, 4 and 5.5 $R_{\rm eff}$) and established that the halo stars have magnitudes and colours consistent with the red giant branch (RGB) stars spanning a wide range of metallicities. Adopting an empirical calibration of $(V-I)$ vs. [M/H] (colour vs. metallicity) based on $(V-I)$ RGB fiducial lines for Galactic globular clusters with spectroscopically measured metallicities, the first metallicity distribution function (MDF) for stars in a halo of an early type galaxy was determined. It showed an asymmetric distribution that peaks around −0.5 dex and has a long tail towards the metal-poor end (Harris *et al.* 1999, Harris & Harris 2000, 2002).

The ACS data in F4, at projected distance of 38 kpc (7 $R_{\rm eff}$) are the deepest observations of an early type galaxy, reaching the core-helium burning red clump evolutionary phase (Rejkuba *et al.* 2005). A comparison of the observed vs. simulated colour-magnitude diagrams (CMDs) and luminosity functions showed that 70-80% of the halo stars formed 12±1 Gyr ago already spanning the entire range of model metallicities used in simulations (Z = 0.0001–0.04). The 20-30% of the younger stars were best fitted with the models having higher minimum metallicity ($\sim 1/10 - 1/4 Z_{\odot}$) and ages between 2-4 Gyr. In other words, the bulk of the halo stars in NGC 5128 formed at redshift $z \gtrsim 2$ and the chemical enrichment was very fast, reaching solar or even twice-solar metallicity already 11-12 Gyr ago, while the minor younger component, contributing $\sim$ 10% of the halo mass formed 2-4 Gyr ago (Rejkuba *et al.* 2011).

The aim of the latest study, which used ACS and WFC3 cameras in parallel, was to address the question how far the halo extends and to measure the metallicity gradients along the minor and major axis of the outer galaxy halo (Rejkuba *et al.* 2014). Two minor axis locations, F8 at 40 kpc and F9 at 90 kpc were observed, while 3 fields were placed along the major axis: F5 at 60 kpc projected distance, F6 at 90 kpc, and F7 at 140 kpc. The expectation was that the remote location, 140 kpc away (25 $R_{\rm eff}$), of F7 might be used as background/foreground control field if no significant RGB population belonging to NGC 5128 halo is identified, while the F5 location at 11 $R_{\rm eff}$ was selected close to expected transition distance between the inner and outer halo, based on observed halo

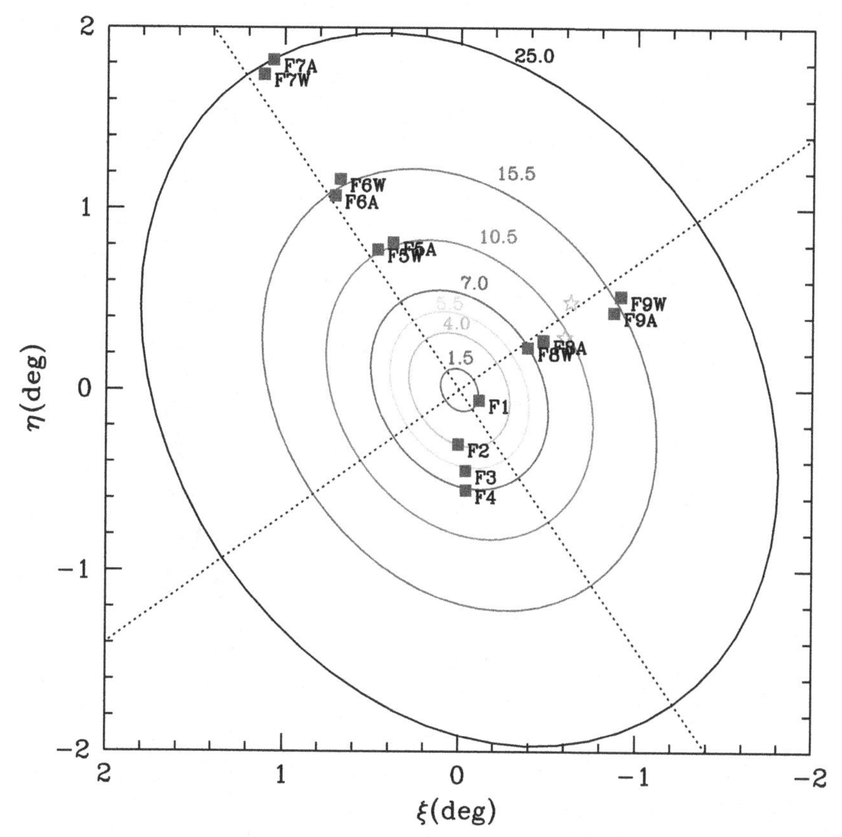

Figure 1. Distribution of HST studies in the halo of NGC 5128 relative to the galaxy centre. The elliptical contours with an axis ratio of 0.77 and PA=35°, measured from the galaxy light profile in the inner $2R_{\rm eff}$, indicate distance from the centre in units of effective radii ($R_{\rm eff}$).

density profile break in NGC 3379 (Harris *et al.* 2007b). The location of F8 was along the same projected elliptical radius (7 $R_{\rm eff}$) as F4, while the minor and major axis fields F9 and F6 were located along the ellipse corresponding to the extrapolation to 15.5 $R_{\rm eff}$. These were chosen to facilitate the comparison of the stellar density profiles along the minor and major axis and thus the overall halo shape.

3. Removing the foreground Milky Way contamination

The relative fraction of the Milky Way (MW) foreground vs. RGB stars in NGC 5128 becomes important, as the density of halo decreases. At some point the density of upper RGB stars in the halo becomes comparable to that of the MW stars and it is impossible to further trace the NGC 5128 halo. Hence, to establish the galaxy and halo extent, it is important to accurately subtract the MW contamination from the observed CMDs.

As a first step to estimate the foreground contamination TRILEGAL (Girardi *et al.* 2005) and Besançon (Robin *et al.* 2003) MW models were examined. The foreground contamination by MW stars consists mainly of dwarf stars in the thin, thick disk, and the halo according to these models. Rejkuba *et al.* (2014) found large differences between the two models: Besançon simulations yielded more than 3 times fewer stars, in particular having fewer faint stars than TRILEGAL, while their colour distribution better fit the observed distribution of MW stars brighter than the tip of the RGB. However, since

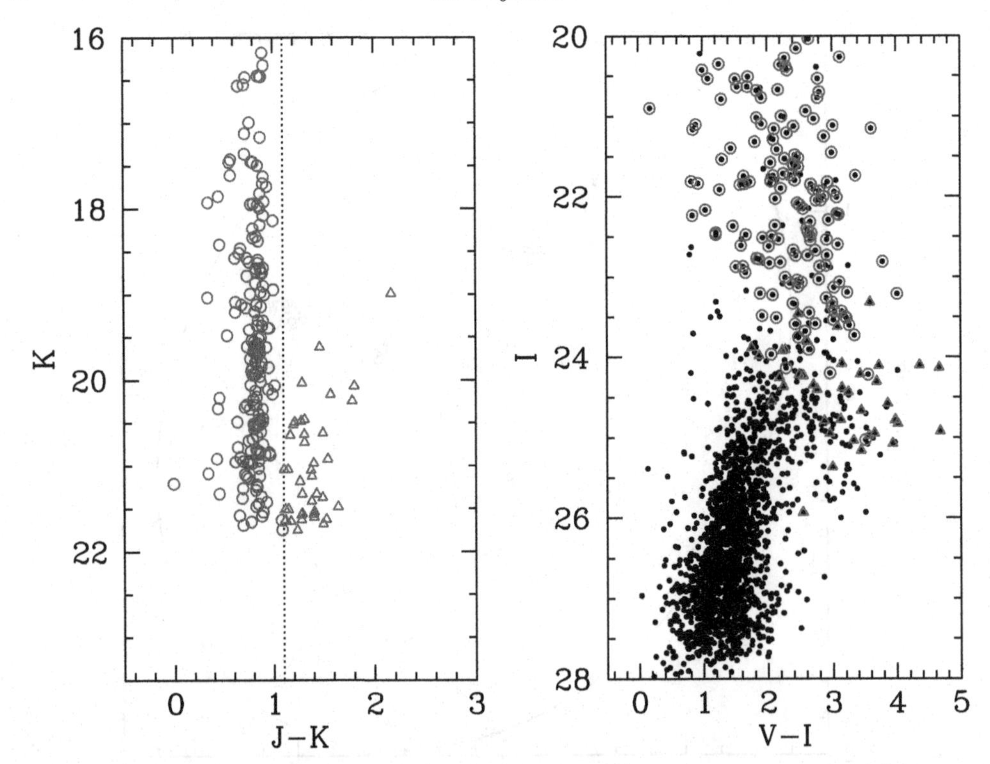

Figure 2. Left: F5 JK_s CMD based on PSF photometry of stars detected on ISAAC images. Stars bluer than $J - K_s < 1.1$ (blue circles) are likely MW foreground contaminants, while the redder stars (red triangles) belong to NGC 5128 halo. Right: VI CMD of the same area (F5) based on WFC3 data. The stars found in common on the ISAAC images are shown as blue or red dots according to their $J - K_s$ colour measured from ISAAC data.

TRILEGAL colour distribution well matched the observed one for fainter ($I > 24.5$) part of the CMD, where most of the NGC 5128 halo stars are located, and given that it had higher number of stars, this simulation was preferred in order to avoid underestimating the MW contamination.

The number of stars observed in the RGB area of the CMD in F7 at 140 kpc distance is 2.14 times higher than the number of expected MW stars according to the TRILEGAL simulation, demonstrating that the NGC 5128 halo extends beyond the most distant field observed. After statistical decontamination of the observed CMDs using TRILEGAL MW model, the RGB colours were interpolated on a grid of empirical RGB fiducial lines in order to derive MDF in each observed field (F5–F9), for ACS and WFC3 fields independently. In each field the mean and median metallicity and stellar density were computed – these profiles are presented in the next section.

Before discussing the resulting gradients (constructed based on foreground decontamination with models), an alternative, purely empirical, way to decontaminate the CMDs is presented. It is based on combination of near-IR and optical data. ISAAC camera on the ESO VLT was used to obtain J and K_s images coinciding with the WFC3 observations of F5–F9 fields. The size of the ISAAC field of view matches very well that of the WFC3. Fig. 2 shows on the left the $J - K_s$ vs. K_s CMD constructed with the PSF photometry of stellar sources on the ISAAC images of F5. On the right the ISAAC detections in common with the ones from WFC3 are overplotted on the WFC3 VI CMD. Stars detected on ISAAC images with $J - K_s < 1.1$ (blue circles) most likely belong to

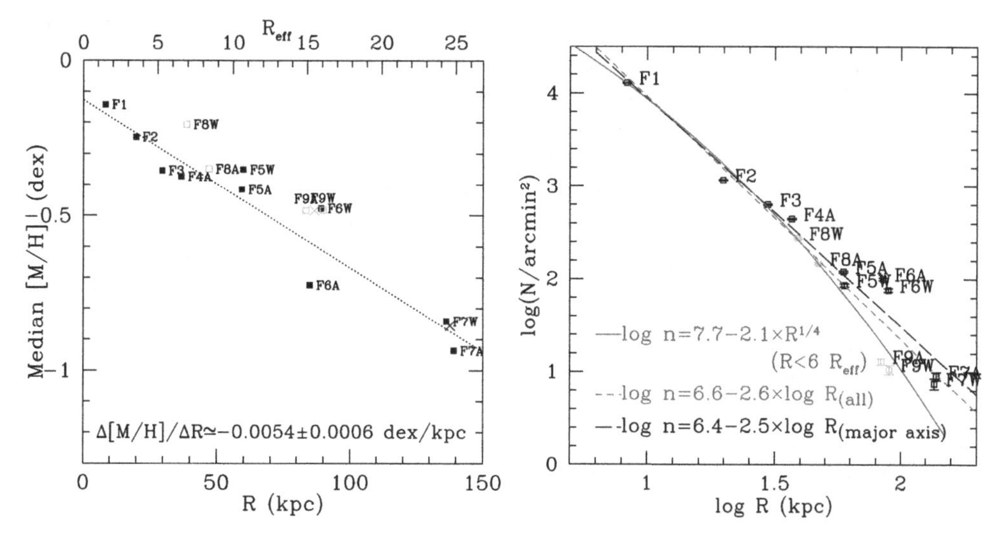

Figure 3. Left: median metallicity gradient. Right: stellar density gradient. The measurements in the minor axis fields (F8 and F9) are shown with the green open symbols and their distances are scaled to the major axis by multiplying with the inverse axis ratio (a/b=1/0.77).

the MW foreground population, while those redder are likely members of the NGC 5128 halo. This figure shows that $J-K_s$ colour can be used to effectively decontaminate the CMDs above the RGB tip and therefore investigate a possible presence of bright AGB stars. At fainter magnitudes it is also possible to get $J-K_s$ colours by adopting positions of stars from WFC3 data. These measurements are somewhat noisier and the separation between foreground and NGC 5128 halo stars is not as clean (Rejkuba *et al.*, in prep).

4. Results and Conclusions

Based on the observed HST CMDs of fields F1–F9, from which the MW foreground contamination was statistically removed, the mean metallicity, metallicity dispersion and stellar density was obtained from the upper RGB, where observations do not suffer from incompleteness. In all fields the metallicity distribution was obtained in a homogeneous way by interpolating the $V-I$ colours on a grid of α-enhanced BASTI isochrones (Pietrinferni *et al.* 2006). While the absolute metallicity scale may depend on the selected set of stellar evolutionary models, the relative metallicity between different fields is robust given the uniform measurement procedure. The MDFs for fields within ~ 15 $R_{\rm eff}$ distance are similar to that of F4 (Rejkuba *et al.* (2005)), presenting a single peak, quick drop on the metal-rich side, and long, less populated metal-poor tail. The median metallicity gradient is shown on the left panel in Figure 3 and its slope $\Delta[M/H]/\Delta R = -0.0054 \pm 0.0006$ dex/kpc is obtained from a linear fit to the data. The minor axis fields have on average higher metallicity than the corresponding fields along the major axis. However, similar size field-to-field scatter is also observed. The gradient is quite shallow, such that even in the most remote field in the halo, at 140 kpc ($25R_{\rm eff}$) distance from the galaxy centre, the median metallicity is close to or higher than $[M/H] = -1$ dex.

The stellar density gradient is shown in the right panel of Fig. 3. Already from the inspection of the observed CMDs it was obvious that there is a larger density along the major axis - this is confirmed by the star counts. The outer halo density along the minor axis is consistent with the extrapolation of the $R^{-1/4}$ profile fitted to the inner three fields, while the major axis fields present an excess of stars and their density gradient is well fitted with a power law. The increasing ellipticity in the outer halo of NGC 5128 confirms a tentative result from the ground based study of Crnojević *et al.* (2013). The halo extends along the major axis at least to 140 kpc and likely also beyond. To find out how far it extends and determine its shape further observations are needed.

So far only a handful of galaxies have been investigated and their halo coverage is very uneven (Tab. 1). In all of them quite metal-rich ($[M/H] > -1$ dex) mean (median) metallicity is measured that is unlike the mean metallicity of the MW halo. In NGC 3379 a transition from the centrally concentrated metal-rich to a more extended metal-poor halo is detected beyond $\sim 12\ R_{\rm eff}$ (Harris *et al.* 2007b). The galaxy with the most extended coverage, after that of NGC 5128, is NGC 3115. It shows a similarly shallow metallicity gradient, and Peacock *et al.* (2015) estimated that the metal-poor component ($[M/H] < -1.3$dex) would only start dominating over the metal-rich one beyond ~ 200 kpc. No galaxy halo was however traced that far. The future facilities, such as JWST and WFIRST will enable deep and wide area mapping of the extended stellar halos in 100s of galaxies up to 10 Mpc, opening the possibility for statistical comparison of their properties with the models (see the review by Johnston in this volume, and references therein).

References

Bird, S., Harris, W. E., Blakeslee, J. P., & Flynn, C. 2010, *A&A* 524, A71
Bird, S., Flynn, C., Harris, W. E., & Valtonen, M. 2015, *A&A* 575, A72
Crnojević, D., Ferguson, A. M. N., & Irwin, M. J. 2013, *MNRAS* 432, 832
Davidge, T. J. 2002, *AJ* 124, 2012
Davidge, T. J. & van den Bergh, S. 2001, *ApJ* (Letters) 553, L133
Elson, R.A.W. 1997, *MNRAS* 286, 771
Ferrarese, L., Mould, J. R., & Stetson, P. B. 2007, *ApJ* 654, 186
Girardi, L., Groenewegen, M. A. T., Hatziminaoglou, E., & da Costa, L. 2005, *A&A* 436, 895
Gregg, M. D., Ferguson, H. C., Minniti, D., *et al.* 2004, *AJ*, 127, 1441
Harris, G. L. H., Harris, W. E., & Poole, G. B. 1999, *AJ* 117, 855
Harris, G. L. H. & Harris, W. E. 2000 *AJ* 120, 2423
Harris, G. L. H., Rejkuba, M., & Harris, W. E. 2010, *PASA* 27, 475
Harris, W. E. & Harris, G. L. H. 2002, *AJ* 123, 3108
Harris, W. E., Harris, G. L. H., Layden, A. C., & Stetson, P. B. 2007a, *AJ* 134, 43
Harris, W. E., Harris, G. L. H., Layden, A. C., & Wehner, E. M. H. 2007b, *ApJ* 666, 903
Neumayer, N. 2010, *PASA* 27, 449
Peacock, M. B., Strader, J, Romanowsky, A. J., & Brodie, J. P. 2015, *ApJ* 800, 13
Peng, E. W., Ford., H. C. & Freeman, K. C. 2004, *ApJ* 602, 685
Pietrinferni, A., Cassisi, S., Salaris, M., & Castelli, F. 2006, *ApJ* 642, 797
Rejkuba, M., Minniti, D., Silva, D. R., & Bedding, T. 2003, *A&A* 411, 351
Rejkuba, M., Harris, W. E., Greggio, L., *et al.* 2005, *ApJ* 631, 262
Rejkuba, M., Greggio, L., Harris, W. E., & Harris, G. L. H. 2011, *A&A* 526, A123
Rejkuba, M., Harris, W. E., Greggio, L., *et al.* 2014, *ApJ* (Letters) 791, L2
Robin, A. C., Reylé, C., Derriére, S., & Picaud, S. 2003, *A&A*, 409, 523
Sakai, S., Madore, B. F., Freedman, W. L., *et al.* 1997, *ApJ* 478, 49
Soria, R., Mould, J. R., Watson, A. M., *et al.* 1996, *ApJ* 465, 79
Tal, T., *et al.* 2009, *AJ* 138, 1417
Wu, P-F., Tully, R. B., Rizzi, L. *et al.* 2014, *AJ* 148, 7

The General Assembly of Galaxy Halos: Structure, Origin and Evolution
Proceedings IAU Symposium No. 317, 2015
A. Bragaglia, M. Arnaboldi, M. Rejkuba & D. Romano, eds.

doi:10.1017/S1743921316000090

Stellar halos around Local Group galaxies

Alan W. McConnachie

NRC Herzberg, Dominion Astrophysical Observatory, 5071 West Saanich Road, Victoria, British Columbia, Canada, V(E2E7
email: alan.mcconnachie@nrc-cnrc.gc.ca

Abstract. The Local Group is now home to 102 known galaxies and candidates, with many new faint galaxies continuing to be discovered. The total stellar mass range spanned by this population covers a factor of close to a billion, from the faintest systems with stellar masses of order a few thousand to the Milky Way and Andromeda, with stellar masses of order $10^{11} M_{\odot}$. Here, I discuss the evidence for stellar halos surrounding Local Group galaxies spanning from dwarf scales (with the case of the Andromeda II dwarf spheroidal), though to intermediate mass systems (M33) and finishing with M31. Evidence of extended stellar populations and merging is seen across the luminosity function, indicating that the processes that lead to halo formation are common at all mass scales.

Keywords. galaxies: dwarf, galaxies: evolution, galaxies: formation, galaxies: general, galaxies: halos, Local Group, galaxies: stellar content, galaxies: structure

---

1. Introduction

As of September 2015, there are approximately 102 candidate Local Group galaxies known. Many of the newest discoveries are exceptionally faint, with literally only a handful of stars amenable to observations from which we can derive the properties of the stellar systems. It remains ambiguous how many of these new discoveries are dwarf galaxies, as opposed to globular clusters, tidal remains, or even asterisms. Figure 1 shows a luminosity function of all Local Group (candidate) galaxies. Red corresponds to the Milky Way and its satellite system ($D_{MW} < 300$kpc), blue corresponds to Andromeda and its satellites ($D_A < 300$ kpc) and green corresponds to isolated galaxies in the Local Group (beyond 300kpc from Andromeda and the Milky Way but within the zero-velocity surface of the Local Group). Data for this luminosity function is given in McConnachie (2012)†.

Figure 2 shows a map of the spatial distribution of red giant branch star candidates in the vicinity of Andromeda and Triangulum as observed by the Pan-Andromeda Archaeological Survey (PAndAS; McConnachie *et al.* 2009), color-coded by RGB color (redder RGB stars are probably more metal-rich in general than bluer RGB stars). Clearly, the halo of Andromeda is highly extended and highly substructured. However, whereas the majority of work on stellar halos is focused towards $L_{\star}$ galaxies like the Milky Way and Andromeda, it is worth highlighting that evidence for stellar halos can be found in galaxies at various positions in the luminosity function shown in Figure 1. Here, we discuss

† An updated version of this catalog is available online at http://www.astro.uvic.ca/~alan/Nearby_Dwarf_Database.html

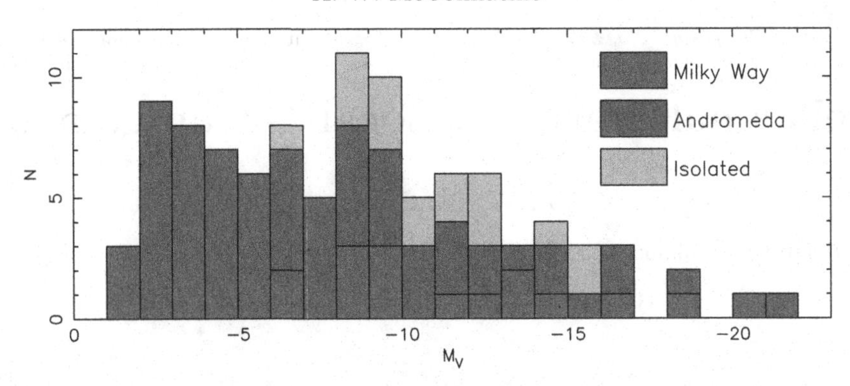

Figure 1. Luminosity function of all galaxies in the Local Group, using the September 2015 update from McConnachie (2012).

Figure 2. Spatial distribution of RGB stars surrounding Andromeda from PAndAS (McConnachie *et al.* 2009). Figure from Martin *et al.* (2013).

examples of stellar halos found in the Local Group across a range of luminosities. All the objects under discussion are visible in Figure 2.

2. Stellar halos in dwarf galaxies: Andromeda II

Using the Isaac Newton Telescope Wide Field Camera, McConnachie & Irwin (2006) derive the surface brightness profiles of all the then-known dwarf spheroidal satellites of M31. These are all generally well described by King or exponential profiles, with the exception of Andromeda II. Andromeda II has a luminosity of $M_V \simeq -12.6$. For this galaxy, a significant discontinuity at ~ 2 arcmins means that a single profile that fits the majority of the radial profile grossly underestimates the number of stars at smaller radii. Subsequent follow-up work with the Subaru SuprimeCam (McConnachie *et al.* 2007) shows that, when split by stellar population, the structure of Andromeda II could

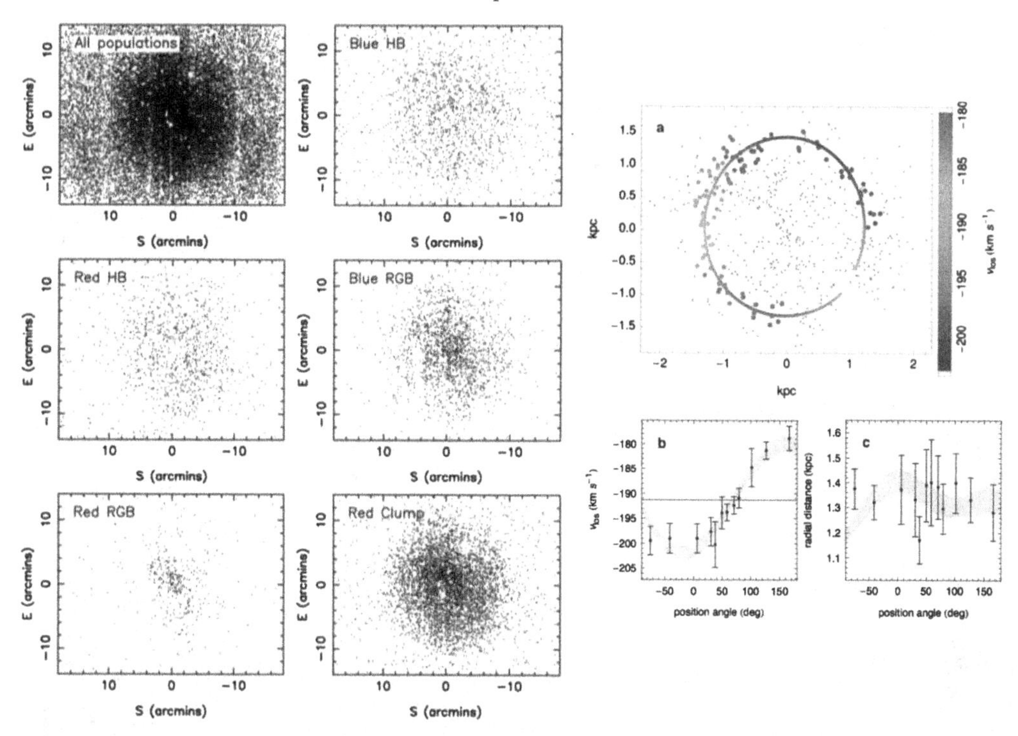

Figure 3. Left panels: stellar distribution of various stellar populations in Andromeda using Subaru/SuprimeCam (figure from McConnachie *et al.* 2007). Right panel: kinematics of stars in Andromeda II, highlighting a coherent "stream-like" structure in the outskirts of this galaxy. Figure from Amorisco *et al.* (2014)

vary dramatically depending on which tracer population was used. In particular, the horizontal branch population shows an extremely extended distribution, whereas redder RGB stars reveal a more centrally concentrated, exponential-style distribution. The left panels of Figure 3 show maps of Andromeda II using these different stellar populations as identified from the color-magnitude diagram.

One interpretation of these results is that Andromeda II has at least two distinct components to its structure, each with slightly different characteristic stellar populations, the more centrally concentrated of which is well traced by redder RGB stars, and the more extended of which is well traced by horizontal branch stars. However, while it is clear that Andromeda II does possess a spatially extended stellar component, its origin is unclear based only on the wide-field Subaru data. More recently, Amorisco *et al.* (2014) use a kinematic survey of Andromeda II by Ho *et al.* (2012) to identify a population of stars that have a kinematically distinct signature in Andromeda II and that trace a potential stellar stream in this dwarf galaxy (right panels of Figure 3). They conclude that Andromeda II has undergone a merger. When interpreted in the context of the earlier photometric work, Andromeda II shows evidence of a spatially extended secondary component whose origin is likely through a merger with a less luminous system, i.e., a stellar halo. Deep HST imaging of Andromeda II from Weisz *et al.* (2014) also points to a merger origin, since here the RGB can be seen to bifurcate, indicating at least two distinct populations, that perhaps formed originally in different systems.

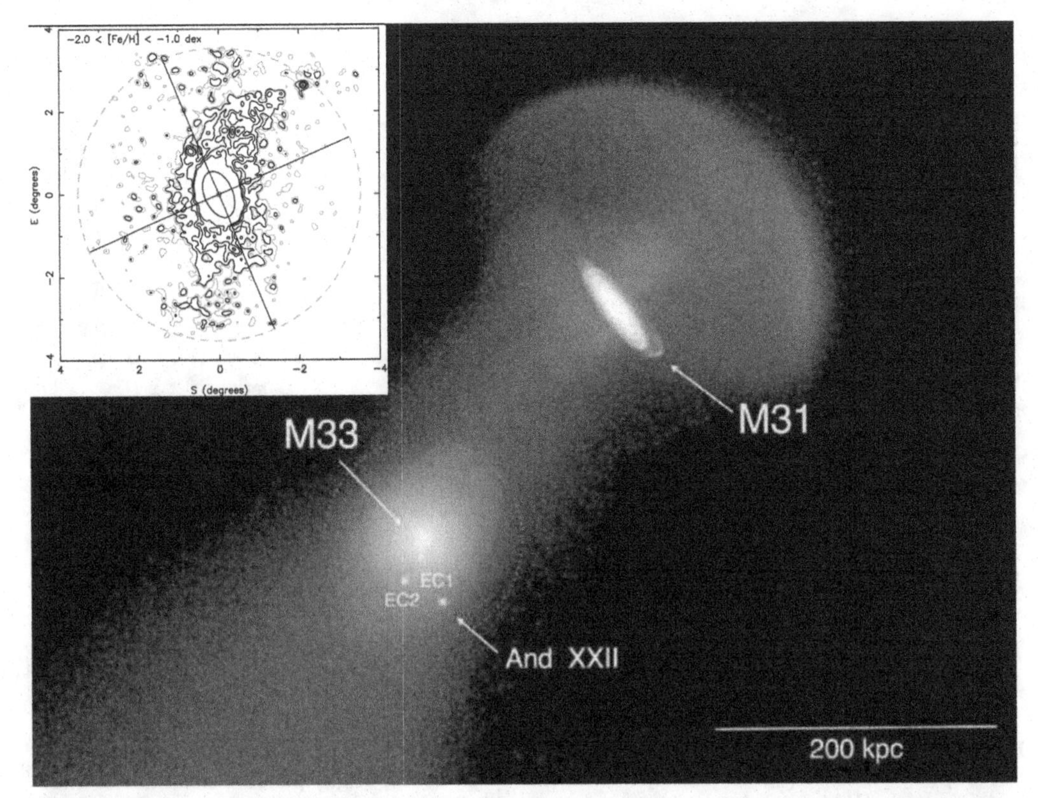

Figure 4. Inset panel: Distribution of metal poor RGB stars surrounding M33, showing a very large, stream-like distortion surrounding this galaxy. Figure from McConnachie *et al.* (2010). Large panel: dynamical model of M33 in orbit around M31, showing the potential stripping of the dark matter halo of M33 in its interaction with M31. Figure from Chapman *et al.* (2013).

3. Stellar halos in sub-L⋆ galaxies: M33

M33 is the third most massive galaxy in the Local Group, but is still an order of magnitude less luminous than either M31 or the Milky Way. There have been several different claims to have detected a stellar halo in this galaxy over the years. Chandar *et al.* (2002) conducted an analysis of the star clusters in this galaxy and concluded that the oldest star clusters have distinct kinematics from the youngest clusters, and also from the HI disk. Their analysis suggested that the oldest clusters are consistent with having an 85-15 % halo-disk contribution. Subsequently, Sarajedini *et al.* (2006) show that the metallicitiy distribution of M33 as implied from the period of RR Lyrae stars has a bimodal distribution, and they attribute the metal-poor peak to a contribution from the stellar halo.

Both of these analyses are based on the study of fields within the main disk of M33, and the presence of an extended population at large radius (i.e., a stellar halo) had not been explicitly identified. McConnachie *et al.* (2006) conduct an analysis of the kinematics of two major axis fields surrounding M33 using Keck/DEIMOS located beyond the edge of the main disk ($> 0.5°$). The majority of stars that they identify have kinematics consistent with the stellar disk, but they identify two additional features, one of which they attribute to a halo component, and another peak that they suggest could be some unidentified substructure. Subsequently, Ibata *et al.* (2007) used incomplete imaging using CFHT/MegaCam surrounding M33 to identify a significant population of stars at

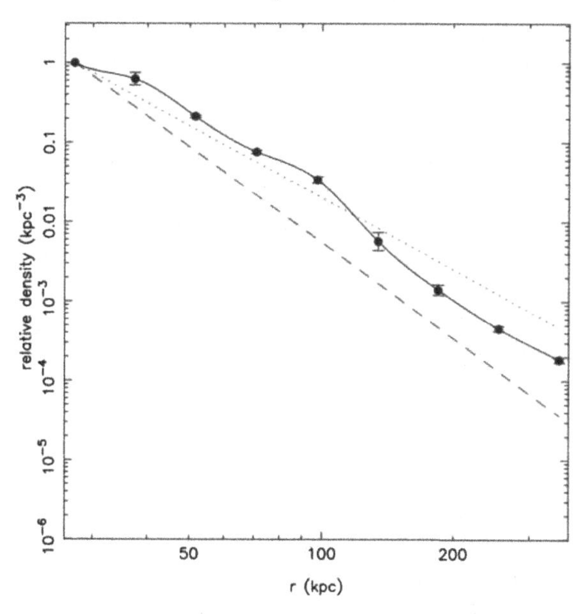

Figure 5. Derived profile of the "smooth" stellar halo of M31. Figure from Ibata *et al.* (2014).

distances of several degrees, confirming the apparent detection of a stellar halo around this galaxy.

However, the complete CFHT/MegaCam panorama of M33 was not completed until McConnachie *et al.* (2009), and it becomes immediately apparent that the stellar halo of M33 identified by Ibata *et al.* (2007) is not a typical, spheroidal-like population. The inset panel of Figure 4 is from McConnachie *et al.* (2010) and shows the distribution of RGB stars at several degrees from this galaxy. This population traces a very large, extended, "S-like" population that is fully consistent with what is expected by a galaxy that is undergoing tidal disruption in an orbit around a larger host. Dynamical modeling of an M31-M33 encounter in McConnachie *et al.* (2009) and Chapman *et al.* (2013; see Figure 4, main panel) suggests that a moderately-close fly-by of M33 by M31 should be enough to strip a large fraction of the dark matter halo, and presumably outer stellar populations, of M33. It is unclear if the material that is now observed is therefore the remnants of a previously extended, spheroidal halo-like population, or the stripped outskirts of the M33 stellar disk. Cockcroft *et al.* (2011) search for any remaining outer stellar populations in M33 away from the main debris, and find only tentative evidence for an underlying population. Having defined a stellar halo as a spatially-extended, outer stellar population, it is clear that M33 indeed possesses such a structure, but its formation mechanism could be quite different compared to other stellar halos, for example around M31.

4. Stellar halos in L⋆ galaxies: Andromeda

Figure 2 shows the extent and complexity of the stellar halo of M31, and this remains the best panoramic image of any stellar halo. There are large numbers of vast substructures and streams, and a significant number of dwarf galaxies (visible in Figure 2 as blue overdensities). Characterisation of this stellar halo is ongoing. Recently, Ibata *et al.* (2014) attempt to quantify the properties of the halo assuming that there is a component that is well described as a "smooth" population. Here, "smooth" implies no obvious substructure down to the surface-brightness limits of the data, and is a convenient way of characterising the pòpulation. Based on a masking procedure, they find that the fraction of the halo population that is smoothly distributed varies significantly as a function of metallicity. Overall, only 1% of the halo is in a smooth distribution, but at the most metal-poor end, over 60% of the stars appear smoothly distributed. Figure 5 shows the radial profile of the smooth M31 halo. It has a power-law distribution (with $\gamma = -3 - -4$) and extends to beyond 300kpc (deprojected). This is an extremely extended population that is also consistent with the very extended spatial distribution of satellite galaxies (Richardson *et al.* 2011). Indeed, it remains to be seen if simulated stellar halos are able to match this type of profile. A companion paper, detailing the statisitical properties of the substructure in M31, is in preparation (McConnachie *et al.* in preparation).

5. Conclusions

Stellar halos are rich environments for understanding the evolutionary and formation processes of galaxies. Merging is a key process by which at least some of the population is formed, both in terms of galaxies merging into the main system and being disrupted , with the remnants being identified with the stellar halo (e.g., Andromeda II and M31), and in terms of the main galaxy itself being disrupted (e.g., M33). Such processes occur in all luminosity regimes in the Local Group and are not confined to $L*$ galaxies.

References

Amorisco, N. C., Evans, N. W., & van de Ven, G. 2014, *Nature*, 507, 335
Chandar, R., Bianchi, L., Ford, H. C., & Sarajedini, A. 2002, *ApJ*, 564, 712
Chapman, S. C., Widrow, L., Collins, M. L. M., *et al.* 2013, *MNRAS*, 430, 37
Cockcroft, R., McConnachie, A. W., Harris, W. E., *et al.* 2013, *MNRAS*, 428, 1248
Ho, N., Geha, M., Munoz, R. R., *et al.* 2012, *ApJ*, 758, 124
Ibata, R. A., Lewis, G. F., McConnachie, A. W., *et al.* 2014, *ApJ*, 780, 128
Ibata, R., Martin, N. F., Irwin, M., *et al.* 2007, *ApJ*, 671, 1591
Martin, N. F., Ibata, R. A., McConnachie, A. W., *et al.* 2013, *Apj*, 776, 80
McConnachie, A. W. 2012, *AJ*, 144, 4
McConnachie, A. W., Chapman, S. C., Ibata, R. A., *et al.* 2006, *ApJ*, 647, L25
McConnachie, A. W., Ferguson, A. M. N., Irwin, M. J., *et al.* 2010, *ApJ*, 723, 1038
McConnachie, A. W., Arimoto, N., & Irwin, M. 2007, *MNRAS*, 379, 379
McConnachie, A. W. & Irwin, M. J. 2006, *MNRAS*, 365, 1263
McConnachie, A. W., Irwin, M. J., Ibata, R. A., *et al.* 2009, *Nature*, 461, 66
Richardson, J. C., Irwin, M. J., McConnachie, A. W., *et al.* 2011, *ApJ*, 732, 76
Sarajedini, A., Barker, M. K., Geisler, D., Harding, P., & Schommer, R. 2006, *AJ*, 132, 1361
Weisz, D. R., Skillman, E. D., Hidalgo, S. L., *et al.* 2014, *ApJ*, 789, 24

The General Assembly of Galaxy Halos: Structure, Origin and Evolution
Proceedings IAU Symposium No. 317, 2015
A. Bragaglia, M. Arnaboldi, M. Rejkuba & D. Romano, eds.

doi:10.1017/S1743921315008376

Resolving the extended stellar halos of nearby galaxies: the wide-field PISCeS survey †

D. Crnojević[1], D. J. Sand[1], N. Caldwell[2], P. Guhathakurta[3], B. McLeod[2], A. Seth[4], J. D. Simon[5], J. Strader[6] and E. Toloba[1,3]

[1]Texas Tech University, Physics Department, Box 41051, Lubbock, TX 79409-1051, USA
email: denija.crnojevic@ttu.edu

[2]Harvard-Smithsonian Center for Astrophysics, Cambridge, MA 02138, USA

[3]UCO/Lick Observatory, University of California, Santa Cruz, 1156 High Street, Santa Cruz, CA 95064, USA

[4]Department of Physics and Astronomy, University of Utah, Salt Lake City, UT 84112, USA

[5]Observatories of the Carnegie Institution for Science, 813 Santa Barbara Street, Pasadena, CA 91101, USA

[6]Michigan State University, Department of Physics and Astronomy, East Lansing, MI 48824, USA

Abstract. In the wide-field Panoramic Imaging Survey of Centaurus and Sculptor (PISCeS), we investigate the resolved stellar halos of two nearby galaxies (the elliptical Centaurus A and the spiral Sculptor, D $\sim$ 3.7 Mpc) out to a projected galactocentric radius of 150 kpc with Magellan/Megacam. The survey has led to the discovery of $\sim$20 faint satellites to date, plus prominent streams and substructures in two environments that are substantially different from the Local Group, i.e. the Centaurus A group dominated by an elliptical and the loose Sculptor group of galaxies. These discoveries clearly attest to the importance of past and ongoing accretion processes in shaping the halos of these nearby galaxies, and provide the first census of their satellite systems down to an unprecedented $M_V < -8$. The detailed characterization of the stellar content, shape and gradients in the extended halos of Sculptor, Centaurus A, and their dwarf satellites provides key constraints on theoretical models of galaxy formation and evolution.

Keywords. galaxies: groups: individual (CenA, NGC253) — galaxies: halos — galaxies: dwarf — galaxies: photometry - galaxies: evolution - galaxies: luminosity function - galaxies: stellar content - galaxies: interaction

---

1. Introduction

The past decade has witnessed the advent of wide-field instrumentation, which has quickly led to the discovery of a wealth of stellar streams and faint satellites in our own Galaxy's halo (Ibata *et al.* 2001, Belokurov *et al.* 2006). Our nearest massive neighbor, M31, has been the next to be systematically surveyed out to large galactocentric radii, uncovering a similarly rich amount of substructures and satellites (McConnachie *et al.* 2009). The widely accepted Λ CDM model for hierarchical structure assembly does indeed predict that the remnants of past/ongoing accretion and interaction events should populate the outskirts of galaxy halos, testifying their evolutionary history. Simulations also show a significant halo-to-halo scatter in the properties of galaxy halos, due to a

† This paper includes data gathered with the 6.5 meter Magellan Telescopes located at Las Campanas Observatory, Chile.

wide variety in their assembly histories. However, the physical processes that regulate star formation and galaxy evolution (e.g. supernova feedback, reionization, environmental effects, etc.) remain poorly understood. While the detection of substructures in the outer halos of virtually all galaxies observed in great depth (Tal *et al.* 2009, Martinez-Delgado *et al.* 2010, Atkinson *et al.* 2013, Duc *et al.* 2015) agree well with theoretical predictions, to date they can only qualitatively confirm this picture: the intrinsic faintness of these features ($\mu_V \gtrsim 28$ mag/arcsec^2) poses a challange to their detailed characterization beyond the Local Group (LG). To make things worse, there are obvious discrepancies between the predicted number and baryonic content properties of the smallest galaxies in our own LG (the "missing satellite" and "too big to fail" problems, e.g., Moore *et al.* 1999, Boylan-Kolchin *et al.* 2012). The faint end of the satellite luminosity function has recently started to be explored beyond the LG (e.g., M81, Chiboucas *et al.* 2013; M101, Merritt *et al.* 2014), and yet far fewer galaxies with $M_V \gtrsim -12$ are observed than predicted by simulations.

In order to test and put quantitative constraints on theoretical predictions, it is imperative to observe and characterize a larger sample of galaxies, with a range of morphologies and living in different environments: this has been the motivation for our Panoramic Imaging Survey of Centaurus and Sculptor (PISCeS), which we introduce in the next section.

2. The PISCeS survey

The PISCeS survey targets two nearby galaxies ($D \sim 3.7$ Mpc): Centaurus A (Cen A, or NGC 5128), the closest elliptical to us and the dominant galaxy of a dense group of galaxies, and Sculptor (NGC 253), a spiral located in a much looser and elongated group. Sculptor has a mass comparable to our own Milky Way (MW; Karachentsev *et al.* 2005), while Cen A is slightly more massive (Woodley *et al.* 2007), and due to their proximity they can be resolved into individual stars, thus allowing for a detailed comparison with the extant surveys of the MW, M31 and M81.

We use the optical Megacam imager at the 6.5-m Magellan II Clay telescope (McLeod *et al.* 2015): with a field-of-view of 24×24 arcmin^2 and a binned pixel scale of 0.16", this is an ideal instrument to deliver a deep, wide-field survey of our target galaxies. With the final goal of reaching projected galactocentric radii of ~ 150 kpc (or ~ 16 deg^2) in the halos of Cen A and Sculptor, since 2010 we have covered ~ 13 deg^2 for each of them in the g and r filters (with 6×300 s exposures). The survey is expected to be completed in 2016.

The standard image reduction and processing is performed by the Smithsonian Astrophysical Observatory Telescope Data Center, and subsequently resolved point spread function photometry is obtained from the stacked images with the DAOPHOT and ALLFRAME packages (Stetson 1987, 1994). Standard stars from SDSS are observed during clear nights to calibrate the data, and adjacent pointings are designed to overlap at the edges to ensure a good calibration under non photometric conditions as well. Finally, we perform artificial star tests in order to assess photometric uncertainties and incompleteness for each individual pointing (this is crucial for all such surveys given the non-uniformity in observing conditions from pointing to pointing).

The deep, resolved Magellan/Megacam images allow us to obtain color-magnitude diagrams (CMDs) such as the one shown in Fig. 1 (for Cen A's survey to date). Our survey reaches limiting magnitudes of ~26.5–27 mag in r-band, which corresponds to ~1.5–2 mag below the tip of the red giant branch (TRGB). A red giant branch (RGB) population at the distance of the target galaxy is clearly identified, as underlined by the

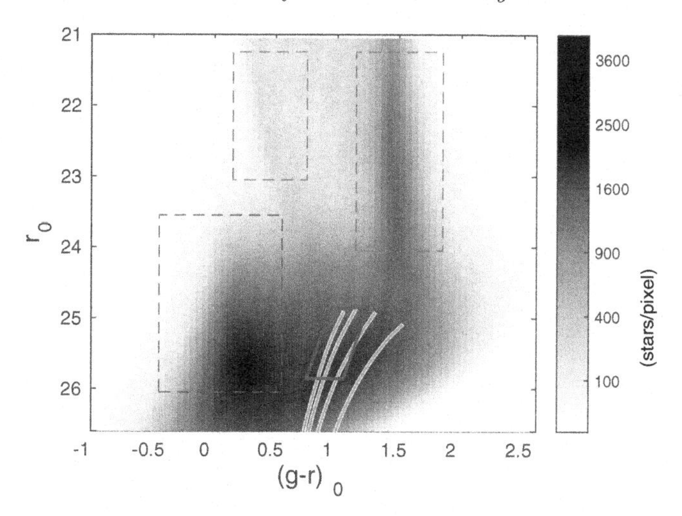

Figure 1. Dereddened Hess color-magnitude diagram (0.05 × 0.05 mag bins) for the Cen A PISCeS stellar catalogue to date. Dartmouth isochrones (Dotter *et al.* 2008) of 12 Gyr with a range of metallicities ([Fe/H]= −2.5 to −1.0 in 0.5 dex steps) are shifted to Cen A's distance and overlaid. The RGB selection box is shown in red, while the contaminants' sequences are approximately delineated by dashed boxes (blue for unresolved galaxies at $(g-r)_0 \sim 0.1$, magenta for MW halo and disk stars, at $(g-r)_0 \sim 0.5$ and $(g-r)_0 \sim 1.5$, respectively).

old isochrones with a range of metallicities, and is clearly separated from the contaminants' sequences (see Fig. 1).

3. Preliminary results

To illustrate the potential of our survey, we present the metal-poor RGB stellar density map of Cen A in Fig. 2, which includes the PISCeS pointings obtained to date (see Fig. 1 for the RGB selection box). The central regions of Cen A clearly show a variety of resolved features including shells and streams, which highlight the high degree of interaction recently experienced by this elliptical.

3.1. *Newly discovered satellites*

We use both visual inspection and the detection of overdensities in the resolved RGB stellar density maps in order to look for previously unknown faint satellites of Cen A and Sculptor. The discoveries of the first robust candidate satellites for both hosts are reported in Crnojević *et al.* (2014) and Sand *et al.* (2014). In total, to date we have identified 13 candidate Cen A satellites and 4 candidates for Sculptor, down to absolute magnitudes of $M_V \sim -8.0$. Resolved populations are clearly significantly more powerful than integrated light alone, given the otherwise prohibitively low surface brightness we are able to reach (down to $\mu_V \sim 32$ mag/arcsec^2). Nevertheless, some of our candidate satellites appear as surface brightness enhancements coupled with only a few resolved stars, and they may represent satellites even fainter than our detection limit. The newly discovered satellites resemble the properties (half-light radius, central surface brightness) of LG dwarfs with comparable luminosities (Crnojević *et al.* 2014, Sand *et al.* 2014, Crnojević *et al.* 2016, Toloba *et al.* 2016a).

Previously known Cen A satellites are easily recognizable in the stellar density map (blue circles in the lower panel of Fig. 2): ESO324-24 at $(\xi \sim 0.3, \eta \sim 1.6)$, NGC5237 at $(2.2, 0.1)$, KKs55 at $(-0.7, 0.3)$, KK197 at $(-0.75, 0.5)$, as well as KK203 and KK196 in

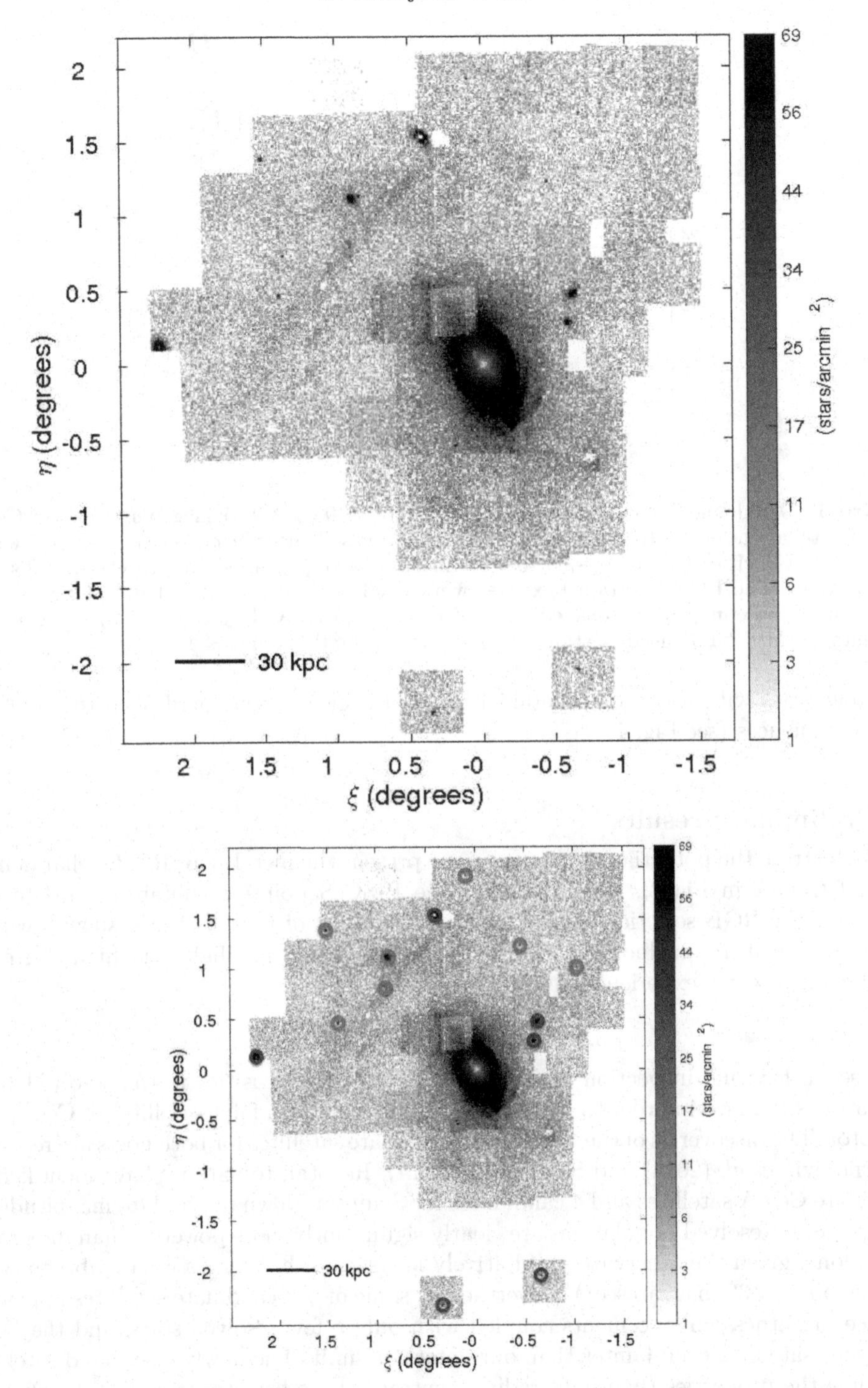

Figure 2. Stellar density map of metal-poor RGB stars (selected from the box in Fig. 1) for the halo of Cen A surveyed by PISCeS to date. Standard coordinates are centered on Cen A (N is up and E is left), and the central regions of the galaxy are replaced by a color image (credit: http://www.eso.org/public/images/eso0903a/); the star-count map in this region suffers from incompleteness due to high stellar crowding. The density scale as well as the physical scale are shown. In the lower panel, we indicate the position of the previously known (blue circles) and newly discovered satellites (red circles).

the two bottom (S) pointings. In the same map, we circle in red the nine robust new Cen A satellites, i.e. those that are firmly detected as RGB overdensities (we do not plot the four remaining candidates that are only seen as surface brightness enhancements). These newly discovered satellites are fainter than the known ones in terms of both absolute magnitude and central surface brightness, see e.g. CenA-MM-Dw1 and CenA-MM-Dw2 at $(0.9, 1.1)$ (Crnojević *et al.* 2014), or even fainter candidates at $(1.5, 1.4)$ (CenA-MM-Dw8), $(-0.5, 1.2)$ (CenA-MM-Dw4), $(0.2, -0.4)$ (CenA-MM-Dw7) and $(0.9, 0.8)$ (CenA-MM-Dw3). The latter is the remnant of a galaxy caught in the midst of tidal disruption, with stunning tails extending in both directions over 1 deg across the survey footprint (from E to NW). We stress that CenA-MM-Dw3 has extreme properties (very low central surface brightness coupled with large extent) with respect to LG satellites within the same luminosity range, and proves how sensitive our strategy is to such elusive objects. A similar potentially disrupting satellite, Scl-MM-Dw2, has also been uncovered in the vicinity of Sculptor (Toloba *et al.* 2016a).

4. Conclusions and future work

The PISCeS survey represents the next observational step in the future of near-field cosmology beyond the LG. PISCeS will charachterize in depth the resolved stellar halos of two $\sim$MW-sized galaxies with different morphology and surrounding environments, and their newly discovered faint satellites and substructures. We have already secured follow-up imaging and spectroscopy of the newly discovered candidate satellites, streams and substructures with a multi-wavelength approach, namely optical with HST, near-infrared with Gemini/FLAMINGOS2, co-added spectroscopy (see also Toloba *et al.* 2016b) with both VLT/VIMOS and Keck/DEIMOS. Data have already started to be collected and will be in hand by the beginning of 2017. This strategy will allow us to derive: the profile, shape and extent of the targets' smooth halos and possible gradients in their resolved populations; the orbital properties of the uncovered streams which will constrain the host halo's mass; the relative mass contribution from in situ versus accreted stellar components; the faint end of the satellite luminosity function down to $M_V \sim -8.0$, as well as the star formation histories of the individual satellites discovered through our survey. Our results will be compared to extant wide- and narrow-field surveys of M31 (PAndAS; SPLASH, Gilbert *et al.* 2012) and of other nearby galaxies (e.g., M81, Chiboucas *et al.* 2013, Okamoto *et al.* 2015; GHOSTS, Monachesi *et al.* 2016). This will ultimately advance the observational census of galaxy outer halos and of their inhabitants, thus providing crucial constraints for current and future theoretical predictions of galaxy evolution.

References

Atkinson, A. M., Abraham, R. G., & Ferguson, A. M. N. 2013, *ApJ*, 765, 28
Belokurov, V., Zucker, D. B., Evans, N. W., *et al.* 2006, *ApJL*, 647, L111
Boylan-Kolchin, M., Bullock, J. S., & Kaplinghat, M. 2012, *MNRAS*, 422, 1203
Chiboucas, K., Jacobs, B. A., Tully, R. B., & Karachentsev, I. D. 2013, *AJ*, 146, 126
Crnojević, D., Sand, D. J., Caldwell, N., *et al.* 2014, *ApJL*, 795, L35
Crnojević, D., *et al.* 2016, *ApJ* in press, ArXiv e-prints, arXiv:1512.05366
Dotter, A., Chaboyer, B., Jevremović, D., *et al.* 2008, *ApJS*, 178, 89
Duc, P.-A., Cuillandre, J.-C., Karabal, E., *et al.* 2015, *MNRAS*, 446, 120
Gilbert, K. M., Guhathakurta, P., Beaton, R. L., *et al.* 2012, *ApJ*, 760, 76
Ibata, R., Irwin, M., Lewis, G. F., & Stolte, A. 2001, *ApJL*, 547, L133
Karachentsev, I. D. 2005, *AJ*, 129, 178
Martínez-Delgado, D., Gabany, R. J., Crawford, K., *et al.* 2010, *AJ*, 140, 962

McConnachie, A. W., Irwin, M. J., Ibata, R. A., *et al.* 2009, *Nature*, 461, 66
McLeod, B., Geary, J., Conroy, M., *et al.* 2015, *PASP*, 127, 366
Merritt, A., van Dokkum, P., & Abraham, R. 2014, *ApJL*, 787, L37
Moore, B., Ghigna, S., Governato, F., *et al.* 1999, *ApJL*, 524, L19
Monachesi, A., Bell, E. F., Radburn-Smith, D. J., *et al.* 2016, *MNRAS*, 457, 1419
Okamoto, S., Arimoto, N., Ferguson, A. M. N., *et al.* 2015, *ApJL*, 809, L1
Sand, D. J., Crnojević, D., Strader, J., *et al.* 2014, *ApJL*, 793, L7
Stetson, P. B. 1987, *PASP*, 99, 191
Stetson, P. B. 1994, *PASP*, 106, 250
Tal, T., van Dokkum, P. G., Nelan, J., & Bezanson, R. 2009, *AJ*, 138, 1417
Toloba, E., *et al.* 2016b, submitted
Toloba, E., Sand, D. J., Spekkens, K., *et al.* 2016a, ApJ, 816, L5
Woodley, K. A., Harris, W. E., Beasley, M. A., *et al.* 2007, *AJ*, 134, 494

The General Assembly of Galaxy Halos: Structure, Origin and Evolution
Proceedings IAU Symposium No. 317, 2015
A. Bragaglia, M. Arnaboldi, M. Rejkuba & D. Romano, eds.

doi:10.1017/S1743921315006857

Intragroup and Intracluster Light

J. Christopher Mihos

Department of Astronomy, Case Western Reserve University

Abstract. The largest stellar halos in the universe are found in massive galaxy clusters, where interactions and mergers of galaxies, along with the cluster tidal field, all act to strip stars from their host galaxies and feed the diffuse intracluster light (ICL) and extended halos of brightest cluster galaxies (BCGs). Studies of the nearby Virgo Cluster reveal a variety of accretion signatures imprinted in the morphology and stellar populations of its ICL. While simulations suggest the ICL should grow with time, attempts to track this evolution across clusters spanning a range of mass and redshift have proved difficult due to a variety of observational and definitional issues. Meanwhile, studies of nearby galaxy groups reveal the earliest stages of ICL formation: the extremely diffuse tidal streams formed during interactions in the group environment.

Keywords. galaxies: clusters, galaxies: halos, galaxies: evolution

---

1. Introduction

In considering the connection between accretion and the formation of galaxy halos, perhaps nowhere is the process more dramatically illustrated than in the assembly of the most massive halos – the extended BCG envelopes and diffuse intracluster light (ICL) that is found in the centers of massive galaxy clusters. Unlike quiescent field galaxies whose major accretion era lies largely in the past, under hierarchical accretion scenarios, clusters of galaxies are the most recent objects to form (e.g. Fakhouri *et al.* 2010); their massive central galaxies continue to undergo active assembly and halo growth even at the current epoch, and may have accreted as much as half their mass since a redshift of $z = 0.5$ (e.g. de Lucia & Blaizot 2007). Thus the cluster environment presents an ideal locale for studying the accretion-driven growth of massive galaxy halos.

As galaxy clusters assemble, their constituent galaxies interact with one another, first within infalling groups, then inside the cluster environment itself. Over the course of time, a variety of dynamical processes liberate stars from their host galaxies, forming and feeding the growing population of intracluster stars. This complex accretion history is illustrated in Figure 1, using the collisionless simulations of Rudick *et al.* (2011). At early times, individual galaxies are strewn along a collapsing filament of the cosmic web. Gravity quickly draws these galaxies into small groups, which then fall together to form larger groups. In the group environment, slow interactions between galaxies lead to strong tidal stripping and the formation of discrete tidal tails and streams. As the groups fall into the cluster, this material is efficiently mixed into the cluster ICL (Rudick *et al.* 2006, 2009). Concurrently, mergers of galaxies in the cluster core expel more stars into intracluster space (Murante *et al.* 2007), as does ongoing stripping of infalling galaxies due to interactions both with other cluster galaxies and with the cluster potential itself (Conroy *et al.* 2007, Purcell *et al.* 2007, Contini *et al.* 2014). Additionally, even *in-situ* star formation in the intracluster medium, from gas stripped from infalling galaxies, may contribute some fraction of the ICL as well (Puchwein *et al.* 2010). All these processes lead to a continual growth of the intracluster light over time, as clusters continue to be fed by infalling groups and major cluster mergers. This evolution predicts that ICL properties should be linked to the dynamical state of the cluster – early in their formation

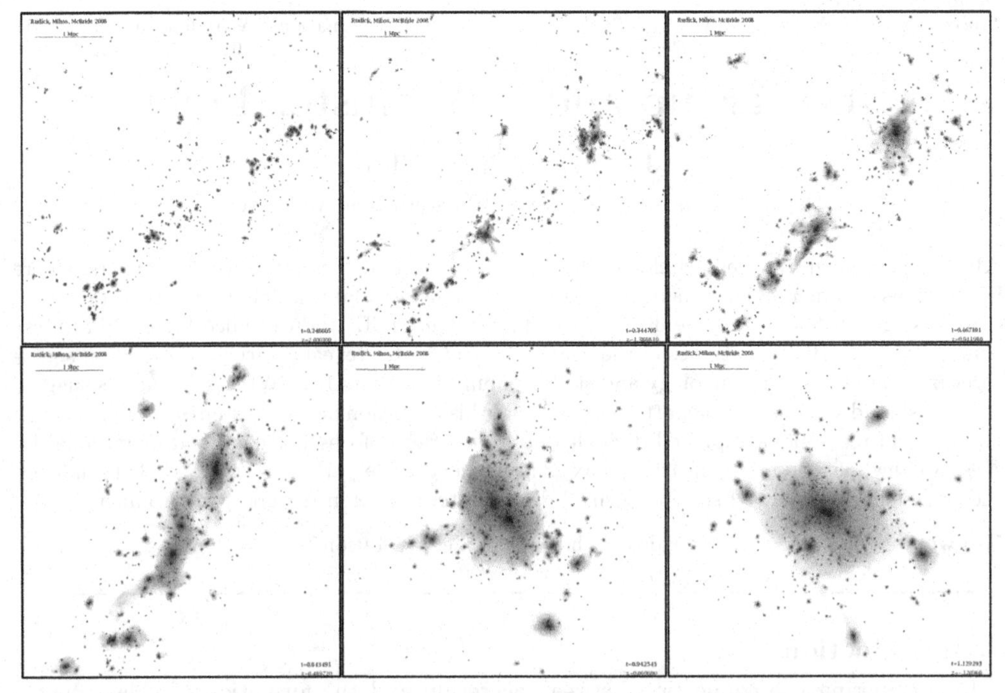

Figure 1. Formation of intracluster light during the assembly of a $10^{15} M_{\odot}$ galaxy cluster. The panels run from $z = 2$ (upper left) to the present day (lower right). From Rudick *et al.* (2011).

history, clusters should be marked by a low *total* ICL fraction but with a high proportion of light in cold (and more easily visible tidal streams), while more evolved clusters would have higher ICL fractions found largely in a smooth, diffuse, and well-mixed state.

The fact that these various processes all operate concurrently makes it difficult to isolate their individual contributions to the ICL, and computational studies differ on whether group accretion, major mergers, or tidal stripping dominate the ICL. Fortunately, these processes imprint a variety of observable signatures in the ICL. The morphology and color of the diffuse light as well as the spatial distribution and kinematics of discrete ICL tracers (red giant branch (RGB) stars, planetary nebulae (PNe), and globular clusters (GCs)) all have potential to disentangle the ICL formation channels. For example, the galaxy mass-metallicity relationship predicts that stripping of low mass satellites would deposit preferentially metal-poor stars into the ICL, while mergers of massive galaxies would lead to more metal-rich ICL. Similarly the age distribution of ICL populations may differentiate between stripping of old stellar systems versus that from star-forming galaxies, or even contributions from *in-situ* ICL production. Thus, observational studies of the morphology, colors, kinematics, and stellar populations in the ICL are well-motivated to track the detailed accretion histories of massive clusters.

2. ICL in the Virgo Cluster

We can use the nearby Virgo Cluster to illustrate the wealth of information locked in the ICL. Figure 2 shows deep, wide-field imaging of the Virgo Cluster taken using CWRU Astronomy's Burrell Schmidt telescope (Mihos *et al.* in prep). Covering 16 square degrees down to a surface brightness of $\mu_V \sim 28.5$, the imaging reveals the complex web of diffuse light spread throughout the core of Virgo. A number of tidal streams are visible, most

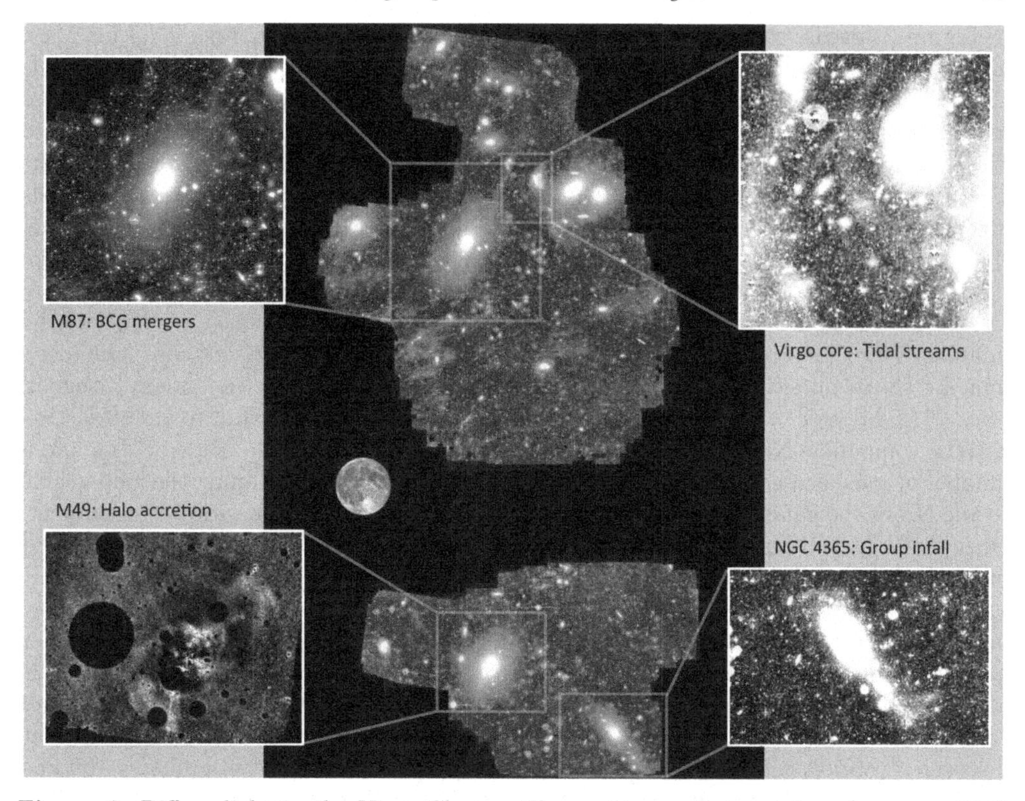

Figure 2. Diffuse light in the Virgo Cluster. The center panel shows deep ($\mu_{V,\rm lim} \approx 28.5$) wide-field imaging of Virgo taken using CWRU's Burrell Schmidt telescope (Mihos *et al.* in prep); the inset moon shows a 30′ scale. Panels show M87's extended halo (upper left; Mihos *et al.* 2005), tidal streams in the Virgo core (upper right; Mihos *et al.* 2005, 2015), M49's system of accretion shells (lower left; Janowiecki *et al.* 2010, Mihos *et al.* 2013) and the diffuse intragroup light surrounding NGC 4365 (lower right; Bogdán *et al.* 2012, Mihos *et al.* in prep).

notably two long ($>$ 100 kpc) thin streams NW of M87 (Mihos *et al.* 2005, Rudick *et al.* 2010). Smaller streams are also found around the Virgo ellipticals M86 and M84, likely due to stripping of low mass satellite galaxies, as well as a system of shells and plumes around M89 suggestive of one or more major mergers (Malin 1979, Janowiecki *et al.* 2010). However, the total luminosity contained in these discrete streams is only $\sim 1-2\times10^9$ $L_\odot$; the bulk of the ICL is likely found in more diffuse form, locked in the extended halo of M87 or strewn throughout the cluster at lower surface brightness.

Indeed the deep imaging reveals not only the thin ICL streams but also the large radial extent of the halos of Virgo ellipticals. In particular, M87's halo is traced beyond 150 kpc, where a variety of signatures indicative of past accretion events can be seen. The outermost regions of M87's halo are extremely boxy (Mihos *et al.* in prep), a behavior reflected in the spatial distribution of its GC system as well (Durrell *et al.* 2014). This combination of boxy isophotes and low halo rotation (Romanowsky *et al.* 2012) hints at a major merger event in M87's past, and indeed, both the GC and PNe systems around M87 show kinematic substructure (Romanowsky *et al.* 2012 and Longobardi *et al.* 2015a respectively), suggesting the recent accretion of one or more $\sim 10^{10}$ $L_\odot$ systems.

Signatures of past accretion are also found in other Virgo ellipticals as well. Located south of the Virgo core, M49 has long been known to have a dynamically complex halo, as traced by kinematic substructure in its GC system (Côté *et al.* 2003). The deep imaging in Figure 2 reveals the cause: after subtraction of a smooth isophotal model for M49, an extensive set of accretion shells (Janowiecki *et al.* 2010, Arrigoni Battaia *et al.* 2011, Capaccioli *et al.* 2015) can be seen, spanning $\sim$ 150 kpc in extent and containing close to 10^9 $L_\odot$ of light (Janowiecki *et al.* 2010). The shells are morphologically similar to those formed during the radial accretion of a low mass satellite, and may be linked to the tidally disturbed dwarf companion VCC 1249 (Arrigoni Battaia *et al.* 2011). The shells are also distinctly *redder* than M49's surrounding halo (Mihos *et al.* 2013), suggesting that the accretion event is building up both the mass *and* metallicity of M49's outer halo.

Figure 2 also illustrates the efficacy of the group environment in driving ICL formation. Lying 5.3° to the SW of the Virgo core (and $\sim$ 7 Mpc behind; Mei *et al.* 2007) is the infalling Virgo W′ group, with the massive elliptical NGC 4365 at its core. Our deep imaging shows an extended, diffuse tidal tail emanating SW from the galaxy (Bogdán *et al.* 2012; Mihos *et al.* in prep), and GC kinematics clearly link the tail to an interaction with its companion NGC 4342 (Blom *et al.* 2014). The tail contains $\sim 1.5 \times 10^9$ $L_\odot$, and a number of other streams are visible in NGC 4365's halo as well (including the loop visible to the NE of the galaxy), all indicative of cold tidal stripping in the group environment. Once the W′ group eventually falls into the main body of Virgo, this diffuse and extended intragroup light will be easily mixed into Virgo's diffuse ICL.

Finally, the imaging contains a dramatic example of the complex dynamical interplay between tidal stripping, ICL formation, and the destruction and formation of cluster galaxy populations. Lying at the center of the "Tidal streams" panel of Figure 2 is a large and extremely dim ultra-diffuse galaxy; with a half light radius of 9.7 kpc and central surface brightness $\mu_V = 27.0$ it is the most extreme ultradiffuse cluster galaxy yet discovered (Mihos *et al.* 2015). The galaxy also sports a long tidal tail arcing $\sim$ 100 kpc to the north, as well as a compact nucleus whose photometric properties are well-matched to those of ultracompact dwarf galaxies (UCDs) found in Virgo (e.g. Zhang *et al.* 2015, Liu *et al.* 2015). In this object, we are clearly seeing the tidal destruction of a low mass, nucleated galaxy which is both feeding Virgo's ICL population and giving rise to a new Virgo UCD.

To go beyond morphology and study the stellar populations in Virgo's ICL in detail, a variety of tools are available. The colors of the streams around M87 ($B - V = 0.7 - 1.0$; Rudick *et al.* 2010) are well-matched to those of the Virgo dE population and of M87's halo itself, suggesting M87's halo may be built at least in part from low mass satellite accretion. HST imaging of *discrete* RGB populations in Virgo intracluster fields shows the ICL to be predominantly old and metal-poor ($t > 10$ Gyr, [Fe/H] ≈ -1; Williams *et al.* 2007), but with an additional population of stars with intermediate ages and higher metallicities ($t \approx 4 - 8$ Gyr, [Fe/H] $\gtrsim -0.5$). These younger populations may arise either from stripped star forming galaxies or from ICL formed *in-situ*. The inference that stripping of late-type galaxies has contributed to the Virgo ICL is also supported by the luminosity function of PNe in M87's outer halo, which shows a "dip" characteristic of lower mass galaxies with extended star formation histories (Longobardi *et al.* 2015b). The diversity of stellar populations seen in Virgo's ICL almost certainly reflects the diversity of processes that create diffuse light in clusters.

3. ICL Systematics: Challenges and Metrics

While Virgo's proximity gives us a detailed view of intracluster stellar populations, to gain a wider census of ICL in galaxy clusters we must move beyond Virgo. Going to greater distances opens up the ability to study ICL in a wider sample of clusters which span a range of mass, dynamical state, and redshift, allowing us to connect ICL

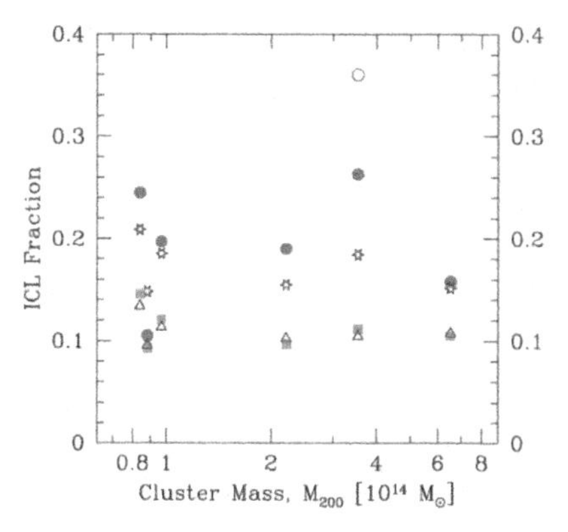

Figure 3. ICL fractions in simulated galaxy clusters, taken from Rudick *et al.* (2011). Five simulated clusters of varying mass are shown; at a given mass, the symbols show ICL fractions calculate for a single cluster using different ICL definitions: stars at low surface brightness ($\mu_V > 26.5$, green squares), stars in low density intracluster space now (open triangles) or ever (blue stars), stars unbound from galaxies (filled red circles), including stars kinematically separated from the central cD galaxy (open red circle). See Rudick *et al.* (2011) for details.

properties with cluster evolution. This comes at a cost, however; beyond Virgo, current generation telescopes cannot directly image intracluster stars, and even studies of more luminous tracers such as PNe and GC become more difficult. At higher redshifts, one becomes limited to broadband imaging, where the strong cosmological $(1+z)^4$ surface brightness dimming makes the already diffuse ICL even more difficult to observe.

Aside from these observational difficulties, a second major problem is the ambiguous definition of intracluster light itself. Since much of the ICL is formed via the mergers that build up the central BCG, there is often no clear differentiation between the BCG halo and the extended ICL – the two components blend smoothly together (and indeed may not be conceptually distinct components at all). In attempts to separate BCG halos from extended ICL, a variety of photometric definitions have been proposed, which typically adopt different functional forms (such as multiple $r^{1/4}$ or Sersic profiles) for each component when fitting the total profile (e.g. Gonzalez *et al.* 2005, Krick & Bernstein 2007, Seigar *et al.* 2007). However, such definitions are very sensitive to the functional forms adopted for the profiles. For example, M87's profile is reasonably well-fit by either a single Sersic or a double $r^{1/4}$ model (Janowiecki *et al.* 2010); the former fit would imply little additional ICL, while the latter fit puts equal light into the inner and outer profiles. To avoid this ambiguity, alternate non-parametric measures have also been employed to characterize the ICL luminosity, defining the ICL as diffuse light fainter than some characteristic surface brightness (e.g. Feldmeier *et al.* 2004, Burke *et al.* 2015). While these present systematic uncertainties of their own, simulations suggest that thresholds of $\mu_V \gtrsim 26.5$ do a reasonable job of separating out an extended and perhaps unrelaxed ICL component from the central BCG light (Rudick *et al.* 2011, Cooper *et al.* 2015).

Still other methods propose kinematic separation of the ICL from the central galaxy light. Dolag *et al.* (2010) used simulated clusters to show that separate kinematic populations exist in cluster cores, well-characterized by distinct Maxwellian distributions. These kinematic populations then separate out spatially into two Sersic-like profiles plausibly identified as the BCG galaxy and the cluster ICL (perhaps reflecting different accretion events as well; Cooper *et al.* 2015). And indeed these definitions have some observational support. Longslit spectroscopy of the BCG galaxy in Abell 2199 shows a velocity dispersion profile that first falls with radius, then increases in the outer halo to join smoothly onto the cluster velocity dispersion (Kelson *et al.* 2002). Meanwhile in Virgo the velocities of the PNe around M87 show a double Gaussian distribution (Longobardi *et al.* 2015b), suggesting distinct BCG and ICL components. However, observational constraints make accessing kinematic information for the ICL in distant clusters a daunting task.

A comparison of these different metrics is shown in Figure 3 (from Rudick *et al.* 2011), which shows that the inferred ICL fraction in simulated clusters can vary by factors of 2–3 depending on the adopted metric (see also Puchwein *et al.* 2010). Kinematic separation leads to higher ICL fractions, as a significant amount of starlight found within the BCG galaxy belongs to the high-velocity ICL component. In contrast, density-based estimates yield systematically lower ICL fractions, as material at high surface brightness is typically assigned to the cluster galaxies independent of its kinematic properties.

Given both the ambiguity in defining the ICL and the observational difficulties in studying it, attempts to characterize ICL in samples of clusters spanning a range of mass and redshift have led to varying results. An early compilation of results for local clusters by Ciardullo *et al.* (2004) showed ICL fractions ranging from $\sim 15 - 40\%$, with no clear dependence on cluster velocity dispersion or Bautz-Morgan type. Recent imaging of more distant clusters probes the connection between cluster evolution and ICL more directly, but again yields mixed results. While Guennou *et al.* (2012) find no strong difference between the ICL content of clusters between at $z \sim 0.5$ and today, Burke *et al.* (2015) find rapid evolution in the ICL fraction of massive clusters over a similar redshift range. Other studies of clusters at $z \sim 0.3 - 0.5$ yield ICL fractions of 10–25% (Presotto *et al.* 2014, Montes & Trujillo 2014, Giallongo *et al.* 2014), similar to $z = 0$ results. However, these studies use different ICL metrics and are limited to only a handful of clusters; clearly a large sample of clusters with ICL fractions measured in a consistent manner is needed to tackle the complex question of ICL evolution.

A similar story holds for recent attempts to constrain ICL stellar populations as well. Using HST imaging of distant CLASH clusters, DeMaio *et al.* (2015) infer moderately low metallicities ([Fe/H] ~ -0.5) from the ICL colors, in contrast to the case of Abell 2744, where Montes & Trujillo (2014) use colors to argue for a dominant population of intermediate age stars with solar metallicity. Meanwhile, spectroscopic population synthesis studies show similarly diverse results. For example, in the Hydra I cluster, Coccato *et al.* (2011) find old ICL populations with sub-solar metallicities, while in the massive cluster RX J0054.0−2823, Melnick *et al.* (2012) find similarly old but metal-rich ICL stars ([Fe/H] $\gtrsim 0$). However, while intriguing, all these studies are subject to strong photometric biases, limited largely to the brightest portions of the ICL which may not be representative of the ICL as a whole and may also include substantial fraction of what would normally be considered BCG light as well.

4. Diffuse Light in Nearby Galaxy Groups

The evolution shown in Figure 1 argues that interactions in the group environment should be particularly effective at stripping stars from galaxies and redistributing them into the diffuse intragroup light, an important precursor to the ICL in massive clusters. Curiously, though, these arguments have not always been borne out observationally. In the nearby Leo I group, searches for intragroup light using both PNe (Castro-Rodriguez *et al.* 2003) and broadband imaging (Watkins *et al.* 2014) have come up empty, particularly notable given that the system is contains a large ($\sim$ 200 kpc) HI ring thought to be collisional in origin (Michel-Dansac *et al.* 2010). Similarly, the M101 group also show little sign of extended diffuse light (Mihos *et al.* 2013), despite the tidal disturbances evident in M101 and its nearby companions. Even in the clearly interacting M81/M82

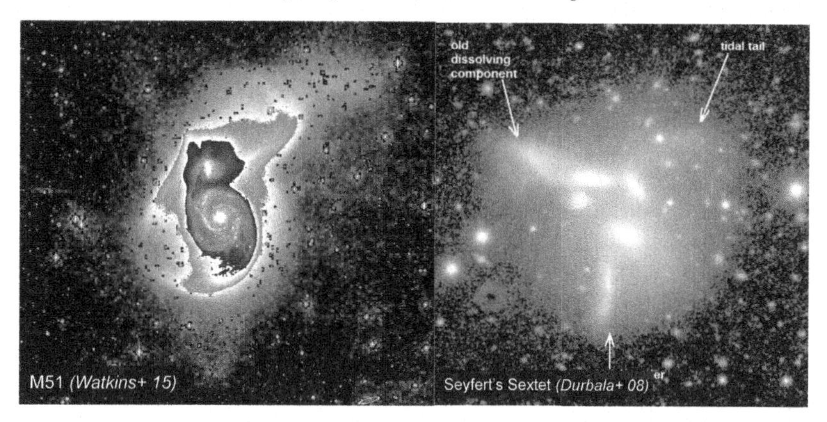

Figure 4. Left: Deep imaging of M51 showing diffuse tidal debris extending to nearly 50 kpc (Watkins *et al.* 2015). Right: Diffuse light in Seyfert's Sextet (Durbala *et al.* 2008).

group, early searches for orphaned RGB stars (Durrell *et al.* 2004) and PNe (Feldmeier *et al.* 2003) could only place upper limits on the intragroup light fraction ($\lesssim 2\%$).

In contrast, intragroup light is quite evident in dense, strongly interacting groups. The ability for these strong interactions to expel diffuse material to large distances is shown in recent deep imaging of the M51 system by Watkins *et al.* (2015; Fig 4a), where several extremely low surface brightness plumes extend nearly 50 kpc from the center. Similarly, many compact groups are awash in diffuse light (e.g. Da Rocha *et al.* 2005, 2008), including the archetypal groups Seyfert's Quintet (Mendes de Oliveira *et al.* 2001) and Seyfert's Sextet (Durbala *et al.* 2008, Figure 4b). The contrast between the copious diffuse light seen in these dense systems and the dearth of light in loose groups is striking, arguing either that the tidal debris is rapidly dispersed to even lower (undetectable) surface brightnesses, or that close interactions in loose groups are relatively uncommon.

More recently, the ability to probe discrete stellar populations in external galaxies provides a powerful new tool for studying intragroup light. Probing stellar densities far below the capabilities of wide-area surface photometry, these techniques are now revealing the diffuse light contained even in loose groups. Deep imaging by Okamoto *et al.* (2015) has uncovered the previously undetected and very extended stellar tidal debris field in the M81 group, while imaging of M31 and M33 by the PAndAS team (e.g. Ibata *et al.* 2014) has mapped the myriad of tidal streams that characterize Andromeda's extended stellar halo and trace its past interaction with M33. While at low surface brightness and containing only a small amount of the total light of their parent groups, the diffuse starlight found in these studies of nearby loose groups represent the important first step in building the intragroup and intracluster light in dense galaxy environments.

References

Arrigoni Battaia, F., Gavazzi, G., Fumagalli, M., *et al.* 2012, *A&A*, 543, A112
Blom, C., Forbes, D. A., Foster, C., *et al.* 2014, *MNRAS*, 439, 2420
Bogdán, Á., Forman, W. R., Zhuravleva, I., *et al.* 2012, *ApJ*, 753, 140
Burke, C., Collins, C. A., Stott, J. P., & Hilton, M. 2012, *MNRAS*, 425, 2058
Burke, C., Hilton, M., & Collins, C. 2015, *MNRAS*, 449, 2353
Côté, P., McLaughlin, D. E., Cohen, J. G., & Blakeslee, J. P. 2003, *ApJ*, 591, 850
Capaccioli, M., Spavone, M., Grado, A., *et al.* 2015, *A&A*, 581, A10
Castro-Rodríguez, N., Aguerri, J. A. L., Arnaboldi, M., *et al.* 2003, *A&A*, 405, 803

Ciardullo, R., Mihos, J. C., Feldmeier, J. J., Durrell, P. R., & Sigurdsson, S. 2004, Recycling Intergalactic and Interstellar Matter, 217, 88
Coccato, L., Gerhard, O., Arnaboldi, M., & Ventimiglia, G. 2011, *A&A*, 533, A138
Conroy, C., Wechsler, R. H., & Kravtsov, A. V. 2007, *ApJ*, 668, 826
Contini, E., De Lucia, G., Villalobos, Á., & Borgani, S. 2014, *MNRAS*, 437, 3787
Cooper, A. P., Gao, L., Guo, Q., *et al.* 2015, *MNRAS*, 451, 2703
Da Rocha, C. & Mendes de Oliveira, C. 2005, *MNRAS*, 364, 1069
De Lucia, G. & Blaizot, J. 2007, *MNRAS*, 375, 2
DeMaio, T., Gonzalez, A. H., Zabludoff, *et al.* 2015, *MNRAS*, 448, 1162
Dolag, K., Murante, G., & Borgani, S. 2010, *MNRAS*, 405, 1544
Durbala, A., del Olmo, A., Yun, M. S., *et al.* 2008, *AJ*, 135, 130
Durrell, P. R., Decesar, M. E., Ciardullo, R., Hurley-Keller, D., & Feldmeier, J. J. 2004, Recycling Intergalactic and Interstellar Matter, 217, 90
Durrell, P. R., Côté, P., Peng, E. W., *et al.* 2014, *ApJ*, 794, 103
Fakhouri, O., Ma, C.-P., & Boylan-Kolchin, M. 2010, *MNRAS*, 406, 2267
Feldmeier, J. J., Durrell, P. R., Ciardullo, R., & Jacoby, G. H. 2003, Planetary Nebulae: Their Evolution and Role in the Universe, 209, 605
Feldmeier, J. J., Mihos, J. C., Morrison, H. L., *et al.* 2004, *ApJ*, 609, 617
Ferrarese, L., Côté, P., Cuillandre, J.-C., *et al.* 2012, *ApJS*, 200, 4
Giallongo, E., Menci, N., Grazian, A., *et al.* 2014, *ApJ*, 781, 24
Gonzalez, A. H., Zabludoff, A. I., & Zaritsky, D. 2005, *ApJ*, 618, 195
Guennou, L., Adami, C., Da Rocha, C., *et al.* 2012, *A&A*, 537, A64
Ibata, R. A., Lewis, G. F., McConnachie, A. W., *et al.* 2014, *ApJ*, 780, 128
Janowiecki, S., Mihos, J. C., Harding, P., *et al.* 2010, *ApJ*, 715, 972
Kelson, D. D., Zabludoff, A. I., Williams, K. A., *et al.* 2002, *ApJ*, 576, 720
Krick, J. E. & Bernstein, R. A. 2007, *AJ*, 134, 466
Liu, C., Peng, E. W., Cote, P., *et al.* 2015, arXiv:1508.07334
Longobardi, A., Arnaboldi, M., Gerhard, O., & Mihos, J. C. 2015, *A&A*, 579, L3
Longobardi, A., Arnaboldi, M., Gerhard, O., & Hanuschik, R. 2015, *A&A*, 579, A135
Malin, D. F. 1979, *Nature*, 277, 279
Mei, S., Blakeslee, J. P., Côté, P., *et al.* 2007, *ApJ*, 655, 144
Melnick, J., Giraud, E., Toledo, I., Selman, F., & Quintana, H. 2012, *MNRAS*, 427, 850
Mendes de Oliveira, C., Plana, H., Amram, P., Balkowski, C., & Bolte, M. 2001, *AJ*, 121, 2524
Michel-Dansac, L., Duc, P.-A., Bournaud, F., *et al.* 2010, *ApJ Letters*, 717, L143
Mihos, J. C., Durrell, P. R., Ferrarese, L., *et al.* 2015, *ApJ Letters*, 809, L21
Mihos, J. C., Harding, P., Feldmeier, J., & Morrison, H. 2005, *ApJ Letters*, 631, L41
Mihos, J. C., Harding, P., Rudick, C. S., & Feldmeier, J. J. 2013, *ApJ Letters*, 764, L20
Mihos, J. C., Harding, P., Spengler, C. E., Rudick, C. S., & Feldmeier, J. J. 2013, *ApJ*, 762, 82
Montes, M. & Trujillo, I. 2014, *ApJ*, 794, 137
Murante, G., Giovalli, M., Gerhard, O., *et al.* 2007, *MNRAS*, 377, 2
Okamoto, S., Arimoto, N., Ferguson, A. M. N., *et al.* 2015, *ApJ Letters*, 809, L1
Presotto, V., Girardi, M., Nonino, M., *et al.* 2014, *A&A*, 565, A126
Puchwein, E., Springel, V., Sijacki, D., & Dolag, K. 2010, *MNRAS*, 406, 936
Purcell, C. W., Bullock, J. S., & Zentner, A. R. 2007, *ApJ*, 666, 20
Romanowsky, A. J., Strader, J., Brodie, J. P., *et al.* 2012, *ApJ*, 748, 29
Rudick, C. S., Mihos, J. C., Frey, L. H., & McBride, C. K. 2009, *ApJ*, 699, 1518
Rudick, C. S., Mihos, J. C., Harding, P., *et al.* 2010, *ApJ*, 720, 569
Rudick, C. S., Mihos, J. C., & McBride, C. 2006, *ApJ*, 648, 936
Rudick, C. S., Mihos, J. C., & McBride, C. K. 2011, *ApJ*, 732, 48
Seigar, M. S., Graham, A. W., & Jerjen, H. 2007, *MNRAS*, 378, 1575
Watkins, A. E., Mihos, J. C., & Harding, P. 2015, *ApJ Letters*, 800, L3
Watkins, A. E., Mihos, J. C., Harding, P., & Feldmeier, J. J. 2014, *ApJ*, 791, 38
Williams, B. F., Ciardullo, R., Durrell, P. R., *et al.* 2007, *ApJ*, 656, 756
Zhang, H.-X., Peng, E. W., Côté, P., *et al.* 2015, *ApJ*, 802, 30

The General Assembly of Galaxy Halos: Structure, Origin and Evolution
Proceedings IAU Symposium No. 317, 2015
A. Bragaglia, M. Arnaboldi, M. Rejkuba & D. Romano, eds.

doi:10.1017/S1743921315007267

New axes for the stellar mass fundamental plane

Paul L. Schechter

MIT Kavli Institute Cambridge, MA 02139, USA
email: schech@mit.edu

Abstract. We argue that the stellar velocity dispersion observed in an elliptical galaxy is a good proxy for the halo velocity dispersion. As dark matter halos are almost completely characterized by a single scale parameter, the stellar velocity dispersion tells us the virial radius of the halo and the mass contained within. This permits non-dimensionalizing of the stellar mass and effective radius axes of the stellar mass fundamental plane by the virial radius and halo mass, respectively.

Keywords. galaxies: elliptical, scaling relations

1. Introduction

With 450 papers that include the words "fundamental plane" in their titles, our purpose is less to say something new than to reassemble some of what has been said before and frame it in a way that renders the fundamental plane somewhat less mysterious.

If one measures stellar velocity dispersions, σ_*, effective radii, r_e, and effective surface brightnesses, as measured in some filter, I_e, for elliptical galaxies, the observations lie close to a plane when plotted in the space spanned by $\log\sigma_*$, $\log r_e$ and $\log I_e$ (Djorgovski & Davis 1987). Fitting a plane to the data yields three coefficients, as shown in equation 1.1.

The effective radius and surface brightness may be combined to produce a luminosity L. The ellipticals lie along a corresponding plane in the space spanned by $\log\sigma_*$, $\log r_e$ and $\log L$. If one has observations in several filters, one can calculate stellar masses, M_*, subject to considerable uncertainty in the initial mass function and its trend with velocity dispersion, and stellar surface densities, Σ_*. These also have associated planes (Hyde and Bernardi 2009). For the present discussion we consider the fundamental plane in the space spanned by $\log\sigma_*$, $\log r_e$ and $\log M_*$.

$$\underbrace{a * \log \begin{pmatrix} \text{surface brightness } I_e \\ \textit{or} \text{ luminosity } L \\ \textit{or} \text{ stellar mass } M_* \\ \textit{or} \text{ surface density } \Sigma_* \end{pmatrix} + b * \log(\text{half light radius } r_e)}_{\text{stellar (baryonic) matter}} \tag{1.1}$$
$$\underbrace{+\, c * \log(\text{ stellar velocity dispersion } \sigma_* \,)}_{\text{thesis: proxy for dark matter halo dispersion } \sigma_{\rm DM}} = 1$$

Two of these quantities, the effective radius and the stellar mass, describe a manifestly baryonic component of elliptical galaxies – the stars. We argue here that the third quantity, the stellar velocity dispersion, is a proxy for the velocity dispersion in the galaxy's dark matter halo.

2. The virial theorem

Astronomers almost always use the virial theorem in the form appropriate to the global properties of equilibrium systems,

$$2T = -U. \tag{2.1}$$

By contrast, physics texts (e.g. Goldstein 1980) often present the virial theorem in a form appropriate to the orbit of a single star,

$$\left\langle v^2 \right\rangle_{\texttt{any object}} = \left\langle \vec{r} \cdot \vec{\nabla}\Phi \right\rangle_{\texttt{the orbit}}, \tag{2.2}$$

where $\left\langle v^2 \right\rangle$ is the twice the orbit averaged kinetic energy per unit mass, $\vec{\nabla}\Phi$ is the gradient of the gravitational potential and $\vec{r}$ is position. The orbit averaged kinetic energies for stars and dark matter particles, are respectively,

$$\left\langle v^2 \right\rangle_{\texttt{any star}} \approx \left\langle \vec{r} \cdot \vec{\nabla}\Phi \right\rangle_{10^0 r_e} \quad \text{and} \tag{2.3}$$

$$\left\langle v^2 \right\rangle_{\texttt{dark particle}} \approx \left\langle \vec{r} \cdot \vec{\nabla}\Phi \right\rangle_{10^1 r_e}. \tag{2.4}$$

The stellar and dark matter velocity dispersions are the product of the *same* gravitational potential and differ *only* insofar as the quantity $\vec{r} \cdot \vec{\nabla}\Phi$ varies with radius.

But there are several lines of evidence that point to little or no variation of $\vec{r} \cdot \vec{\nabla}\Phi$ with radius.

3. The spheroid-halo conspiracy

It was the observation that spiral galaxies have circular velocities in excess of what was expected from their observable baryons that first lead to the conclusion that they were embedded in dark matter halos. But beyond that, H I rotation curves are remarkably flat, leading van Albada and Sancisi (1986) to conclude that there was a "disk-halo conspiracy." The baryons dissipate to the point where they just compensate for what would otherwise be a decline in the rotation curve in the absence of baryons.

Gavazzi *et al.*(2007) subsequently used a combination of strong and weak lensing to show that there is likewise a "spheroid-halo conspiracy" for early-type systems. The gravitational potentials are isothermal giving $\left\langle v^2 \right\rangle \sim$ constant. Humphrey and Buote (2010) use X-ray observations and likewise find a conspiracy producing isothermal potentials. Stellar velocity dispersions are therefore telling us the velocity dispersion of the dark matter.

We have marked up the fundamental plane equation 1.1 to show that while the effective radius and stellar mass are telling us about the baryons, the stellar velocity dispersion is telling us the velocity dispersion of the dark matter halo, σ_{DM}.

4. A fundamental *line* for dark matter halos

Dark matter halos are often characterized by their virial radii, r_{200} and the mass within that radius, M_{200}. It is natural to ask whether these, along with σ_{DM}, likewise organize themselves into a plane. They do not. Instead they lie along a tight line (Diemer *et al.*2013). They are governed by a single parameter, an overall scale. Parameterizing the line by the maximum observed circular velocity, one has

$$M_{200} \sim V_{max}^3 \quad \text{and} \quad r_{200} \sim V_{max}. \tag{4.1}$$

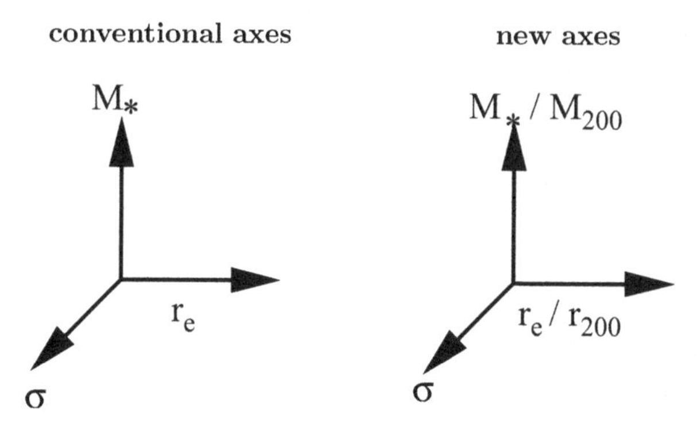

Figure 1. Axes for the stellar mass fundamental plane, with dimensioned effective radii and stellar masses unscaled on the left and scaled by the virial radius and mass of the halo (determined from the stellar velocity dispersion) on the right.

The first of these is a consequence of defining halos in terms of densities while the second is consequence of the ongoing nature of virialization and the measurement of the virial radius at the same time for all halos.

Now if a stellar velocity dispersion tells us the velocity dispersion of the dark matter halo, and if that in turn tells us the mass and radius of the dark matter halo, then we can scale the stellar mass by the mass in dark matter (or alternatively, the total mass), and we can likewise scale the effective radius by the dark matter radius,

5. New axes for the stellar mass fundamental plane

This gives us new axes for the fundamental plane, as shown in Figure 1. One axis gives us the overall scale of the system, and a second axis tells us what fraction of the the baryons that have been incorporated into stars, on the assumption that the system started out with the cosmological baryon fraction.

While the above argument invokes the spheroid-halo conspiracy in its strongest form, with $\sigma_{DM} = \sigma_*$, the argument can still be made as long as there is a well defined relation between stellar velocity dispersion and halo dispersion.

6. Why not a fundamental line for ellipticals?

The stellar components of elliptical galaxies are clearly not as simple as the halos. While the baryons and dark matter were once nearly uniformly distributed, some baryons dissipated and formed stars at the centers of the halos. The M_*/M_{200} axis tells the fraction of baryons that did so. The r_e/r_{200} axis is governed both by the dissipation prior to star formation and by the subsequent remixing of stars by mergers.

If these three processes – dissipation, star formation and remixing – were strictly governed by the scale of the halo, the ellipticals would lie along a line. If there were a stochastic component to each of them, the scatter would thicken the line into a sausage. The fact that the ellipticals are confined to a plane indicates that the three processes are coupled.

7. Antecedents

Zaritsky *et al.*(2008) anticipated our representation of the fundamental plane. They defined an "efficiency", $M_*/M_{\rm baryonic}$, which differs from our choice of coordinate by a factor of the cosmic baryon fraction. And they defined a "concentration", r_{200}/r_e, which is just the inverse of our coordinate. They plot both, separately, against stellar velocity dispersion, but do not examine the three dimensional distribution and do not argue that the planarity of the elliptical distribution indicates coupling. But it is clear from their discussion that a coupling is both possible and likely.

8. Two postdictions

There is considerable evidence that the gravitational potential inside the effective radius, and hence the velocity dispersion, is dominated by stars rather than dark matter (e.g. Treu & Koopmans, 2004, Mediavilla *et al.* 2009). Measurement of the stellar velocity dispersion beyond the effective radius might therefore reflect the halo dispersion yet more closely than the central measurements that are typically used.

Schechter *et al.*(2014) found that if one used Einstein ring radii to calculate equivalent velocity dispersions for the SLACS (Auger *et al.*2010) sample, the resulting fundamental plane is substantially tighter than the one constructed using measured stellar velocity dispersions. It was this result that led to the present formulation, so it cannot count as confirmation.

Along the same line, Falcón-Barosso and van de Ven (2011) find that the they get a tighter fundamental plane when they use velocity dispersions measured out to r_e than the one they get using dispersions interior to $r_e/8$. More extensive samples, and samples extending out to yet larger radii, are on the horizon.

References

Auger, M. W., Treu, T., Bolton, A. S., *et al.* 2010, *ApJ*, 724, 511
Diemer, B., Kravtsov, A. V., & More, S. 2013, *ApJ*, 779, 159
Djorgovski, S. & Davis, M. 1987, *ApJ*, 313, 59
Falcón-Barroso, J., van de Ven, G., Peletier, R. F., *et al.* 2011, *MNRAS*, 417, 1787
Gavazzi, R., Treu, T., Rhodes, J. D., *et al.* 2007, *ApJ*, 667, 176
Goldstein, H. 1980, *Classical Mechanics, 2nd ed.* (Addison-Wesley), p. 85
Hyde, J. B. & Bernardi, M. 2009, *MNRAS*, 396, 1171
Humphrey, P. J., & Buote, D. A. 2010, *MNRAS*, 403, 2143
Mediavilla, E., Muñoz, J. A., Falco, E., *et al.* 2009, *ApJ*, 706, 1451
Schechter, P. L., Pooley, D., Blackburne, J. A., & Wambsganss, J.*ApJ*, 793, 96
Treu, T., & Koopmans, L. V. E. 2004, *ApJ*, 611, 739
van Albada, T. S., & Sancisi, R. 1986, *Royal Society of London Philosophical Transactions Series A*, 320, 447
Zaritsky, D., Zabludoff, A. I., & Gonzalez, A. H. 2008, *ApJ*, 682, 68

The General Assembly of Galaxy Halos: Structure, Origin and Evolution
Proceedings IAU Symposium No. 317, 2015
A. Bragaglia, M. Arnaboldi, M. Rejkuba & D. Romano, eds.
doi:10.1017/S1743921315007164

Direct imaging of haloes and truncations in face-on nearby galaxies

J. H. Knapen[1,2], S. P. C. Peters[3], P. C. van der Kruit[3], I. Trujillo[1,2], J. Fliri[1,2], M. Cisternas[1,2] and L. S. Kelvin[1,2,4,5]

[1]Instituto de Astrofísica de Canarias, E-38205 La Laguna, Tenerife, Spain

[2]Departamento de Astrofísica, Universidad de La Laguna, E-38205 La Laguna, Tenerife, Spain

[3]Kapteyn Astronomical Institute, University of Groningen, P.O.Box 800, 9700 AV Groningen, the Netherlands

[4]Institut für Astro- und Teilchenphysik, Universität Innsbruck, Technikerstrasse 25, 6020 Innsbruck, Austria

[5]Astrophysics Research Institute, Liverpool John Moores University, IC2, Liverpool Science Park, 146 Brownlow Hill, Liverpool, L3 5RF, UK

Abstract. We use ultra-deep imaging from the IAC Stripe 82 Legacy Project to study the surface photometry of 22 nearby, face-on to moderately inclined spiral galaxies. The reprocessed and co-added SDSS/Stripe 82 imaging allows us to probe down to $29 - 30\,r'$-mag/arcsec^2 and thus reach into the very faint outskirts of the galaxies. We find extended stellar haloes in over half of our sample galaxies, and truncations in three of them. The presence of stellar haloes and truncations is mutually exclusive, and we argue that the presence of a stellar halo can hide a truncation. We find that the onset of the halo and the truncation scales tightly with galaxy size. We highlight the importance of a proper analysis of the extended wings of the point spread function (PSF), finding that around half the light at the faintest levels is from the inner regions of a galaxy, though not the nucleus, re-distributed to the outskirts by the PSF. We discuss implications of this effect for future deep imaging surveys, such as with the LSST.

Keywords. Galaxies: spiral, Galaxies: haloes, Galaxies: structure

---

1. Introduction

The expected surface brightness levels of haloes in face-on galaxies are similar to those of truncations. Truncations were discovered in edge-on stellar disks by van der Kruit (1979), who noticed that in successively deeper photographic exposures the radial extent of edge-on galaxies does not grow, in contrast to their vertical extent (see Fig. 10 in van der Kruit & Freeman 2011 for an illustration). We now know that truncations occur in most edge-on galaxies, are sharp with scalelengths of less than 1 kpc, and typically occur at $4-5$ times the exponential scalelength of the inner disk. They seem to be fundamental features of disks, and as the evolutionary timescales in the outer regons of galaxies are relatively long, studying truncations can have immediate impact on our understanding of the early evolution of disks.

The face-on counterparts of the truncations observed in most edge-on galaxies have remained elusive. Mostly because of the shorter line-of-sight integration through face-on disks, the surface brightness where truncations might be expected is faint, at around 27 mag/arcsec^2. Some face-on truncations have been reported as such in the literature but van der Kruit (2008) has argued that these are in fact breaks in the inner disk similar to those found by Freeman (1970)—a plausible interpretation in light of the view

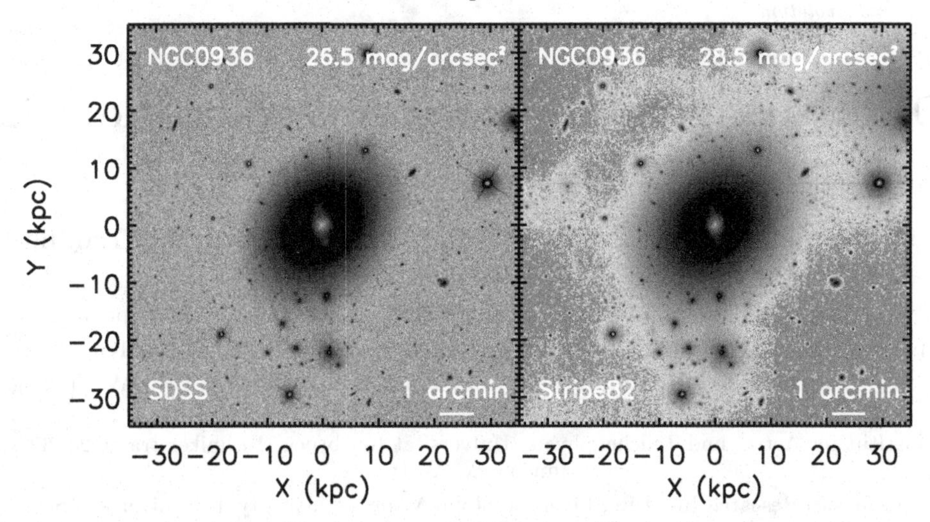

Figure 1. Standard SDSS (left) and re-reduced Stripe 82 (right) r'-band images of NGC 936, highlighting the faintest surface brightness levels. From Fliri & Trujillo (2016).

presented by Martín-Navarro *et al.* (2012) of breaks and truncations as well separated common features in disks.

We have used deep images of relatively face-on galaxies to look for truncations at surface brightness levels of 27 or below (see Peters *et al.* 2016 for full details of our study, which is only summarised in the present paper). These levels are most interesting in the context of the subject of the current Symposium, because stellar haloes are expected to become visible at roughly similar surface brightnesses. Such stellar haloes are the remants of the merger process now thought to be the dominant process shaping galaxies in ΛCDM cosmology.

As reported in detail elsewhere in these proceedings, the faint stellar haloes around galaxies can be studied in the most nearby galaxies by observing their resolved stars. In order to study haloes also beyond the nearest galaxies, and in many more objects, one needs to image the haloes directly in integrated light. Our images allow us to do this, as described below. We will also discuss to what extent light redistributed by the point spread function (PSF) can contaminate these imaged haloes, and how this informs us of difficulties ahead in future deep imaging studies.

2. Data and Analysis

We use imaging from the Stripe 82 subset of the Sloan Digital Sky Survey (SDSS), as reprocessed by us in the context of the IAC Stripe 82 Legacy Project (Fliri & Trujillo 2016). We selected a sample of 22 galaxies from the Stripe 82 area, with a diameter larger than 1 minute of arc, face-on or moderately inclined, and without signs of distortions, mergers or nearby foreground objects which might cause problems with the photometry.

The Stripe 82 images allow us to reach surface brightness levels down to $29 - 30\,r'$-mag/arcsec^2, a few magnitudes deeper than standard SDSS images. This is illustrated in Fig. 1 for one of our galaxies, NGC 936, as a comparison of a standard SDSS and a new Stripe 82 image. For our sample of 22 galaxies, we added g, r' and i images, destroying all colour information but gaining another half a magnitude in depth. We then used various techniques to identify truncations, as explained in detail in Peters *et al.* (2016). Here, we concentrate on the results obtained from standard ellipse fitting to derive radial surface brightness profiles. We inspected those profiles by eye and identified truncations as increases and haloes as decreases of the steepness of the exponential profiles (see Peters *et al.* 2016), and measured the radial distance and the surface brightness levels at which these changes, indicating the onset of the truncation or halo, occur. We also identify breaks in the profiles occurring well within the disk.

3. Effects of the PSF

Before continuing with a description of our results, we first consider whether the effects of an extended PSF can artificially flatten a radial profile, mimicking a halo signature. A PSF describes the response of an imaging system to a point source, and can be either modelled or empirically derived. We do the latter, to quantify the extent to which extended wings to a PSF, often of Gaussian shape or similar, can redistribute light from the central regions of a galaxy to the outskirts.

For the faint outer regions we are interested in, we need to combine the observed profiles of relatively faint stars in the Stripe 82 area (which yield a good representation of the central part of the PSF) with those of the very brightest stars in the area. These latter show large saturated cores, but are the only ones which allow to characterise the PSF at distances of tens of arcsec from the point source, and thus where we observe our faint structure.

After combining the PSFs from all these stars, we create a combined PSF with which we can then convolve a model of our galaxy. The results are shown for the galaxy IC 1515, a rather typical case, in Fig. 2, in terms of original and convolved models to components of the light of the galaxy, namely the central, bulge, component, and a double exponential which describes the disk. From our PSF modelling we can draw the following most interesting conclusions. (1) Indeed light from the central regions of a galaxy can be redistributed by the PSF and add to the surface brightness at levels of below 27 mag/arcsec^2, and at radial distances from the centre of the galaxy of over one minute of arc. (2) This PSF component contributes typically a few tens of per cent (some 50% in the example of Fig. 2) but is not enough to explain the flattening of the profile. We thus claim that we indeed detect haloes (see next Section). (3) The bulk of the redistributed light adding to the outer regions does *not* originate in the nuclear point source or even the central bulge component, but in the spiral arm region of the disk of a galaxy.

4. Results

Having established that the flattening observed in many of our radial surface brightness profiles at levels of 26 – 28 mag/arcsec^2 is indeed indicative of a halo component, we then proceed to characterise the profile shapes. The two main results are that we indeed detect truncations in face-on or moderately inclined disk galaxies, and that we detect the halo component in most other galaxies. In particular, we find truncations in three of our 22 sample galaxies, and haloes in 15 of them (see Peters *et al.* 2016 for full details). We also find breaks in the inner disk region in most of our galaxies.

Truncations and haloes are mutually exclusive, in the sense that we either observe a truncation, or a halo, but never both in the same galaxy (some galaxies have neither). This is easily understandable, but does imply that a truncation can only be observed if a halo is either absent, or very faint. We have studied the three galaxies in which we found this to occur, but could not find any parameter that might cause such a faint or absent halo (Peters *et al.* 2016). Most galaxies do have a halo, and we can image that directly using our very deep Stripe 82 data.

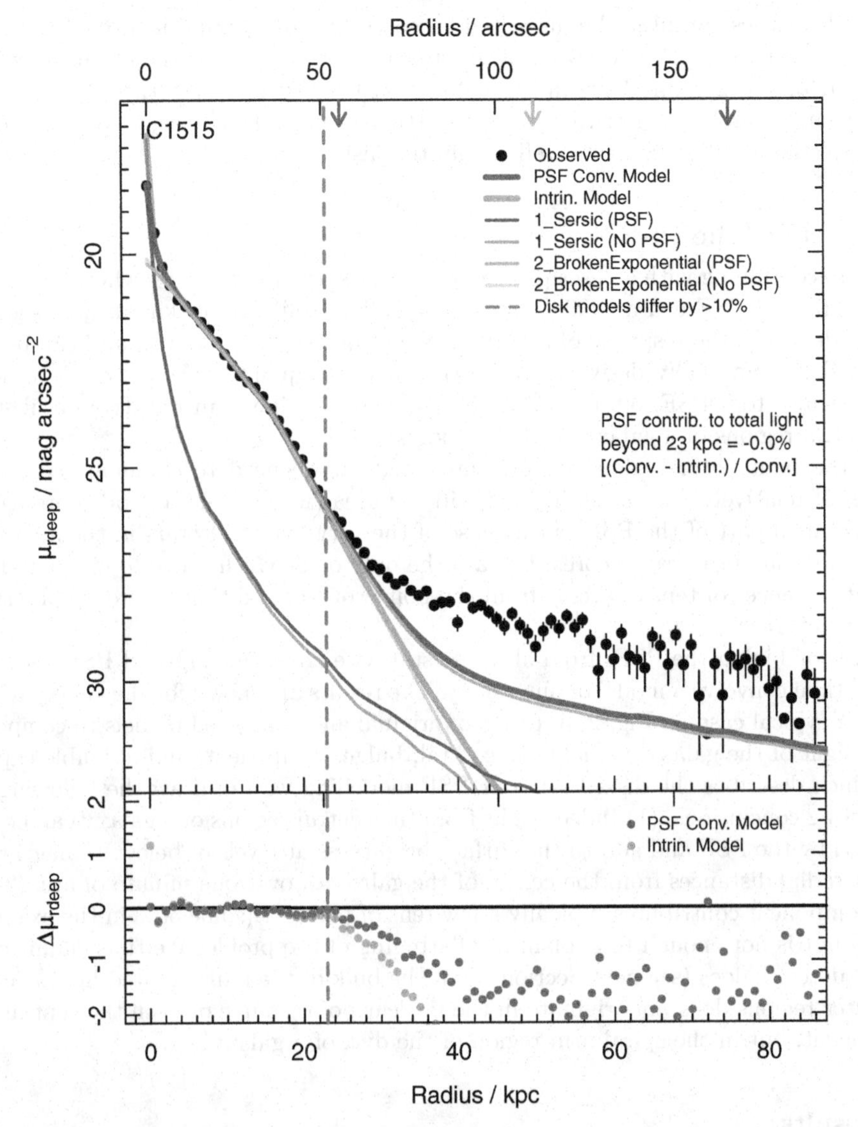

Figure 2. Graphical description of modelling whether the PSF can create an artificial halo, showing radial surface brightness profiles for IC 1515. The red curves depict a Sersic fit to the inner galaxy, corresponding to the bulge, the green curves a broken exponential, and the purple curves the sum of the two. Black dots denote the observations. In all cases, the light curves are before and the darker ones after the convolution with the PSF model. The lower panel shows the differences between the observations and the PSF-convolved and intrinsic models. The conclusion is that half the light at large radii can be PSF-scattered light, but that a halo is indeed observed. From Peters *et al.* (2016).

The physical sizes of the various features truncation, halo, and break correlate rather well with the size of their host galaxy. This is shown in Fig. 3 where we plot the radius of the onset of the feature (location for the breaks) versus the size of the galaxy, and contains some surprising aspects. First, the break radius correlates with the size of the galaxy. As the break seems to be related to the zones where spiral arms occur and bars may end (e.g., Martín-Navarro *et al.* 2012) this is probably related to structural components of galaxies being related to their host galaxy.

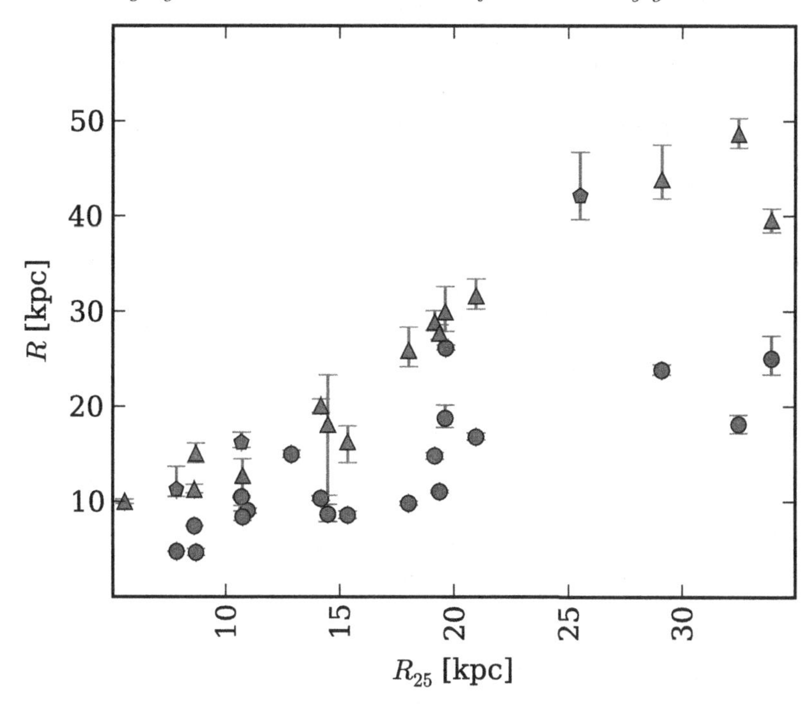

Figure 3. The correlation of R_{25}, indicative of the size of a galaxy, with the radii R of the various features. Red pentagons represent the truncations, blue circles the breaks, and green triangles the haloes. From Peters *et al.* (2016).

We also see in Fig. 3 how the onset of both truncations and haloes is tightly correlated with host galaxy size, but, in fact, following the exact same correlation. So not only do truncations and halo know about the size of their host galaxy, also do they occur (or start to become apparent) at exactly the same relative location. Correlations between other parameters of breaks, truncations, or haloes and a variety of parameters of the galaxy or its bulge or disk are not statistically significant (Peters *et al.* 2016).

5. Implications for future studies

Perhaps surprisingly to many workers in the field, the effects of the PSF extend to the faint outer regions of relatively nearby galaxies (see also Sandin 2014, 2015 for a detailed analysis). As we show here (and in Peters *et al.* 2016, see also L. Kelvin, I. Trujillo & J. Fliri, in prep.), light from the inner disk of a galaxy which has been redistributed by the PSF can contribute up to half of the light which we identify here as originating in a halo component. For the purpose of our study, and going down to some $29 - 30\,\mathrm{mag/arcsec^2}$, this cannot fully explain the observed flattening of the profile, allowing us to claim the successful detection of the halo component in 15 galaxies. But in deeper imaging the problem will become worse.

The extended wings of the PSF are primarily caused by turbulence in the atmosphere, and are also time-variable. Space-based imaging is less affected by PSF wings, and they are easier to model thanks to the lack of the varying atmosphere component. But all ground-based imaging will be affected, and the deeper one goes the more pronounced the effect. In particular surveys such as those with the LSST (Large Synoptic Survey Telescope) will be affected. Stacking multiple images will allow one, in principle, to reach limits in surface brightness of around $33\,\mathrm{mag/arcsec^2}$, but whether faint extended structure in galaxies can be detected and believed at those depths critically depends on the ability to model the PSF.

Another problem with such deep imaging is background subtraction. As described in detail in Peters *et al.* (2016, see also Fliri & Trujillo 2016) significant modelling and subsequent subtraction of the residual background was necessary before we could use our Stripe 82 images. This involves extrapolating over the area of the galaxy. But a typical Virgo galaxy such as NGC 4321 (at a distance of some 17 Mpc) will have a diameter of 50 arcmin (240 kpc) at a depth of 33 mag/arcsec^2, extrapolating the profile of its exponential disk. In the presence of a flatter halo profile, it will become even larger. So the background modelling will not only have to be done outside this huge region, but will then have to be extrapolated across the galaxy. These are serious challenges that must be overcome before the far outer regions of galaxies can be imaged to such extreme depth.

6. Conclusions

We report on our detection of truncations in three relatively face-on galaxies, found in a sample of 22 galaxies for which we analysed deep Stripe 82 images. In another 15 galaxies, we found a flattening of the radial surface brightness profile which we interpret as the direct detection of a stellar halo. The presence of a truncation or a halo is mutually exclusive. Light from the inner galaxy which has been redistributed to the outskirts by the PSF can add up to half the light in the outer, halo, regions, but the remaining excess light (above an extrapolated exponential profile) is due to a stellar halo. We caution that the combined difficulties in modelling the PSF and modelling and subtracting the background will make deeper direct imaging of outer regions of galaxies extremely challenging. This is particularly true for ground-based imaging, including stacked images from the LSST.

7. Acknowledgments

We acknowledge financial support to the DAGAL network from the People Programme (Marie Curie Actions) of the European Union's Seventh Framework Programme FP7/2007-2013/ under REA grant agreement number PITN-GA-2011-289313, and from the Spanish Ministry of Economy and Competitiveness (MINECO) under grant numbers AYA2013-48226-C3-1-P (IT) and AYA2013-41243- P (JHK). PCvdK thanks the IAC for financial support through their Severo Ochoa visitors programme. We also acknowledge use of SDSS imaging data.

References

Fliri, J. & Trujillo, I. 2016, *MNRAS*, 456, 1359
Freeman, K. C. 1970, *ApJ*, 160, 811
Martín-Navarro, I., Bakos, J., Trujillo, I., *et al.* 2012, *MNRAS*, 427, 1102
Peters, S. P. C., van der Kruit, P. C., Knapen, J. H. *et al.* 2016, *MNRAS*, submitted
Sandin, C. 2014, *A&A*, 567, A97
Sandin, C. 2015, *A&A*, 577, A106
van der Kruit, P. C. 1979, *A&AS*, 38, 15
van der Kruit, P. C. 2008, *Formation and Evolution of Galaxy Disks*, 396, 173
van der Kruit, P. C. & Freeman, K. C. 2011, *ARAA*, 49, 301

The General Assembly of Galaxy Halos: Structure, Origin and Evolution
Proceedings IAU Symposium No. 317, 2015
A. Bragaglia, M. Arnaboldi, M. Rejkuba & D. Romano, eds.

doi:10.1017/S174392131500959X

Very Low-Mass Stars with Extremely Low Metallicity in the Milky Way's Halo

Wako Aoki[1], Timothy C. Beers[2], Takuma Suda[3], Satoshi Honda[4] and Young Sun Lee[5]

[1]National Astronomical Observatory of Japan,
Mitaka, Tokyo, Japan
email: aoki.wako@nao.ac.jp

[2]University of Notre Dame, Notre Dame, IN, United States

[3]University of Tokyo, Tokyo, Japan

[4]University of Hyogo, Sayo, Hyogo, Japan

[5]Chungnam National University, Daejeon 305-764, Republic of Korea

Abstract. Large surveys and follow-up spectroscopic studies in the past few decades have been providing chemical abundance data for a growing number of very metal-poor ([Fe/H]<-2) stars. Most of them are red giants or main-sequence turn-off stars having masses near 0.8 solar masses. Lower mass stars with extremely low metallicity ([Fe/H]<-3) are yet to be explored. Our high-resolution spectroscopic study for very metal-poor stars found with SDSS has identified four cool main-sequence stars with [Fe/H]<-2.5 among 137 objects (Aoki *et al.* 2013). The effective temperatures of these stars are 4500–5000 K, corresponding to a mass of around 0.5 solar masses. Our standard analysis of the high-resolution spectra based on 1D-LTE model atmospheres has obtained self-consistent chemical abundances for these objects, assuming small values of micro-turbulent velocities compared with giants and turn-off stars. The low temperature of the atmospheres of these objects enables us to measure their detailed chemical abundances. Interestingly, two of the four stars have extreme chemical-abundance patterns: one has the largest excesses of heavy neutron-capture elements associated with the r-process abundance pattern known to date (Aoki *et al.* 2010), and the other exhibits low abundances of the α-elements and odd-Z elements, suggested to be signatures of the yields of very massive stars (>100 solar masses; Aoki *et al.* 2014). Although the sample size is still small, these results indicate the potential of very low-mass stars as probes to study the early stages of the Milky Way's halo formation.

Keywords. stars:abundances, stars:fundamental parameters, stars:individual(SDSS J001820.51-093939.2, stars:low-mass, stars:population II, stars:subdwarfs)

---

1. Introduction

Very metal-poor stars are believed to be ancient objects that were born in the early phases of the Milky Way's formation. They should be low-mass ($M<1$ $M_{\odot}$) stars with long lifetimes (>10 billion years), containing rich information about the nucleosynthetic yields of the first generations of stars and early chemical evolution, as well as the assembly history of the Milky Way. The large-scale surveys of very metal-poor stars and follow-up high resolution spectroscopy in the past few decades have accumulated chemical abundance data for a large number of stars (Frebel & Norris 2015). According to the SAGA database (Suda *et al.* 2008), which is collecting information on metal-poor stars, chemical compositions have been reported for about 1000 stars with [Fe/H]<-2.5. Most of the stars studied so far are red giants or main-sequence turn-off stars. By comparison,

Table 1. Cool main-sequence, very metal-poor stars found with SDSS and Subaru

Object	Object Name	$T_{\rm eff}$ (K)	log g	[Fe/H]	$v_{\rm Helio}$ (km s^{-1})	M ($M_\odot$)	d (pc)
SDSS J0018−0939	SDSS J001820.51−093939.2	4600	5.0	−2.5	−122.9	0.47	300
SDSS J0259+0057	SDSS J025956.45+005713.3	4550	5.0	−3.3	+35.7	0.47	380
SDSS J1703+2836	SDSS J170339.60+283649.9	5000	4.8	−3.2	−147.9	0.52	370
SDSS J2357−0052	SDSS J235718.91−005247.8	5000	4.8	−3.4	−9.4	0.52	440

Notes:
Stellar name, atmospheric parameters, and radial velocities are taken from Aoki *et al.* (2013). Stellar mass and luminosity are estimated from isochrones (Kim *et al.* 2002), and the distance is estimated from the luminosity and the apparent magnitude.

studies of cool main-sequence stars with very low metallicity are quite limited (e.g., Yong *et al.* 2003a).

This is not surprising, given the faintness of the cool main-sequence stars. According to isochrones for very low metallicity ([Fe/H]= −3.5; e.g., Kim *et al.* 2002), red giants with effective temperature ($T_{\rm eff}$) of 5000 K (0.8 $M_\odot$) have luminosity (L) as high as 100-200 $L_\odot$, and main-sequence stars with $T_{\rm eff}$=6000 K have 0.7 $L_\odot$. The luminosity of cool main-sequence stars with $T_{\rm eff}$ of 5000 K (0.5 $M_\odot$) is only 0.1 $L_\odot$. Hence, the small sample of cool main-sequence stars with extremely low metallicity obtained so far does not indicate that very low-mass stars are not formed at low metallicity, just that they are difficult to find. Future surveys of metal-poor stars including cool main-sequence stars have the potential to constrain the initial mass function of low-mass stars as a function of metallicity.

Cool main-sequence stars are also useful to study molecular features that provide unique opportunities to measure elemental and isotopic abundances. For instance, MgH features have been systematically measured by Yong *et al.* (2003b) to investigate Mg isotope ratios.

2. Low-mass very metal-poor stars found with SDSS and Subaru

We have conducted high-resolution spectroscopy for candidate very metal-poor stars found with Sloan Digital Sky Survey (SDSS; York *et al.* 2000). The results obtained for the full sample (137 stars) were reported by Aoki *et al.* (2013). The sample includes four main-sequence stars with $T_{\rm eff}$ = 4500–5000 K (Table 1).

The surface gravities of these stars are estimated by demanding that the Fe abundances derived from neutral and ionized species (i.e., Fe I and Fe II lines) are consistent. For SDSS J 0018-0939, the value obtained by this assumption is slightly larger than $\log g$ = 5.0. For this star $\log g$ = 5.0, the value expected from the isochrone, is adopted.

Estimates of Fe abundances from SDSS spectra are compared with the determination from the high-resolution spectra obtained with Subaru/HDS in Figure 1 of Aoki *et al.* (2013). The comparison shows no significant offset in general, although the scatter is as large as 0.5 dex for extremely metal-poor stars. The Fe abundances estimated from the SDSS spectra for the four cool main-sequence stars are, however, systematically lower than the values obtained by the analyses of high-resolution spectra. One reason for this discrepancy is the difference in the gravity estimates: the high-resolution spectra provide clear evidence for high gravity ($\log g \sim 5$), such as the weakness of ionized Fe features and the broad wings of strong absorption lines. Such information was not available in the pipeline analyses for SDSS data (Lee *et al.* 2008), and relatively low gravities ($\log g \sim 3$–4) was derived. Further calibration will be useful for the analysis of SDSS spectra of cool main-sequence stars.

3. A signature of very massive first stars

Chemical-abundance ratios obtained for three of the four cool main-sequence stars studied by Aoki *et al.* (2013) are similar to those found for main-sequence turn-off stars and red giants (Fig. 1). Aoki *et al.* (2014) also confirmed that our analysis for the comparison star G 39-36, which is a bright object with similar stellar parameters, obtains similar results. The abundance patterns of these stars are well-explained by the yields of core-collapse supernovae with mass of several tens of solar masses (Fig. 1).

However, among the four cool main-sequence stars found in our SDSS/Subaru sample, SDSS J 0018–0039 exhibits peculiar chemical abundance patterns (Aoki *et al.* 2014). While the Fe abundance of this star is not extreme ([Fe/H]~ -2.5), light elements such as C and Mg are deficient ([C/Fe]$= -0.8$ and [Mg/Fe]$= -0.5$). The low abundance of Co ([Co/Fe]$= -0.7$) is also remarkable. This abundance pattern is not explained by nucleosynthesis models for the usual core-collapse supernovae. Instead, Aoki *et al.* (2014) discussed that an explosion of a very massive star (> 100 $M_\odot$) is a possible source of this peculiar abundance pattern (Fig. 2).

In order to investigate the abundance pattern in more detail, we obtained a high-resolution ($R = 60,000$) spectrum, with higher S/N ratio, with Subaru/HDS in August 2014. A preliminary analysis of the new spectrum confirms the abundance properties of this star. New results obtained from the high-S/N spectrum are as follows;

- Fe abundances from ionized species: The Fe abundances obtained from absorption lines of neutral and singly ionized Fe exhibit a discrepancy in the previous analysis if $\log g = 5.0$ is adopted. In order to obtain consistent Fe abundances from both species, a higher $\log g$ value needs to be assumed, which is not realistic according to isochrones for metal-poor main-sequence stars (Kim *et al.* 2002). This problem is not solved by the analysis of the new spectrum. The number of Fe II lines detected is, however, still quite small, and the result is not very conclusive. It should be noted that the Ti abundances derived from Ti I and Ti II show good agreement if $\log g = 5.0$ is adopted.
- Si abundance: The Si abundance, estimated from the saturated Si I line at 3906 Å in the previous study (Aoki *et al.* 2014), is quite low ([Si/Fe]$= -0.4$). The preliminary analysis of another Si line (Si I 4102 Å) for our new spectrum confirms the under-abundance of Si. The result is slightly dependent on the treatment of contamination of the SiH molecular feature, and more careful analysis is required to obtain a final result.
- MgH feature: Although molecular features are strong, in general, in the cool and dense atmospheres of low-mass main-sequence stars, only weak MgH features are found in the new spectrum of SDSS J0018–0939. This confirms the low abundance of Mg of this object.
- New upper limits: We obtained new and stronger upper limits for Zn, Cu, and Ba, which provide even tighter constraints on the origin of the peculiar abundances of this star.

The abundance pattern derived by the preliminary analysis for the new spectrum confirms the results obtained by our previous study. The low Si abundance suggests that the abundance pattern of this star is not well-explained by pair-instability supernovae, expected to be explosions of very massive stars (140-300 $M_\odot$), and core-collapse supernovae of more massive objects (Ohkubo *et al.* 2006) might be preferable for the progenitor (Fig. 2). The upper limit of Zn abundance would be another constraint on the progenitor.

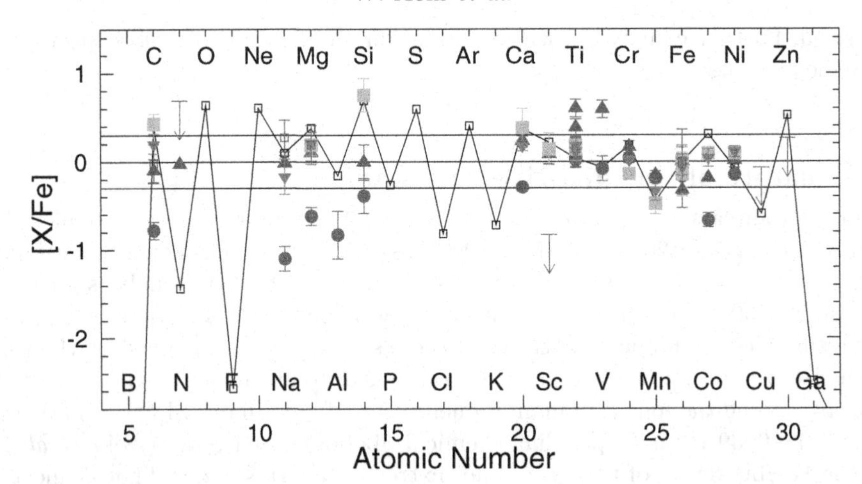

Figure 1. Chemical abundance pattern of the four cool main-sequence stars found with SDSS and Subaru observations (Aoki *et al.* 2010; 2013; 2014), and a comparison star G 39-36 (blue triangles: Aoki *et al.* 2014). The result for SDSS J0018−0939 (red filled circles) is updated by the preliminary analysis for the new spectra reported in the present work. The solid line is predictions of a core-collapse supernova model for usual massive stars (Tominaga *et al.*, private communication). The abundance patterns of stars other than SDSS J0018−0939 are well-explained by the model, whereas SDSS J0018−0939 exhibits clear under-abundances of C, Mg, and Co.

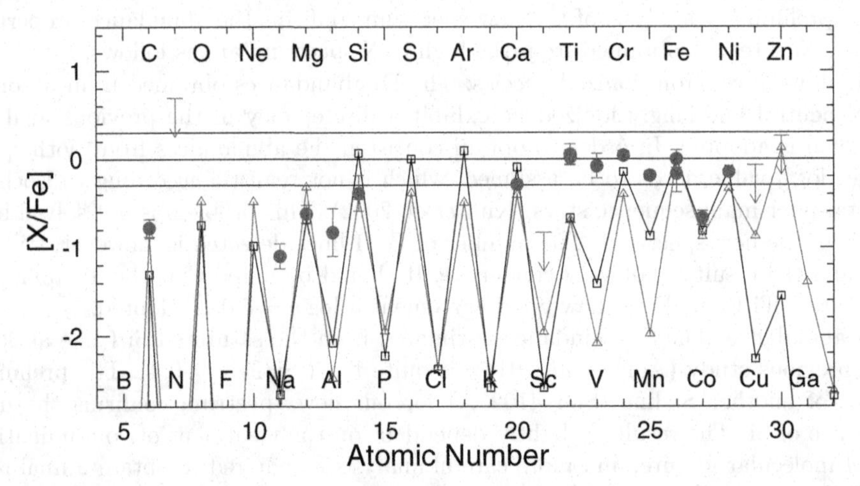

Figure 2. Chemical abundance pattern of SDSS J0018−0939 (red filled circles), compared with model predictions for a pair instability supernova of a ∼300 M$_\odot$ star (the solid blue line with squares), and for a core collapse supernova of a more massive object (∼ 1000 M$_\odot$; the solid line with triangles). See text for more details.

4. An r-process-element enhanced star

Another interesting object among the four cool main-sequence stars found with SDSS/Subaru is SDSS J2357−0052, as it exhibits large excesses of heavy neutron-capture elements (Aoki *et al.* 2010). Eu II lines are clearly detected in this object, and abundance ratios of neutron-capture elements (e.g., [Ba/Eu], [La/Eu]) indicate that their origin is the r-process. The star is classified as an "r-II", which have Eu ([Eu/Fe] $> +1.0$). This is the first (and so far, unique) example of a cool main-sequence star among the r-II stars known to date.

The metallicity of this star ([Fe/H] = −3.4) is the lowest among the r-II stars. As a result, the excess of r-process elements with respect to Fe is the highest yet found. This means that r-process nucleosynthesis has occurred even at such low metallicity, and the products have been preserved in some low-mass objects. Exploring the astrophysical sites of the r-process is currently a hot topic in nuclear astrophysics, and the distribution of metallicity and abundance ratios of heavy elements in metal-poor stars are regarded as useful constraints on the models (e.g., Wehmeyer *et al.* 2015; Ishimaru *et al.* 2015). SDSS J2357−0052 could provide one of the strongest constraints.

5. Summary and future prospects

Cool main-sequence stars appear to be common among very/extremely metal-poor stars, although the number of objects that have been well-studied to date is small, due to their intrinsic faintness. The sample size should increase with the advent of new surveys of very metal-poor stars in the near future, e.g., LAMOST. The fraction of cool main-sequence stars among very metal-poor stars will help determine the initial mass function for $M < 0.8$ $M_\odot$ stars at very low metallicity.

Our abundance analysis of four cool main-sequence stars with [Fe/H]< -2.5 discovered with SDSS confirms that typical abundance ratios derived for main-sequence turn-off stars and red giants with very low metallicity are also found in most cases. Large excesses of heavy neutron-capture elements are found for one of the four stars, and the abundance pattern is also similar to that obtained for previously reported r-II stars. Another star, however, exhibits very peculiar abundance patterns for the α and Fe-peak elements that is not well-explained by the usual core-collapse supernovae of massive stars.

There is no clear reason why stars with such remarkable abundance ratios are contained in the small sample of cool main-sequence stars. Future studies for larger samples of such objects will enable us to discuss the statistics of these stars.

Acknowledgments. This work is based on data collected at the Subaru Telescope, which is operated by the National Astronomical Observatory of Japan. We would like to thank N. Tominaga for providing us with plots shown in Figure 1 with his model calculations. W.A. and T.S. are supported by the JSPS Grant-in-Aid for Scientific Research (S:23224004). S. H. is supported by the JSPS Grant-in-Aid for Scientific Research (C:26400231). T.C.B. acknowledges partial support from grant PHY 08-22648: Physics Frontiers Center/Joint Institute for Nuclear Astrophysics (JINA), and PHY 14-30152; Physics Frontier Center/JINA Center for the Evolution of the Elements (JINA-CEE), awarded by the U.S. National Science Foundation. Y.S.L. acknowledges support provided by the National Research Foundation of Korea to the Center for Galaxy Evolution Research (No. 2010-0027910) and the Basic Science Research Program through the National Research Foundation of Korea (NRF) funded by the Ministry of Science, ICT & Future Planning (NRF-015R1C1A1A02036658).

References

Aoki, W., Beers, T. C., Honda, S., & Carollo, D. 2010, *ApJL*, 723, L201
Aoki, W., Beers, T. C., Lee, Y. S., *et al.* 2013, *AJ*, 145, 13
Aoki, W., Tominaga, N., Beers, T. C., Honda, S., & Lee, Y. S. 2014, *Science*, 345, 912
Frebel, A., & Norris, J. E. 2015, *ARAA*, in press, arXiv:1501.06921
Ishimaru, Y., Wanajo, S., & Prantzos, N. 2015, *ApJL*, 804, L35
Kim, Y.-C., Demarque, P., Yi, S. K., & Alexander, D. R. 2002, *ApJS*, 143, 499

Lee, Y. S., Beers, T. C., Sivarani, T., *et al.* 2008, *AJ*, 136, 2022
Ohkubo, T., Umeda, H., Maeda, K., *et al.* 2006, *ApJ*, 645, 1352
Suda, T., Katsuta, Y., Yamada, S., *et al.* 2008, *PASJ*, 60, 1159
Wehmeyer, B., Pignatari, M., & Thielemann, F.-K. 2015, *MNRAS*, 452, 1970
Yong, D., & Lambert, D. L. 2003a, *PASP*, 115, 796
Yong, D., Lambert, D. L., & Ivans, I. I. 2003b, *ApJ*, 599, 1357
York, D. G., Adelman, J., Anderson, J. E., *et al.* 2000, *AJ*, 120 1579

The General Assembly of Galaxy Halos: Structure, Origin and Evolution
Proceedings IAU Symposium No. 317, 2015
A. Bragaglia, M. Arnaboldi, M. Rejkuba & D. Romano, eds.

doi:10.1017/S1743921315010789

Searching for chemical relics of first stars with LAMOST and Subaru

Haining Li[1,2], Wako Aoki[3,4], Gang Zhao[1], Satoshi Honda[5], Norbert Christlieb[6] and Takuma Suda[7]

[1]Key Lab of Optical Astronomy, National Astronomical Observatories, Chinese Academy of Sciences, Beijing 100012, China
[2]email: lhn@nao.cas.cn

[3] National Astronomical Observatory of Japan, 2-21-1 Osawa, Mitaka, Tokyo, 181-8588, Japan

[4] School of Physical Sciences, The Graduate University of Advanced Studies (SOKENDAI), 2-21-1 Osawa, Mitaka, Tokyo 181-8588, Japan

[5] University of Hyogo, 407-2, Nishigaichi, Sayo-cho, Sayo, Hyogo, 679-5313, Japan

[6] Zentrum für Astronomie der Universität Heidelberg, Landessternwarte, Königstuhl 12, D-69117 Heidelberg, Germany

[7] Research Center for the Early Universe, The University of Tokyo, Hongo 7-3-1, Bunkyo-ku, Tokyo 113-0033, Japan

Abstract. We report progresses of a joint project on searching for extremely metal-poor (EMP) stars based on LAMOST survey and Subaru follow-up observation. Follow-up high-resolution snapshot spectra have been obtained for 70 objects, resulting in 42 EMP stars. A number of chemically interesting objects have already been identified, including (1) Two UMP (ultra metal-poor) stars with [Fe/H] ~ -4.0. One of them is the second UMP turnoff star with Li detection. (2) A super Li-rich (A(Li) ~ 3.1) EMP giant. This is the most metal-poor and extreme example of Li enhancement in giants known to date, and will shed light on Li production during the evolution of red giants. (3) A few EMP stars showing extreme overabundance in heavy elements. Detailed abundances of these extreme objects and statistics obtained by the large sample of EMP stars will provide important constraints on the Galactic halo formation.

Keywords. stars:abundances, stars: Population II, nucleosynthesis

1. Introduction

Extremely metal-poor ([Fe/H]† < -3.0, EMP) stars are believed to record chemical and dynamical features of the early universe, since their atmosphere preserves signature of the gas at the time and place that they were born. EMP stars and stars with even lower metallicities such as ultra metal-poor ([Fe/H] < -4.0, UMP) and hyper metal-poor ([Fe/H] < -5.0, HMP) stars provide fundamental knowledge to the formation of first generation of stars (Frebel & Norris 2015), the nucleosynthesis yields of first supernovae (Nomoto *et al.* 2013), and the primordial nucleosynthesis model (Boesgaard & Steigman 1985).

It is now clear that a large fraction of stars with low metallicities present significant enhancements of carbon and are usually referred to carbon-enhanced metal-poor (CEMP) stars. There are also several subclasses of CEMP stars based on different abundance patterns of heavier elements, especially the neutron-capture elements (e.g., defined by Beers & Christlieb 2005), including CEMP-s (enriched in $s-$process elements), CEMP-

† $[A/B] = \log(N_A/N_B)_\star - \log(N_A/N_B)_\odot$, where N_A and N_B are the number densities of elements A and B respectively, and $\star$ and $\odot$ refer to the star and the Sun respectively

rs (enriched in both $s-$ and $r-$process elements), and CEMP-no (no enhancement in neutron-capture elements). Statistical studies on CEMP stars have revealed that at lower metallicities, CEMP-no becomes dominant (Norris *et al.* 2013; Carollo *et al.* 2014). However, the origin of this subclass is not yet well understood (Masseron *et al.* 2010), although various models try to explain the observed abundance pattern, such as mass transfer from the AGB companion (Suda *et al.* 2004), carbon-rich winds of massive rotating EMP stars (Cescutti *et al.* 2013), and faint supernovae associated with first stars (Umeda & Nomoto 2005; Nomoto *et al.* 2013).

Abundances of the slow ($s-$) and rapid ($r-$) neutron-capture elements among metal-poor stars are very important to constrain early nucleosynthesis. Previous studies reveal that in the early Galaxy, the $r-$process is the primary contributor to the production of elements heavier than the iron group (Sneden *et al.* 1996). Only at higher metallicities (i.e., later time), the effect of $s-$process starts to emerge (Burris *et al.* 2000). A few stars have been found to exhibit extreme enhancements in $r-$process elements, e.g., the first $r-$process enhanced EMP giant found by Sneden *et al.* 1994, CS 22892−052 with a [Eu/Fe] $\sim +1.6$. Stars with [Eu/Fe] $> +1$ and [Ba/Eu] < 0 are referred to $r-$II stars (Beers & Christlieb 2005). An important result is that, the abundance pattern from Ba through Dy of the $r-$II stars is similar to the scaled solar system $r-$process (SSr) pattern. The astrophysical site of the $r-$process is not yet clear, but it is linked to explosive conditions of neutron star mergers (Goriely *et al.* 2013), or massive-star core-collapse supernovae (Woosley *et al.* 1994). Therefore, $r-$II stars are regarded as the best candidates to explore the nature of the $r-$process and its site.

Lithium abundances of metal-poor main-sequence turnoff stars provide important observational constraints on a number of basic questions, including the origin of different CEMP subclasses (e.g., Masseron *et al.* 2012; Hansen *et al.* 2014), the primordial Li production in the Big Bang nucleosynthesis (Asplund *et al.* 2006; Spite *et al.* 2013), etc. A plateau of lithium abundances around A(Li)$\sim$ 2.2 † has been observed among metal-poor main-sequence turnoff stars, and it seems to "meltdown" when it goes down to [Fe/H] < -3.0 (Sbordone *et al.* 2010). On the other hand, red giants are expected to show low lithium abundances due to the first dredge-up which mixes the surface with internal Li-depleted material. However, a few metal-poor giants are found to show significant excess of lithium, which has raised challenges to the standard stellar evolution theory. Various modifications have been adopted to the standard model to explain the observed features of Li-rich giants at different evolutionary stages, including Charbonnel & Balachandran (2000) for the RGB bump, Nollett *et al.* (2003) for the AGB, Sackmann & Boothroyd (1999) for the RGB as well as extra mixing, etc.

2. Opportunity with LAMOST

Large scale survey including Hamburg/ESO survey (HES, Christlieb *et al.* 2008), and Sloan Digital Sky Survey and Sloan Extension for Galactic Understanding and Exploration (SDSS/SEGUE, Yanny *et al.*2009) has tremendously increased the number of candidate metal-poor stars. High-resolution spectroscopic follow-up observations of these candidates have resulted in detailed chemical abundances of more than 300 EMP stars (e.g., Cayrel *et al.* 2004; Norris *et al.* 2013; Aoki *et al.* 2013; Roederer *et al.* 2014). However, EMP stars and metal-poor stars with peculiar chemical abundances are quite rare, e.g., the number of stars with [Fe/H] < -4.0 is no larger than 20, and the number of identified metal-poor $r-$II stars is only 12. Hence additional EMP stars with extremely

† $A(\mathrm{Li}) = 12 + log[n(Li)/n(H)]$ where n is the number density of atoms

low metallicities or peculiar abundance patterns are very important to further explore stellar evolution and the enrichment of the earliest Milky Way.

The Large sky Area Multi-Object fiber Spectroscopic Telescope, LAMOST † (also known as Wang-Su Reflecting Schmidt Telescope or Guoshoujing Telescope, Cui *et al.* 2012) started its 5-year regular survey in 2012 (Zhao *et al.* 2012). The unique design of LAMOST combines a 4-meter large aperture, a 5-degree field of view, and 4000 fibers on the focal plane. It can averagely observe about 3400 targets at one exposure, and thus allows to carry out large scale spectroscopic surveys of the Milky Way. Based on LAMOST low-resolution (R=1800) spectra covering 3700−9100 Å, one can reliably identify candidate metal-poor stars in the survey mode, and hence notably enhance the searching efficiency. LAMOST has already obtained more than 3 million stellar spectra in the first two years, which is a huge database compared to previous stellar spectroscopic survey. More importantly, LAMOST survey is designed in such a way that the selection bias on spectral types (i.e., colors) are negligible (Carlin *et al.* 2012). Therefore it provides a great opportunity to enlarge the sample size of metal-poor stars and is also suitable for statistical studies.

We have selected more than 200 candidate metal-poor stars from the second data release of LAMOST spectroscopic survey. EMP candidates have been selected based on stellar parameters including metallicities determined from LAMOST spectra for all stars with S/N higher than 20 in the g-band, adopting methods by Li *et al.* (2015a).

3. Follow-up with Subaru and early results

For 70 of the selected candidate metal-poor stars, "snapshot" high-resolution spectra were acquired with the resolving power R=36,000 and exposure times of 10−20 minutes during the two and half clear nights in May 2014 and March 2015 with the High Dispersion Spectrograph (HDS) at Subaru. This is similar to the observing mode as made by Aoki *et al.* (2013). After a quick visual check of the snap-shot spectra, a few interesting objects including two ultra metal-poor stars and a super Li-rich EMP star were selected out of the observed sample. Then for these targets, spectra with higher resolving power (R=60,000) and higher signal-to-noise ratio covering 4000−6800 Å were further obtained for detailed abundance analysis.

Due to different signal-to-noise ratio and spectral quality of the snapshot spectra, stellar parameters have been estimated for 52 stars out of the 70 observed targets (with their location in $T_{\rm eff}$ vs. $\log g$ diagram shown in Figure 1). The resultant metallicities confirm that all the measured 52 candidates are metal-poor, and there are 42 EMP stars with [Fe/H] < -3.0, including 17 with [Fe/H] < -3.5 and 2 with [Fe/H] < -4.0. Such a result infers a very efficient selection, reaching a fraction of about 60% for the observed candidates and 80% for the targets whose parameters can be determined in identifying truly extremely metal-poor stars. It should be noted that the metallicity estimation depends on the determination of the effective temperature. The effective temperature is determined here using the spectroscopic method, i.e., by adjusting the temperature to make the Fe abundances derived from individual Fe I lines independent of the excitation potentials. For targets with fewer Fe I lines, more detailed analysis might result in different values of effective temperatures and metallicities.

The two UMP stars which were further observed with higher-resolution and higher-S/N include LAMOST J1253+0753, one new discovery to the other dozen of such rare objects, and LAMOST J1313−0552, an independent discovery of HE 1310−0520 (Hansen *et al.*

† See http://www.lamost.org for more detailed information, and the progress of the LAMOST surveys.

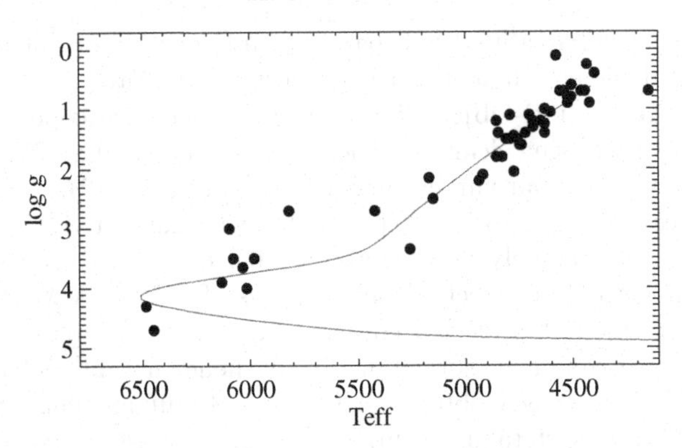

Figure 1. Distribution in the $T_{\rm eff}$ vs. $\log g$ diagram of the 52 observed targets whose metallicities can be accurately determined from the Subaru snapshot spectra. A theoretical isochrone corresponding to a metallicity [Fe/H] ~ -3.0 and an age of 13 Gyr (Demarque *et al.* 2004) is plotted for reference.

2014). Detailed abundance analysis shows that both UMP stars are carbon-enhanced, consistent with the scheme of higher fraction of CEMP stars at lower metallicities. The fact that neither of them shows an enhancement in neutron-capture elements support the result of previous studies that CEMP-no stars are dominant in the extremely low-metallicity region. Moreover, for the UMP turnoff LAMOST J1253+0753, we could determine its lithium abundance from the Li 6707 line (A(Li) $\sim$ 1.80) and thus make it the second UMP turnoff star with accurate Li abundance. As can be seen from the top left panel of Figure 2, this newly discovered UMP turnoff star is located below the observed Li plateau of metal-poor main-sequence turnoff stars, and provides unique evidence (the filled circle) to the so-called "meltdown" of the Li plateau around [Fe/H] ~ -4.2 where there used to be no observational point before. The Li abundance of LAMOST J1253+0753 is very close to the value of another UMP main-sequence turnoff star HE 0233−0343 with A(Li) $\sim$ 1.80. For the measured elements, both of the two UMP turnoff stars generally follow the "normal" pattern of metal-poor stars (Li *et al.* 2015b).

Another peculiar object which was selected from snapshot spectra and observed with R=60,000 is LAMOST J0705+2552, a super Li-rich EMP star. This EMP giant shows extremely strong and saturated Li 6707 line with a large equivalent width of about 290 mÅ. Moreover, the Li 6103 line is also detectable in the spectra of this object. The lithium abundance of LAMOST J0705+2552 has thus been determined from the unsaturated Li 6103 line, resulting in A(Li) $\sim$ 3.10 (the filled star symbol in Figure 2). The lithium content of LAMOST J0705+2552 far exceeds the average lithium abundance of normal metal-poor giants (i.e., A(Li) $\sim$ 1.0, the dashed line in the top left panel) and the Li plateau of main-sequence turnoff stars (A(Li) $\sim$ 2.2, the dash-dotted line), and makes it the most Li-rich giant with such low metallicity ([Fe/H] ~ -3.3). Except for Li, LAMOST J0705+2552 also shows excess in abundances of N ([N/Fe] $\sim$ 2.97), Na ([Na/Fe] $\sim$ 1.67) and Mg ([Mg/Fe] $\sim$ 0.96). It is quite difficult to enhance nitrogen by the first dredge-up in EMP stars, and extra mixing would be necessary to enrich the surface of LAMOST J0705+2552 to reach such extreme enhancement in nitrogen (Suda & Fujimoto 2010). More interestingly, no nitrogen-enhanced metal-poor star has been reported to date showing large Li excess, and further detailed investigation on this object will help to understand the evolutionary status of this first super Li-rich nitrogen-enhanced EMP giant.

Among the observed 52 targets, we have also found an EMP star LAMOST J1109+0754 showing extreme enhancements in heavy elements. Although this object was only

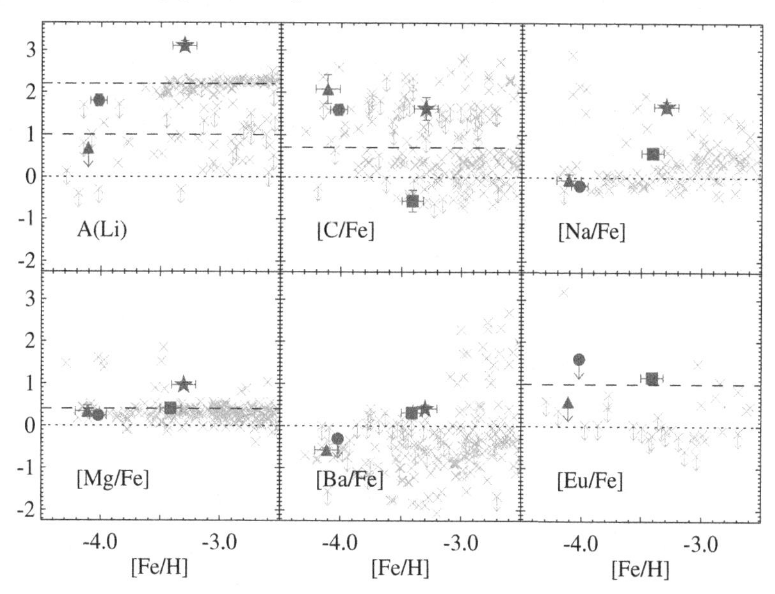

Figure 2. A(Li) vs. [Fe/H] and [X/Fe] vs. [Fe/H] for C, Na, Mg, Ba and Eu. For lithium, the dashed line refers to the average A(Li) among normal red giants with A(Li)$\sim$1.0, and the dash-dotted line refers to the observed Li plateau around A(Li)$\sim$2.2. For carbon, the dashed line refers to the division of carbon-enhanced and carbon-normal stars with [C/Fe] $\sim$ +0.7. For Mg, the dashed line refers to the canonical value of [α/Fe] $\sim$ +0.4 for the halo stars. Crosses correspond to literature metal-poor stars. Filled symbols correspond to the peculiar low-metallicity stars discovered and studied with LAMOST and Subaru, with the circle referring to LAMOST J1253+0753, the triangle referring to LAMOST J1313−0552, the square referring to LAMOST J1109+0754 and the star referring to LAMOST J0705+2552. Arrows correspond to upper or lower limits.

observed in snapshot mode, its being relatively bright enables us to obtain relatively high quality Subaru/HDS spectrum and to determine accurate parameters and elemental abundances for this object as well. Among the measured species, there are 11 elements in the nuclear charge range of $Z = 38 - 66$, covering the light trans-iron and the second $r-$process peak elements. The abundance pattern confirms that LAMOST J1109+0754 is a strongly $r-$process enhanced EMP star, e.g., [Eu/Fe] = +1.16 (filled square in Figure 2), and is very similar to that of literature cool $r-$II giants in the range from Sr through Dy (Li *et al.* 2015c). The abundance pattern of heavy elements of LAMOST J1109+0754 can also be well explained by the classical $r-$process. Therefore it is the thirteenth $r-$II star, with the lowest metallicity of [Fe/H] ~ -3.4 among all $r-$II giants. The abundance pattern of LAMOST J1109+0754 from Ba through Dy well matches the theoretical SSr pattern, which provides additional evidence of a universal production ratio of these elements during the evolution of the Galaxy.

About 6 million stellar spectra will be obtained through LAMOST spectroscopic survey, and the joint collaboration with Subaru will certainly enable us to significantly enlarge the sample of EMP stars, and to explore the nature of the nucleosynthesis and chemical enrichment at the very beginning of the Universe.

Acknowledgements

H.N.L. and G.Z. acknowledge supports by NSFC grants No. 11573032, 11233004, and 11390371. W.A. and T.S. are supported by the JSPS Grant-in-Aid for Scientific Research (S:23224004). S.H. is supported by the JSPS Grant-in-Aid for Scientific Research (c:26400231). N.C. acknowledges support from Sonderforschungsbereich 881 "The Milky Way System" (subproject A4) of the German Research Foundation (DFG). Guoshoujing Telescope (the Large Sky Area Multi-Object Fiber Spectroscopic Telescope, LAMOST) is a National Major Scientific Project built by the Chinese Academy of Sciences. Funding for the project has been provided by the National Development and Reform Commission. LAMOST is operated and managed by the National Astronomical Observatories, Chinese Academy of Sciences. This work is based on data collected at the Subaru Telescope, which is operated by the National Astronomical Observatory of Japan.

References

Aoki, W., Beers, T. C., & Lee, Y. S., *et al.* 2013, *AJ*, 145, 13
Asplund, Martin, Lambert, David L. & Nissen, Poul Erik, *et al.* 2006, *ApJ*, 664, 229
Beers, T. C. & Christlieb, N. 2005, *ARA&A*, 43, 531
Boesgaard, A. M. & Steigman, G. 1985, *ARA&A*, 23, 319
Bromm, V. & Yoshida, N. 2011, *ARA&A*, 49, 373
Burris, D. L., Pilachowski, C. A., & Armandroff, T. E., *et al.* 2000, *ApJ*, 544, 302
Carlin, J. L., Lpine, S., & Newberg, H. J., *et al.* 2012, *RAA*, 12, 775
Carollo, D., Freeman, K., & Beers, T. C., *et al.* 2014, *ApJ*, 788, 180
Cayrel, R., Depagne, E., Spite, M., *et al.* 2004, *A&A*, 416, 1117
Cescutti, G., Chiappini, C., & Hirschi, *et al.* 2013, *A&A*, 553, A51
Charbonnel, C. & Balachandran, S. C. 2000, *A&A*, 359, 563
Christlieb, N., Schrck, T., Frebel, A., *et al.* 2008, *A&A*, 484, 721
Cui, X.-Q., Zhao, Y.-H., & Chu, Y.-Q., *et al.* 2012, *RAA*, 12, 1197
Frebel, A. & Norris, J. E. 2015, *ARA&A*, in press, arXiv:1501.06921
Goriely, S., Sida, J.-L., & Lematre, J.-F., *et al.* 2013, *Physical Review Letters*, 111, 242502
Hansen, T., *et al.* 2014, *ApJ*, 787, 162
Li, H. N., Zhao, G., Christlieb, N., *et al.* 2015a, *ApJ*, 798, 110
Li, H. N., Aoki, W., Zhao, G., *et al.* 2015b, *PASJ*, in press, arXiv:1506.05684
Li, H. N., Aoki, W., Honda, S., *et al.* 2015c, *RAA*, 15, 1264
Masseron, T., Johnson, J. A., Plez, B., *et al.* 2010, *A&A*, 509, A93
Masseron, T., Johnson, J. A., Lucatello, S., *et al.* 2012, *ApJ*, 751, 14
Nollett, Kenneth M., Busso, M., & Wasserburg, G. J. 2003, *ApJ*, 582, 1036
Nomoto, K., Kobayashi, C., & Tominaga, N. 2013, *ARA&A*, 51, 457
Norris *et al.* 2013a, *ApJ*, 762, 28
Norris *et al.* 2013b, *ApJ*, 762, 25
Roederer, I. U., Cowan, J. J., & Preston, G. W., *et al.* 2014, *MNRAS*, 445, 2970
Sackmann, I.-Juliana & Boothroyd, Arnold I. 1999, *ApJ*, 510, 217
Sbordone, L., Bonifacio, P., Caffau, E., *et al.* 2010, *A&A*, 522, 26
Sneden, C., Preston, G. W., McWilliam, A., & Searle, L. 1994, *ApJ*, 431, L27
Sneden, C., McWilliam, A., & Preston, G. W., *et al.* 1996, *ApJ*, 467, 819
Spite, M., Caffau, E., Bonifacio, P., *et al.* 2013, *A&A*, 552, A107
Suda, T., Aikawa, M., & Machida, M. N., *et al.* 2004, *ApJ*, 611, 476
Suda, T. & Fujimoto, M. Y. 2010, *MNRAS*, 405, 177
Umeda, H. & Nomoto, K. 2005, *ApJ*, 619, 427
Woosley, S. E., Wilson, J. R., & Mathews, G. J., *et al.* 1994, *ApJ*, 433, 229
Yanny, B., Rockosi, C., & Newberg, H. J., *et al.* 2009, *ApJ*, 137, 4377
Yong, D., *et al.* 2013, *ApJ*, 762, 27
Zhao, G., Zhao, Y. H., & Chu, Y. Q., *et al.* 2012, *RAA*, 12, 723

The General Assembly of Galaxy Halos: Structure, Origin and Evolution
Proceedings IAU Symposium No. 317, 2015
A. Bragaglia, M. Arnaboldi, M. Rejkuba & D. Romano, eds.

doi:10.1017/S1743921315009783

Inhomogeneous chemical enrichment in the Galactic Halo

Chiaki Kobayashi

Centre for Astrophysics Research, Science and Technology Research Institute, University of Hertfordshire, Hertfordshire, AL10 9AB, UK; email: `c.kobayashi@herts.ac.uk`

Abstract. In a galaxy, chemical enrichment takes place in an inhomogeneous fashion, and the Galactic Halo is one of the places where the inhomogeneous effects are imprinted and can be constrained from observations. I show this using my chemodynamical simulations of Milky Way type galaxies. The scatter in the elemental abundances originate from radial migration, merging/accretion of satellite galaxies, local variation of star formation and chemical enrichment, and intrinsic variation of nucleosynthesis yields. In the simulated galaxies, there is no strong age-metallicity relation. This means that the most metal-poor stars are not always the oldest stars, and can be formed in chemically unevolved clouds at later times. The long-lifetime sources of chemical enrichment such as asymptotic giant branch stars or neutron star mergers can contribute at low metallicities. The intrinsic variation of yields are important in the early Universe or metal-poor systems such as in the Galactic halo. The carbon enhancement of extremely metal-poor (EMP) stars can be best explained by faint supernovae, the low $[\alpha/\mathrm{Fe}]$ ratios in some EMP stars naturally arise from low-mass ($\sim 13-15M_\odot$) supernovae, and finally, the $[\alpha/\mathrm{Fe}]$ knee in dwarf spheroidal galaxies can be produced by subclasses of Type Ia supernovae such as SN 2002cx-like objects and sub-Chandrasekhar mass explosions.

Keywords. Galaxy: abundances, Galaxy: evolution, Galaxy: halo, supernovae: general

1. Introduction

Different elements and isotopes are produced from stars on different timescales. Elemental and isotopic abundance ratios of interstellar medium (ISM) evolve as a function of time in a galaxy. Therefore, stars are fossils that retain information on the formation and evolutionary history of the galaxy. This approach is called *galactic archaeology*. Nucleosynthesis yields have recently been updated and a complete yield table is provided for $1-300M_\odot$ (Nomoto, Kobayashi & Tominaga 2013, hereafter NKT13). Galactic chemical evolution models (Kobayashi *et al.* 2006, Kobayashi *et al.* 2011a) are in excellent agreement with observations except for some minor elements. The ν-process plays an essential role in the production of F (Kobayashi *et al.* 2011b), while multi-dimensional calculation is necessary for Ti (Kobayashi *et al.* 2006).

Galactic chemical evolution has been studied with so-called *one-zone models* (also called monolithic models in the cosmological context) for many years. In one-zone models, it is assumed that the interstellar medium (ISM) of the region under consideration is well mixed instantaneously and has a uniform chemical composition. In reality, the ISM is not well mixed and more-realistic models have been proposed. *Stochastic models* can involve inhomogeneous mixing statistically. *Hierarchical (semi-analytic) models* include cosmological mass accretion (see NKT13 for the references).

Thanks to the development of high-performance computers and numerical techniques, it became possible to calculate chemical enrichment combined with 3D hydrodynamical simulations, which are called *chemodynamical simulations* (e.g., Kobayashi & Nakasato 2011, hereafter KN11, and references therein). With such chemodynamical simulations, positions, kinematics, ages, metallicities, and elemental abundances of star particles are obtained. Thus, the scatter of observational data and the effect of inhomogeneous mixing may be studied. These are the predictions that can be, and should be, compared with future observations with high-resolution multi-object spectrographs such as APOGEE, HERMES, and GAIA-ESO surveys.

2. Dependence on locations in chemodynamical simulations

We use a hydrodynamical code that includes relevant physical processes such as radiative cooling, star formation, supernova feedback, and chemical enrichment from core-collapse supernovae (SNe II and hypernovae), Type Ia supernovae (SNe Ia), and asymptotic giant branch (AGB) stars. The progenitor model of SNe Ia is based on the single degenerate scenario with the metallicity effects of white dwarf (WD) winds, and the lifetime distribution functions are calculated as in Kobayashi & Nomoto (2009). See Kobayashi (2004), Kobayashi, Springel & White (2007), and KN11 for the details. We use ΛCDM initial conditions, so that any galaxy forms through the successive merging of subgalaxies with various masses. The initial condition is chosen in order to avoid major mergers at $z \lesssim 2$, which is necessary to retain the disk structure. In this simulated galaxy, the bulge is formed by the initial starburst that is induced by the assembly of gas-rich sub-galaxies. Because of the angular momentum, the gas accretes onto the plane forming a rotationally supported disk that grows from inside out. In the disk, star formation takes place with a longer timescale, which is maintained not by the slow gas accretion, but by the self-regulation due to supernova feedback. Many satellite galaxies successively come in and disrupt, and half of thick disk stars have formed in merging galaxies.

In a real galaxy, the star formation history is not so simple and in particular the interstellar medium (ISM) is not homogeneous at any time, which is different from one-zone chemical evolution models. The effects of inhomogeneous enrichment can be summarized as follows. (I) There is a local variation in star formation and metal flow by the inflow and outflow of the ISM. (II) Heavy elements are distributed via stellar winds and supernovae, and the elemental abundance ratios depends on the metallicity and mass of progenitor stars. (III) The ISM may be mixed before the next star formation by other effects such as diffusion and turbulence. (IV) There is a mixing of stars due to dynamical effects such as merging and migration.

In chemodynamical simulations (KN11; Kobayashi 2014, hereafter K14), hydrodynamics and chemical enrichment are solved self-consistently throughout the galaxy formation. Star formation and chemical enrichment depend on the local density. With our feedback scheme, star particles obtain heavy elements from the gas particles from which the stars form. Gas particles obtain heavy elements only when they pass through within a feedback radius of dying star particles. The mass and metallicity dependencies on nucleosynthesis yields are included by looking at the age and metallicity of progenitor star particles, although the energy dependence has not been included. Because it is not possible to resolve supernova ejecta, the size of feedback region, or the number of gas particle that receive feedback, is given by a parameter. This will account for some of gas-phase mixing. Therefore, (I), (II), (IV) are naturally included but (III) is probably not enough.

Because star formation and chemical enrichment histories depend on locations inside the galaxy, elemental abundances also depend on locations. Figure 1 shows [O/Fe]-[Fe/H] relations as a function of Galactocentric radius and latitude. Because of the delayed enrichment of SNe Ia, which produce more iron-peak elements than α elements (O, Mg, Si, S, and Ca), there is a plateau at [Fe/H] ~ -1, and then a decreasing trend with [Fe/H]. The α-enhanced population is more prominent at the center of galaxy and at higher latitudes, and very similar figures are obtained with APOGEE and RAVE.

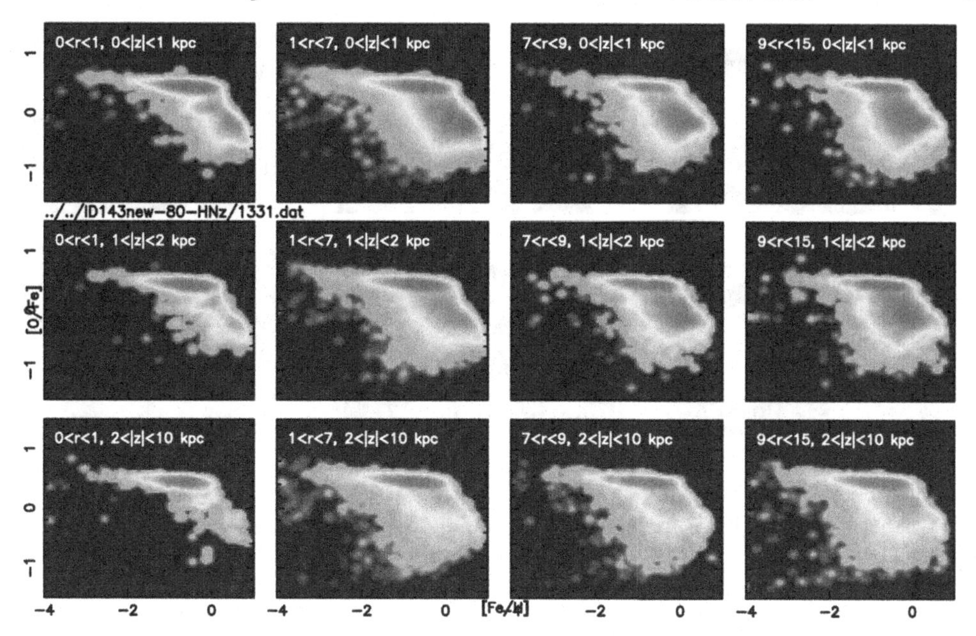

Figure 1. [O/Fe]-[Fe/H] relations as a function of Galactocentric radius ($r < 1, 1-7, 7-9, 9-15$ kpc) and latitude ($z < 1, 1-2, 2-10$ kpc). The contours show the frequency distribution of stars in the simulated galaxies, where red is for the highest frequency.

In the solar neighborhood, the trends and the scatter of elemental abundance ratios are in good agreement with observations (see KN11 for the [X/Fe]-[Fe/H] diagrams of the solar neighborhood, thick disk, and bulge). At the same [Fe/H], [Mn/Fe] also shows an increasing trend with [Fe/H] because Mn is more produced by SNe Ia than Fe. Odd-Z elements (Na, Al, and Cu) show increasing trends at [Fe/H] $\lesssim -1$ because of the metallicity dependence of nucleosynthesis yields. In the bulge, the star formation timescale is so short that the [α/Fe] plateau continues to [Fe/H] $\sim +0.3$. Because of the smaller contribution from SNe Ia, the majority of stars shows high [α/Fe] and low [Mn/Fe]. [(Na, Al, Cu, Zn)/Fe] are also high because of the high metallicity in the bulge. The stellar population of the thick disk is neither disk-like nor bulge-like. For thick disk stars, [α/Fe] is higher, and [Mn/Fe] is lower than thin disk stars because of the short formation timescale. However, [(Na, Al, Cu, Zn)/Fe] are lower than bulge stars because of the lower chemical enrichment efficiency. This is because half of the thick disk stars have already formed in satellite galaxies before they accrete onto the disk, and the metals have been ejected from the satellite galaxies by galactic winds.

In the disk, metallicity radial and vertical gradients exist, but no [α/Fe] radial gradient, which seems consistent with observations by SDSS and RAVE. These suggest that the chemical enrichment efficiency is higher in the center but the star formation timescale does not so much depend on the radius. In the bulge, metallicity and [α/Fe] vertical gradients exist, which is caused by the increase of metal-rich and low [α/Fe] populations at lower latitudes. For the thick disk stars, there is metallicity vertical, weak metallicity radial, and no [α/Fe] radial gradients (see K14 for the radial profiles and histograms).

The time evolution of these gradients are important to determine the major physical process of the disk formation. In chemodynamical simulations like our model, metallicity radial gradients are steeper at higher redshifts (see KN11 for the figure). This time evolution is consistent with the observations of lensed disk galaxies at high redshifts (Pilkington *et al.* 2012). On the other hand, classical monolithic collapse models often

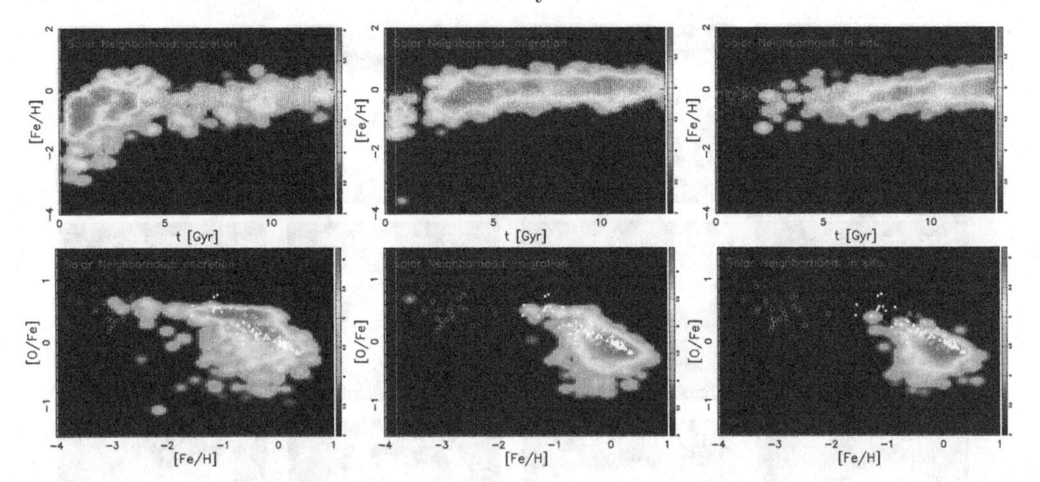

Figure 2. The age-metallicity and [O/Fe]-[Fe/H] relations for stars that formed in-situ (right panels), and originate from migration (middle panels) and stellar accretion of mergers (left panels).

predict opposite evolution, and migration may have much faster evolution. Note that an inverse gradient is shown at $z \sim 3$ by Cresci *et al.* (2010), but there is no stable disk at such a high redshift in chemodynamical simulations from cosmological initial conditions, and this observation is not suitable for the discussion of Milky Way-size disks.

3. *Local* inhomogeneous effect - the origin of the scatter in [X/Fe]

In this section, we discuss three processes that cause a scatter of stellar abundances. The first two processes are in the effect (IV), and the third process consists of the effects (I)-(III). By selecting stars from the locations when the stars formed, we show how each process affects on stellar populations in the solar neighborhood in Figure 2.

(1) Accretion of stars in merging satellites that have a different evolutionary history. Usually chemical enrichment efficiency is low because of the small potential well. No age-metallicity relation is expected. In Figure 2, there are not many stars at [Fe/H] $\gtrsim 0$, and no time evolution. There is not very much contribution from SNe Ia. The most metal-poor population arise from this process.

(2) Migration of stars along the disk plane. Stars formed in the center can be located in the solar neighborhood at present. Because of the metallicity gradients, metallicity can be high. No age-metallicity relation is expected. In Figure 2, [Fe/H] is as high as ~ 0, and there is no time evolution. Most of these stars are disk stars and enriched by SNe Ia.

(3) There is an intrinsic scatter in the stars formed in the solar neighborhood because star formation and chemical enrichment depend on local density (in situ). In Figure 2, there is no old stars at $t \lesssim 2$ Gyr because the disk grow inside-out. The age-metallicity relation is clearly seen from (t, [Fe/H])=(6 Gyr, -0.5) to (13 Gyr, $+0.5$). The [O/Fe]-[Fe/H] relation is seen from [Fe/H] ~ -1 to $+0.5$. These are no metal-poor stars with [Fe/H] $\lesssim -1.4$, which suggests that there is pre-enrichment from pre-existing supernovae. However, there is still a significant scatter in both relations. This is the *local* inhomogeneous enrichment effect.

In summary, the following phenomena occur in the case of inhomogeneous enrichment: i) The age-metallicity relation is weak. In other words, the most metal-poor stars are not always the oldest stars. In our simulations, the metallicity of the first enriched stars reaches [Fe/H] ~ -3. At later times, star forming regions become denser, and both metal richer and poorer stars than [Fe/H] ~ -3 appear. ii) Some stars can be affected by SNe Ia at [Fe/H] $\lesssim -1$ even with our metallicity inhibition of SNe Ia. The SN Ia contribution is characterised by low [α/Fe] and high [Mn/Fe]. iii) The scatter of elemental abundance ratios becomes large if the supernova yield depends on progenitor metallicity such as Na. iv) Some CEMP/NEMP stars can be explained with the local enrichment from AGB stars even without the binary effect. In fact, the observed [N/O]-[O/H] trend can be reproduced with our simulation without including the effect of rotating massive stars (see K14 for the figure).

4. The first chemical enrichment

Despite more than ten years of surveys, no stars that have an elemental abundance pattern consistent with pair-instability supernovae (PISNe) have been found. The characteristics of PISN enrichment are: i) The odd-Z effect is much larger than ~ 1 dex. ii) [(Si, S, Ar, Ca)/Fe] are much larger than [(O, Mg)/Fe] because of more extensive explosive oxygen burning. iii) [Cr/Fe] is much larger because of the larger incomplete Si-burning region. iv) [(Co, Zn)/Fe] are much smaller because of the much larger ratio between the complete and incomplete Si-burning regions. The star reported by Aoki *et al.* (2014) does not show the Si enhancement ([Si/Mg] ~ 0.1) and [Co/Fe] is extremely low (~ -0.7), although the [Mg/Fe] ratio is very low (~ -0.5). This abundance pattern may be explained with a anisotropic enrichment by jet-like explosions.

Instead, the elemental abundance patterns of extremely metal-poor (EMP) stars (Ishigaki *et al.* 2014, including the [Fe/H]<-7 star) and metal-poor damped Lyman α systems (Kobayashi *et al.* 2011c), in particular the large carbon enhancement, are well reproduced by faint supernovae. The central parts of supernova ejecta that contain most of the iron fall back onto the black hole, while the stellar envelopes that contain carbon are ejected as in normal supernovae. Therefore, the [C/Fe] ratio of faint supernovae is as large as that of carbon-enhanced metal-poor (CEMP) stars. Among α elements, O and Mg are synthesized during hydrostatic burning and are located in the outskirts of ejecta. Therefore, faint supernovae often have high [(O, Mg)/Fe] ratios, depending on the mixing-fallback processes. The mixing-fallback effect is naturally expected in the case of hypernovae, which are jet-induced explosions (of rotating stars) followed by black hole formation. In supernovae, it is also possible that some degree of mixing occurs through Rayleigh-Taylor instability.

Some EMP stars show low [α/Fe] ratios, which can be explained with low-mass ($\sim 13-15M_\odot$) supernovae (Kobayashi *et al.* 2014). $\sim 10-20M_\odot$ supernovae have a smaller mantle mass that contains α elements than more massive stars, and thus give lower [α/Fe] ratios than the initial mass function (IMF) weighted values of core-collapse supernova yields, i.e., the plateau values of [α/Fe]-[Fe/H] relations. These supernovae will leave a neutron star behind, and should be very common for the standard IMF weighted for the low-mass end. A couple of stars also show carbon enhancement, which is interesting. If there is a CEMP-low-α star with [Zn/Fe] $\sim$ [Co/Fe] ~ 0, that would require a new population, i.e., $\sim 10-20M_\odot$ faint supernovae that form black holes. For $\lesssim 20M_\odot$ stars, they are believed to form neutron stars, and thus black-hole forming faint supernovae have not been discussed in previous works.

In summary, an intrinsic variation of [α/Fe] ratios can be caused by the following enrichment sources: (1) SNe Ia, (2) low-mass supernovae, (3) hypernovae ($E/10^{51}\mathrm{erg} \gtrsim 10$

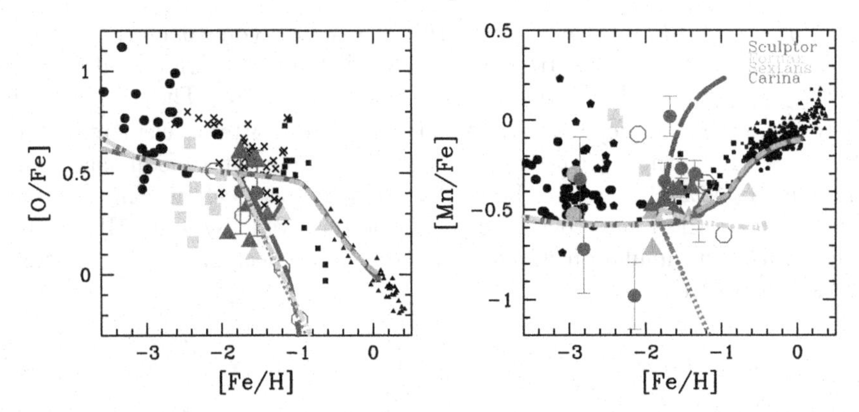

Figure 3. Evolution of elemental abundance ratios for the solar neighborhood with only normal SNe Ia (red solid lines) and a model with 50% SNe Iax and 50% sub-Ch SNe Ia (green short–dashed lines). The blue long-dashed, cyan dotted, and magenta dotted-dashed lines are for dSph galaxies with 100% SN Iax, 100% sub-Ch SNe Ia, and equal mix of the two, respectively. See Kobayashi *et al.* (2011, black) and Venn *et al.* (2012, color) for the observational data sources of the solar neighborhood and dSphs, respectively.

for $\gtrsim 25 M_\odot$, NKT13), (4) faint supernovae, and (5) PISNe. (1) and (2) are fully, (3) is partially, included in my chemodynamical simulations.

5. Dwarf Spheroidal Galaxies

The low [α/Fe] ratios and their trends in dwarf spheroidal galaxies can be explained by subclasses of SNe Ia, such as SN 2002cx-like objects and sub-Chandrasekhar (Ch) mass explosions (Kobayashi *et al.* 2015), that have been discovered with recent extensive observations of supernovae. The [α/Fe] ratios of stars in dSphs such as Fornax and Sagittarius are lower than those of the Galactic halo stars. This difference is often interpreted to indicate a larger contribution of SNe Ia in dSphs than that of the solar neighborhood. It is also possible to find a quite clear knee at lower metallicities; [Fe/H] ~ -2 in Sculptor, ~ -1.5 in Fornax, and so on (e.g., Tolstoy *et al.* 2009). If this knee is caused by SNe Ia, [Mn/Fe] ratios should show an increasing trend with [Fe/H] from this [α/Fe] knee. However, in dSph galaxies, [Mn/Fe] ratios are as low as in the Galactic halo stars for a wide range of metallicity (e.g., North *et al.* 2012). One possible scenario is the lack of contributions from massive core-collapse supernovae due to the incomplete sampling of IMF (e.g., Venn *et al.* 2012). Less-massive supernovae ($\sim 20 M_\odot$) give low [α/Fe] ratios without changing [Mn/Fe] ratios.

We propose another scenario. Figure 3 shows our new chemical evolution models that includes two possible channels in the single degenerate scenario: (1) double detonations in sub-Ch mass CO WDs, where a thin He envelope is developed with relatively low accretion rates after He novae even at low metallicities, and (2) carbon deflagrations in Ch-mass *possibly* hybrid C+O+Ne WDs, where WD winds occur at [Fe/H] ~ -2.5 at high accretion rates. These subclasses of SNe Ia are rarer than 'normal' SNe Ia and do not affect the chemical evolution in the solar neighborhood, but can be very important in metal-poor systems with stochastic star formation. In dwarf spheroidal galaxies in the Local Group, the decrease of [α/Fe] ratios at [Fe/H] ~ -2 to -1.5 can be produced depending on the star formation history. SNe Iax give high [Mn/Fe], while sub-Ch-mass SNe Ia give low [Mn/Fe], and thus a model including a mix of the two is favored by the available observations.

References

Aoki, W, Tominaga, N., Beers, T. C., Honda, S., & Lee, Y. S. 2014, Science, 345, 912
Cresci, G., Mannucci, F., Maiolino, R., *et al.* 2010, *Nature*, 467, 811
Ishigaki, M. N., Tominaga, N., Kobayashi, C., & Nomoto, K. 2014, *ApJ*, 792, L32
Kobayashi, C. 2014, in IAU Symposium 298, 298, 167 (K14)
Kobayashi, C., Ishigaki, M. N., Tominaga, N., & Nomoto, K. 2014, *ApJ*, 5, L5
Kobayashi, C., Karakas, I. A., & Umeda, H. 2011a, *MNRAS*, 414, 3231
Kobayashi, C., Izutani, N., Karakas, A. I. *et al.* 2011b, *ApJ*, 739, L57
Kobayashi, C. & Nakasato, N. 2011, *ApJ*, 729, 16 (KN11)
Kobayashi, C. & Nomoto, K. 2009, *ApJ*, 707, 1466
Kobayashi, C., Nomoto, K., & Hachisu, I. 2015, *ApJ*, 804, L24
Kobayashi, C., Springel, V., & White, S. D. M. 2007, *MNRAS*, 376, 1465
Kobayashi, C., Tominaga, N., & Nomoto, K. 2011c, *ApJ*, 730, L14
Kobayashi, C., Umeda, H., Nomoto, K., Tominaga, N., & Ohkubo, T. 2006, *ApJ*, 653, 1145
Nomoto, K., Kobayashi, C., & Tominaga, N. 2013, *ARA&A*, 51, 457 (NKT13)
North, P., Cescutti, G., Jablonka, P., *et al.* 2012, *A&A*, 541, 45
Pilkington, K., Few, C. G., Gibson, B. K., *et al.* 2012, *A&A*, 540, A56
Tolstoy, E., Hill, V., & Tosi, M. 2009, *ARA&A*, 47, 371
Venn, K. A., Shetrone, M. D., Irwin, M. J., *et al.* 2012, *ApJ*, 751, 102

The General Assembly of Galaxy Halos: Structure, Origin and Evolution
Proceedings IAU Symposium No. 317, 2015
A. Bragaglia, M. Arnaboldi, M. Rejkuba & D. Romano, eds.

doi:10.1017/S1743921315009576

Exploring the early Universe with extremely metal-poor stars.

Terese T. Hansen[1], Norbert Christlieb[1], Camilla J. Hansen[2] and Timothy C. Beers[3]

[1]Landessternwarte, Heidelberg University
Königstuhl 12, 69117 Heidelberg, Germany
email: thansen@lsw.uni-heidelberg.de
[2]Dark Cosmology Center, Copenhagen University
Juliane Maries vej 30, 2100 Copenhagen, Denmark.
email: cjhansen@dark-cosmology.dk
[3]Department of Physics and JINA Center for the Evolution of the Elements, University of Notre Dame
Notre Dame, IN 46556, USA
email: tbeers@nd.edu

Abstract. The earliest phases of Galactical chemical evolution and nucleosynthesis can be investigated by studying the old metal-poor stars. It has been recognized that a large fraction of metal-poor stars possess significant over-abundances of carbon relative to iron. Here we present the results of a 23-star homogeneously analyzed sample of metal-poor candidates from the Hamburg/ESO survey. We have derived abundances for a large number of elements ranging from Li to Pb. The sample includes four ultra metal-poor stars ([Fe/H] < -4.0), six CEMP-no stars, five CEMP-*s* stars, two CEMP-*r* stars and two CEMP-*r*/*s* stars. This broad variety of the sample stars gives us an unique opportunity to explore different abundance patterns at low metallicity.

Keywords. Galaxy: formation – Galaxy: halo – Stars: abundances – Stars: chemically peculiar Binaries: spectroscopic – ISM: structure

1. Introduction

The earliest stages of galaxy formation and chemical evolution can be explored by means of metal-poor stars. These very old stars are believed to hold the fossil record of the nucleosynthesis products of the first stars that formed shortly after the Big Bang. The vast majority of the metal-poor stars observed today are found in the Milky Way halo system.

Recent studies, such as Carollo *et al.*(2012), Lee *et al.*(2013), and Norris *et al.*(2013) confirm that carbon-enhanced metal-poor (CEMP) stars† constitute a large fraction of the most metal-poor stars known, and that the fraction of CEMP stars increases dramatically with decreasing metallicity, accounting for ~40% of all stars with [Fe/H] $\leqslant -3.5$. The CEMP stars can be divided into four sub-classes defined by Beers & Christlieb (2005), the first being the CEMP-no stars which exhibit no over abundances of neutron-capture elements ([Ba/Fe] $\lesssim 0.0$). Several progenitors have been suggested for these stars such as fast rotating massive metal-free stars, the so-called "Spinstars" (Meynet *et al.* 2006, Hirschi 2007, Chiappini *et al.* 2008), supernovae with mixing and fallback (Umeda & Nomoto 2003,2005), and mass transfer from an asymptotic giant branch (AGB) binary companion (Masseron *et al.* 2010). The CEMP-no stars dominate at the lowest metallicity, hence these stars are most likely associated with elemental-abundance patterns that were produced by the very first generation of massive stars to form in the Galaxy.

† Originally defined by Beers & Christlieb (2005) as metal-poor ([Fe/H] $\leqslant -1.0$) stars with [C/Fe] $\geqslant +1.0$; a level of carbon enrichment [C/Fe] $\geqslant +0.7$ is used in most contemporary work.

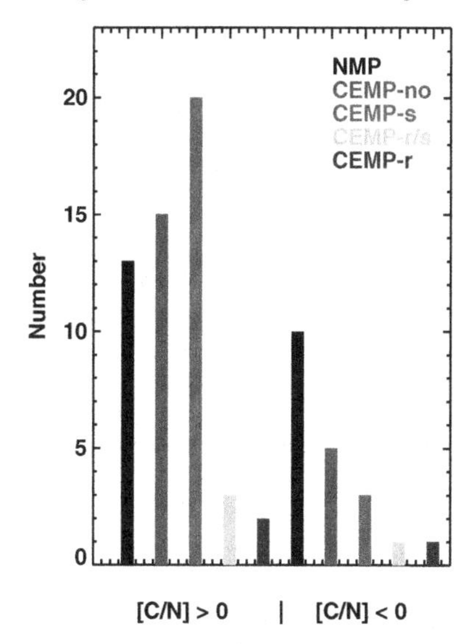

Figure 1. Number of stars with either [C/N] > 0 or [C/N] < 0 for the different types of metal-poor stars. Black: NMP stars, red: CEMP-*no* stars, green: CEMP-*s* stars, blue: CEMP-*r* stars, yellow: CEMP-*r*/*s* stars (from left to right for each group).

The second sub-class are the CEMP-*s* stars, which show over abundances of neutron-capture elements produced in the slow neutron-capture process ([Ba/Fe] > 1.0). These are widely believed to be the result of mass transfer from an AGB binary companion, a scenario also supported by radial-velocity monitoring of these stars (Lucatello *et al.* 2003 and Hansen *et al.* 2015b submitted). The two last sub-classes of CEMP stars are the CEMP-*r* and CEMP-*r*/*s*. The first show over abundances of element produced in the rapid neutron-capture process ([Eu/Fe] > 1.0), and the latter exhibit over abundances of neutron-capture elements suggesting contributions from both the *s*- and *r*-process (0.0 < [Ba/Eu] < 0.5). Only few stars belonging to these two sub-classes have been detected and analysed. Each of these sub-classes appear to be associated with different element-production histories, thus their study provides insight into the variety of astrophysical sites in the early Galaxy that were primarily responsible for their origin.

2. Mixing in the early Universe

All of the proposed progenitors of CEMP stars have experienced some degree of internal mixing, whether that mixing is due to convection driven by rapid rotation in the spinstars, or convection in AGB stars during their evolution (Herwig 2005). When this mixing occurs the carbon is transported from the core (spinstars) or from the surface (AGB stars) to the H-burning shell where the CNO cycle is active, where it is transformed into ^{13}C and ^{14}N. Thus the mixing creates a signature we can detect in the C and N abundances and in the ^{12}C/^{13}C isotopic ratio of the un-evolved CEMP star. High ^{12}C/^{13}C and [C/N] ratios indicate only partial hydrogen burning by the CNO cycle, while low ^{12}C/^{13}C and [C/N] ratios are a signature of more complete burning by the CNO cycle (Maeder *et al.* 2015).

We have therefore divided the stars into two groups; one with [C/N] > 0 ([C/Fe] > [N/Fe]) and one with [C/N] < 0 ([C/Fe] < [N/Fe]). Figure 1 shows the number of each type of metal-poor star with either [C/N] > 0 or [C/N] < 0. There are clearly more stars with [C/N] > 0 than with [C/N] < 0. For the "normal" metal-poor (NMP) stars the numbers are roughly equal, while the CEMP-no and especially the CEMP-*s* stars are of the [C/N] > 0 variety.

Figure 2 shows the C and N abundances and the [C/N] ratios, of the sample of stars along with those from Yong *et al.*(2013), as a function of metallicity. Circles represent stars with [C/N] > 0 and triangles stars with [C/N] < 0. From inspection of the bottom panel of Figure 2, none of the CEMP-no stars with [C/N] < 0 are found with metallicities above [Fe/H] > −3.4; all CEMP-no

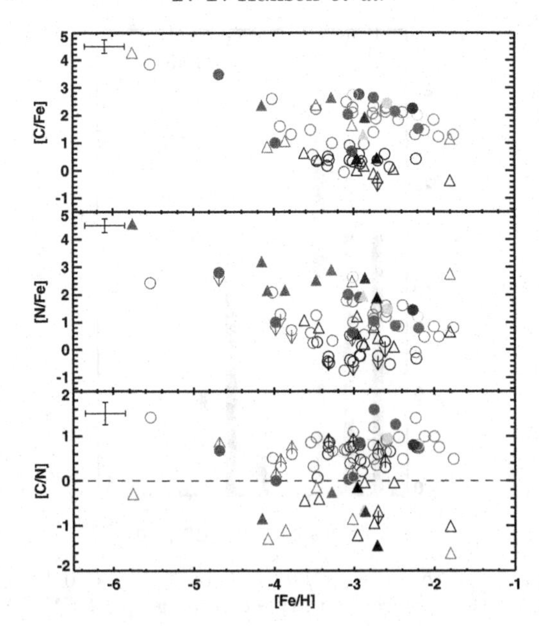

Figure 2. C and N abundances and [C/N] ratios, filled symbols – Hansen *et al.*(2015a), unfilled symbols – Yong *et al.*(2013). Circles represent stars with [C/N] > 0; triangles are stars with [C/N] < 0. Symbols are color-coded (black: NMP stars, red: CEMP-*no* stars, green: CEMP-*s* stars, blue: CEMP-*r* stars, yellow: CEMP-*r*/*s* stars). An approximate error bar for the sample stars is shown in the upper left of each panel.

stars with [C/N] < 0 are at the extremely low-metallicity end. Some CEMP-no stars are found to have large enhancements in Na and Mg. In our sample these stars are only found below this extremely low metallicity, as can be seen in Figure 3, where Na, Mg and Al abundances for the stars are plotted against metallicity. This indicates that the large degrees of internal mixing and processing required to produce the abundance pattern seen in such CEMP-no stars was only operating at the very earliest times.

As mentioned above the mixing also alters the ^{12}C/^{13}C isotopic ratio. We find low ($\sim$ 5) ^{12}C/^{13}C isotopic ratios for all of our CEMP-no stars, consistent with the equilibrium value for CNO-cycle processed material (^{12}C/^{13}C $\sim$ 4). This shows that the material from which these stars formed has undergone mixing, whether in spinstars or in some pre-supernova evolution. The ^{12}C/^{13}C isotopic ratios found in the CEMP-*s* stars of our sample are generally higher ($\sim$ 13). This value is low enough to be a signature of H-burning via the CNO cycle, which is also expected if the carbon excess found in CEMP-*s* stars is due to mass transfer from an AGB companion, where multiple dredge up events mix the material in the star. However, according to Bisterzo *et al.* (2012), current AGB models do not include sufficient mixing to replicate the low ^{12}C/^{13}C isotopic ratios found in CEMP-*s* stars.

3. Carbon bands and Ba floor

It has been suggested by Spite *et al.* (2013) that the absolute C abundances of CEMP stars fall into two bands. The C abundances for stars with metallicities [Fe/H] > −3 appeared to cluster around the solar carbon abundance (A(C) $\sim$ 8.5), while those with [Fe/H] < −3 cluster around a lower C abundance, A(C) $\sim$ 6.5. The higher band is mainly populated by CEMP-*s* stars that are thought to have gained their carbon from an extrinsic source (i.e., mass transfer from an AGB binary companion), whereas the lower band is mainly populated by CEMP-no stars that are believed to be born with their carbon excess. The recent paper by Bonifacio *et al.*(2015) confirms the existence of the two carbon bands for a larger sample, including the stars from Yong *et al.*(2013).

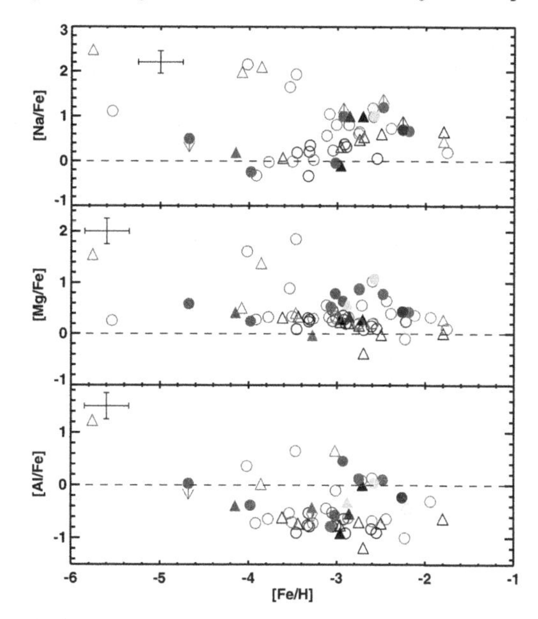

Figure 3. [Na/Fe], [Mg/Fe], and [Al/Fe] for the stars from Hansen *et al.*(2015a) (filled symbols) and Yong *et al.*(2013) (unfilled symbols). Color-coding of the stellar classes is as in Figure 2. An approximate error bar for the sample stars is shown in the upper left of each panel.

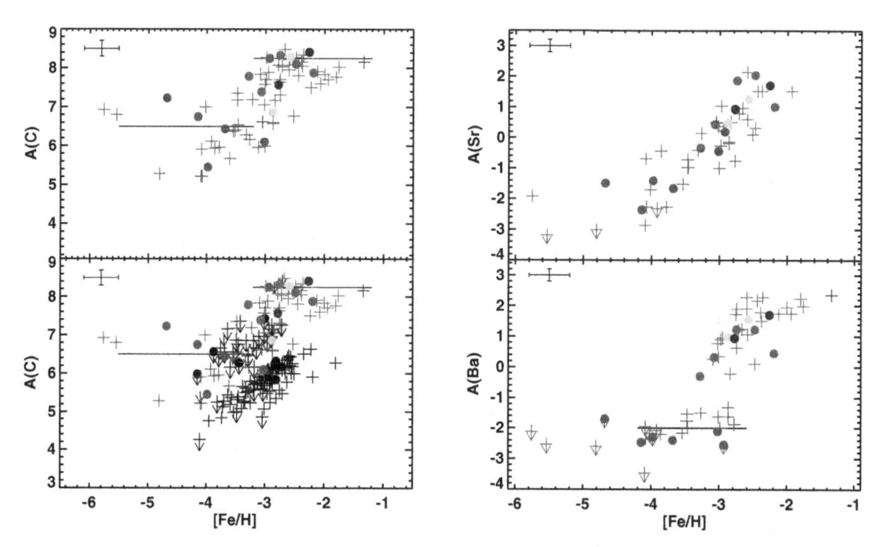

Figure 4. Absolute C, Sr, and Ba abundances for the sample stars (dots: Hansen *et al.*(2015a); plusses: Yong *et al.*(2013)), colour coding as in Figure 2. C-bands and Ba floor are indicated by solid lines.

The left panels of Figure 4 show the distribution of absolute C abundances for our stars (Hansen *et al.* 2015a) and those of Yong *et al.* (2013). Our data also support the presence of two carbon "bands" that comprise the distribution of the absolute carbon abundances for CEMP stars, although with a smoother transition between the bands than was found by Spite *et al.*(2013). However a few of the CEMP-no stars with metallicities above $[Fe/H] > -3.0$ have carbon abundances on the higher band, indicating a different origin for the carbon found in these stars as opposed to those with carbon abundances on the lower band.

It can also bee seen from Figure 4 that the CEMP-r and CEMP-r/s stars are found at both the high and the low levels of C enhancement. There is as such no indication of a possible dominance of either high or low carbon-abundance stars for either of these CEMP sub-classes.

The right panels of Figure 4 show the absolute Sr and Ba abundances for the CEMP stars in our sample (Hansen *et al.* 2015a) and those of Yong *et al.* (2013). There is clear grouping of the different classes of stars considered in our study. Recall that [Ba/Fe] is used to differentiate the CEMP-no stars from the CEMP-s and CEMP-r/s stars. For the Sr abundances we see that the individual classes of the stars in our sample are mixed together in a band showing decreasing A(Sr) with decreasing [Fe/H], but with a possible change in the trend at the lowest metallicity (around [Fe/H] ~ -4.2). In contrast, the behaviour of Ba for the CEMP-no stars in our sample is substantially different, the stars for which we have Ba detections exhibit Ba abundances of A(Ba) ~ -2.0, independent of metallicity. It should be emphasized though, that the area below [Fe/H] $= -4$ is only sparsely populated, with most stars only having an upper limit on Sr and Ba. Never the less the current data suggest the presence of a floor in Ba at extremely low metallicity.

4. Summary

We have here presented some of the key results of an homogeneous abundance analysis of 23 extremely metal-poor stars found in the Milky Way halo. A large fraction of these are carbon enhanced, and from inspection of their C and N abundances along with $^{12}C/^{13}C$ ratios, we see signs of mixing having occurred in the progenitors of these stars. For the CEMP-no stars the results indicate that this mixing was most efficient at the lowest metallicities. Our sample confirms the presence of two bands in the absolute C abundances of CEMP stars, as suggested by Spite *et al.*(2013). However, the presence of CEMP-no stars on both bands challenges the interpretation as purely intrinsic/extrinsic C origin. Finally we detect different behavior for the absolute abundances for the two neutron-capture elements Sr and Ba, where an indication of a floor is seen in the Ba abundances of CEMP-no stars.

Acknowledgements

T.T.H. and N.C acknowledge support from Sonderforschungsbereich SFB 881 "The Milky Way System" (subproject A4) of the German Research Foundation (DFG), C.J.H acknowledge support from research grant VKR023371 from the Villum Foundation, and T.C.B acknowledge partial support for this work from grants PHY 08-22648; Physics Frontier Center/Joint Institute or Nuclear Astrophysics (JINA), and PHY 14-30152; Physics Frontier Center/JINA Center for the Evolution of the Elements (JINA-CEE), awarded by the US National Science Foundation.

References

Beers, T. C., & Christlieb, N. 2005, *A&A*, 43, 531
Bisterzo, S., Gallino, R., Straniero, O., Cristallo, S., & Käppeler, F. 2012, *MNRAS*, 422, 849
Bonifacio, P., Caffau, E., Spite, M., & *et al.*, 2015, *A&A*, 579, A28
Chiappini, C., Ekström, S., Meynet, G., & *et al.*, 2008, *A&A*, 479, L9
Hansen, T. T., Hansen C. J., Christlieb, N., & *et al.*, 2015a, *ApJ*, 807, 173
Hansen, T. T., Andersen, J., & Nordström, B *et al.*, 2015b, *A&A*, submitted
Herwig, F, 2005, *ARAA*, 43, 435
Hirschi, R, 2007, *A&A*, 461, 571
Lee, Y. S., Beers, T. C., Masseron, T., *et al.*, 2013, *AJ*, 146, 132
Lucatello, S., Tsangarides, S., Beers, T. C., & *et al.*, 2005, *ApJ*, 625, 825
Maeder, A., Meynet, G., & Chiappini, C, 2015., *A&A*, 576, A56
Masseron, T., Johnson, J. A., Plez, B., & *et al.*, 2010, *A&A*, 509, A93
Meynet, G., Ekström, S., & Maeder, A, 2006, *A&A*, 447, 623
Norris, J. E., Yong, D., Bessell, M. S., & *et al.*, 2013, *ApJ*, 762, 28
Spite, M., Caffau, E., Bonifacio, P., & *et al.*, 2013, *A&A*, 552, A107
Umeda, H. & Nomoto, K. 2003, *Nature*, 422, 871
Yong, D., Norris, J. E., Bessell, M. S., & *et al.*, 2013, *ApJ*, 762, 26

The General Assembly of Galaxy Halos: Structure, Origin and Evolution
Proceedings IAU Symposium No. 317, 2015
A. Bragaglia, M. Arnaboldi, M. Rejkuba & D. Romano, eds.

doi:10.1017/S1743921315008340

Planetary Nebulae and their parent stellar populations. Tracing the mass assembly of M87 and Intracluster light in the Virgo cluster core

Magda Arnaboldi[1,2], Alessia Longobardi[3] and Ortwin Gerhard[3]

[1]ESO, K. Schwarzschild Str. 2, 85748 Garching, Germany
email: marnabol@eso.org

[2]INAF, Oss. Astr. di Pino Torinese, 10025 Pino Torinese, Italy

[3]Max-Planck-Institut für Extraterrestrische Physik, Postsach 1312, 85741 Garching, Germany

Abstract. The diffuse extended outer regions of galaxies are hard to study because they are faint, with typical surface brightness of 1% of the dark night sky. We can tackle this problem by using resolved star tracers which remain visible at large distances from the galaxy centers. This article describes the use of Planetary Nebulae as tracers and the calibration of their properties as indicators of the star formation history, mean age and metallicity of the parent stars in the Milky Way and Local Group galaxies. We then report on the results from a deep, extended, planetary nebulae survey in a 0.5 deg^2 region centered on the brightest cluster galaxy NGC 4486 (M87) in the Virgo cluster core, carried out with SuprimeCam@Subaru and FLAMES-GIRAFFE@VLT. Two planetary nebulae populations are identified out to 150 kpc distance from the center of M87. One population is associated with the M87 halo and the second one with the intracluster light in the Virgo cluster core. They have different line-of-sight velocity and spatial distributions, as well as different planetary nebulae specific frequencies and luminosity functions. The intracluster planetary nebulae in the surveyed region correspond to a luminosity of four times the luminosity of the Large Magellanic Cloud. The M87 halo planetary nebulae trace an older, more metal-rich, parent stellar population. A substructure detected in the projected phase-space of the line-of-sight velocity vs. major axis distance for the M87 halo planetary nebulae provides evidence for the recent accretion event of a satellite galaxy with luminosity twice that of M33. The satellite stars were tidally stripped about 1 Gyr ago, and reached apocenter at a major axis distance of 60 – 90 kpc from the center of M87. The M87 halo is still growing significantly at the distances where the substructure is detected.

Keywords. Stars: AGB and post-AGB. (ISM) Planetary nebulae: general. Galaxies: general, abundances, Virgo cluster, elliptical and lenticular, cD, halos, formation, NGC 4486, kinematics and dynamics, structure.

---

1. Introduction

Accretion events are believed to be responsible for the build up of stellar halos in elliptical galaxies (Delucia & Blaizot 2007) at relatively low redshift ($z < 2$; Oser *et al.* 2010). In the dynamical centers of galaxy clusters, brightest cluster galaxies (BCGs) are expected to have the majority of their stars accreted (Cooper *et al.* 2015). The galaxy NGC 4486 (M87) is one of the nearest BCGs (at D=14.5 Mpc) in the densest region of the Virgo cluster (Binggeli *et al.* 1987). Its halo represents a benchmark for studies of the assembly history of extended halos in high density environments. Because of the large orbital time in their outer regions, these halos may still contain fossil records of the mass accretion events that lead to their hierarchical build up.

M87 has been the target of several imaging surveys and its close proximity made it possible to identify planetary nebulae (PNs) with 8 meter class telescopes (Arnaboldi *et al.* 2003, Aguerri *et al.* 2005, Castro-Rodriguez *et al.* 2009). The goal is to use PN as kinematic tracers and their general PN population properties as probes for star formation history, age and metallicity of the parent stellar population in those regions of M87 where the surface brightness is too low to carry out absorption line spectroscopy.

In the next sections, we present a concise summary of the general characteristics of PN population as tracers of stars and motions in galaxies and then describe the results from the extended PN survey in M87.

2. General properties of PN population

PNs are the final evolutionary stage for most stars in the mass range $1 - 8\,\mathrm{M}_{\odot}$. In the Milky Way (MW), about 95% of the stars will end their lives as PNs, while only 5% explode as supernovae. The PN phase lasts $\tau_{PN} \sim 3\times10^4$ years at most, and its duration depends on the age and metallicity of the parent stellar population (Buzzoni *et al.* 2006). τ_{PN} is also related to the expansion time of a nebular shell $\tau_{PN} = D_{PN}/V_{exp}$, where D_{PN} is the diameter and V_{exp} is the expansion velocity of a PN shell; typical expansion velocities for the brightest PNs are in the range $11 - 22$ kms^{-1} (Arnaboldi *et al.* 2008). τ_{PN} could be shortened by the presence of a hot interstellar medium (Dopita *et al.* 2000, Villaver & Stanghellini 2005) which may remove the gaseous shell during its expansion.

Because the diffuse nebula around the core is very efficient in re-emitting $\sim$ 15% of the UV energy radiated by the central star in the optical Oxygen forbidden line [OIII] at 5007 Å [which is the brightest optical emission of a PN (Dopita *et al.* 1992)], PN stars can be efficiently selected via narrow band imaging centered on the Oxygen line.

There are about 2000 PNs known out of 200 billion stars in the MW, and they are mostly concentrated towards the MW plane. A typical Galactic PN has an average shell diameter of about 0.3 pc. Hence when a sample of PNs similar to those in the MW are detected in external galaxies at distances larger than 1 Mpc, they are identified as spatially unresolved emissions of monochromatic green light at 5007Å.

The integrated [OIII] flux F_{5007} of a spatially unresolved PN can be expressed as m_{5007} magnitude via the formula:

$$m_{5007} = -2.5\log(F_{5007}) - 13.74 \tag{2.1}$$

(Jacoby 1989). Narrow band imaging of external galaxies provide m_{5007} magnitudes for the entire PN population of the surveyed galaxy, down to a given limiting flux. It is then possible to derive the Planetary Nebulae luminosity function (PNLF) $N(m_{5007})$ for that PN population at the galaxy distance. The PNLF has been used extensively as secondary distance indicator in early and late-type galaxies within $10-15$ Mpc distance (see Ciardullo *et al.* (2002) for a review). The PNLF is often approximated by an analytical formula given by

$$N(M) \propto e^{0.307M} \times \left(1 - e^{3(M^*-M)}\right) \tag{2.2}$$

as introduced by Ciardullo *et al.* (1989), where $M^* = -4.51$ is the absolute magnitude of the bright cut-off of the PNLF (Ciardullo *et al.* 1998). This analytical formula is the product of two exponential terms: the first term can be thought of as the dimming of the [OIII] flux as the shell expands at uniform speed (Heinze & Westerlund 1963), and the second term models the cut-off at bright magnitudes. Hence the formula in Eq. 2.2 describes a PN population as an ensemble of diffuse expanding shells powered by unevolving massive cores, all at about $M_{core} \simeq 0.7 M_{\odot}$, which are emitting a total luminosity of $L \simeq 6000 L_{\odot}$ (Ciardullo *et al.* 2002).

Simple stellar population theory predicts that PN cores should become fainter as the stellar population ages, with core masses as low as $M_{core} \leqslant 0.55 M_{\odot}$ in a 10 Gyr old, solar-enriched, stellar population (Buzzoni *et al.* 2006). Marigo *et al.* (2004) computed a relative dimming of 4 magnitudes for the bright cut-off M^* of the PNLF for a 10 Gyr old population with respect to that of a 1 Gyr old population. Such a strong dependence of M^* on age is **not** observed though: empirically, the absolute magnitude of the PNLF bright cut-off is the same for PN populations in star forming disks and in ellipticals (Longobardi *et al.* (2013)). A possible explanation is that binary stars are progenitors to the brightest PNs in old populations (Ciardullo *et al.* 2005). Still there may be systematic variations of the PNLF that correlate with the star formation history, mean age and metallicity of the parent stellar population, which can be used to constrain their values and any spatial variations in extended stellar halos.

2.1. *PN visibility lifetime and luminosity functions in the Milky Way and Local Group galaxies*

PN specific frequency - The total number of PNs associated with the bolometric luminosity of a parent stellar population is expressed as $N_{PN} = \alpha L_{\odot,bol}$, where α is the PN specific frequency. The value of the α parameter is related to the normalization in Eq. 2.2; i.e. to the observed total number of PNs of a given detected population. The value of α is related to the PN visibility lifetime τ_{PN} by the equation

$$\alpha = \frac{N_{PN}}{L_{\odot,bol}} = \mathrm{B}\tau_{PN} \tag{2.3}$$

where B (independent of the stellar population) is the PN formation rate (stars/yr/$L_{\odot}$; Buzzoni *et al.* 2006). The measured values for the α parameter show strong scatter for stellar populations redder than $(B - V) \geqslant 0.8$ (Coccato *et al.* 2009, Cortesi *et al.* 2013) with an inverse correlation with the far ultraviolet (FUV) color excess, that is stellar populations with a strong FUV excess are *PN-starved.* Stellar populations with an UV up-turn or FUV color excess are the old and metal rich populations in massive elliptical galaxies. Differently, stellar populations in irregular galaxies like the LMC are *PN-rich.* In general PN populations show systematic variations of the α values with the integrated photometric properties of the parent stellar population, hence variations of the measured α values as function of radius in an extended stellar halo can be used as a signal for different stellar populations in the halo.

PNLF morphology - The PNLF shows systematic variations that correlate with the average age and metallicity of the parent stellar population. In the $\log(N)$ vs. m_{5007} plot, the gradient of the PNLF within 2.5 magnitude below the brightest can be steeper or shallower than 0.307 (as in Eq. 2.2). The PN population associated with the MW bulge has a steeper PNLF than that derived for the M31 PNs (Arnaboldi *et al.*, in prep.), while those in the outer regions of star-forming disks have a shallower or no gradient (Ciardullo *et al.* 2004). Longobardi *et al.* (2013) proposed a generalized formula for the PNLF according to the equation

$$N(M) = c_1 e^{c_2 M} \times \left(1 - e^{3(M^* - M)}\right); M^* = -4.51 \tag{2.4}$$

where c_1 is related to the value of the α parameter to first order and c_2 is related to the gradient of the $\log(N)$ at faint magnitudes. We are planning to apply the generalized formula to complete and extended PN populations and to correlate the c_1, c_2 derived values

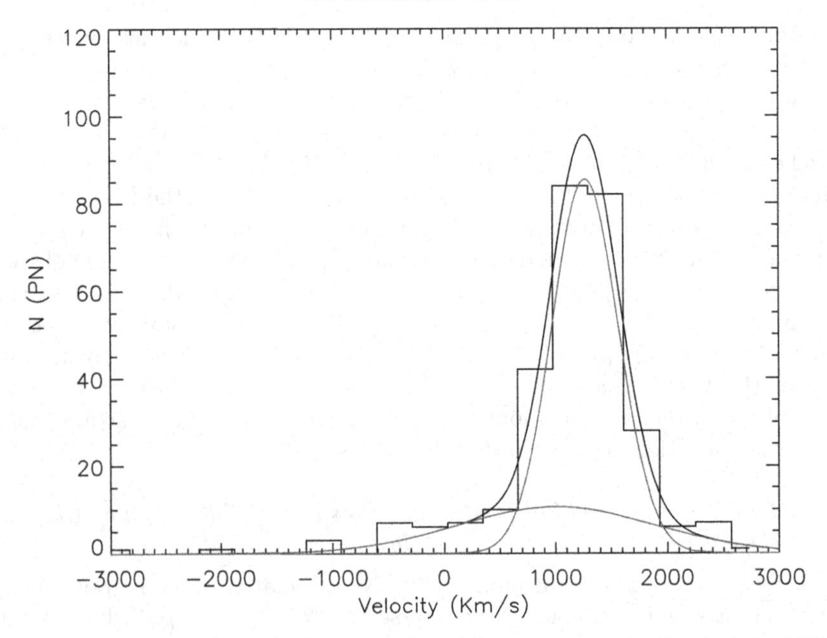

Figure 1. Histogram of the line-of-sight velocities of the spectroscopically confirmed PNs (black histogram) fitted with a double Gaussian (black curve). Red and blue lines represent the two Gaussian associated with the M87 halo and the IC components. From Longobardi *et al.* (2015a).

with the mean ages and metallicities of the parent stellar populations from spectroscopic measurements, so that we can have a better calibration of the PN probes.

In addition to the gradient, the PNLF of stellar population in low luminosity - metal poor galaxies - shows the presence of a *dip*. This feature of the PNLF is measured with high significance in the well sampled PNLF for the LMC (Reid & Parker 2010), the SMC (Jacoby & De Marco 2002), NGC 6822 (Hernández-Martínez & Peña 2009) and in the outer regions of M33 (Ciardullo *et al.* 2004). This dip falls within an interval of 2 to 4 magnitudes below M^*, but the magnitude at which the dip is detected varies between the PN populations sampled in the Local Group.

Strategy - we can use the global properties of the PN populations, their PNLFs, the gradients, dips and α values to signal transition from old/metal-rich to fading/metal-poor populations when the individual stars cannot be resolved or their surface brightness is too low to carry out integrated light photometry or absorption line spectroscopy.

3. The PN populations in the Virgo cluster core

In 2010 we started an imaging survey with SuprimeCam@Subaru to cover 0.5 deg^2 in the M87 halo; at the distance of the Virgo cluster (D=14.5 Mpc) this is equivalent to an area of (130 kpc)^2. We wanted to use the general properties described in Sec. 2 to study its stellar population and kinematics.

We acquired deep narrow band images centered on the [OIII] emission redshifted to the systemic velocity of M87 (V_{sys} = 1275 kms^{-1}) and deep off-band images in the V-band. We identified PN candidates as spatially unresolved [OIII] sources with no continuum following the procedure described in Arnaboldi *et al.* (2002). The final magnitude limited catalog consisted of 688 PN candidates down to 2.5 magnitude from the brightest PN (Longobardi *et al.* 2013). We then carried out the spectroscopic follow-up with

FLAMES/GIRAFFE@VLT to acquire spectra for these candidates. We obtained spectra for 289 confirmed PNs (Longobardi *et al.* 2015a) which we analyzed together with 12 previous identified PNs from Doherty *et al.* (2009), for a total combined sample of 301 PNs.

We built the line-of-sight velocity distribution (LOSVD) for the whole PN sample: this is shown in Fig. 1. The LOSVD is characterized by a strong peak at 1275 kms^{-1} and a second moment $\sigma_n \simeq 300\ \mathrm{kms}^{-1}$, with large asymmetric wings, with a tail extended towards zero and negative LOS velocities. These wings can be modeled with a broad Gaussian component, centered at $v_b = 995\ \mathrm{kms}^{-1}$ and $\sigma_b = 900\ \mathrm{kms}^{-1}$. The narrow component is consistent with the LOSVD for the stars in the M87 halo, while the broad component has a LOSVD similar to that of Virgo cluster galaxies. We thus selected the M87 halo PNs as those associated with the narrow component in the LOSVD by means of a robust iterative procedure (Longobardi *et al.* 2015a). The PNs in the broad asymmetric wings of the LOSVD are thereby tagged as Virgo intracluster (IC) PNs. In Figure 2 we show the projected phase space diagram for the 301 PNs around M87 and the additional ICPNs from Doherty *et al.* (2009).

We studied the spatial distribution of the two selected PN samples and compared their profiles with the V band photometry of M87 by Kormendy *et al.* (2009). The number density profile of the M87 halo PNs follows the Sersic $n = 11.8$ surface brightness profile, while that of the ICPNs follows a much flatter radial profile, which is consistent with a power law, $N_{ICPN}(R) \propto R^{\gamma}$ with $\gamma = [-0.34, -0.04]$. By scaling the M87 PN density profile to the inner regions we derived the value of the α parameter for the M87 PN halo population: $\alpha_{halo} = (1.06 \pm 0.12) \times 10^{-8} PNL_{\odot,bol}^{-1}$. The value of the α parameter for the intracluster light is different: the same procedure applied to the ICPN number density profile returns $\alpha_{ICL} = (2.72 \pm 0.72) \times 10^{-8} PNL_{\odot,bol}^{-1}$.

We then compared our α_{halo} and α_{ICL} values with those for PN populations in nearby galaxies. Galaxies with $(B - V) \leqslant 0.8$ have PN specific frequency values similar to that of α_{ICL}. For redder galaxies, the scatter increases; the value of α_{halo} is the same as for populations with an FUV excess that are PN-starved. Hence the IC component contributes three times more PN per unit bolometric luminosity than the M87 halo light, signaling a change of population from halo to ICL. This transition is consistent with the existence of a color gradient towards bluer colors in M87 at large radii (Rudick *et al.* 2010), and an ICL population that is mostly old (age $\simeq$ 10 Gyr) with a mean metallicity of $[M/H] \simeq -1.0$ (Williams *et al.* 2007). On the basis of the kinematic separation of the M87 halo and intracluster PNs, we built the two independent PNLFs, and examined them in turn. The PNLF for the M87 halo PNs is steeper than in Eq. 2.2. We used the generalized formula from Eq. 2.4 and derived $c_2 = 0.72$ and a distance modulus of $m - M = 30.8$. The PNLF of the ICPNs shows a dip at $1 - 1.5$ mag below M^*, reminiscent of the morphology of the PNLF observed in PN populations associated with low luminosity/metal poor galaxies. We fitted the generalized PNLF formula to the PNLF of the ICPNs, and found $c_2 = 0.66$ and a distance modulus of $m - M = 30.8$.

We could estimate the luminosity associated with the ICPNs by integrating the ICL surface brightness profile over the surveyed area, which amounts to $L_{ICL} = 5.3 \times 10^9 L_{\odot,bol}$. From the presence of the "dip" of the PNLF, the parent stellar population is similar to that of the Magellanic Clouds or in the outer regions of the M33 disk. Thus the sampled luminosity of the ICL is equivalent to four time the luminosity of the LMC or 1.5 times that of $M33$, over the whole surveyed area of $(130\ \mathrm{kpc})^2$.

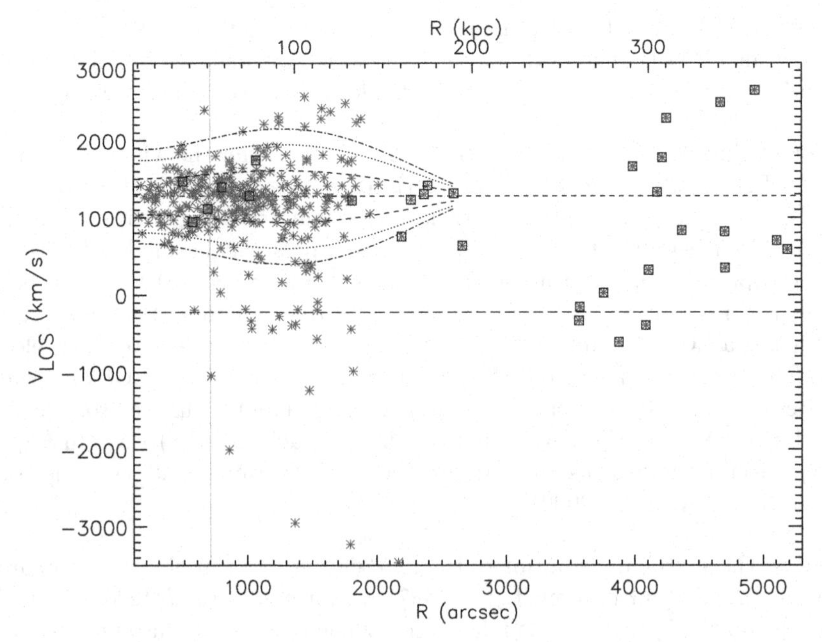

Figure 2. Projected phase-space diagram, V_{LOS} vs. major axis distance from the center of M87, for all spectroscopically confirmed PNs in the Virgo core.The major axis distance is given both in arcsec (bottom axis), and in kpc (top axis), where 73 pc= 1″. The PNs are classified as M87 halo PNs (red asterisks) and ICPNs (blue asterisks), respectively. Black squares identify spectroscopically confirmed PNs from Doherty *et al.* (2009). The dashed horizontal line shows the M87 systemic velocity $V_{sys} = 1275$ kms^{-1}, while the continuous green line shows the effective radius $R_e = 703.914''$ (Kormendy *et al.* 2009). From Longobardi *et al.* (2015a).

4. The late mass assembly of the M87 halo

We then turned to the M87 halo PNs and studied the projected phase-space diagram v_{LOS} vs. R_{Maj} for this component. The density of points is not uniform in the range of velocities covered by the M87 halo. There is a clearly identifiable **V**-shaped over-density, or *"Chevron"*, between 30 and 90 kpc radii, with its vertex, or edge, culminating at 90 kpc or 1200″ (Longobardi *et al.* 2015b) . These high density substructures in phase-space are likely to be associated with the disruption of a satellite galaxy in the deeper potential of a massive host (Quinn 1984).

We looked at the LOSVD in three radial bins over the radial range of the Chevron, and employed a Gaussian Mixture Model to assign each M87 halo PN a probability to belong to the **V** substructure or to the smooth halo. A total of 54 PNs are thereby associated to the **V** substructure and 200 PNs to the smooth halo component (Longobardi *et al.* 2015b). We then looked at the spatial distribution of the Chevron PNs: the highest density occurs at 1200″ NW along the M87 major axis, and is spatially correlated with a diffuse, extended substructure, labeled "the crown" of M87 (Longobardi *et al.* 2015b), see Figure 3. The luminosity of the substructure is about 60% of the light at the location where the crown is found. We computed the total luminosity of the disrupted satellite from the 54 PNs: it amounts to $L_{sat} = 2.8 \times 10^9 L_{\odot,bol}$. We also looked at correlation between the high density of the chevron PNs and the (B-V) color in the M87 halo. A strong correlation is found, with bluer color (B-V = 0.76) measured at the position of the "crown". From the luminosity and the color we infer that the luminosity of the dissolved satellite is equivalent to about twice the luminosity of M33.

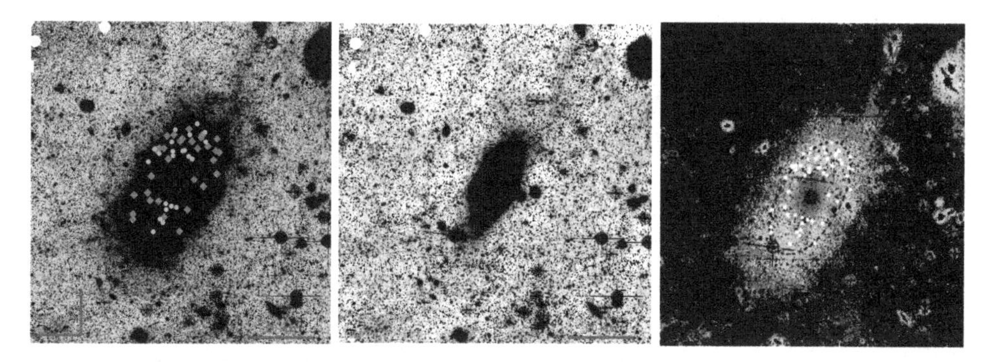

Figure 3. Spatial and color distribution associated with the kinematic substructure identified in the phase space of the M87 halo PNs. Left panel: V-band image of a 1.6 × 1.6 deg^2 centered on M87 from Mihos *et al.* (in prep.). Full circles, and diamonds indicate the spatial positions of the M87 halo PNs in the chevron substructure. Magenta and green colors indicate PN LOSVs above and below V_{LOS} = 1254 kms^{-1}, the LOSV at the end of the Chevron. Central panel: unsharp masked image of M87 median binned to enhance faint structures. The crown-shaped substructure is visible at distance of 800″ – 1200″(60 – 90 kpc) along the major axis, NW of M87. The blue line measures 90 kpc. Right panel: (B – V) color image of M87 from Mihos *et al.* (in prep.) with chevron PNs overplotted (white dots). The dashed ellipse indicates the isophote at a major axis distance of 1200″. The crown is found in a region where the (B – V) color is on average 0.8, bluer than on the minor axis. From Longobardi *et al.* (2015b).

From the distribution and velocities of Chevron PNs in Figure 3, a possible interpretation of the satellite orbit could be that it was first disrupted entering M87 from the south (along the green dots), with the debris then moving up north, turning around in the crown region, coming back down on both sides across M87 (magenta dots). The velocities would then imply that the northern side of M87 is closer to the observer. The dynamical time for such an orbit is of the order of 1 Gyr (Weil *et al.* 1997).

5. Conclusions

Using PNs as tracers we showed that the stellar halo of the BCG M87 is distinct from the surrounding ICL in its kinematics, density profile, and parent stellar population, consistent with the halo of M87 being redder and more metal-rich that the ICL. The ICL in our surveyed fields corresponds to about four times the luminosity of the LMC, spread out over a region of (130 kpc)^2. It is remarkable that population properties can be observed for such a diffuse component. Based on its population properties we propose that the progenitors of the Virgo ICL were low-mass star forming galaxies.

We also presented kinematic and photometric evidence for an accretion event in the halo of M87. This event is traced by PNs whose phase space shows a distinct chevron-like feature, which is the result of the incomplete phase-space mixing of a disrupted galaxy. At a major axis distance of $R \sim 69-90$ kpc where the width of the chevron goes to zero, a deep optical image shows the presence of a crown-like substructure that contribute more than 60% of the light in this area. The luminosity of the satellite corresponds to about two times M33 with color $(B-V) = 0.76$. The similar colors of the accreted satellite and ICL suggest that the halo of M87 is presently growing through the accretion of similar star-forming systems as those that build up the diffuse ICL component. The newly discovered substructure within the halo of M87 demonstrates that beyond a distance of 60 kpc its halo is still assembling.

6. Acknowledgments

The authors wish to thank J.C Mihos, R. Hanuschik for their contribution, and the time allocation committees of the Subaru Telescope and the ESO OPC for the opportunity to carry out this exciting project. Based on observations made with the VLT at Paranal Observatory under programmes 088.B-0288(A) and 093.B-066(A), and with the Subaru Telescope under programme S10A-039.

References

Aguerri, J. A. L., Gerhard, O. E., Arnaboldi, M., Napolitano, N. R., *et al.* 2005, *AJ*, 129, 2585
Arnaboldi, M., Doherty, M., Gerhard, O. E., Ciardullo, R. *et al.* 2008, *ApJ*, 674, L17
Arnaboldi, M., Freeman, K. C., Okamura, S., Yasuda, N. *et al.* 2003, *AJ*, 125, 514
Arnaboldi, M., Aguerri, J. A. L., Napolitano, N. R., Gerhard, O. E. *et al.* 2002, *AJ*, 123, 760
Binggeli, B., Tammann, G. A., & Sandage, A. 1987, *AJ*, 94, 251
Buzzoni, A., Arnaboldi, M., & Corradi, R. 2006, *MNRAS*, 368, 877
Castro-Rodriguez, N., Arnaboldi, M., Aguerri, J. A. L., Gerhard, O. *et al.* 2009, *A&A*, 507, 621
Ciardullo, R., Sigurdsson, S., Feldmeier, J. J., Jacoby, G. H. 2005 *ApJ* 629, 499
Ciardullo, R., Durrell, P. R., Laychak, M. B. *et al.* 2004, *ApJ*, 614, 167
Ciardullo, R., Feldmeier, J. J., Jacoby, G. H., Kuzio de Naray, R. *et al.* 2002, *ApJ*, 577, 31
Ciardullo, R., Jacoby, G. H., Feldmeier, J. J., & Bartlett, R. E. 1998, *ApJ*, 492, 62
Ciardullo, R., Jacoby, G. H., Ford, H. C., & Neill, J. D. 1989, *ApJ*, 339, 53
Coccato, L., Gerhard, O., Arnaboldi, M. *et al.* 2009, *MNRAS*, 394, 1249
Cooper, A. P., Gao, L., Guo, Q., Frenk, C. S., *et al.* 2015, *MNRAS*, 451, 2703
Cortesi, A., Arnaboldi, M., Coccato, L. *et al.* 2013, *A&A*, 549, 115
De Lucia, G.,& Blaizot, J. 2007, *MNRAS*, 375, 2
Doherty, M., Arnaboldi, M., Das, P., Gerhard, O. *et al.* 2009, *A&A*, 502, 771
Dopita, M., Massaglia, S., Bodo, G., Arnaboldi, M. *et al.* 2000, *ASPC*, 199, 423
Dopita, M., Jacoby, G. H., & Vassiliadis, E. 1992, *ApJ*, 389, 27
Jacoby, G. H. & De Marco, O. 2002, *AJ*, 123, 269
Jacoby, G. H. 1989, *ApJ*, 339, 39
Kormendy, J., Fisher, D. B., Cornell, M. E., & Bender, R. 2009, *ApJS*, 182, 216
Hernández-Martínez, L. & Peña, M. 2009, *A&A*, 495, 447
Henize, K. G. & Westerlund, B. E. 1963, *ApJ*, 137, 747
Longobardi, A., Arnaboldi, M., Gerhard, O., & Hanuschik, R. 2015a, *A&A* 579, 135
Longobardi, A., Arnaboldi, M., Gerhard, O., & Mihos, J. C. 2015b, *A&A* 579L, 3
Longobardi, A., Arnaboldi, M., Gerhard, O., Coccato, L. *et al.* 2013, *A&A* 558, 42
Marigo, P., Girardi, L., Weiss, A., Groenewegen, M. A. T., *et al.* 2004, *A& A*, 423,
Oser, L., Ostriker, J. P., Naab, T., Johansson, P. H., *et al.* 2010, *ApJ*, 725, 2312
Quinn, P. J. 1984, *ApJ*, 279, 596
Reid, W. A. & Parker, Q. A. 2010, *MNRAS*, 405, 1349
Rudick, C. S., Mihos, J. C., Harding, P., Feldmeier, J. J. *et al.* 2010, *ApJ*, 720, 569
Villaver, E. & Stanghellini, L. 2005, *ApJ*, 632, 854
Weil, M. L., Bland-Hawthorn, J., & Malin, D. F. 1997, *ApJ* 490, 664
Williams, B. F., Ciardullo, R., Durrell, P. R., Vinciguerra, M. *et al.* 2007, *ApJ*, 656, 756

The General Assembly of Galaxy Halos: Structure, Origin and Evolution
Proceedings IAU Symposium No. 317, 2015
A. Bragaglia, M. Arnaboldi, M. Rejkuba & D. Romano, eds.
doi:10.1017/S1743921315009710

RR Lyrae to understand the Galactic halo

Giuliana Fiorentino

INAF-Osservatorio Astronomico di Bologna, via Ranzani 1, 40127, Bologna, Italy
email: giuliana.fiorentino@oabo.inaf.it

Abstract. We present recent results obtained using old variable RR Lyrae stars on the Galactic halo structure and its connection with nearby dwarf galaxies. We compare the period and period-amplitude distributions for a sizeable sample of fundamental mode RR Lyrae stars (RRab) in dwarf spheroidals (~1300 stars) with those in the Galactic halo (~16'000 stars) and globular clusters (~1000 stars). RRab in dwarfs –as observed today– do not appear to follow the pulsation properties shown by those in the Galactic halo, nor they have the same properties as RRab in globulars. Thanks to the OGLE experiment we extended our comparison to massive metal–rich satellites like the dwarf irregular Large Magellanic Cloud (LMC) and the Sagittarius (Sgr) dwarf spheroidal. These massive and more metal–rich stellar systems likely have contributed to the Galactic halo formation more than classical dwarf spheroidals.

Finally, exploiting the intrinsic nature of RR Lyrae as distance indicators we were able to study the period and period amplitude distributions of RRab within the Halo. It turned out that the inner and the outer Halo do show a difference that may suggest a different formation scenario (in situ vs accreted).

Keywords. Galaxy: structure, Galaxy: halo, stars: variables: RR Lyrae

1. Introduction

"How did the Galactic halo (Halo) form and evolve?" remains a fundamental open question in astrophysics. Among the different components of our Galaxy the Halo plays a crucial role. The Halo is one of the oldest Galactic component and as a such it is a direct witness of the infancy of the Milky Way (MW). Trapped in its fine and large structure (globular clusters, inner/outer Halo, tidal debris of accreted satellites) there is the key information on how it was formed.

The early suggestion by Searle & Zinn (1978), based on about 20 GCs, that the outer Halo formed from the aggregation of protogalactic fragments is nowadays supported by *1)* theoretical lambda-Cosmological Dark Matter (CDM) simulations of galaxy formation that predict small galaxies form first and then cluster to form larger galaxies, and *2)* the observation of stellar streams and merging satellites in the MW and in other galaxies. Currently there are two main open issues concerning the Halo structure and formation:

–*Duality of the Halo*– Although the Halo was once considered a single component, evidence for its duality is emerging. The existence of an inner and outer (galactocentric distance d_G ~15 Kpc) component has been recently supported by several works. In particular Carollo *et al.* (2007, 2010) used a large sample (~16'000) of local stars (within 4 Kpc) included in recent data release of the Sloan Digital Sky Survey (SDSS, see York *et al.* 2000). In order to perform an accurate kinematical analysis and to use a simple model for the Halo potential, the authors carefully selected a sample of local stars. Using this sample, they found a kinematic signature of counter rotating and spherically distributed stars that they have attributed to the outer Halo. This is in contraposition with a more flattened and concentrated inner halo component with a zero or slightly prograde rotation around the Galactic center. The inner and outer Halo seem also to differ in their chemical compositions. In fact the outer Halo results to be metal poorer (<[Fe/H]>∼-2.2) than the mean peak of the inner Halo (<[Fe/H]>∼-1.6). However, a possible strong bias in the distance determination (up to 50%) may be the culprit of the "artificial counter-rotating component" (Schonrich *et al.* 2011). Thus, we still lack a general consensus concerning the Halo large structure;

–*The nature of the Halo building blocks* –if any– is still a matter of debate. In particular, the question of whether the current dwarf spheroidal satellites (dSphs) of the MW are surviving representatives of the Halo building blocks has been explored in several works but none of them is firmly conclusive (see Tolstoy *et al.* 2009). Among these, Helmi *et al.* 2006 found a significant discrepancy between the metallicity distribution function of Halo stars when compared with that observed in dSphs. In particular, the Halo metallicity distribution function shows a very pronounced metal–poor tail which is not present in dSphs. This disagreement seems to be alleviated (but not removed) when selection effect that privileged metal–poor stars are accounted for in the Galactic halo sample (Schorck *et al.* 2009). Emerging possible candidates as protogalactic fragments are the ultra faint dwarfs that have been recently discovered (L $\lesssim 10^5$ $L_\odot$, e.g. Belokurov *et al.* 2006). It is worth noting that most of the studies comparing chemical abundances in Halo and dSphs are based on red giant stars (Venn *et al.* 2004) that suffer the age-metallicity degeneracy. On the other hand the unknown distance (and thus luminosity) of individual red giant Halo stars further complicate a proper comparison. This is why the possibility to compare similar, and preferably old, stellar tracers is crucial in the comparison between the Halo and dSphs. The ideal stellar tracers are old globular clusters (GCs) and field stars (e.g. RR Lyrae, RRL; blue horizontal branch, BHB).

2. Constraints on the Galactic halo structure

The use of GCs to constrain the time scale of the Halo formation dates back to half century ago (Eggen *et al.* 1962, Searle & Zinn 1978). Stars in GCs have the key advantage to have similar iron abundance (with few exceptions, e.g. Ω Cen), to be roughly coeval and all at the same distance to us. Thanks to these properties and to the improving accuracy in performing photometry in dense environments, nowadays we are able to estimate very accurate relative ages (δ age $\lesssim$ 1 Gyr). Marin-Franch *et al.* 2009, measured relative ages for a sample of 64 GCs observed with HST optical imaging. They found a dichotomy in the distribution of GCs. A group of young GCs that follows an age–metallicity relation (AMR) and a group of coeval old clusters with an age dispersion of ∼5%. These results suggest a scenario where the Halo formed in two steps: 1) a rapid event (duration less than 0.8 Gyr) that originated the old coeval group; 2) a slow event that may consist of the accretion of nearby satellites and their GCs. A different result has been discussed in Leaman *et al.* (2013) using a similar sample of 61 GCs. They found that GCs define two parallel AMRs. On the basis of kinematical properties of GCs, this evidence is interpreted in the following way: 1) The metal–poor branch is formed by Halo clusters that were accreted by dwarfs ; 2) The metal–rich branch belongs to the Galactic disk. These two results are controversial and leave the main question of how much Halo was formed from accretion or in situ still open. Here, it is worth noting that also GCs have some drawbacks: i) most of them are within 20 Kpc from the Galactic centre (the Halo extends out to ∼100 Kpc, Deason *et al.*2012); ii) they contribute only to few percents of the entire Halo mass (∼10^9 M⊙); iii) they are more complex than previously believed in terms of their chemical enrichment histories (Carretta *et al.* 2009, see also Carretta's

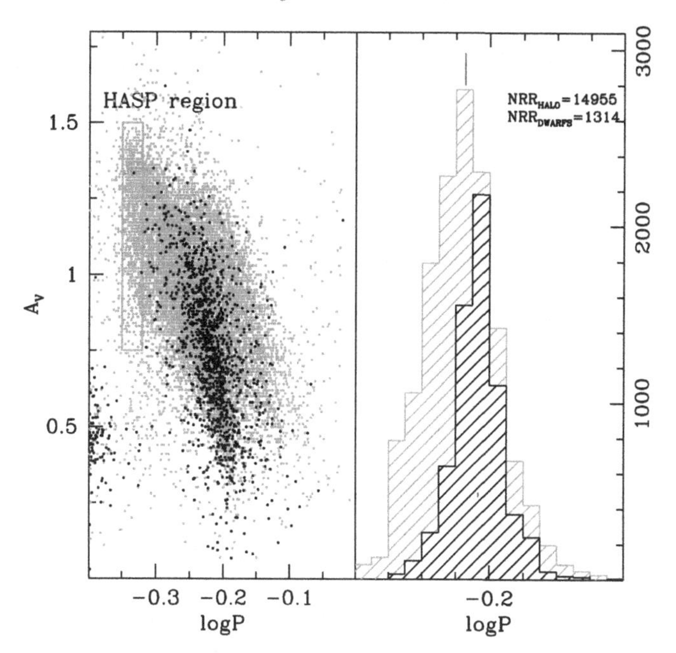

Figure 1. The V-Amplitude vs Period (left) and Period (right) distributions are shown for fundamental RRab stars belonging to the Halo sample (grey) and to the dwarf spheroidal sample (black). The HASP region has been highlighted in green.

review paper, this symposium) and their photometric multiple sequences (e.g., Piotto *et al.* 2015 and reference therein, see also Piotto's review paper, this symposium). This means that GCs may not be fairly representative of the Halo field stars, thus picturing only part of the story.

Increasing attention has been payed to BHB and RRL stars that can be easily recognised and observed over all across the Halo field. Kinman *et al.*(2012, and references therein) derived kinematic properties from radial velocities and proper motions of field Halo stars. These authors estimated distances with several different methods for both BHB and RRL stars. Their main conclusion is that a signature for a different kinematics between inner and outer Halo does exist. In particular the outer Halo shows a retrograde motion confirming the result obtained by Carollo and collaborators. Even thought that this is a very carefully done analysis, it is based on a quite limited sample of BHB and RRL stars, i.e. about 100 objects. Gaia will allows us to drastically increase the statistics of the sample where this kind of analysis will be possible returning a full description of the large and fine structure of the Halo. However, waiting for the Gaia data release, much can be learned using the RRLs detected and characterised by the on going photometric and spectroscopic surveys.

3. New insights from RR Lyrae stars

There are several good reasons to use RRL stars in this context, they are: 1) old ($\gtrsim$ 10 Gyr) stellar population tracers; 2) excellent distance indicators, thus they can trace the different components of the Galaxy; 3) roughly ubiquitous, they have been observed in several kind of different stellar systems (with enough statistics and the proper metallicity), e.g. globular clusters, classical and ultra–faint dwarf galaxies, Halo, bulge, thick disk ; 4) observationally very easy to pick up and to separate from not (or other kind of) variable stars. Hence they are a unique tool to directly compare the ancient stellar population of the Halo at different Galactocentric distance (d_G) with surrounding stellar systems. Thanks to extensive photometric variability surveys, today we have at disposal robust light curves and hence periods, mean magnitudes and amplitudes for more than 15'000 RR Lyrae belonging to the Halo (QUEST, Zinn *et al.* 2014; NSVS, Wozniak *et al.* 2004; ASAS, Szczygiel *et al.* 2009 and CATALINA, Drake *et al.* 2013). Using a magnitude vs [Fe/H] relation (Cacciari & Clementini 2003), where the mean metallicity is assumed [Fe/H]$\sim$-1.6, one can derive d_G for each individual RRL, they span from $\sim$5 to 80 Kpc. This is the largest, deepest, and most homogeneous catalog at our disposal for the Galactic Halo. Because of time-sampling and completeness problems, the fundamental mode RRLs (hereinafter RRab) stars are the most reliable.

-RRLs and the Halo duality- The period and the period-amplitude distributions, that are **reddening and distance** independent, can be analysed at varying the d_G using a Kolmogorov-Smirnov (KS) test (see Stetson *et al.* 2014 and Fiorentino *et al.* 2015 for details). This predicts a low likelihood that the inner ($d_G \lesssim$ 15 Kpc) and the outer ($d_G \gtrsim$ 15 Kpc) Halo come from the same parent population. Furthermore, within this catalogue $\sim$3000 RRab (with $5 \lesssim d_G \lesssim 50$ Kpc) have metallicity estimations from low resolution spectra of SDSS (σ[Fe/H]$\sim$0.3). They provide the most robust evidence of a metallicity gradient of the ancient population in the Halo (Bono *et al.* in preparation), similarly to what is found in other massive spiral galaxies. This metallicity gradient varies from [Fe/H]$\sim$-1.4 (at 8 Kpc) to [Fe/H]$\sim$-1.8 (at 50 Kpc). Amplitudes and periods depend on metallicity, hence the KS test prediction of a difference between inner and outer Halo may be only highlighting the presence of the hidden [Fe/H]–gradient.

-RRLs and the Halo Building blocks- RRLs can also tell us something about the role that classical and ultrafaint dwarf galaxies may have played in the formation of the Halo. Historically, the average properties of RRLs in individual dSphs and UFDs -such as the mean periods of both RRab and RRc stars- are compared to those observed in GCs as representative of the Galactic halo (Smith, Catelan & Clementini 2009). However, today we have at disposal this unique huge sample of Halo stars that can be directly compared for the first time with the dSph RRL population. To make a proper statistical comparison one can decide to build-up from the literature a homogeneous and complete sample of RRLs in dSphs, this includes Carina, Cetus, Tucana, Leo I, Draco, Ursa Minor and eleven UFDs (see Stetson *et al.* 2014 and Fiorentino *et al.* 2015 for the references of the full compilation). It contains more than 1300 RRab stars. This choice is supported by the similarity of the period and period-amplitude distributions of the RRLs in these galaxies. First of all the period distribution of RRab is well peaked around 0.61 d and it can be well approximated with a gaussian function whereas the Halo distribution is very broader with an average of $\sim$0.58 d (see Fig.1). Second, a net difference is clear: stars with High Amplitudes and Short Periods (HASP, Av $\gtrsim$ 0.75 mag and P $\lesssim$ 0.48 d) that are observed in the Halo distribution are not found in any of the considered seventeen dwarfs. The result is solid and indeed it does not change even when the sample of RRab in dSph is almost doubled including more than 1000 RRab belonging to the massive Fornax dSph (Fiorentino *et al.* in preparation). Covering a broad range in metallicity (-2.3 $\lesssim$ [Fe/H] $\lesssim$ -1.1, see Clement *et al.* 2001), GCs can be used to interpret the HASP evidence. Only GCs hosting more than 35 RRL are accounted for to avoid statistical bias. The metallicity turns out to be the main parameter driving the HASP occurrence, since only GCs more metal–rich than [Fe/H]$\sim$-1.5 host RRL in the HASP region.

4. Conclusions

As discussed in Fiorentino *et al.* (2015), the missing HASP RRab suggest that dSphs and UFDs did not have a metallicity enrichment fast enough to build–up alone the Halo. Scaling the dSphs distribution to the Halo one, an extreme upper limit of 50% can be estimated to the contribution of current dSphs to the Galactic Halo formation. This is in quite good agreement with recent results of Cosmological simulations (Tissera *et al.* 2014). Even though the above results rely on rough preliminary estimates, they pose a serious question: "Where does the rest (in fact most) of the Halo mass come from?" There are two main alternatives: 1) from few large and metal-rich stellar systems LMC or Sgr–like (e.g., Zinn *et al.* 2014); 2) from in situ stellar formation (e.g., Vincenzo *et al.* 2014). Thanks to the extensive OGLE survey (Soszynski *et al.* 2009, 2014), more than ∼17'000 and 1600 RRLs in the LMC and Sgr dSph respectively, can be used to investigate the first option. These galaxies both show RRLs in the HASP region. This is reasonable due to the existence of a mass-metallicity relation for the galaxies of the nearby Universe (McConnachie *et al.* 2012). More quantitatively, a KS test performed on these two galaxies, as compared with the Halo, returns a likelihood of ∼10% for Sgr and few % for LMC. Although the exceptionally complete LMC sample may affect the statistical test, it is clear that more massive, and thus metal–richer MW satellites may have had a major role in the Halo assembly.

Acknowledgements

GF is in debt with Monica Tosi and Eline Tolstoy for organizing the meeting "From Dwarfs to Giants" (2013, Sexten, Italy), where the idea of this paper arose from useful discussions. GF has been supported by the FIRB 2013 (grant RBFR13J716).

References

Carollo, D., *et al.* 2007, *Nature*, 450, 1020
Carollo, D., Beers, T. C., Chiba, M., *et al.* 2010, *ApJ*, 712, 692
Clement, C. M., *et al.* 2001, *AJ*, 122, 2587
Drake, A. J., *et al.* 2013, *ApJ*, 763, 32
Eggen, O. J., Lynden-Bell, D., & Sandage, A. R. 1962, *ApJ*, 136, 748
Fiorentino, G., Bono, G., Monelli, M., *et al.* 2015, *ApJ* (Letters), 798, L12
Helmi, A., *et al.* 2006, *ApJ* (Letters), 651, L121
Leaman, R., VandenBerg, D. A., & Mendel, J. T. 2013,*MNRAS* , 436, 122
Marín-Franch, A., Aparicio, A., Piotto, G., *et al.* 2009,*ApJ*, 694, 1498
McConnachie, A. W. 2012, *AJ*, 144, 4
Schönrich, R., Asplund, M., & Casagrande, L. 2011, *MNRAS*, 415, 3807
Schörck, T., Christlieb, N., Cohen, J. G., *et al.* 2009, *A&A*, 507, 817
Searle, L., & Zinn, R. 1978, *ApJ*, 225, 357
Smith, H. A., Catelan, M., & Clementini, G. 2009, *AIP-CP*, 1170, 179
Soszyński, I., *et al.* 2009, *AcA*, 59, 1
Soszyński, I., Udalski, A., Szymański, M. K., *et al.* 2014, *AcA*, 64, 177
Stetson, P. B., Fiorentino, G., Bono, G., Bernard, E. J., Monelli, M., Iannicola, G., Gallart, C., & Ferraro, I. 2014, *PASP*, 126, 616
Szczygieł, D. M., Pojmański, G., & Pilecki, B. 2009, *AcA*, 59, 137
Tissera, P. B., Beers, T. C., Carollo, D., & Scannapieco, C. 2014, *MNRAS*, 439, 3128
Tolstoy, E., Hill, V., & Tosi, M. 2009, *A&AR*, 47, 371
Venn, K. A., Irwin, M., Shetrone, M. D., Tout, C. A., Hill, V., & Tolstoy, E. 2004, *AJ*, 128, 1177
Vincenzo, F., Matteucci, F., Vattakunnel, S., & Lanfranchi, G. A. 2014, *MNRAS*, 441, 2815

Vivas, A. K., *et al.* 2004, *AJ*, 127, 1158
Woźniak, P. R., Vestrand, W. T., Akerlof, C. W., *et al.* 2004, *AJ*, 127, 2436
York, D. G., Adelman, J., Anderson, J. E., Jr., *et al.* 2000, *AJ*, 120, 1579
Zinn, R., Horowitz, B., Vivas, A. K., Baltay, C., Ellman, N., Hadjiyska, E., Rabinowitz, D., & Miller, L. 2014, *ApJ*, 781, 22

The General Assembly of Galaxy Halos: Structure, Origin and Evolution
Proceedings IAU Symposium No. 317, 2015
A. Bragaglia, M. Arnaboldi, M. Rejkuba & D. Romano, eds.

doi:10.1017/S1743921315007292

PN populations in the local group and distant stellar populations

Warren Reid[1,2]

[1]Department of Physics and Astronomy, Macquarie University
Sydney, NSW, 2109, Australia
email: warren.reid@outlook.com

[2]University of Western Sydney,
Locked Bag 1797, Penrith South DC, NSW 1797, Australia

Abstract. Our understanding of galactic structure and evolution is far from complete. Within the past twelve months we have learnt that the Milky Way is about 50% wider than was previously thought. As a consequence, new models are being developed that force us to reassess the kinematic structure of our Galaxy. Similarly, we need to take a fresh look at the halo structure of external galaxies in our Local Group. Studies of stellar populations, star-forming regions, clusters, the interstellar medium, elemental abundances and late stellar evolution are all required in order to understand how galactic assembly has occurred as we see it. PNe play an important role in this investigation by providing a measure of stellar age, mass, abundances, morphology, kinematics and synthesized matter that is returned to the interstellar medium (ISM). Through a method of chemical tagging, halo PNe can reveal evidence of stellar migration and galactic mergers. This is an outline of the advances that have been made towards uncovering the full number of PNe in our Local Group galaxies and beyond. Current numbers are presented and compared to total population estimates based on galactic mass and luminosity. A near complete census of PNe is crucial to understanding the initial-to-final mass relation for stars with mass >1 to <8 times the mass of the sun. It also allows us to extract more evolutionary information from luminosity functions and compare dust-to-gas ratios from PNe in different galactic locations. With new data provided by the Gaia satellite, space-based telescopes and the rise of giant and extra-large telescopes, we are on the verge of observing and understanding objects such as PNe in distant galaxies with the same detail we expected from Galactic observations only a decade ago.

Keywords. Planetary nebulae: general, Galaxies: halos, Surveys, Kinematics.

1. Introduction

Access to large telescopes and innovative identification techniques are contributing to an increased number of planetary nebulae (PNe) being discovered and used to trace stellar populations regardless of the type of galaxy. Although distant PN populations have been used as distance indicators by means of the invariant bright cut-off of the PN luminosity function (PNLF, eg. Ciardullo *et al.* 1989) for some time now, they are more recently being used to trace stellar luminosity (eg. Buzzoni *et al.* 2006), galaxy kinematics (eg. Cortesi *et al.* 2013; Longobardi *et al.* 2015a) and undertake chemical tagging (eg. Gonçalves *et al.* 2014; Corradi *et al.* 2015). PNe, through their bright [O III] and H lines have the advantage of being detected and measured across the 3.1 Mpc diameter of our local group and identified up to distances of 30 Mpc in the local universe, providing a snapshot of stellar populations, their luminosity, age, metallicity and their kinematic migrations over periods up to 8 Gyr ago. This makes them very useful tools with which to test a number of theories concerning the evolution of stars and galaxies.

Tidal encounters between galaxies have occurred in many places such as between M31 and M33 (Bekki 2008) and between the LMC and SMC (Bekki *et al.* 2008b), both about 3 Gyr ago. Such encounters can result in vast extended disks, halos and streams where measurements of metallicity are able to reveal progenitor origins. Low stellar surface brightness beyond projected radii of $\sim 2R_e$ makes it very difficult to study the kinematics and individual properties of stars in elliptical galaxies. The study of the structural, chemical and kinematic properties of PNe in the outer regions of galaxies such as their halos are now revealing previously unseen clues as to galaxy formation and evolution. The properties of stellar halos in the Milky Way and M31 as well as several dwarf galaxies in the LG have already been well investigated (eg. Reitzel, Guhathakura & Gould, 1998; Davidge, 2002). More recently, however, observations have begun extending the physical properties of stellar halos to giant elliptical galaxies such as NGC 2768, M87 and NGC 5128, beyond the LG (eg. Cortesi *et al.* 2013; Longobardi *et al.* 2015a).

High resolution spectroscopic observations of PNe in LG galaxies and beyond permit the measurement of radial velocities and allow the estimate of global mass distributions (eg. Arnaboldi *et al.* 1998; Méndez *et al.* 2001; Romanowsky *et al.* 2003; Napolitano *et al.* 2004; Peng *et al.* 2004; Cortesi *et al.* 2013; Longobardi *et al.* 2015a,b). We can explore the 2D distribution of PNe and the stellar regions they populate, providing clues to the triaxial shapes of galaxies and formation processes. The structure and origin of the outer halo regions of ellipticals is only now beginning to become clear. We now have well constrained methods to separate the halo components from the inter-cluster light (ICL) (eg. Dolag *et al.* 2010; Cui *et al.* 2014; Longobardi *et al.* 2015a).

2. PN populations

There are a number of good reasons for the high increase in the number of PN studies in external galaxies. The PNLF from these objects serves to test distance estimates made through other means and estimate 'α', the luminosity specific number of PNe expected according to the bolometric luminosity of the galaxy (see next section). Visible number densities provide information about past kinematic evolution such as possible ram pressure stripping of PNe by a hot intercluster medium. In addition, the α parameter and the PNLF correlate with the age, colour and metallicity of the progenitor stellar population, providing an effective means of measurement where other tracers may be too faint. The line of sight velocity dispersion (LOSVD) of PNe in a galaxy can indicate tidal disruptions or the relative mass and luminosity of the halo component, where the PN histogram may range from Gaussian to highly non-gaussian and multi-peaked. Table 1 provides an up-to-date estimate of the number of PNe now available for study within the MW and the LG. Table 2 provides an estimate of the number of PNe now available at the outskirts of the LG and beyond.

2.1. *The α parameter*

The PN population size, when scaled to the available completeness limits, directly correlates with the visual magnitude of a galaxy (the number of PNe $L_{\odot bol}^{-1}$) or "α ratio" (Magrini *et al.* 2003). This relation, however, is weakly dependent on the age, metallicity, morphological type of the galaxy (Buzzoni *et al.* 2006) and the degree of star formation at the time the PN progenitors were forming. Please see Reid (2012) for further details. In short, the expected number of PNe (N_{PNe}) for a simple stellar population of total luminosity (L_{bol}) is:

$$N_{PN} = ß\, L_{bol}\, \tau_{PN} \tag{2.1}$$

Table 1. Known planetary nebulae populations in the local group in the years 2011 and 2015 showing the improvement. Local group galaxies newly identified/discovered in the past seven years are indicated by an asterisk next to the name in column 1.

Name	Type	Mv	Dist. [kpc]	PNe 2011	PNe 2015	Ref (old) 2011	Ref (new) 2015
M 31	Sb	-21.5	785	2766	2779	Merrett 2006	Jacoby *et al.* 2013
Milky Way	Sbc	-20.9		~3000	3288	Parker *et al.* 2006	Parker *et al.* 2015
M 33	Sc	-18.9	795	152	152	Ciardullo *et al.* 2004	
LMC	Ir	-18.5	50	740	740	Reid *et al.* 2006, 2010	Reid *et al.* 2013; 2014
SMC	Ir	-17.1	59	89	105	Jacoby *et al.* 2002	Drašković *et al.* 2015
M 32 (NGC221)	E2	-16.5	760	30	45	Ciardullo *et al.* 1989	Sarzi *et al.* 2011
NGC 205 (M110)	Sph	-16.4	760	35	35	Corradi *et al.* 2005	
IC 10	Ir	-16.3	660	27	35	Kniazev, *et al.* 2008	Gonçalves *et al.* 2012a
NGC 6822	dIr	-16.0	500	26	26	HM[2] *et al.* 2009	
NGC 185	Sph	-15.6	660	5	8	Corradi *et al.* 2005	Gonçalves *et al.* 2012b
IC 1613	dIr	-15.3	725	3	3	Magrini *et al.* 2005	
NGC 147	Sph	-15.1	660	9	9	Corradi *et al.* 2005	
WLM	dIr	-14.4	925	1	1	Magrini *et al.* 2005	
Sagittarius	dSph/E7	-13.8	24	3	4	Zijlstra *et al.* 2006	
Fornax (E351-G30)	dSph	-13.1	138	1	2	Larsen 2008	
Pegasus (DDO 216)	dIr	-12.3	760	1	1	Jacoby *et al.* 1981	
Leo I (DDO 74)	dSph	-11.9	250				
Andromeda I	IDsPH	-11.8	810				
Andromeda II	dSph	-11.8	700				
Leo A (Leo III)	dIr	-11.5	690	1	1	Magrini *et al.* 2003	
DD 210	dIr	-11.3	1025				
KKs 3*	dSph	-10.8	2146				
Sag DIGD	dIr	-10.7	1300				
Pegasus II (An VI)	dSph	-10.6	830				
Pisces (LGS3)	dIr	-10.4	810				
Andromeda V	dSph	-10.2	810				
Andromeda III	dSph	-10.2	760				
Leo II (Leo B)	dSph	-10.1	210				
Cetus	dSph	-9.9	755				
Phoenix	dSph	-9.8	395		1	Saviane *et al.* 2009	
Sculptor (E351-G30)	dSph	-9.8	87				
Cassiopeia (An VII)	dSph	-9.5	690				
Tucana	dSph	-9.6	870				
Sextans	dSph	-9.5	86				
Carina (E206-G220)	dSph	-9.4	100				
Draco (DDO 208)	dSph	-8.6	79				
Ursa Minor	dSph	-8.5	63				
Canes Venatici I*	dSph	-7.8	220				
Leo T*	dSph	-7.1	420				
Ursa Major*	dSph	-6.7	100				
Hercules*	dSph	-6.6	147				
Eridanus II*	dSph	-6.6	380				
Canis Major Dwarf*	Irr		7.6				
Canes Venatici II*	dSph	-5.8	150				
Boötes I*	dSph	-5.8	60				
Boötes III*	dSph	-5.8	46				
Leo IV	dSph	-5.5	154				
Leo V	dSph	-5.2	175				
Pisces II*	dSph	-5.0	180				
Coma Berenices*	dSph	-4.1	44				
Ursa Major II*	dSph	-3.8	30				
Tucana II*	dSph	-3.8	57				
Indus I*	dSph	-3.5	100				
Horologium I*	dSph	-3.4	79				
Grus I*	dSph	-3.4	120				
Pictoris I*	dSph	-3.1	114				
Phoenix II*	dSph	-2.8	83				
Boötes II*	dSph	-2.7	42				
Reticulum II*	dSph	-2.7	30				
Segue 2*	dSph	-2.5	35				
Eridanus III*	dSph	-2.0	87				

Notes:
[2]Hernández-Martínez *et al.* (2009).

Table 2. Known planetary nebulae populations in the outer local group and outlying galaxies in the years 2011 and 2015 showing the improvement. Local group galaxies newly identified/discovered in the past seven years are indicated by an asterisk next to the name in column 1.

Name	Type	Mv	Dist. [kpc]	PNe 2011	PNe 2015	Ref (old) 2011	Ref (new) 2015
LG outskirts							
GR8	dSph	-11.8	2200	0		Magrini *et al.* 2005	
Antlia	dSph	-15.8	1330				
NGC3109	dIr	-15.8	1330	18	20	Peña *et al.* 2007	
Sextans B	dIr	-14.3	1600	5	5	Magrini *et al.* 2000	
Sextans A	dIr	-14.2	1320	1	1	Magrini *et al.* 2003	
EGB0427+63	sIr	-10.9	2200				
Beyond LG							
NGC 1316	SAB9s0	-22.0	19000	43	796	Coccato *et al.* 2009	McNeil-Moylan *et al.* 2012
NGC 2768	E6	-21.8	22400		289		Cortesi *et al.* 2013
NGC 5846	E0	-21.7	23100	123		Coccato *et al.* 2009	
NGC 1399	E1	-21.4	18500	37	187	McNeil *et al.* 2010	
M 84 (NGC 4374)	E1	-21.3	18400	450		Coccato *et al.* 2009	
NGC3311	E/S0	-21.3	41000		56		Ventimiglia *et al.* 2011
M 86 (NGC 4406)	E3	-22.1	15000	16		Arnaboldi *et al.* 1996	
NGC 821	E6	21.55	22400		167		Teodorescu *et al.* 2011
M 87	E0p	-21.5	16400		300		Longobardi *et al.* 2015a,b
NGC 5128	gE/S0	-21.0	3800	1141	1267	Peng *et al.*2004	Walsh *et al.* 2015
NGC 1344	E5	-20.92	18400	194		Teodorescu *et al.* 2005	
NGC 3608	E2	-20.7	21300	87		Coccato *et al.* 2009	
NGC 4494	E1	-20.4	15800	255		Napolitano *et al.* 2009	
NGC 3489	SB0-(rs)	-20.2	12100		75		Cortesi *et al.* 2013
NGC 3115	SO	-20.0	9700		186		Cortesi *et al.* 2013
NGC 1023	SB0-(rs)	-19.9	11400		236		Cortesi *et al.* 2013
NGC 3377	E5	-19.8	10400	151		Coccato *et al.* 2009	
NGC 4564	E6	-19.7	13900	49		Coccato *et al.* 2009	
NGC 1023	S0	-19.7	10600	183		Noordermeer *et al.* 2008	
M 105 (NGC 3379)	E1	-19.7	9800	186		Douglas *et al.* 2007	
NGC 7457	SA0-(rs)	-19.6	13200		156		Cortesi *et al.* 2013
NGC 3394	SB0-(s)	-19.4	11600		77		Cortesi *et al.* 2013
NGC 3384	SO	-19.2	10800	68		Tremblay *et al.* 1995	
NGC 4697	E6	-19.2	10900	535		Méndez *et al.* 2001	

where ß is the so-called "specific evolutionary flux" (which approximates to 2×10^{-11} $L_{\odot bol}^{-1}yr^{-1}$); and τ_{PN} is the PN (emission-detectable) lifetime in years.

In addition, the stellar core mass evolution directly affects the PN lifetime since it depends on the central star temperature ($T_{eff} \simeq 10^5$ K) during the hot post AGB phase. With an increase in the hot post AGB lifetime with increasing dynamical ages and metallicities, the τ_{HPAGB} is somewhat equivalent to Fuel/l_{HPAGB}, where l_{HPAGB} is the luminosity of the central star at the onset of the PN stage. The luminosity-specific PN density then becomes:

$$\alpha = \frac{N_{PN}}{L_{SSP}} = ß\,\tau_{PN} = ß\min\{\tau_{HPAGB}, \tau_{dyn}\} \qquad (2.2)$$

The luminosity-specific PN number density presented here is calibrated to the near-complete number density and metallicity found in both the LMC and SMC. This calibration is then applied to the estimate PN population expected in similar dwarf or irregular galaxies of varying low metallicity. Data supplied by Reid & Parker (2013) and Reid (2014) suggest there may not be more than 740 genuine PNe in the LMC. Likewise, data and on-going analysis supplied by Drašković *et al.* (2015) suggest the SMC may not support more than 105 PNe. For large galaxies with accordingly higher metallicities, we calibrate to the theoretical luminosity-specific PN density found by Buzzoni *et al.* (2006).

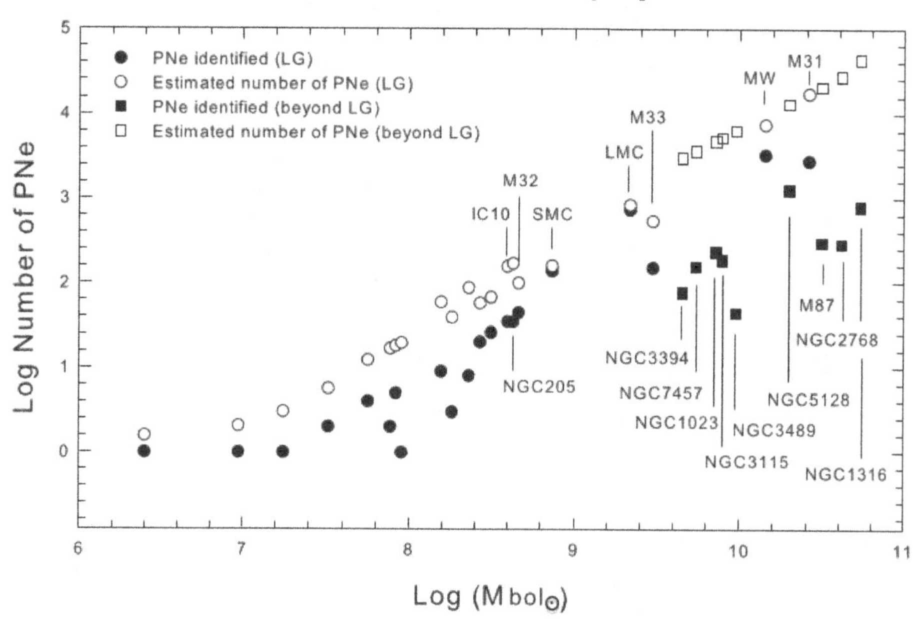

Figure 1. The number of PNe currently known in each galaxy (filled circles for LG and filled squares for galaxies beyond) as a function of the absolute estimated luminosity of their host galaxies. Open circles for LG and open squares beyond represent the possible number of PNe predicted using α, based on the theory of simple stellar populations calibrated to well assessed populations and metallicities.

For calibration we use $\alpha = 1$ PN / 1.5×10^6 $L_\odot$ for M31 (metallicity $\sim$ –0.6), $\alpha = 1$ PN / 2.5×10^6 $L_\odot$ for the LMC (metallicity $\sim$ –1.15) and $\alpha = 1$ PN / 4.6×10^6 $L_\odot$ for the SMC (metallicity $\sim$ –1.25).

In addition, Longobadi *et al.* (2015a) found that galaxies with (B-V) colours <0.8 have small α whereas galaxies with B-V colours >0.8 have higher α with more spread. With a small bolometric correction applied to the L_v of the galaxies, it is possible to use α to estimate the number of PNe that may exist in each galaxy. This bolometric correction does not amount to much more than -0.2 Mag. In order to compare the number of PNe currently uncovered and the likelihood of further discoveries, the known number of PN for each galaxy is plotted on the same scale and shown in Figure 1.

3. PNe provide information on stellar populations

The vast majority of studies based on extragalactic PNe surveys include a luminosity function (eg. Hernández-Martínez & Peña 2009; Reid & Parker 2010; Cortesi *et al.* 2013; Longobardi *et al.* 2015a). Even if the ultimate goal of the research is aimed at kinematics, the role of the PNLF is key to understanding the characteristics of the sample at hand. For example, the PNLF will quickly suggest the relative completeness of the sample. It will show if the brightest possible PNe in the population have been included and provide a distance estimate. The presence of a dip and its position between two and four magnitudes below the brightest possible PN will indicate the star-forming history of the galaxy and suggest the metallicity by inference. The presence of a peak in the function, 6$\sim$7 mag below the brightest indicates the luminosity at which most PNe reside during their 20 to 30 K yr lifetimes (Reid & Parker 2010).

Using PNe in the Large Magellanic Cloud (LMC) Badenes *et al.* (2015) have compiled a statistical analysis of the relationship between PNe and their parent stellar population. To achieve this they use a delay time distribution (DTD) which is the rate of PN production as a function of time after a hypothetical brief burst of star formation. Given a set of observed PNe in each region of the LMC and a stellar age distribution (SAD) map of the galaxy, the DTD can be recovered for each discrete SAD cell. The mean lifetime of the PNe produced in each time bin of the DTD was calculated by dividing the recovered number of stars that turn off the main sequence by the rate at which stars from a coeval population with a specific same age leave the MS. The study shows that most PN progenitors in the LMC have main-sequence lifetimes in a narrow range between 5 and 8 Gyr, corresponding to masses between 1.2 and 1.0 $M_\odot$ producing PNe visible for $26 \pm 4\,6$ kyr. There is also a distinct second population of PN progenitors in the LMC with main-sequence lifetimes between 35 and 800 Myr, corresponding to masses between 8.2 and 2.1 $M_\odot$ and average PN lifetimes of $11\,^{+6}_{-8}$ kyr but no PNe that appear to result from stars in the 800 Myr - 2 Gyr MS lifetime bracket. The DTD-derived PN lifetimes result in an integrated PN formation rate of $\sim$0.02 yr^{-1} in the surveyed area, which includes $\sim$80% of the stellar mass of the LMC. This implies a bolometric luminosity specific PN formation rate of $\sim 7 \times 10^{-12}$ PNe yr^{-1} $L_\odot{}^{-1}$. In the LMC today, $40\,^{+23}_{-29}$% of PNe are generated by young progenitors and 25 ± 8% by older progenitors.

Using the CTIO 4-m telescope and MOSAIC-2 wide-field camera on [O III]5007 and Hα, Hernandez-Martinez and Pena (2009) found 26 PNe in NGC 6822, increasing the known number by 8. The PNLF has a similar dip to the one found for the Small Magellanic Cloud at 2.5mag below the maximum. The best fit to the observed PNLF gives a rough estimate of 23.64 $\pm$0.23 mag for the distance modulus. The authors have estimated the number of PNe in the brightest 0.5 mag normalized to the galactic bolometric luminosity ($\alpha_{0.5}$) as $3.8\,^{+0.9}_{-0.71}$ E-9. This value is similar to the values derived from small galaxies (M_B fainter than -18 mag) and galaxies with recent star formation (larger than values obtained for early-type galaxies).

PNLFs were constructed for nine galaxies with various Hubble types by Rodríguez-González *et al.* (2015). They report that all the galaxies except one have a two-mode population where the PNe represent two episodes of star formation in which the second episode is significantly stronger. These authors approach the analysis in an unusual way by using cumulative PNLFs and estimating progenitor mass. Peak starbursts at $t_1 =$ 8 Gyr and $t_2 = 11.5$ Gyr and the estimated age of $\sim$13.5 Gyr for NGC 6822 are interesting results if one has confidence in the methods and analysis used.

4. Kinematic studies

A paper by Cortesi *et al.* (2013) uses the kinematics of PNe to very large radii to help explain the origins of S0 galaxies. The galaxies surveyed are divided into three groups, the first are isolated S0s (NGC 3115 & NGC 7457), those in poor groupings (NGC 1023 & NGC 2768) and those in the richer Leo group (NGC 2284 & NGC 3489). The authors find that the kinematics are largely dominated by rotational motion with significant random velocities. The spatial distribution of PNe is compared to that of the underlying stellar population. To do this, elliptical isophotes are fitted to images of the galaxies and surface brightness profiles extracted from the fits. The α value varies slightly as the redder ellipticals are poorer in PNe per unit L_{gal} than spirals. A comparison of α to the B-V mag and UV excess show that S0s are indistinguishable from ellipticals but suggest that bluer galaxies have higher specific frequencies of PNe.

A spectroscopic study of 287 PNe associated with M 87 in Virgo A by Longobardi *et al.*(2015a) found that 211 are located between 40 kpc and 150 kpc from the galaxy center. The velocity dispersion is found to be bimodal with a narrow component centered on the systematic velocity of M87 (234 PNe) and a broader off-center component identified as halo and ICL (44 PNe). The number density of the galaxy halo PNe and ICL PNe have different spatial distributions where halo PNe follow the galaxy's surface brightness profile and ICL PNe have a shallower power-law profile $l_{ICL} \propto R^y$ with y in the range [-0.34, -0.04]. PN number densities confirm different PN populations in M87 and the ICL, each with a different value for α. The stellar halo is found to extend out to ~150 kpc while PNe belonging to the ICL overlap outwards from a radius of ~60 kpc. Both populations are separated on the basis of their velocity relative to the systematic velocity of 1275 km s^{-1} where the V_{halo} = 1270.4 km s^{-1} and has a dispersion of 298.4 km s^{-1} compared to the V_{ICL} which is centered at 999.5 km s^{-1} and has a dispersion of 881 km s^{-1}. Statistically, 10% or 30 PNe from all halo sample bins are likely associated with the IC component.

An imaging and spectroscopic survey of PNe in NGC 5128 (Cent A) by Walsh *et al.* (2015) presented velocities for kinematic and dynamical studies. NTT was used to image 15 fields with EMMI and [O III] together with off-band filters. New discoveries were added to the existing count and observed with FLAMES in MEDUSA. This resulted in 1118 PN candidates and 1267 spectroscopiclly confirmed PNe.

5. Chemistry in PNe

Spectroscopic observations of nine newly discovered PNe in the outskirts of M31 obtained with the 10.4-m GTC telescope by Corradi *et al.* 2015 has extended their previous study to a galactocentric radii of 100pc. None of the observed PNe are members of a classical metal-poor and ancient halo. Two of the outermost PNe have solar O abundances and kinematics similar to the extended disk of M31. Other PNe have slightly lower O content (O/H]~-0.4) and sometimes large derivations from disk kinematics. In other studies, Hernández-Martínez *et al.*(2009) analysed the chemical behavior of the irregular galaxy NGC 6822 and more recently, Gonçalves *et al.* (2014) analysed the universal mass-metallicity relation and the chemical abundances in dwarf galaxy NGC 205.

6. Summary: The future

- Many more PNe need to be identified and observed in a larger number and range of galaxies in order to measure and separate the halo and IC properties.
- Different stellar populations will be exposed through PN densities and kinematics.
- More abundances need to be determined for external PNe and used to reveal tidal stripping, streams, extended stellar disks and tidal encounters with other galaxies.
- Work is required for a better theoretical understanding of the way in which metallicity, age and different star-fiormaing histories affect post-AGB phases of stellar evolution.
- More theoretical work needs to be done on the physics behind PNLF dips and peaks.

References

Arnaboldi, M., Freeman, K. C., & Méndez, R. H. 1996, *ApJ*, 472, 145

Arnaboldi, M., *et al.* 1998, *ApJ*, 507, 759

Badenes, C., Maoz, D., & Ciardullo, R. 2015, *ApJ*, 804, 25

Bekki, K., 2008, *MNRAS*, 390L, 24

Bekki, K., 2008b, *ApJ*, 684L, 87

Buzzoni, A., Arnaboldi, M., & Corradi, R. L. M. 2006, *MNRAS*, 368, 877

Ciardullo, R., Jacoby, G., Ford, H. C., & Neill, J. D., 1989, *ApJ*, 339, 53 1989, *ApJ*, 339, 53

Ciardullo, R., *et al.* 2004, *ApJ*, 614, 167

Corradi, R. L. M., *et al.* 2005, *A&A*, 431, 555

Corradi, R. L. M., Kwitter, K. B., Balick, B., Henry, R. B. C., & Hensley, K. 2015, *ApJ*, 807, 181

Coccato, L., *et al.* 2009, *MNRAS*, 1249, 1283
Cortesi, A., Arnaboldi, M., Coccato, L., *et al.* 2013, *A&A*, 549, 115
Cui, W., Murante, G., & Monaco, P., 2014, *MNRAS*, 437, 816
Davidge, T. J., 2002, *AJ*, 124, 2012
Dolag, K., Murante, G., & Borgani, S., 2010, *MNRAS*, 405, 1544
Douglas, N. G., *et al.* 2007, *ApJ*, 664, 257
Drašković, D., Parker, Q. A., Reid, W. A., & Stupar, M., 2015, *MNRAS*, 452, 1402
Gonçalves, D., Teodorescu, A., Alver-Brito, A., *et al.* 2012a, *MNRAS*, 425, 2557
Gonçalves, D., Magrini, L., Martins, L. P., Teodorescu, A., & Quireza, C. 2012b, *MNRAS*, 419, 854
Gonçalves, D., Magrini, L., Teodorescu, A. M., & Carneiro, C. M. 2014, *MNRAS*, 444, 1705
Hernández-Martínez, L. & Peña, M. 2009, *A&A*,495, 447
Jacoby, G. H. & Lesser, M. P. 1981, *AJ*, 86, 185
Jacoby, G. H. & De Marco, O. 2002, *AJ*, 123, 269
Jacoby, G., *et al.* 2013, *ApJ*, 769, 10
&Kniazev, A. Y., Pustilnik, S. A., Zucker, D. B. 2008, *MNRAS*, 384, 1045
Larsen, S. S. 2008, *A&A*, 477, L17
Longobardi, A., Arnaboldi, M., & Gerhard, O., Hanuschik R. 2015a, *A&A*, 579A, 135
Longobardi, A., Arnaboldi, M., Gerhard, O., & Mihos, J. C. 2015b, *A&A*, 579L, 3
Magrini, L., Corradi, R. L. M., Mampaso, A., & Perinotto, M., 2000, *A&A*, 355, 713
Magrini, L., Corradi, R. L. M., Greimel, R., Leisy, P., *et al.* 2003, *A&A*, 407, 51
Magrini, L., Corradi, R. L. M., Greimel, R., Leisy, P., *et al.* 2005, *MNRAS*, 361, 517
Merrett, H. *et al.* 2006, Procedings of the ESO workshp *"Planetary Nebulae beyond the Milky Way"*, Eds. L.Stanghellini, J.R. Walsh, N.G. Douglas, p. 281
Magrini, L., Corradi, R. L. M., Greimel, R., Leisy, P., *et al.* 2005, *MNRAS*, 361, 517
McNeil, E. K., Arnaboldi, M., & Freeman, K. C. 2010, *A&A*, 518, A44
McNeil-Moylan, E. K., Freeman, K. C., Arnaboldi, M., & Gerhard, O. E. 2012, *A&A*, 539, A11
Méndez, R. H. *et al.* 2001, *ApJ*, 563, 135
Napolitano, N. R., *et al.* 2004, , in: S.D. Ryder, D.J. Pisano, M.A. Walker & K.C. Freeman, K.C. (eds.), IAU Symp. 220 *Dark Matter* (Astron. Soc. Pac.), San Francisco, p. 173
Napolitano, N. R., *et al.* 2009, *MNRAS*, 393, 329
Noordermeer, E., *et al.* 2008, *MNRAS*, 384, 943
Parker, Q. A., Acker, A., Frew, D. J., *et al.* 2006, *MNRAS*, 373, 79
Parker, Q. A., Bojici, I., Frew, D., Acker, A., & Ochsenbein, F., 2015, *AAS*, 22510806P
Peña, M., Richer, M. G., & Stasińska, G. 2007, *A&A*, 466, 75
Peng, E. W., Ford, H. C., & Freeman, K. C. 2004, *ApJ*, 602, 685
Reid, W. A. & Parker, Q. A. 2006, *MNRAS*, 373, 521
Reid, W. A. & Parker, Q. A. 2010, *MNRAS*, 405, 1349
Reid, W. A. & Parker, Q. A. 2013, *MNRAS*, 436, 604
Reid, W. A. 2012, in Manchado A., Stanghellini L., Schoenberner D., (eds.), IAU Symp. 283 *Planetary Nebulae: An eye to the Future* (Cambridge Univ. Press), Cambridge, p. 227
Reid, W. A. 2014, *MNRAS*, 438, 2642
Reitzel, D. B., Guhathakurta, P., & Gould, A., 1998, *AJ*, 116, 707
Rodríguez-González, A. & Hernández-Martínez, L., *et al.* 2015, *A&A*, 575, 1
Romanowsky A. J., *et al.* 2003, *Sci*, 301, 1696
Sarzi, M., Mamon, G. A., Cappellari, M., *et al.* 2011, *MNRAS*, 415, 2832
Saviane, I., Exter, K., Tsamis, Y., Gallart, C., & Péquignot, D. 2009, *A&A*, 494, 515
Teodorescu, A. M., Méndez, R. H., Saglia, R. P., *et al.* 2005, *ApJ*, 635, 290
Teodorescu, A. M., Mendez, R. H., Bernardi, F., *et al.* 2011, *ApJ*, 736, 65
Tremblay, B., Merrit, D., & Williams, T. B. 1995, *ApJ*, 443, 149
Ventimiglia, G., Arnaboldi, M., & Gerhard, O., 2011, *A&A*, 528, A24
Walsh, J. R., Rejkuba, M., Walton, N. A. 2015, *A&A*, 574A, 109
Zijlstra, A. A., Gesicki, K., Walsh, J. R., *et al.* 2006, *MNRAS*, 369, 875

The General Assembly of Galaxy Halos: Structure, Origin and Evolution
Proceedings IAU Symposium No. 317, 2015
A. Bragaglia, M. Arnaboldi, M. Rejkuba & D. Romano, eds.

doi:10.1017/S174392131500976X

The Outer Galactic Halo As Probed By RR Lyr Stars From the Palomar Transient Facility + Keck

Judith Cohen[1], Branimir Sesar[1,2], Sophianna Banholzer[1], the PTF Collaboration[1]

[1]California Institute of Technology, Palomar Observatory,
Mail Code 249-17, Pasadena, Ca., 91125, USA
email: jlc@astro.caltech.edu

[2]Max Planck Institute for Astronomy,
Konigstühl 17, D-69117,
Heidelberg, Germany
email: bsesar@mpia.de

Abstract. We present initial results from our study of the outer halo of the Milky Way using a large sample of RR Lyr(*ab*) variables datamined from the archives of the Palomar Transient Facility. Of the 464 RR Lyr in our sample with distances exceeding 50 kpc, 62 have been observed spectroscopically at the Keck Observatory. v_r and $\sigma(v_r)$ are given as a function of distance between 50 and 110 kpc, and a very preliminary rather low total mass for the Milky Way out to 110 kpc of $\sim 7 \pm 1.5 \times 10^{11}$ $M_\odot$ is derived from our data.

Keywords. The Galaxy, Galaxy: halo, stars: variables: RR Lyrae

1. Introduction

We present initial results from our study of the outer halo of the Milky Way (MW) using a large sample of RR Lyr(*ab*) variables datamined for the archives of the Palomar Transient Facility (PTF) (Law *et al.* 2009, Rau *et al.* 2009) (P.I. S. R. Kulkarni of Caltech). RR Lyr are old low-mass pulsating stars with distinctive light curves, amplitudes at V of $\sim$ 1 mag, and periods of $\sim$0.5 days. These characteristics makes them fairly easy to distinguish in a wide field, multi-epoch optical imaging survey of moderate duration if the survey cadence is suitable. Their most desirable characteristic is that they can be used as standard candles. Accurate luminosities, which have only a small metallicity dependence, can be inferred directly from the light curves, and these stars, with $M_V \sim +0.6$ mag, are fairly luminous and hence can be detected at large distances.

Using machine learning techniques we have isolated a sample of 464 RR Lyr from the PTF. These were found by searching for PTF fields which had enough epochs of observation (25 minimum) in the R filter. A total of roughly 10,000 deg^2 on the sky met this requirement and was searched. The regions of known outer halo objects (i.e. dwarf galaxies, globular clusters, and known halo streams) were excluded. The search criterion was variability, and the sample was refined by requiring the derived period and amplitude to have values appropriate for RR Lyr. Since no observing time was assigned for this purpose, we are effectively datamining the PTF archive, and our sample lies in random pencil beams through the halo, each a PTF tile (area on sky: 7.6 deg^2).

Our sample begins at a brightness corresponding to a distance of 50 kpc, and extends out to $\sim$110 kpc, after which RR Lyr are too faint to pick out with PTF data. The selection procedure was carried out in 2012; at present the PTF+iPTF variable star database

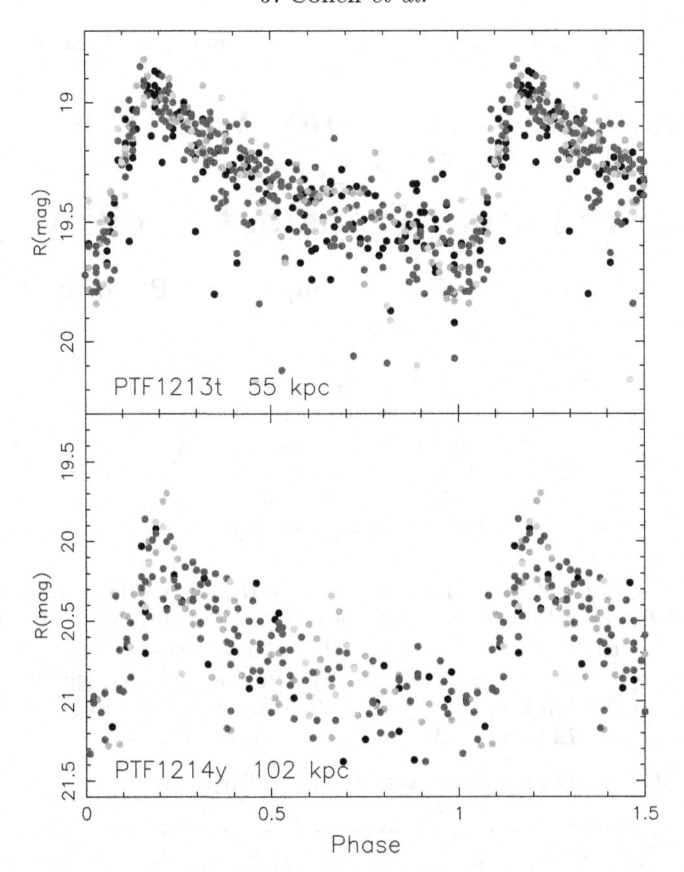

Figure 1. Examples of phased PTF light curves: **Top:** A RR Lyr at a distance of 55 kpc with 468 measurements with the R filter in the PTF database. **Bottom:** A RR Lyr at a distance of 102 kpc with 219 R measurements in the PTF database. Note that prior to the addition of 2 years of data to the PTF variable star database in early 2015, this star had only 61 detections at R in the PTF database. The color scheme denotes the year of observation, with the first day set to the first night the object was observed by the PTF. Black = year 1, red = year 2, green = year 3, blue = year 4, etc using the PGPLOT color routine. The sample selection was carried out using only the first two years of PTF data.

contains photometry through Dec 2014, i.e. two more years of data, none of which was used to select the current sample. A more detailed description of the search method is given in Sesar *et al.* (2014). Given this is a variability search, the only contaminants with colors similar to those of RR Lyr are quasars (QSOs).

Fig. 1 (top) shows the light curve of one of the brighter RR Lyr in our sample (r = 55 kpc) as well as that of one of the most distant RR Lyr found to date (r = 102 kpc) (bottom).

2. Radial Velocity Measurements

A spectroscopic campaign to obtain radial velocities for RR Lyr candidates began at the Keck Observatory with the Deimos spectrograph (Faber *et al.* 2003) in the spring of 2014. RR Lyr are pulsating variable stars, hence their observed radial velocities need to be corrected for the motion of the atmosphere. To measure the center-of-mass velocity v_r of RR Lyr stars, we use the Balmer Hα line and the method described by Sesar (2012).

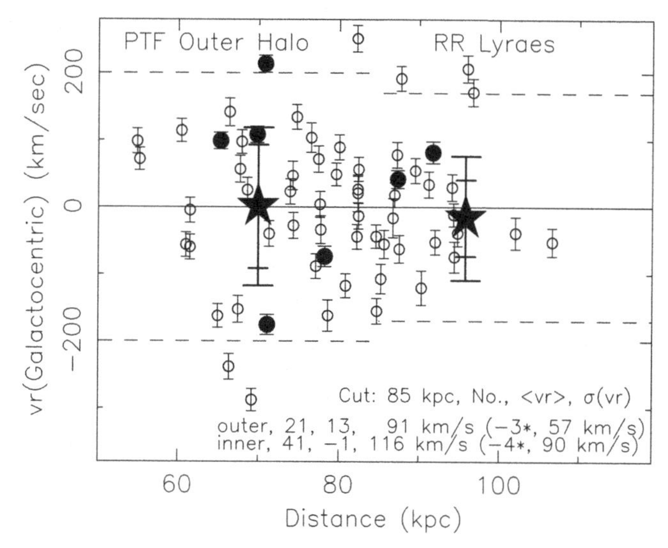

Figure 2. Radial velocities as a function of heliocentric distances are shown for our sample of 62 RR Lyr selected from the PTF with Keck/Deimos moderate resolution spectra. The borders of the regions (both high and low v_r) considered outliers are indicated by the dashed horizontal lines. The text at the bottom right gives the number of RR Lyr, mean and σ (with 20 km s^{-1} removed in quadrature for observational uncertainties) for the inner and for the outer sample, cut at 85 kpc, as well as these with the largest outliers removed (3 stars for the outer region, and 4 for the closer sample). Filled circles denote stars with two Deimos spectra, open circles have only one spectrum. One σ error bars are shown for each RR Lyr. The two large stars denote the means for the inner and outer sample with the outliers removed; their two error bars correspond to σ for the full sample, and that for the cropped sample.

Typical uncertainties in center-of-mass velocities, including both observational and phase correction, are 15 to 20 km s^{-1} for a single Deimos spectrum.

Fig. 2 shows v_r vs r for the 62 RR Lyr with Keck/Deimos spectra. We cut the sample at 85 kpc, and assume an uncertainty in each individual measurement of 20 km s^{-1}. We find a mean v_r for the inner 41 RR Lyr of -1 km s^{-1} with $\sigma = 116$ km s^{-1}, while for the outer 21 stars, the mean is $+13$ km s^{-1} with $\sigma(v_r) = 91$ km s^{-1}. The mean of v_r falls close to 0, suggesting that our corrections to observed radial velocities are appropriate.

When 4 major outliers (2 high and 2 low) are deleted from the inner sample, σ falls to 90 km s^{-1} for the inner sample. Deleting three high outliers from the outer sample leaves 18 R Lyr beyond 85 kpc, and σ falls to 57 km s^{-1}, a remarkably low dispersion. We are not yet certain whether the outliers arise from substructures within the outer halo.

The velocity dispersion as a function of radius for the Milky Way halo is shown in Fig. 3, where we compare our work to selected values from the recent literature. The small stars denote our entire sample, cut at 85 kpc, and the larger stars show our sample of outer halo RR Lyr cleaned of outliers; recall that the outliers comprise only about 10% of our total sample. The star symbols are plotted at the median distance of the inner and of the outer sample. The velocity dispersions found for the inner halo from recent studies are also shown: from SDSS blue horizontal branch (BHB) stars (Xue *et al.* 2004), from SDSS K giants (Xue *et al.* 2014) as well as a sample of halo high velocity stars from Brown *et al.* (2010). Most of these other samples are confined to distances less than 80 kpc. The only attempt to reach the distances probed by the outer part of our sample is that of Deason *et al.* (2012). Our results agree quite well with the value they published of $\sigma(v_r)$ of 50 to 60 km s^{-1} at distances of $\sim$110 kpc, rising to $\sim$90 km s^{-1} at distances

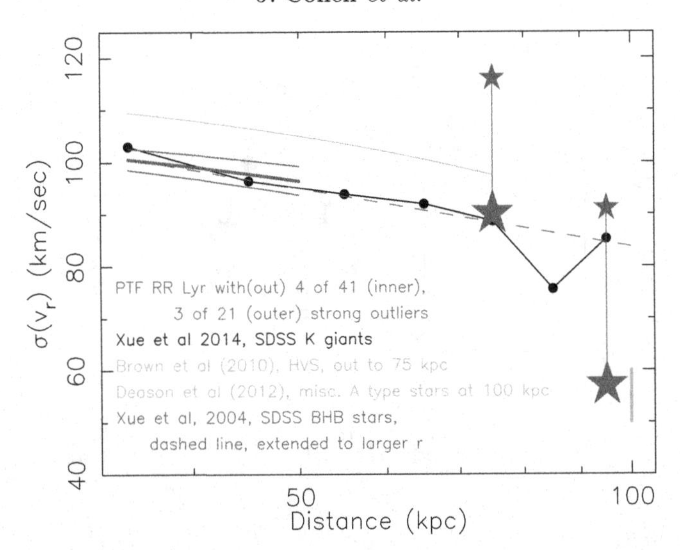

Figure 3. $\sigma(v_r)$ is shown as a function of $\log(r)$ for our inner and outer sample of RR Lyr (split at 85 kpc), with and without eliminating the single strong outliers. Values from Xue *et al.* 2008), whose extrapolation to larger r is indicated by a dashed line, Xue *et al.* (2011, 2014), Brown, Geller & Kenyon (2014), and Deason *et al.* 2012b), are shown. The text within the figure indicates the colors used for each reference and the nature of the sample in each case.

of 70 kpc. We agree reasonably well with the result of Brown *et al.* (2010) for $\sigma(v_r)$ as a function of r from the MMT high velocity star survey, extending out to 75 kpc, when their non-parametric method to eliminate outliers is used (see their Fig. 6).

3. Implications

One of our goals is to determine the radial distribution of the outer halo of the Milky Way using RR Lyrae as tracers of its stellar population, and to use this information to derive the total mass of the MW. We calculate distances assuming a fixed M_R of +0.6 mag as the flux-averaged mean over the period, then correct for interstellar reddening assuming that the variable is so distant that the full reddening from the maps of Schlegel, Finkbeiner & Davis (1998) applies.

Fig. 4 shows a histogram of the full sample of 464 candidate RR Lyr beyond 50 kpc as a function of $\log(r)$, where r is the heliocentric distance. Both axes have logarithmic scales. The vertical axis is the number of variables in bins equally spaced in distance starting at 50 kpc, with 2.8 kpc/bin. These counts, accumulated over a solid angle on the sky of $\Omega \sim 10{,}000$ deg^2, represent $\rho(r)$ $(\Omega/4\pi)$ $4\pi r^2 \Delta r$. We fit a power law $\rho \propto r^{-\gamma}$, over the range of r we cover, i.e. out to just over 100 kpc. Therefore a linear fit of the log(counts) vs $\log(R_{GC})$ will yield a slope that is $-\gamma + 2$. Our best fit slope is $\gamma = 3.8 \pm 0.3$.

The completeness corrections are very difficult to calculate because the number of epochs observed varies greatly among the PTF fields, depending on which (other) project requested observations of the field; the Galactic halo project has until recently been allocated no P48 imaging time at all; we have simply used observations acquired for other PTF subprojects and thus have no control over the cadence of observing nor the fields observed. Furthermore, PTF observations at the P48 are carried out as long as the dome can safely remain open irrespective of clouds. Thus the limiting magnitude of each exposure of a given field has a very wide range, with clouds (i.e. a bright limiting magnitude) more common than exceptional nights which are perfectly clear and with

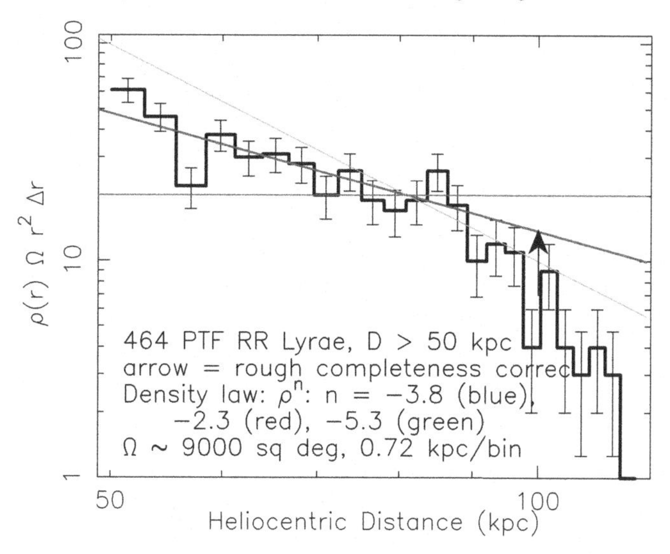

Figure 4. The number density of RR Lyr as a function of distance is shown together with several power law fits which suggest $\gamma \approx 3.8$. A lower limit to the incompleteness correction at a distance of 100 kpc is shown as a vertical arrow.

good seeing, which have the faintest limiting magnitudes. An RR Lyr at 100 kpc has a mean R of 20.6, close to the limit of the PTF survey, so that in many PTF images it will not be detected. To aquire a specified minimum number N of R detections, where N is set by our search procedure to identify a candidate RR Lyr, will thus require many more epochs of observation with PTF of its field than would be the case for a RR Lyr at 50 kpc in the same field. We indicate our best guess of a lower limit to the completeness correction arising from this issue at 100 kpc by the vertical arrow in Fig. 4. Other such issues will further increase the completeness correction above 90 kpc.

The only effort that reaches out to the radial range covered by our RR Lyr sample is that of Deason *et al.* (2014), who claim there is a very steep outer halo profile, with $\gamma \sim 6$ beyond $r = 50$ kpc and even steeper slopes $\gamma \sim 6 - 10$ at larger radii. Their sample of ~5200 stars is contaminated by QSOs; a photometric separation is not sufficient and cuts out many BHB stars.

We cannot reproduce the extremely steep decline in $n(r)$ that they claim to observe. We believe that with our RR Lyr sample selected through variability, the precise distances we obtain for our RR Lyr stars, and our low QSO contamination, that our result that $\rho \propto r^{-3.8\pm0.3}$ for $50 < r < 100$ kpc is correct. Most other recent analyses, i.e. Xue *et al.* (2015), Brown *et al.* 2010, find γ between 3.5 and 4.5 from $r = 20$ kpc out to the limit of their data, between 60 and 80 kpc; see e.g. Fig. 1 of Gnedin *et al.* (2010).

As a rough indication of the total mass of the MW, we assume a spherical halo, ignoring the subtleties in the inner part of the MW of the thick and thin disk. We also ignore the difference between our heliocentric distances and galactocentric distances. The standard way to obtain the total enclosed mass given a set of tracers, be they globular clusters or RR Lyr or any other low mass objects in the outer halo of a massive galaxy, is to solve the Jeans equation. Watkins, Evan & An (2010) have solved the Jeans equation for the case of set of mass tracers with both distances (not projected distances) and radial velocities located in the outer halo of a massive galaxy such as the Milky Way. To accomplish this they assume that the tracers follow a power-law density distribution

$\rho \propto r^{-\gamma}$. They further assume that a NFW halo (Navarro, Frenk & White 1996) is an adequate representation of the outer part of the Milky Way halo.

Their result then simplifies (see Evans, An, & Deason 2011) to

$$M_{vir} \approx \frac{r_{vir}^{0.5}\,(0.5 + |\gamma| - 2\beta)}{GN} \sum_{i=1}^{N} r_i^{0.5} v_{r,i}^2. \qquad \text{(eqtn. 1)}$$

Lacking proper motions, we cannot evaluate the velocity anisotropy β; we assume isotropic orbits ($\beta = 0$) for the outer halo RR Lyr. Attempts to measure β in the inner halo are more successful as much larger samples of tracers can be assembled, usually SDSS BHB stars. Kafle *et al.* (2012), among others, suggest $\beta = +0.5$ from $\sim$25 kpc out to the limit of their sample at $\sim$60 kpc, a value similar to that deduced by Williams & Evans (2015), who apply a new theoretical analysis to previously published data, while Deason *et al.* (2011) suggest from CDM simulations that $\beta = 0$.

Using the above formula with $\beta = 0.0$ we find a total mass out to 110 kpc of $\sim 7{\pm}1.5{\times}10^{11}\ M_{\odot}$, where the uncertainty corresponds to a range in β from -0.5 to $+0.5$. This suggests a rather low total mass for the MW, but is in reasonable agreement with several recent determinations based on studying halo stars at smaller radii (see the compiled recent measurements in Fig. 5 of Williams & Evans 2015), although analyses including the outermost MW satellites and M31 timing arguements continue to suggest a higher total mass.

Future work should expand our sample of outer halo RR Lyr significantly. Another two years of PTF imaging (2012–2014) has recently been added to the PTF variable star database, suggesting it is time to conduct another search for candidate RR Lyr in the PTF+iPTF database, which is a rather daunting task given the limited available human resources.

References

Brown, W. R., Geller, M. J., Kenyon, S. J., & Diaferio, A., 2010, *AJ*, 139, 59
Deason, A. J., McCarthy, I. G., Font, A. S. *et al.* 2011, *MNRAS*, 415, 2607
Deason, A. J., Belokurov, V., Evans, N. W. *et al.* 2012b, *MNRAS*, 425, 2840
Deason, A. J., Belokurov, V., Koposov, S. E., & Rockosi, C., 2014, *ApJ*, 787, 30
Evans, N. W., An, J., & Deason, A. J., 2011, *ApJL*, 730, L26
Faber, S., Phillips, A. C., Kibrick, R. I. *et al.* 2003, *SPIE*, 4841, 1657
Gnedin, O. Y., Brown, W. R., Geller, M. J., & Kenyon, S., 2010, *ApJL*, 720, L108
Kaffle, P. R., Sharma, S., Lewis, G. F., & Bland-Hawthorn, J., 2012, *ApJ*, 761, A98
Kaffle, P. R., Sharma, S., Lewis, G. F. *et al.* 2014, *ApJ*, 794, A59
Law, N. M., Kulkarni, S. R., Dekany, R. G. *et al.* 2009, *PASP*, 121, 1395
Navarro, J. F., Frenk, C. S., & White, S. M., 1996, *ApJ*, 462, 563
Rau, A., Kulkarni, S. R., Law, N. M. *et al.* 2009, *PASP*, 121, 1334
Schlegel, D., Finbeiner, D. P., & Davis, M., 1998, *ApJ*, 500, 525
Sesar, B., 2012, *AJ*, 144, A114
Sesar, B., Banholzer, S. R., Cohen, J. G. *et al.* 2014, *ApJ*, 793, A135
Watkins, L. L., Evans, N. W., & An, J. H., 2010, *MNRAS*, 406, 264
Williams, A. A. & Evans, N. W., 2015, *MNRAS*, 454, 698
Xue, X. X., Rix, H.-W., Zhao, G. *et al.* 2008, *ApJ*, 684, 1143
Xue, X. X., Rix, H.-W., Yanny, B. *et al.* 2011, *ApJ*, 738, A79
Xue, X. X., Ma, Z., Rix, H.-W. *et al.* 2014, *ApJ*, 784, 170
Xue, X. X., Rix, H. W., Ma, Z., Morrison, H., Bovy, J., Sesar, B. & Janesh, H., 2015, *ApJ*, 809, 144

The General Assembly of Galaxy Halos: Structure, Origin and Evolution
Proceedings IAU Symposium No. 317, 2015
A. Bragaglia, M. Arnaboldi, M. Rejkuba & D. Romano, eds.

doi:10.1017/S1743921315006730

Globular clusters and their contribution to the formation of the Galactic halo

Eugenio Carretta

INAF-Osservatorio Astronomico di Bologna, via Ranzani 1, I-40127 Bologna, Italy
email: eugenio.carretta@oabo.inaf.it

Abstract. This is a "biased" review because I will show recent evidence on the contribution of globular clusters (GCs) to the halo of our Galaxy seen through the lens of the new paradigm of multiple populations in GCs. I will show a few examples where the chemistry of multiple populations helps to answer hot questions including whether and how much GCs did contribute to the halo population, if we have evidence of the GCs-halo link, what are the strengths and weak points concerning this contribution.

Keywords. stars: abundances, stars: Population II, Galaxy: abundances, (Galaxy:) globular clusters: general, Galaxy: halo

1. Introduction

We obviously know that GCs *do* contribute to the Galactic halo, there are about 100 "smoking guns", i.e. the GCs currently observed in the Milky Way (MW) halo. Furthermore, large surveys like SDSS, Pan-STARRS, Dark Energy Survey, public ESO surveys such as VVV and ATLAS routinely discover new systems in the grey zone between globular clusters and dwarf galaxies, sometimes with controversial classification, see e.g. the case of Crater/Laevens1 considered to be a dwarf galaxy (Bonifacio *et al.* 2015) or a GC (Kirby *et al.* 2015a).

Mass loss is expected for GCs in particular at early phases, during the violent relaxation following gas expulsion (see e.g. Lynden-Bell 1967, Baumgardt *et al.* 2008) but GCs are dynamically evolved systems, hence mass loss is predicted during all their lifetime. It is then possible that many clusters dissolved, and what we see now are the smaller survivors of a potentially much larger population.

Early (e.g. Ibata *et al.* 1994, Fusi Pecci *et al.* 1995) and more recent studies (Forbes and Bridges 2010, Leaman *et al.* 2013) on the accretion history focussed on dwarf galaxies and entire clusters or entire systems of GCs, not on the direct contribution from the clusters themselves. A possible reason is summarized in Fig. 1: the chemical tagging made with $\alpha-$elements may distinguish between clusters and dwarf galaxies, but usually GC stars are superimposed to several Galactic components, formed in situ, accreted or even kicked out (see Sheffield *et al.* 2012). Clearly we need another diagnostic.

2. The Na-O anticorrelation

One of the best tools for chemical tagging of GC stars is the Na-O anticorrelation, which represents the unique DNA of GCs (see Fig. 2 which summarizes the results of our FLAMES survey with $\sim$ 2500 red giant stars in 25 GCs, see Carretta 2015 for references). The huge spreads in Na and O tell us that GCs are not simple stellar populations (by definition, coeval stars with the same initial chemical composition) because GC stars have very different Na, O contents. Hence, their stars are not strictly coeval: huge chemical

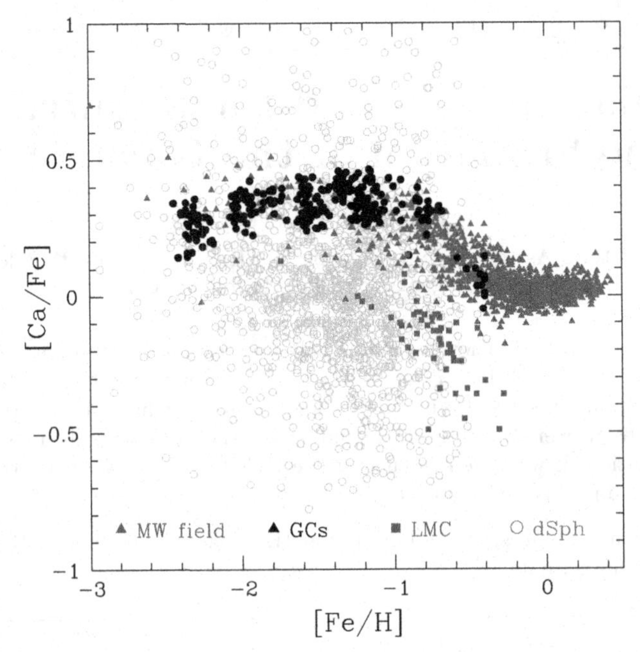

Figure 1. [Ca/Fe] as a function of [Fe/H] for several stellar populations. Stars in GCs (black filled circles) from Carretta *et al.* (2010a), Carretta *et al.* (2010c), Carretta *et al.* (2011), Carretta *et al.* (2013), Carretta *et al.* (2014a), Carretta *et al.* (2014b), Carretta *et al.* (2015); in dwarf galaxies (open grey circles) from Kirby *et al.* (2011); in LMC (filled squares) from Pompeia *et al.* (2008); and in the Milky Way (filled red triangles) from Adibekyan *et al.* (2012), Chen *et al.* (2000), Gratton *et al.* (2003), and Jonsell *et al.* (2005).

differences translate into tiny age differences, $10^6 - 10^7$ yrs, depending on the kind of polluter chosen to release enriched matter in the intracluster gas (see the review by Gratton *et al.* 2012 for details).

This chemistry is explained as the result of proton-capture reactions in H-burning at high temperature, enhancing or destroying different elements according to the temperature stratification (Denisenkov & Denisenkova 1989).

The occurrence of this pattern is only seen in the high density GC environment, whereas in Galactic Pop. II field stars O and Na are basically untouched from the main sequence up to the giant branch (shaded box in Fig. 2, where the field sample is taken from Gratton *et al.* 2003) and Gratton *et al.* (2000).

The Na-O anticorrelation, discovered by the Lick-Texas group (e.g. Kraft 1994), joined other features like the anticorrelation of C and N, observed even in unevolved stars, and of other elements along increasing Coulomb barrier like Al, Mg and even K (Mucciarelli *et al.* 2012, Cohen & Kirby 2012, Carretta 2014). In turn, these abundance variations have an impact on the photometric sequences in the color-magnitude diagrams due to molecular bands, mostly of CNO elements (Sbordone *et al.* 2011, Milone *et al.* 2012, Larsen *et al.* 2014a). For heavier species there is no direct photometric observable.

The anticorrelation of Na and O is the most notable of these features because it is observed in GCs spanning the whole mass range from the tiny Pal 5 (Smith *et al.* 2002) in dissolution phase to ω Cen (e.g. Johnson & Pilachowski 2010), the likely remnant of a nucleated dwarf galaxy; it is observed in GCs of sure or suspected extragalactic origin (e.g. M 54, Carretta *et al.* 2010a) and in those showing a metallicity spread (ω Cen, M 54). This signature does not discriminate between GCs formed in situ or

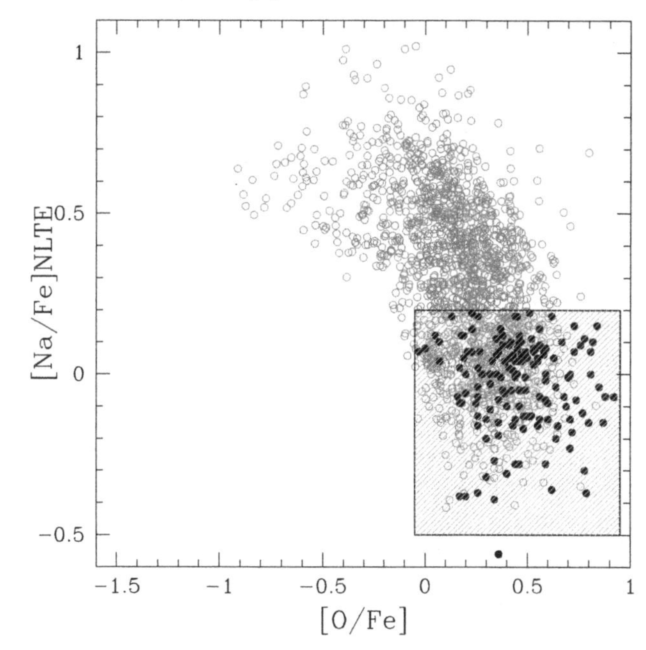

Figure 2. [Na/Fe] as a function of [O/Fe] for GC (empty circles, Carretta *et al.* 2009 and references in Fig.1) and field stars (filled circles, Gratton *et al.* 2003 and Gratton *et al.* 2000). The green shaded box encompasses the field stars.

likely accreted by our Galaxy (Leaman *et al.* 2013). In short, it is a widespread feature among GCs, likely related to the intrinsic mechanism of formation: we conclude that *a genuine old GC is a system that formed in a short time at least two generations of stars with different content of proton-capture elements.*

This signature allows us to put strong constraints on the initial masses of GCs. An almost constant fraction ($\sim 33\%$) of first generation (FG) stars with primordial composition still resides in GCs (Carretta *et al.* 2009). However the bulk of present day GC stars is composed of second generation (SG) stars, with modified composition. Therefore, since SG stars are formed using the ejecta of only a fraction of FG stars, and the ejecta from the main candidate polluters are not enough (de Mink *et al.* 2009), we incur in the so called *m*ass budget problem. This is solved with either a peculiar IMF or assuming that the precursors of GCs were much more massive than present-day GCs, losing about 90% of their stars and becoming good candidates as main contributors to the halo.

3. Multiple populations in GCs: second generation stars in the halo

A selective mass loss is expected because SG stars form more centrally concentrated in many models (e.g. D'Ercole *et al.* 2008, Decressin *et al.* 2007) and also the following dynamical evolution will push FG stars more easily outside tidal radius. Observations show that currently SG stars are usually more concentrated (e.g. Milone *et al.* 2012, Carretta 2015); however, after reaching full dynamical mixing, also SG stars with their unique chemical signature may be lost from GCs. These stars provide a clearcut probe of the contribution of GCs to the halo.

Carretta *et al.* (2010b) made a first attempt by comparing sodium abundances in field stars and SG stars in GCs. They found a small fraction (1.4%) of stars with GC signature evaporated in the halo (the fraction is doubled by considering the current ratio of FG/SG).

A more systematic survey found a similar fraction of SG stars by looking for large N excess in metal-poor halo stars with spectra from the Sloan survey. After cleaning the sample for contamination by thick disk stars Martell *et al.* (2011) found 3% of halo stars with SG composition, in agreement with serendipitous discoveries using O and Na (Ramirez *et al.* 2012), unless one of their two O-poor stars is found to be polluted by mass transfer from a companion AGB star, as suggested by the large abundance of barium and yttrium.

Finally, signatures of SG stars (Na, Al enhancements and O, Mg depletions) are more and more used to retrieve a cluster origin for streams (e.g. Sesar *et al.* 2015: Ophiuchus stream; Wylie-de Boer *et al.* 2012: Aquarius stream)

Using these signatures many studies (Carretta *et al.* 2010b, Vesperini *et al.* 2010, Schaerer & Charbonnel 2011, Martell *et al.* 2011, Gratton *et al.* 2012) provide estimates for the fraction of halo stars originated in GCs. These range from a few percent, using only observations, up to almost 50% of the halo mass, if the initial mass of GCs is assumed to be many times larger than present-day values.

4. Multiple populations in GCs: first generation stars in the halo

We get important information also from the FG component, which is much more elusive, since FG stars share the same chemistry of field stars of similar metallicity. In the homogeneous set of Na abundance from Carretta (2013, empty triangles in Fig. 3 here) there are a few stars clearly below the general trend (filled triangles in Fig. 3): these are objects tagged as accreted stars (e.g. Gratton *et al.* 2003, Nissen & Schuster 2010) and their abundances are clearly similar to those observed in dwarf satellites of Milky Way. Therefore, we may infer that a minority of halo stars has a composition similar to the dwarfs, but the bulk of halo stars looks like the FG component of GCs (red filled squares in Fig. 3).

This confirms the findings from $\alpha-$elements (Tolstoy *et al.* 2009) and carbon (Kirby *et al.* 2015b): the majority of field halo stars cannot have formed in present-day dSphs.

Another supporting evidence comes from the horizontal branch (HB) morphology. We know that the luminosity function of BHB stars is very different in GCs and in the halo (e.g. Kinman & Allen 1996, Gratton *et al.* 2012). We further know that Na-rich stars are also more enriched in He. Carretta *et al.* (2007), Carretta *et al.* (2010b) found that the extension of the Na-O anticorrelation is correlated to the maximum temperature reached at the bluest HB region (see Fig. 4) through the He abundance, increasing blueward. The maximum temperature on the HB of field BHB stars implies that they are not He enhanced. Since He-rich, Na-rich stars are rare in the field this suggests that the field BHB are related to the FG in GCs.

Additional evidence comes from binaries: Goodwin (2010) summarizes how the frequency and separation of binaries in different environments can be used to estimate the past density of their birth place. D'Orazi *et al.* (2010) and Lucatello *et al.* (2015) found that the binary fraction in large samples of GC stars is much lower among SG stars than among FG stars, with evidence of a formation in much denser environment, as predicted by theoretical models (Hong *et al.* 2015), and suggesting a common origin for the field stars and the FG populations.

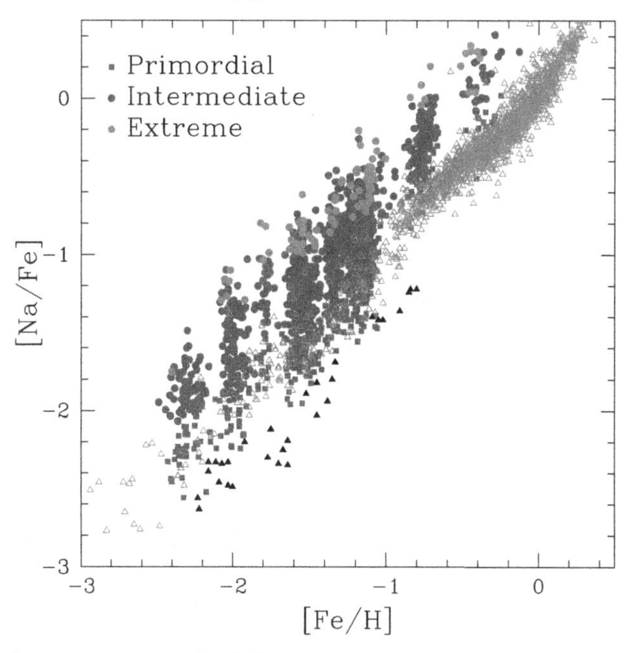

Figure 3. [Na/H] as a function of [Fe/H] for GC stars of different generations (filled squares and circles) and field stars (triangles). Filled triangles are the accreted field component.

5. Challenges and conclusions

The basic assumptions for a large contribution of GC stars to the halo (initial masses much larger, with selective loss of FG stars) must confront with two major challenges.

Evidence is accumulating that in dwarf galaxies GCs cannot have been initially more than 4-5 times today's mass, as their present-day masses account for about 25% of the galaxy mass in metal poor stars (Larsen *et al.* 2012, Larsen *et al.* 2014b, Tudorica *et al.* 2015). However, GCs like those in the Fornax dSph show normal evidence of multiple populations (Larsen *et al.* 2014a).

Furthermore, simulations of mass loss from gas expulsion (Khalaj & Baumgardt 2015) predict a strong anticorrelation between the fraction of SG stars and the cluster mass, that is not observed. Moreover, no variation of the fraction of SG stars as a function of cluster mass or Galactocentric distance or metallicity is observed (Bastian & Lardo 2015), at odds with what expected by mechanisms of mass loss.

The above points directly concern the multiple population scenario. However, independently from this scenario, larger birth cluster masses and large amount of mass loss are predicted, providing a mass comparable with the halo total mass (e.g. Marks & Kroupa 2010, Fall & Zhang 2001). In particular, primordial residual-gas expulsion and infant mortality may account for the bulk of the halo with FG signature (Baumgardt *et al.* 2008).

In summary, we have plenty of evidence for a relevant contribution of GCs to the halo: streams likely originated from GCs (from their size or chemical tagging, see also Grillmair, this volume), tidal tails, direct or indirect evidence from chemistry, HB stars and binary fraction. Considering that a good third of the halo seems due to the so-called "big four" accretion events (Sagittarius, Hercules-Aquila, GASS, Virgo Cloud; see Belokurov 2013), a large contribution of GCs to the general assembly of the Galactic halo seems to be a viable option.

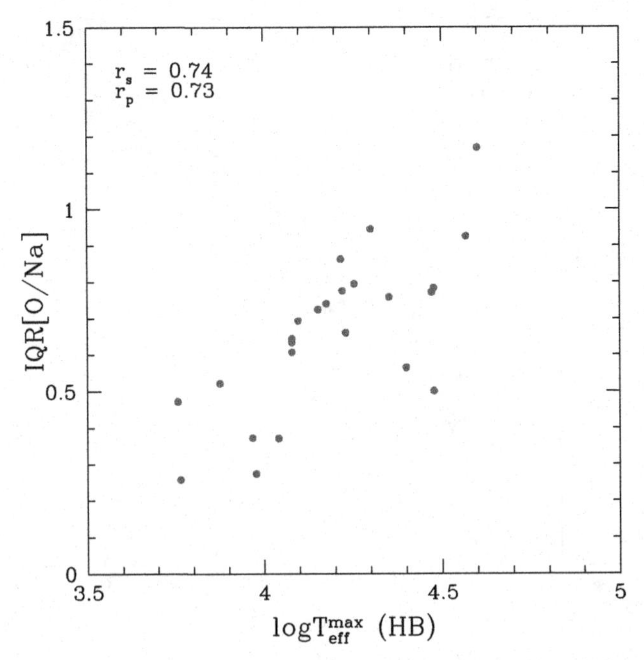

Figure 4. The extension of the Na-O anticorrelation (measured by the interquartile range IQR[O/Na] as a function of the maximum temperature along the HB for GCs in the FLAMES survey (Carretta *et al.* 2007, Carretta *et al.* 2010b).

References

Adibekyan, V. Zh, Sousa, S. G., Santos, N. C., Delgado Mena, E., Gonzales Hernandez, J. I., Israelian, G., Mayor, M., & Khachatryan, G. 2012, *A&A*, 545, A32

Bastian, N. & Lardo, C. 2015, *MNRAS*, 453, 357

Baumgardt, H., Kroupa, P., & Parmentier, G. 2008, *MNRAS*, 384, 1231

Belokurov, V. 2013, *New AR*, 57, 100

Bonifacio, P., Caffau, E., Zaggia, S., Franqis, P., Sbordone, L., Andrievsky, S. M., & Korotin, S. A. 2015, *A&A*, 579, L6

Carretta, E. 2013, *A&A*, 557, A128

Carretta, E. 2014, *ApJ*, 795, L28

Carretta, E. 2015, *ApJ*, 810, 148

Carretta, E., Recio-Blanco, A., Gratton, R. G., Piotto, G., & Bragaglia, A. 2007, *ApJ*, 671, L125

Carretta, E., Bragaglia, A., Gratton, R. G., *et al.* 2009, *A&A*, 505, 117

Carretta, E., Bragaglia, A., Gratton, R. G., *et al.* 2010a, *A&A*, 520, 95

Carretta, E., Bragaglia, A., Gratton, R. G., Recio-Blanco, A., Lucatello, S., D'Orazi, V., & Cassisi, S. 2010b, *A&A*, 516, 55

Carretta, E., Bragaglia, A., Gratton, R., Lucatello, S., Bellazzini, M., & D'Orazi, V. 2010c, *ApJ*, 712, L21

Carretta, E., Lucatello, S., Gratton, R. G., Bragaglia, A., & D'Orazi, V. 2011, *A&A*, 533, 69

Carretta, E., Bragaglia, A., Gratton, R. G. *et al.* 2013, *A&A*, 557, A138

Carretta, E., Bragaglia, A., Gratton, R. G., D'Orazi, V., Lucatello, S., & Sollima, A. 2014a, *A&A*, 561, A87

Carretta, E., Bragaglia, A., Gratton, R. G. *et al.* 2014b, *A&A*, 564, A60

Carretta, E., Bragaglia, A., Gratton, R. G. *et al.* 2015, *A&A*, 578, A116

Chen, Y. Q., Nissen, P. E., Zhao, G., Zhang, H. W., & Benoni, T. 2000, *A&AS*, 141, 491

Cohen, J. G. & Kirby, E. N. 2012, *ApJ*, 760, 86

Decressin, T., Meynet, G., Charbonnel C. Prantzos, N., & Ekstrom, S. 2007, *A&A*, 464, 1029

de Mink, S. E., Pols, O. R., Langer, N., & Izzard, R. G. 2009, *A&A*, 507, L1

Denisenkov, P. A. & Denisenkova, S. N. 1989, *A.Tsir.*, 1538, 11
D'Ercole, A., Vesperini, E., D'Antona, F., McMillan, S. L. W., & Recchi, S. 2008, *MNRAS*, 391, 825
D'Orazi, V., Gratton, R. G., Lucatello, S., Carretta, E., Bragaglia, A., & Marino, A. F. 2010, *ApJ*, 719, L213
Fall, S. M. & Zhang, Q. 2001, *ApJ*, 561, 751
Forbes, D. A. & Bridges, T. 2010, *MNRAS*, 404, 1203
Fusi Pecci, F., Bellazzini, M., & Cacciari, C., Ferraro F. R. 1995, *AJ*, 110, 1664
Goodwin, S. P. 2010, *RSPTA*, 368, 851
Gratton, R. G., Sneden, C., Carretta, E., & Bragaglia, A. 2000, *A&A*, 354, 169
Gratton, R. G., Carretta, E., Claudi, R., Lucatello, S., & Barbieri, M. 2003, *A&A*, 404, 187
Gratton, R. G., Carretta, E., & Bragaglia, A. 2012, *A&AR*, 20, 50
Hong, J., Vesperini, E., Sollima, A., McMillan, S. L. W., D'Antona, F., & D'Ercole, A. 2015, *MNRAS*, 449, 629
Ibata, R. A., Irwin, M. J., & Gilmore, G. 1994, *Nature*, 370, 194
Johnson, C. I. & Pilachowski, C. A. 2010, *ApJ*, 722, 1373
Jonsell, K., Edvardsson, B., Gustafsson, B., Magain, P., Nissen, P. E., & Asplund, M. 2005, *A&A*, 440, 321
Khalaj, P. & Baumgardt, H. 2015, *MNRAS*, 452, 924
Kinman, T. D. & Allen, C. 1996, *ASP-CS*, 92, 36
Kirby, E. N., Cohen, J. G., Smith, G. H., Majewski, S. R., Sohn, S. T., & Guhathakurta, P. 2011, *ApJ*, 727, 79
Kirby, E. N., Simon, J. D., & Cohen, J. G. 2015, *ApJ*, 810, 56
Kirby, E. N., Guo, M., Zhang, A. J., *et al.* 2015, *ApJ*, 801. 125
Kraft, R. P. 1994, *PASP*, 106, 553
Larsen, S. S., Strader, J., & Brodie, J. P. 2012, *A&A*, 544, L14
Larsen, S. S., Brodie, J. P., Forbes, D. A., & Strader, J. 2014a, *A&A*,565, A98
Larsen, S. S., Brodie, J. P., Grundahl, F., & Strader, J. 2014b, *ApJ*, 797, 15
Leaman, R., VandenBerg, D. A., & Mendel, J. T. 2013, *MNRAS*, 436, 122
Lucatello, S., Sollima, A., Gratton, R. G., Vesperini, E., D'Orazi, V., Carretta, E., & Bragaglia, A. 2015, *A&A*, in press, arXiv:1509.05014
Lynden-Bell, D. 1967, *MNRAS*, 136, 101
Marks, M. & Kroupa, P. 2010, *MNRAS*, 406, 2000
Martell, S. L., Smolinski, J. P., Beers, T. C., & Grebel, E. K. 2011, *A&A*, 534, 136
Milone, A., Piotto, G., Bedin, L. *et al.* 2012, *ApJ*, 744, 58
Mucciarelli, A., Bellazzini, M., Ibata, R., Merle, T., Chapman, S. C., Dalessandro, E., & Sollima, A. 2012, *MNRAS*, 426, 2889
Nissen, P. E. & Schuster, W. J. 2010, *A&A*, 511, L10
Pompéia, L., Hill, V., Spite, M., Cole, A. *et al.* 2008 2008, *A&A*, 480, 379
Ramírez, I., Meléndez, J., & Chanamé, J. 2012, *ApJ*, 757, 164
Sbordone, L., Salaris, M., Weiss, A., & Cassisi, S. 2011, *A&A*, 534, A9
Schaerer, D. & Charbonnel, C. 2011, *MNRAS*, 413, 2297
Sesar, B., Bovy, J., Bernard, E. J. *et al.* 2015, *ApJ*, 809, 59
Sheffield, A. A., Majewski, S. R., Johnston, K. V. *et al.* 2012, *ApJ*, 761, 161
Smith, G. H., Sneden, C., & Kraft, R. P. 2002, *AJ*, 123, 1502
Tolstoy, E., Hill, V., & Tosi, M. 2009, *ARA&A*, 47, 371
Tudorica, A., Georgiev, I. Y., & Chies-Santos, A. L. 2015, *A&A*, 581, 84
Vesperini, E., McMillan, S. L. W., D'Antona, F., & D'Ercole, A. 2010, *ApJ*, 718, 112
Wylie-de Boer, E., Freeman, K., williams, M., Steinmetz, M., Munari, U., & Keller, S. 2012, *ApJ*, 755, 35

The General Assembly of Galaxy Halos: Structure, Origin and Evolution
Proceedings IAU Symposium No. 317, 2015
A. Bragaglia, M. Arnaboldi, M. Rejkuba & D. Romano, eds.

doi:10.1017/S1743921315010649

Did globular clusters contribute to the stellar population of the Galactic halo?

Corinne Charbonnel[1,2] **and Martin Krause**[3,4]

[1]Department of Astronomy, University of Geneva
1290 Versoix - Switzerland
email: `Corinne.Charbonnel@unige.ch`
[2]IRAP CNRS UMR 5277, Université de Toulouse III
31400 Toulouse, France
[3] Universitäts-Sternwarte München, Ludwig-Maximilians-Universität
Scheinerstr. 1, 81679 München, Germany
email: `krause@mpe.mpg.de`
[4] Max Planck Institute for extraterrestrial Physics,
PO Box 1312, Giessenbachstr., 85741 Garching, Germany

Abstract. The origin of Galactic halo stars and the contribution of globular clusters (GC) to this stellar population have long been (and still are) debated. The discovery of multiple stellar populations with peculiar chemical properties in GCs both in the Milky Way and in Local Group galaxies recently brought a renewal on these questions. Indeed most of the scenarios that compete to reproduce the present-day GC characteristics call for fast expulsion of both gas and low-mass stars from these clusters in their early infancy. In this framework, the initial masses of GCs could have been 8 to 25 times higher than their present-day stellar mass, and they could have contributed to 5 to 20 % of the low-mass stars in the Galactic halo. Here we revisit these conclusions, which are in tension with observations of dwarf galaxies and of young massive star clusters in the Local Group. We come back in particular on the paradigm of gas expulsion from massive star clusters, and propose an alternative interpretation of the GC abundance properties. We conclude by proposing a major revision of the current concepts regarding the role massive star clusters play in the assembly of galactic haloes.

Keywords. globular clusters: general

---

1. Introduction

Globular clusters (GC) have long been considered as the perfect archetype of single stellar populations of coeval stars born with the same chemical composition. Indeed and although important differences in metallicity from cluster to cluster were long recognized, no significant star-to-star [Fe/H] abundance variations are found within individual clusters except the most massive ones (e.g., Omega Cen, M54, M22, NGC 1851, NGC 5286; see e.g. Simmerer *et al.* 2013, Marino *et al.* 2015 and references therein).

A major paradigm shift has occurred in the domain since the early 2000's, thanks to very high resolution spectroscopy with 8-10m class telescopes, together with high-precision photometry of Galactic GCs especially performed with HST (also ground based photometry is showing the same effects of MP, eg with Stromgren filters, or the various U filters). Those challenging and revolutionary studies have revealed that individual GCs do actually host multiple stellar populations. Those manifest themselves through their different chemical properties, in particular the well-documented O-Na anticorrelation that appears to be ubiquitous in GCs more massive than a few 10^4 $M_\odot$ (e.g. Carretta *et al.* 2009a, Carretta *et al.* 2010, Lind *et al.* 2009, Gratton *et al.* 2012). These peculiarities are associated to exquisite photometric complexity, with the appearance of multimodal sequences in different regions of the GC color-magnitude diagrams (e.g. Piotto *et al.* 2002, Milone *et al.* 2015). These properties have not been found yet in open clusters (Bragaglia *et al.* 2014). However, spectroscopic searches of field halo giant stars with chemical properties similar to that of GCs have opened the possibility that a large fraction of the halo population was originally formed in globular clusters (Martell *et al.* 2011). To improve the statistics, these studies are based on CN and CH bandstrength behavior, which is easily determined, but which might not reflect the initial composition of the stars as C and N are subject to evolution variations induced by internal mixing processes within the red giant stars themselves (e.g. Charbonnel & Zahn 2007a, Charbonnel & Lagarde 2010, Lagarde *et al.* 2012). Such intrinsic variations should be accounted for when estimating the actual number of peculiar stars that have really escaped from GCs. This is extremely important as the percentage of field halo stars with chemical properties similar to those observed in GCs does put, in principle, tight constraints on the actual contribution of GCs to the halo stellar populations.

2. Self-enrichment scenarii and the mass budget problem

Different scenarii were built to explain GCs chemical and photometric patterns. Most call for "self-enrichment" of GCs during their infancy, implying the formation of at least two stellar populations in all GCs. In this framework, first population (1P) stars are thought to be born with the proto-cluster original composition (i.e., that of contemporary field halo stars), while second population (2P) stars formed from original gas polluted to various degrees by hydrogen-burning processed material ejected by more massive, short-lived, 1P GC stars. The nature of the "polluters" is strongly debated, with the most commonly-invoked ones being the so-called Fast Rotating Massive Stars (FRMS; masses higher than $\sim$ 20 $M_{\odot}$; Decressin *et al.* 2007a, Decressin *et al.* 2007b, Krause *et al.* 2013) and massive AGB stars (masses $\sim$ 6 - 6.5 $M_{\odot}$; e.g.Ventura *et al.* 2001, Ventura *et al.* 2013, D'Ercole *et al.* 2010). Variants and combinations of these objects have been proposed as possible solutions, as well as supermassive stars ($\sim 10^4$ $M_{\odot}$ stars; Denissenkov & Hartwick 2014).

When considering a classical stellar IMF for the polluters, both the FRMS and the AGB scenarii face the so-called "mass-budget" problem. Indeed, the ratio between 1P and 2P stars is predicted to be of the order of ten to one, even when assuming that the maximum amount of their H-processed ejecta is entirely recycled into the second stellar population after a 50-50% dilution with pristine gas. To make things even worse, this 10-1 ratio is obtained when assuming that the 2P was composed of low-mass stars only. However, the percentage of 2P GC stars observed today in GCs is $\sim$ 70%, with only slight variations from cluster to cluster (e.g. Prantzos & Charbonnel 2006, Carretta 2013).

3. Gas and star expulsion from GCs in their infancy?

Drastic gas expulsion has been identified as a way to enhance the ratio of 2P to 1P stars (e.g. Prantzos & Charbonnel 2006, D'Ercole *et al.* 2010, Schaerer & Charbonnel 2011). If the 2P was formed in the immediate surrounding of the polluters at the center of a mass-segregated cluster, the less tightly bound 1P would have been lost preferentially as a result of a strong change in the GC potential well induced by gas expulsion (Decressin *et al.* 2010). In order to overcome the mass budget issue within this standard IMF framework, the initial masses of GCs must be $\sim$ 8 - 10 times (or up to 25 times, if 2P stars also escape from GCs, as suggested by the "CN-CH surveys", see § 1) larger than the present-day stellar mass. Therefore, the present-day Galactic GC population would then have contributed to approximately 5 to 8 % (10 to 20 %) of the low-mass stars in the Galactic halo (Schaerer & Charbonnel 2011, Martell *et al.* 2011).

For this to work, N-body simulations require explosive gas expulsion to unbind $\geqslant$ 95 % of 1P stars; more precisely, it must happen on the crossing timescale of the cluster (Decressin *et al.* 2010, Khalaj & Baumgardt 2015). According to gas expulsion models happening via the formation of a superbubble, with kinematics described by a thin-shell model (Krause *et al.* 2012, Krause *et al.* 2013), stellar winds and supernovae explosions fail to produce sufficient power for this process to work. With these sources of energy, the superbubbles are destroyed by the Rayleigh-Taylor instability before they reach escape speed for all but perhaps the least massive and most extended clusters. Rather, energy released by the coherent onset of accretion onto the stellar remnants of 1P massive stars (neutron stars and black holes) might plausibly accomplish the task (Krause *et al.* 2012). However, this fast episode of gas and stars removal would happen at the end of the SNe phase, when turbulence decreases within the ISM and all massive stars have turned into such dark remnants, i.e., typically $\sim$ 35 Myr after cluster formation.

4. Gas expulsion from Young Massive Star Clusters

4.1. *Observational constraints*

These conclusions are in tension with observations of young massive star clusters (YMC) in the Local Group and in dwarf galaxies, that can potentially test the conditions for gas expulsion. In particular, a sample of YMC with ages below 10 Myr, current masses between 8 and 50×10^5 $M_\odot$, and half-mass radii between 1.5 and 18 pc (i.e., very similar to the properties of GC progenitors), appear to be gas free by an age of $\sim$ 3 Myr, and show no evidence for multiple epochs of star-formation (Bastian *et al.* 2014, hereafter B14). In the case GC formation at high redshift happened the same way as YMC formation today, this implies that the dark remnant scenario for fast gas removal occurs too late to explain the disappearance of the gas at such an early stage.

4.2. *Grid of thin shell models for the gas dynamics*

To solve this inconsistency and understand how such YMCs could lose their gas so early in their evolution, we have computed a tailor-made grid of thin shell models for the gas dynamics (Krause *et al.* 2015, hereafter K15). We considered different sources of energy production (metallicity dependent stellar winds, radiation, supernovae, winds of young pulsars, and gamma ray burst related hypernovae). We used standard assumptions for the stellar IMF (Kroupa *et al.* 2013) and for the coupling between the energy produced by stellar feedback and the gas (20% efficiency). We first assumed a local star formation efficiency (SFE) of 30%, which is the preferred value for self-enrichment scenarios, and then we increased this quantity in some of our models.

With this grid, we show that the success of gas expulsion is strongly and mainly dependent on the compactness† of a star cluster as well as on the local SFE. We find that for two out of the eight YMCs in the sample of B14 gas expulsion is possible with a SFE of 30% for both the winds and SNe energy injection cases. However, none of these processes is sufficient to expel the residual gas for six YMCs in the sample, if the SFE was 30%. We find that 10^{52} erg hypernovae would be needed to expel the gas at this SFE

† We compute the compactness index with the formula $C_5 \equiv (M_*/10^5 M_\odot)/(r_h/\mathrm{pc})$, with the initial stellar mass M_* and half-mass radius r_h.

in one case (T352/W38220), and 10^{53} erg hypernovae would be required for the other four.

A possible way to remove residual gas is to assume a higher star formation efficiency. For two of the eight clusters of B14, an increase of the SFE up to $\sim 50\%$ is sufficient to account for gas expulsion. For the four most compact YMCs however, the SFE has to be $\sim 90\%$ to remove the remaining case. However in that case, the associated change of potential well is far too small to lead to a significant loss of stars, and the contribution of the clusters to the field population is minute compared to the standard framework for GC self-enrichment.

5. What if YMCs happen to be the modern counterparts of GCs that formed in the early universe?

YMC formation happening in the local and modern universe might be totally different than GC formation at high redshift. However, in the case where YMCs happen to be the modern counterparts of GCs, we should conclude that gas and star expulsion is a serious issue for the current GC self-enrichment scenarii that make classical assumptions for the SFE and for the stellar IMF of the first and second stellar populations. Rather, gas could be cleared for very high SFE, but in that case the total stellar mass would not be changed significantly enough to affect the 1P-to-2P ratio as requested by the O-Na anticorrelation observed in all GCs. Unfortunatly, the detection of the O-Na anticorrelation in YMCs is not yet feasible, although this would be the definitive test to assess the genetic filiation, or real relashionship, between YMCs and GCs. Indeed and as can be seen in Fig. 1, the distinction between star clusters showing or not the Na-O anticorrelation is clear in the C_5-[Fe/H] plane.

Alternative solutions have been investigated to alleviate the mass budget issue. An interesting option is to assume a flat IMF for the polluters (Smith & Norris 1982, D'Antona & Caloi 2004, Prantzos & Charbonnel 2006, Schaerer & Charbonnel 2011). Enlarging the mass domain of the polluters was also proposed, by including i.e. massive binary stars (9 to 20 $M_\odot$) as an additional source of H-burning ashes (De Mink *et al.* 2009), or by considering hypothetical supermassive stars ($\sim 10^4$ $M_\odot$) as the source of abundance anomalies. Alternatively, a scenario was proposed where a single stellar generation forms, where low-mass stars with proto-planetary discs sweep up material ejected by interacting binaries or rapidly rotating massive stars (Early-Disc Accretion scenario; Bastian *et al.* 2013). However, all these scenarii appear to have serious drawbacks (see e.g. Bastian & Lardo 2015).

6. Do GCs really host 1P stars?

Finally, we proposed an alternative interpretation of the observed sodium distribution, and suggested that stars with low sodium abundance that are counted as 1P members could actually be 2P stars (Charbonnel *et al.* 2014). For this, we computed the number ratio of 2P stars along the Na distribution following the FRMS model using tight constraints from the well-documented case of NGC 6752. We showed that the typical percentage of low-sodium stars usually classified as 1P stars can be reproduced by invoking only secondary star formation from material ejected by massive stars and mixed with original GC material in proportions that account for the Li-Na anti-correlation in this cluster. In other words, GCs could be totally devoid of 1P low-mass stars, and all the low-mass stars we observe today could have formed out of the ejecta of 1P massive stars mixed with original gas in roughly 50 to 50 proportions.

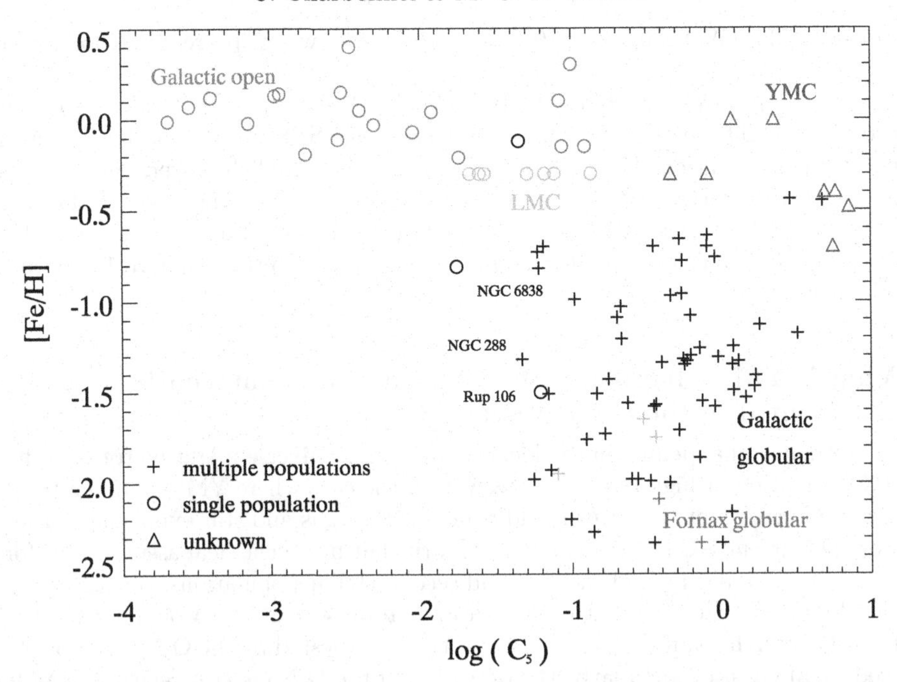

Figure 1. [Fe/H] versus compactness index C_5 for star clusters of various types and origins. Different symbols are used according to the presence or absence of the O-Na anticorrelation (plus and circles respectively; triangles are for clusters where the presence of the anticorrelation has not been assessed). Multiple populations occur at high C_5 where gas expulsion becomes increasingly difficult and eventually impossible due to large binding energies. See K15 for more details and for sample references

If the absence of "true" 1P stars in GCs today is confirmed (this can be tested by deriving the carbon isotopic ratio in GC low-mass main sequence stars; Charbonnel *et al.* 2014), we should find out whether all 1G low-mass stars were lost from the GCs or whether they have not formed, and why. The first case seems very improbable in view of the mass budget and gas expulsion problems, which would be even more exacerbated than discussed in the present paper. Also, this would be in tension with the recent census of halo stars compared to the number of GCs in the dwarf galaxy Fornax (Larsen *et al.* 2014).

In the second case (no formation of 1P low-mass stars), the mass initially locked in 1P GC massive stars could have been only two to four times the present-day stellar mass (instead of 8 to 25 times, as discussed previously). This would definitively release the current tension between the different model predictions and the constraints coming from YMCs as well as from dwarf galaxies haloes. We should then understand why LMS could not form initially. Work is in progress on all these open issues.

Acknowledgements

CC acknowledges support from the Swiss National Science Foundation (FNS) for the project 200020-159543 "Multiple stellar populations in massive star clusters - Formation, evolution, dynamics, impact on galactic evolution" (PI CC). MK acknowledges support from Deutsche Forschungsgemeinschaft under DFG project number PR 569/10-1 in the context of the Priority Program 1573 Physics of the Interstellar Medium. We thank the International Space Science Institute (ISSI, Bern, CH) for welcoming the activities of ISSI Team 271 "Massive star clusters across the Hubble Time (2013 - 2016, team leader

CC), as well as our colleagues, N.Bastian and R.Diehl, co-authors of K15 paper, and F.Primas, W.Chantereau, and Y.Wang, co-authors of C14.

References

Bastian, N. & *et al.* , 2013, *MNRAS*, 436, 2398
Bastian, N., & Hollyhead, K, Cabrera-Ziri, I., 2014, *MNRAS*, 445, 378
Bastian, N., & Lardo, C., 2015, *MNRAS*, 453, 357
Bragaglia, A., *et al.* 2014, *ApJ*, 796, 68
Carretta, E., *et al.* 2009, *A&A* 505, 139
Carretta, E., *et al.* 2010, *A&A*, 516, A55
Carretta, E. 2013, *A&A*, 557, A128
Charbonnel, C. & Lagarde, N. 2010, *A&A*, 522, A10 (C14)
Charbonnel, C. & Zahn, J. P. 2007, *A&A* (Letter), 467, L15
Charbonnel, C., Chantereau, W., Krause, M., Primas, F., & Wang, Y. 2014, *A&A* (Letter), 569, L6
D'Antona, F. & Caloi, V. 2004, *ApJ*, 611, 871
Decressin, T., *et al.* 2007a, *A&A*, 475, 859
Decressin, T., Charbonnel, C., & Meynet, G, 2007b, *A&A*, 464, 1029
Decressin, T., Baumgardt, H., Charbonnel, C., & Kroupa, P., *A&A*, 516, A73
De Mink, S., *et al.* 2009, *A&A* (Letters), 507, L1
Denissenkov, P. A. & Hartwick, F. D. A.. 2014, *MNRAS* (Letter), 437, L21
D'Ercole, *et al.* 2010, *MNRAS*, 407, 854
Khalaj, P. & Baumgardt,H. 2015, *MNRAS*, 492, 924
Gratton, R. G., Carretta, E., & Bragaglia, A. 2012, *A&ARv*, 20, 50
Krause, M. Charbonnel, C., *et al.* 2012, *A&A* (Letter), 546, L5
Krause, M. Charbonnel, C., *et al.* 2013, *A&A*, 552, A121
Krause, M., Charbonnel, C., Bastian, N., & Diehl, R. 2015, submitted to *A&A* (K15)
Kroupa, P., *et al.* 2013, in Planets, Stars and Stellar Systems (Springer Netherlands), 115
Lagarde, N., *et al.* 2012, *A&A*, 543, A108
Larsen, S. S., *et al.* 2014, *A&A*, 565, A98
Lind, K., *et al.* 2009, *A&A*, 503, 545
Marino, A. F., *et al.* 2015, *MNRAS*, 450, 815
Martell, S. L., Molinski, J. P., Beers, T. C., & Grebel, E. K. 2011, *A&A*, 354, A136
Milone, A. P., Marino, A. F., Piotto, G. P., *et al.* 2015, *ApJ*, 808, 51
Piotto, G., *et al.* 2002, *ApJ*, 670, 39
Prantzos, N. & Charbonnel, C. 2006, *A&A*, 458, 135
Schaerer, D. & Charbonnel, C. 2011, *MNRAS*, 413, 2297
Simmerer, J., *et al.* 2013, *ApJ* (Letters), 764, L7
Smith, G. H. & Norris, J. 1982, *ApJ*, 254, 594
Ventura, P., *et al.* 2001, *ApJ* (Letter), 550, L65
Ventura, P., *et al.* 2013, *MNRAS*, 431, 3642

The General Assembly of Galaxy Halos: Structure, Origin and Evolution
Proceedings IAU Symposium No. 317, 2015
A. Bragaglia, M. Arnaboldi, M. Rejkuba & D. Romano, eds.

doi:10.1017/S174392131500678X

Are the globular clusters with significant internal [Fe/H] spreads all former dwarf galaxy nuclei?

G. S. Da Costa

Research School of Astronomy & Astrophysics, Australian National University
email: gary.dacosta@anu.edu.au

Abstract. In this contribution the hypothesis that the Galactic globular clusters with substantial internal [Fe/H] abundance ranges are the former nuclei of disrupted dwarf galaxies is discussed. Evidence considered includes the form of the metallicity distribution function, the occurrence of large diffuse outer envelopes in cluster density profiles, and the presence of ([*s*-process/Fe], [Fe/H]) correlations. The hypothesis is shown to be *plausible* but with the caveat that if significantly more than the current nine clusters known to have [Fe/H] spreads are found, then re-evaluation will be required.

Keywords. Globular Clusters, Dwarf Galaxies, Nuclear Star Clusters

---

1. Introduction

The current standard picture of the formation of our Galaxy, and particularly its stellar halo, postulates that there is a substantial contribution from the disruptive merger and accretion of dwarf galaxies. Dwarf galaxies often possess their own globular cluster systems, and in some cases they also have nuclei/nuclear star clusters (e.g., den Brok *et al.* 2014). Consequently, these dwarf galaxy globular and nuclear star clusters will be accreted into the halo of the Galaxy as the dwarf is tidally disrupted. The on-going disruption of the Sagittarius (Sgr) dwarf by the Milky Way is an example of this process in action. There are a number of globular clusters currently associated with this dwarf (e.g., Law & Majewski 2010), and the cluster M54 lies at the dwarf galaxy's centre. When the Sgr accretion process is complete, these clusters will simply become part of the Galaxy's halo globular cluster population and their origin in Sgr will not be easily established.

The vast majority of Galactic globular clusters have one characteristic in common, and that is that their constituent stars are chemically homgeneous as regards the abundances of the heavier elements such as Fe and Ca. There are, however, a small number of "globular clusters" that show definite intrinsic internal [Fe/H] abundance dispersions. In this contribution the hypothesis that these systems are the former nuclear star clusters of now disrupted dwarf galaxies is investigated (see also Willman & Strader 2012 and Marino *et al.* 2015). Before embarking on this task, however, there are a number of caveats that need to be mentioned. First, Yong *et al.* (2013) have used an exquisite level of analysis to show that in the globular cluster NGC 6752, there are real star-to-star [Fe/H] abundance variations at very low levels (Δ[Fe/H] $\sim$ 0.03 dex). It is possible that all clusters show [Fe/H] abundance variations at this level; here we consider only those where the [Fe/H] abundance variations are substantially larger. Second, while the stellar system Terzan 5 does contain a significant [Fe/H] abundance range, Massari *et al.* (2015) argue that it is a fossil remnant of the formation of the Galactic bulge, and that it does not have an external accretion origin. It will not be considered further here. Further, while Simmerer *et al.*

(2013) have claimed that the globular cluster NGC 3201 contains a significant [Fe/H] range, that result has been questioned. For example, Muñoz *et al.* (2013) did not find any evidence to support an abundance range in their study of the cluster, and Mucciarelli *et al.* (2015a) have raised concerns about the analysis approach used by Simmerer *et al.* (2013). This cluster will also not be considered further. Moreover, in this contribution we will also not consider the complexity of the colour-magnitude diagrams now being revealed by *HST* photometry (e.g., Milone *et al.* 2015a,b), since the complexity does not seem to be restricted solely to the clusters showing [Fe/H] abundance variations. We also note that the light-element abundance variations, collectively known as the O-Na anti-correlation and which are seemingly ubiquitous in the globular cluster population, are also found in the clusters with intrinsic [Fe/H] variations.

The most well-known case of a "globular cluster" with an internal abundance range is the stellar system ω Cen. This system has a large spread in [Fe/H] among its member stars (e.g., Johnson & Pilachowski 2010) and there is evidence for multiple populations in both the abundance distribution and the colour-magnitude diagram. The unusual properties have led to the common speculation that ω Cen is the nuclear remnant of a tidally disrupted dwarf galaxy. Bekki & Freeman (2003) have used dynamical model calculations to show that it is feasible to have the nuclear remnant of a disrupted dwarf galaxy end up in a tightly bound orbit similar to that of the present-day ω Cen. Additional support for the scenario lies in the existence in the solar neighbourhood of field stars whose usually high [*s*-process/Fe] abundance ratios correspond to those for ω Cen stars at similar [Fe/H] values. Analysis of the kinematics indicates that these stars are very likely to be tidal debris from the ω Cen accretion event (Majewski *et al.* 2012).

The most straightforward case of a dwarf galaxy nuclear star cluster with an internal abundance range is the globular cluster M54, which lies at the centre of the Sgr dwarf. As shown by, for example, Carretta *et al.* (2010), this cluster possesses an intrinsic dispersion in [Fe/H] among its member stars: σ_{int}[Fe/H] = 0.18 dex. The Sgr dwarf is currently being tidally disrupted, and once the disruption is complete, M54 will become 'just another globular cluster in the halo'. Its origin as the nuclear star cluster of a dwarf galaxy will then be much less obvious.

In addition to ω Cen and M54, there are currently seven other globular clusters with intrinsic [Fe/H] ranges. These clusters and some of their properties are given in Table 1. Note that the cluster M22 is included in the list. Mucciarelli *et al.* (2015b) have questioned the existence of an [Fe/H] spread in this cluster, as for NGC 3201. However, the Mucciarelli *et al.* (2015b) results do not explain the range in [*s*-process/Fe] among the stars in this cluster, nor do they offer any explanation for the observed range in Ca II triplet line strengths among the giants, which points to the presence of an abundance spread (Da Costa *et al.* 2009). We now discuss the common characteristics of this set of objects and their relevance to a potential connection with dwarf galaxy nuclear star clusters.

1.1. *Metallicity Distribution Functions*

There are now a number of Milky Way dwarf spheriodal companion galaxies for which individual values of [Fe/H] for sizeable samples of red giant stars have been determined, enabling the characterisation of the *Metallicity Distribution Function* (MDF) – the number of stars as a function of [Fe/H]. For example, Leaman *et al.* (2013) give MDFs for six dSphs. One common feature of the dSph MDFs (see also Kirby *et al.* 2011a,b) is that they rise relatively slowly on the metal-poor side of the metallicity peak. In other words, there is a large range (often more than 1 dex) in [Fe/H] between the most metal-poor stars and those at the peak of the distribution. This is in direct contrast to the situation

Table 1. Globular clusters with intrinsic [Fe/H] spreads

Cluster	M_V[1]	R_{gc}[1] (kpc)	[Fe/H] Spread Reference
NGC 1851	–8.33	16.6	Carretta *et al.* (2011)
NGC 5139 (ω Cen)	–10.26	6.4	Johnson & Pilachowski (2010)
NGC 5286	–8.74	8.9	Marino *et al.* (2015)
NGC 5824	–8.85	25.9	Da Costa *et al.* (2014)
NGC 6273 (M19)	–9.13	1.7	Johnson *et al.* (2015)
NGC 6656 (M22)	–8.50	4.9	Marino *et al.* (2009)
NGC 6715 (M54)	–9.98	18.9	Carretta *et al.* (2010)
NGC 6864 (M75)	–8.57	14.7	Kacharov *et al.* (2013)
NGC 7089 (M2)	–9.03	10.4	Yong *et al.* (2014)

[1] Values from the on-line version of the Harris (1996) catalogue.

in the globular clusters with [Fe/H] ranges – in those systems with sufficient stars to form the MDF it always rises very sharply on the metal-poor side of the peak in the distribution. For example, in the sample of 55 NGC 5286 red giants observed with GIRAFFE by Marino *et al.* (2015), there are none with [Fe/H] < -1.90 but 14 with $-1.90 \leqslant$ [Fe/H] < -1.80, an extremely rapid rise in the MDF. Da Costa & Marino (2011) illustrate the same effect in a comparison of the MDFs for M22 and ω Cen. Although the extent of the MDFs on the metal-rich side of the peak abundances are different, with ω Cen having a notably longer tail to higher metallicities, both MDFs show steep rises on the metal-poor side of the peak.

A natural interpretation of the MDF difference is that cluster MDFs represent the outcome of rapid enrichment processes at high star formation rates consistent with high densities at the centre of a dwarf galaxy during the formation of a nuclear star cluster, while the dwarf galaxy MDFs represent the result of a more extended star formation process over a larger physical scale (e.g., Kirby *et al.* 2011a,b).

1.2. *Outer Density Profiles*

In the original photographic based work of Grillmair *et al.* (1995), and in the new DECam based work of Kuzma (ANU PhD thesis), the cluster M2 has been shown to be surrounded by a diffuse halo of stars extending to at least 250pc in radius. This is much larger than the nominal 'tidal' radius of the cluster, which is ~40pc. The outer portion of the surface density profile is well described by a power-law with a slope of -2.0 ± 0.1, and in 2-dimensions, the outer structure is relatively symmetrical with little indication of any 'tidal tails'. The situation is similar to that in NGC 1851 where Olszewski *et al.* (2009) discovered that the cluster is also surrounded by a large diffuse stellar envelope. The diameter is ~500pc, again much more extended than the tidal radius. Recent work by Marino *et al.* (2014) has verified that the NGC 1851 outer diffuse envelope is unambiguously associated with the cluster, and has revealed that it is dominated by stars whose properties match the cluster "1st generation" (i.e., the stars with lower [Fe/H] and lower [*s*-process/Fe] – see following sub-section).

The same situation occurs in a third globular cluster with an internal [Fe/H] range, NGC 5824. Here the analysis of Grillmair *et al.* (1995) reveals that the cluster is also surrounded by an extensive diffuse outer envelope. The outer density profile is a power law (slope -2.2 ± 0.1) and cluster stars are detected to $r \approx 45'$ (see also Carballo-Bello *et al.* 2012). At the distance of NGC 5824 this corresponds to $r \approx 420$pc or a diameter nearly 1kpc in size. This is significantly larger than the outer envelopes surrounding M2 and NGC 1851, which are closer to the centre of the Galaxy and thus potentially more susceptible to tidal stripping. Indeed the size of the NGC 5824 outer envelope approaches the extent of present-day low-luminosity dwarf galaxies.

While it is possible that in each case we are seeing an envelope of escaped cluster stars, it is equally possible that the outer envelopes represent remnant populations from a tidally disrupted dwarf galaxy that have remained bound to the former nuclear star cluster.

As regards the other clusters in Table 1, M54 is embedded in the Sgr dwarf galaxy, so it could be said that this cluster is also surrounded by a 'large diffuse outer envelope'. On the other hand, the density profiles of ω Cen and M22 do not show any significant extra-tidal structure, but given the locations of these clusters relatively close to the Galactic centre, any outer envelope is likely to have been stripped off. As for NGC 5296 and M19, there is little detailed information available on their surface density profiles, but again these clusters are relatively close to the Galactic centre. As regards NGC 6864, the surface density profile in Grillmair *et al.* (1995) hints at the presence of an outer envelope; a modern study based on digital wide-field imaging is required for confirmation.

1.3. *([s-process/Fe], [Fe/H]) correlations*

Perhaps the most intriguing characteristic in the clusters with [Fe/H] ranges is the existence of correlation between [Fe/H] and [s-process element/Fe] abundance ratios: in all clusters where sufficient data exist (seven of the nine objects), the stars with larger [Fe/H] abundances also have higher [s-element/Fe] abundance ratios†. The correlation is unexpected in a nucleosynthetic sense because it requires an additional s-process element contribution, most probably from AGB-stars, over and above the contribution needed to maintain the abundance ratio as the iron abundance increases. An example of the correlation for the cluster NGC 5286 is shown in Fig. 6 of Marino *et al.* (2015), who point out two further pieces of information. First, the relative enhancement of different neutron-capture elements correlates with the s-process element fraction in material with the solar composition. This verifies that the enrichment does indeed involve s-process nucleosynthesis. Second, a comparison of a subset of the clusters (NGC 5286, M2, M22, and ω Cen – see Fig. 19 of Marino *et al.* 2015) shows that the rate of increase in [s-process/Fe] with [Fe/H] is similar but not identical among these clusters. NGC 5286 has the steepest gradient while that for M22 and ω Cen are somewhat shallower. The nucleosynthesis process is therefore apparently similar from cluster-to-cluster, but not identical. It is, however, not by any means clear how this correlation fits into the "[Fe/H abundance range implies a former nuclear star cluster" hypothesis.

2. Discussion

The information presented above for the nine clusters with [Fe/H] spreads can be summarised as follows. (i) M54 is the nuclear star cluster of the Sgr dwarf. (ii) ω Cen is almost certainly the nuclear remnant of a dwarf galaxy that has been accreted and disrupted by the Milky Way. (iii) The distribution of [Fe/H] values in the clusters consistently shows steep rises on the metal-poor side, which is consistent with rapid enrichment at a central location. (iv) At least some of the clusters in question are surrounded by extended stellar envelopes that might represent the remnant population of an accreted dwarf galaxy.

It is then important to note that Georgiev *et al.* (2009) have shown that the nuclear star clusters in current dwarf galaxies have similar properties to those for luminous Milky Way globular clusters as regards luminosity (M_V), and size (half-light radius r_h).

† In the case of ω Cen, which has the largest [Fe/H] range, the effect appears to saturate in the sense that the [s-process/Fe] ratios reach a plateau and do not continue to increase with increasing [Fe/H] beyond [Fe/H] ≈ -1.3.

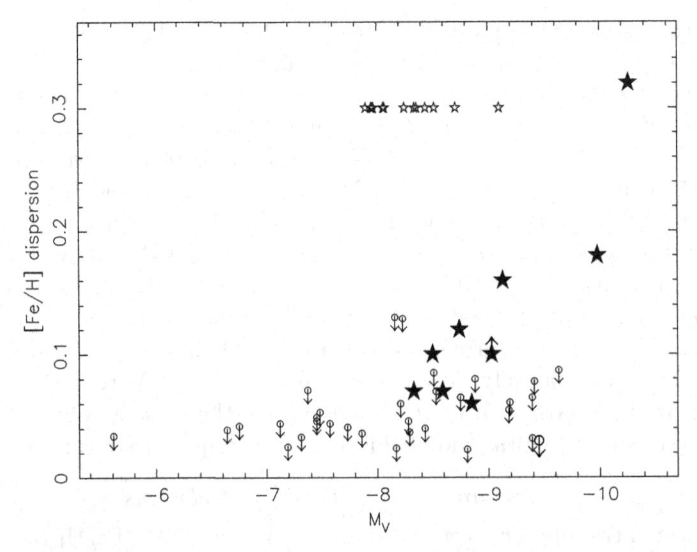

Figure 1. Metallicity dispersion, or upper limit, for Milky Way globular clusters plotted against absolute visual magnitude. The sample of clusters included is complete for $M_V < -7.9$, fainter clusters are taken from Carretta *et al.* (2010). The filled star symbols are the 9 clusters with intrinsic [Fe/H] dispersions. The 13 unstudied (or poorly studied) clusters with $M_V < -7.9$ are shown in the upper part of the plot as open star symbols.

These results then make it plausible that the clusters with internal [Fe/H] ranges are the former nuclear star clusters of dwarf galaxies accreted and disrupted by the Milky Way. However, this inference can only be valid if the number of such former nuclear star clusters is consistent with other approaches to the issue. For example, the current number of clusters with [Fe/H] ranges is broadly consistent with the results of Pffeffer *et al.* (2014), who have used cosmological simulations to suggest 1–3 massive Milky Way globular clusters are former dwarf galaxy nuclei. Another estimate of the expected number of former dwarf galaxy nuclear star clusters can be made as follows. The absolute visual magnitude of the Galactic stellar halo is approximately $M_V \approx -17$ (e.g., Freeman 1993). If we assume that ~50% of this luminosity comes from the disruption of satellite galaxies with the rest formed in-situ, and if we further assume that the disrupted systems had comparable luminosites to the present-day Fornax or Sgr systems, i.e., $M_V \approx -14$, then the disruption of ~15 systems can provide the required luminosity. If we then further assume that ~50% of the disrupted dwarfs had nuclear star clusters, then we arrive at the, admittedly uncertain, estimate of approximately eight such systems in the Milky Way halo. As noted in Table 1 there are nine known "globular clusters" with internal [Fe/H] spreads. The numbers are thus consistent but clearly if substantially more globular clusters are found to possess significant internal [Fe/H] abundance spreads, the hypothesized connection between such clusters and the nuclear star clusters of disrupted dwarfs will require re-consideration.

To assess this question we show in Fig. 1 an updated version of a plot first presented by Carretta *et al.* (2010). In this plot we show either an upper limit on the potential [Fe/H] abundance range present, or abundance dispersion estimates for the clusters in Table 1. The literature has been searched to include upper limits for all Milky Way clusters with $M_V < -7.9$; for the fainter clusters the limits shown are only for the clusters studied by Carretta *et al.* (2010). As indicated in the figure, there are 13 unstudied or poorly studied relatively luminous ($M_V < -7.9$) Milky Way globular clusters. Not surprisingly, these clusters are mostly at large distances from the Sun and/or have large reddenings.

Five have $E(B-V) < 0.3$ mag: NGC 5024, 6541, 5986, 6229 and 6284. In the terminology of Lee *et al.* (2007) NGC 5986 has a strongly extended blue HB (like ω Cen, M22, M54 and M2) and NGC 6629 has a moderately extended blue HB (like NGC 1851, 5824 and 6864). Detailed studies of these clusters would be very worthwhile.

In summary, the hypothesis that the globular clusters with substantial internal [Fe/H] abundance ranges are the former nuclear star clusters of now disrupted dwarf galaxies has to be considered at least plausible. However, further work is required to substantiate the total number of such clusters – if many more are discovered then the hyphothesis would need revision.

References

Bekki, K. & Freeman, K. C. 2003, *MNRAS*, 346, L11
Carballo-Bello, J. A., Gieles, M., Sollima, A., *et al.* 2012, *MNRAS*, 419, 14
Carretta, E., Bragaglia, A., Gratton, R. G., *et al.* 2010, *A&A*, 520, A95
Carretta, E., Lucatello, S., Gratton, R. G., Bragaglia, A., & D'Orazi, V. 2011, *A&A*, 533, A69
Da Costa, G. S. & Marino, A. F. 2011, *PASA*, 28, 28
Da Costa, G. S., Held, E. V., Saviane, I., & Gullieuszik, M. 2009, *ApJ*, 705, 1481
Da Costa, G. S., Held, E. V., & Saviane, I. 2014, *MNRAS*, 438, 3507
den Brok, M., Peletier, R. F., Seth, A., *et al.* 2014, *MNRAS*, 445, 2385
Georgiev, I. Y., Hilker, M., Puzia, T. H., *et al.* 2009, *MNRAS*, 396, 1075
Grillmair, C. J., Freeman, K. C., Irwin, M., & Quinn, P. J. 1995, *AJ*, 109, 2553
Freeman, K. C. 1993, in: G. H. Smith & J. P. Brodie (eds.), *The Globular Cluster – Galaxy Connection*, ASP Conf. Ser. Vol. 48 (San Francisco: ASP), p. 608
Harris, W. E. 1996, *AJ*, 112, 1487
Johnson, C. I. & Pilachowski, C. A. 2010, *ApJ*, 722, 1373
Johnson, C. I., Rich, R. M., Pilachowski, C. A., *et al.* 2015, *AJ*, 150, 63
Kacharov, N., Koch, A., & McWilliam, A. 2013, *A&A*, 554, A81
Kirby, E. N., Lanfranchi, G. A., Simon, J. D., *et al.* 2011a, *ApJ*, 727, 78
Kirby, E. N., Cohen, J. G., Smith, G. H., *et al.* 2011b, *ApJ*, 727, 79
Law, D. R. & Majewski, S. R. 2010, *ApJ*, 718, 1128
Leaman, R., Venn, K., Brooks, A. M., *et al.* 2013, *ApJ*, 767, 131
Lee Y.-W., Gim H. B., Casetti-Dinescu, D. I. 2007, *ApJ*, 661, L49
Majewski, S. R., Nidever, D. L., Smith, V. V., *et al.* 2012, *ApJ*, 747, L37
Marino, A. F., Milone, A. P., Piotto, G., *et al.* 2009, *A&A*, 505, 1099
Marino, A. F., Milone, A. P., Yong, D., *et al.* 2014, *MNRAS*, 442, 3044
Marino, A. F., Milone, A. P., Karakas, A. I., *et al.* 2015, *MNRAS*, 450, 815
Massari, D., Dalessandro, E., Ferraro, F. R., *et al.*, 2015, *ApJ*, 810, 69
Milone, A. P., Marino, A. F., Piotto, G., *et al.* 2015a, *MNRAS*, 447, 927
Milone, A. P., Marino, A. F., Piotto, G., *et al.* 2015b, *ApJ*, 808, 51
Mucciarelli, A., Lapenna, E., Massari, D., Ferraro, F. R., & Lanzoni, B. 2015a, *ApJ*, 801, 69
Mucciarelli, A., Lapenna, E., Massari, D., *et al.* 2015b, *ApJ*, 809, 128
Muñoz, C., Geisler, D., & Villanova, S. 2013, *MNRAS*, 433, 2006
Olszewski, E. W., Saha, A., Knezek, P., *et al.* 2009, *AJ*, 138, 1570
Pfeffer, J., Griffen, B. F., Baumgardt, H., & Hilker, M. 2014, *MNRAS*, 444, 3670
Simmerer, J., Ivans, I. I., Filler, D., *et al.* 2013, *ApJ*, 764, L7
Willman, B. & Strader, J. 2012, *AJ*, 144, 76
Yong, D., Meléndez, J., Grundahl, F., *et al.* 2013, *MNRAS*, 434, 3542
Yong, D, Roederer, I. U., Grundahl, F., *et al.* 2014, *MNRAS*, 441, 3396

The General Assembly of Galaxy Halos: Structure, Origin and Evolution
Proceedings IAU Symposium No. 317, 2015
A. Bragaglia, M. Arnaboldi, M. Rejkuba & D. Romano, eds.

doi:10.1017/S174392131500695X

RR Lyrae stars as probes of the Milky Way structure and formation

Pawel Pietrukowicz and OGLE collaboration

Warsaw University Observatory, Al. Ujazdowskie 4, 00-478 Warszawa, Poland
pietruk@astrouw.edu.pl

Abstract. RR Lyrae stars being distance indicators and tracers of old population serve as excellent probes of the structure, formation, and evolution of our Galaxy. Thousands of them are being discovered in ongoing wide-field surveys. The OGLE project conducts the Galaxy Variability Survey with the aim to detect and analyze variable stars, in particular of RRab type, toward the Galactic bulge and disk, covering a total area of 3000 deg^2. Observations in these directions also allow detecting background halo variables and unique studies of their properties and distribution at distances from the Galactic Center to even 40 kpc. In this contribution, we present the first results on the spatial distribution of the observed RRab stars, their metallicity distribution, the presence of multiple populations, and relations with the old bulge. We also show the most recent results from the analysis of RR Lyrae stars of the Sgr dwarf spheroidal galaxy, including its center, the globular cluster M54.

Keywords. Galaxy: structure, Galaxy: formation, stars: variables: RR Lyrae

---

RR Lyrae stars are core helium-burning giants with theoretically estimated masses in a range from about 0.55 to 0.80 $M_\odot$ and ages >10 Gyr (Marconi *et al.* 2015). These pulsating stars have spectral types from A2 to F6 or effective temperatures between 6500 and 9000 K and V-band absolute magnitudes in a range from +0.3 to +0.9 mag. RR Lyrae stars can be found everywhere in our Galaxy. Thousands of them have been discovered in wide-field surveys such as: ASAS, Catalina, MACHO, NSVS, OGLE, PTF, QUEST, SDSS, SEKBO, VVV. RR Lyrae stars are divided into fundamental-mode (type RRab), first-overtone (type RRc), and rarely found double-mode (type RRd) pulsators. RRab stars are on average intrinsically brighter and have higher amplitudes than RRc stars. More importantly, RRab variables with their characteristic saw-shaped light curves, in comparison to nearly-sinusoidal light curves of RRc stars, are hard to overlook making the searches for this type of variables highly complete. RRab stars have also a very practical photometric property. Based on the pulsation period and shape of the light curve one can estimate the metallicity of the star (Jurcsik 1995; Jurcsik & Kovács 1996; Smolec 2005).

The OGLE project (the Optical Gravitational Lensing Experiment) is a long-term variability survey which started in 1992 with the original aim to detect microlensing events toward the Galactic bulge (Udalski *et al.* 1992). Since the installation of a 32-chip camera with 1.4 deg^2 field of view in 2010, the project has been in its fourth phase (OGLE-IV, Udalski *et al.* 2015) and focuses on large-scale monitoring. Currently, OGLE monitors about 1.3 billion stars located in dense regions of the sky such as the Galactic bulge, Galactic disk, and Magellanic Clouds, by covering a total area of over 3000 deg^2. The survey is conducted with the 1.3-m Warsaw telescope at Las Campanas Observatory, Chile, administrated by the Carnegie Institution for Science.

Recently, Soszyński *et al.* (2014) released a collection of 38,257 RR Lyrae stars detected in the OGLE-IV Galactic bulge fields. An analysis of the subset of 27,258 RRab stars has

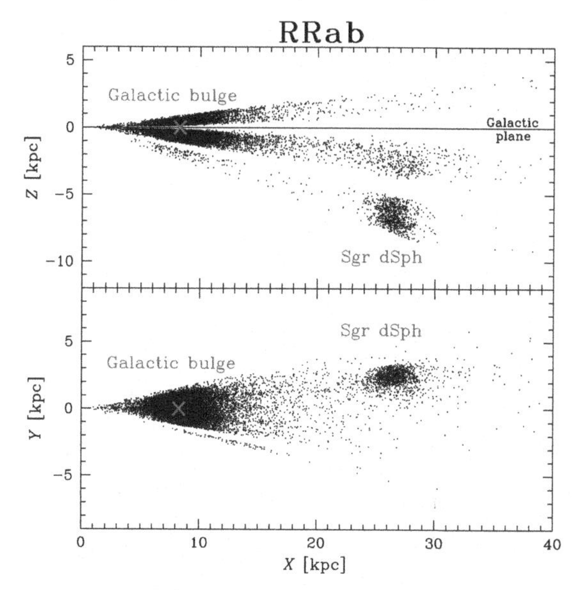

Figure 1. Projection of nearly 23,000 RRab stars observed toward the OGLE-IV bulge fields onto XZ and XY planes. The Sun is located at the origin of the system. The variables are concentrated around the Galactic Center and distributed to the outer halo. The second concentration is formed of variables from the tidally disrupted Sgr dSph galaxy.

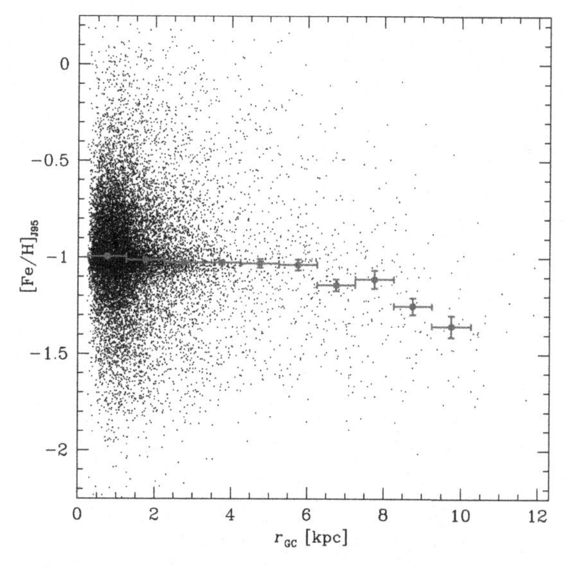

Figure 2. Metallicity distribution on the Jurcsik (1995) scale as a function of distance from the Galactic Center. Note a break around 6 kpc.

been published by Pietrukowicz *et al.* (2015). According to their work, these metal-poor stars trace closely the barred structure formed of intermediate-age red clump giants. The obtained distance to the Galactic center (GC) from the RR Lyrae stars is $R_0 = 8.27 \pm 0.01(\text{stat}) \pm 0.40(\text{sys})$ kpc, in very good agreement with the most recent estimates of R_0 from other methods. They show that the spatial distribution has the shape of a triaxial ellipsoid with proportions 1:0.49(2):0.39(2) and the major axis located in the Galactic plane and inclined at an angle $i = 20° \pm 3°$ to the Sun-GC line of sight. Another

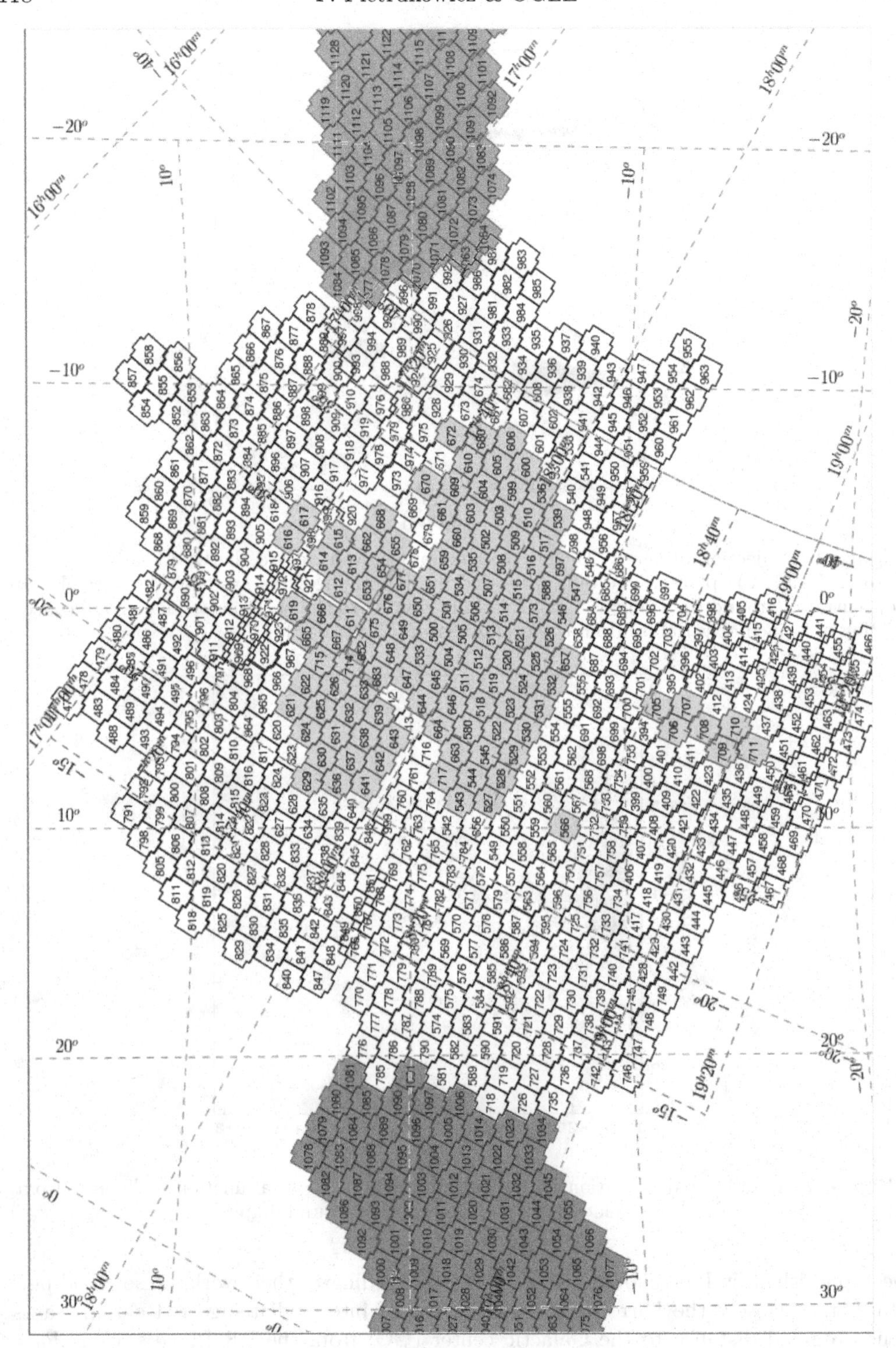

Figure 3. Extended coverage of the OGLE-IV bulge area. RR Lyrae stars from the central fields (in light pink) are analyzed in Pietrukowicz *et al.* (2015) and presented in this contribution. Fields at larger Galactic longitudes (in green and red) represent Galactic disk fields.

discovery is the presence of multiple old populations being likely the result of mergers in the early history of the Milky Way.

In this contribution, we report important results on far RR Lyrae variables, namely stars observed behind the Galactic bulge. Fig. 1 shows that the distribution of stars from the Galactic bulge area to the outer halo is smooth. Pietrukowicz *et al.* (2015) has found that the spatial density profile of bulge RR Lyrae variables can be described as a single power law with an index of -2.96. This value is very similar to indices obtained for halo stars, for example -2.8 for main-sequence stars near turn-off point (Jurić *et al.* 2008).

The prominent structure seen in Fig. 1 at a distance about three times larger than that to the GC is the remnant of the Sagittarius dwarf spheroidal (Sgr dSph) galaxy. Our search for variables within the tidal radius of M54, the globular cluster located at the core of Sgr dSph, has brought the detection of 277 such objects including 182 RR Lyrae stars, of which 23 are new. Based on 65 RRab stars, very likely members of the cluster, we have estimated the distance to M54 as $27.1 \pm 0.2(\text{stat}) \pm 1.3(\text{sys})$ kpc (Hamanowicz *et al.*, in prep.).

In Fig. 2, we present the radial metallicity distribution for RRab stars from the GC out to about 12 kpc. The metallicity clearly decreases with the distance from the center but the decrease is very mild to about 6 kpc and much more steeper farther out. At a distance of 8 kpc from the GC it amounts to about -1.1 dex on the Jurcsik (1995) scale, in agreement with what is observed in the solar vicinity. For instance, a mean metallicity for 28 field RR Lyrae stars listed in Smolec (2005) is -1.05 dex.

Our results indicate that the old bulge and halo form one component of the Galaxy. Recently, in a sample of $\sim$100 RR Lyrae stars from the bulge area, Kunder *et al.* (2015) found a high velocity object on a halo-like orbit. Ongoing and future surveys will complete our knowledge on the shape and properties of the old component. The OGLE survey has extended the coverage of the bulge area with the prime aim to find and characterize RR Lyrae stars to an angular distance of about 22° from the Galactic center (see Fig. 3).

Acknowledgements

The OGLE project has received funding from the National Science Centre, Poland, grant MAESTRO 2014/14/A/ST9/00121 to A. Udalski. This work has been also supported by the Polish Ministry of Sciences and Higher Education grants No. IP2012 005672 under the Iuventus Plus program to P. Pietrukowicz and No. IdP2012 000162 under the Ideas Plus program to I. Soszyński.

References

Marconi, M., Coppola, G., Bono, G., *et al.* 2015, *ApJ*, 808, 50
Jurić, M., Ivezić, Ž., Brooks, A., *et al.* 2008, *ApJ*, 673, 864
Jurcsik, J. 1995, *AcA*, 45, 653
Jurcsik, J. & Kovács, G. 1996, *A&A*, 312, 111
Kunder, A., *et al.* 2015, *ApJL*, 808, L12
Pietrukowicz, P., *et al.* 2015, *ApJ*, 811, 113
Smolec 2005, *AcA*, 55, 59
Soszyński, I., Udalski, A., & Szymański, M. K., *et al.* 2014, *AcA*, 64, 177
Udalski, A., Szymański, M., Kaluzny, J., Kubiak, M., & Mateo, M. 1992, *AcA*, 42, 253
Udalski, A., Szymański, M., & Szymański, G. 2015, *AcA*, 65, 1

The General Assembly of Galaxy Halos: Structure, Origin and Evolution
Proceedings IAU Symposium No. 317, 2015
A. Bragaglia, M. Arnaboldi, M. Rejkuba & D. Romano, eds.

doi:10.1017/S1743921315006821

Globular clusters in M31, Local Group, and external galaxies

Søren S. Larsen

Department of Astrophysics/IMAPP, Radboud University,
Postbus 9010, NL-6500GL, Nijmegen, the Netherlands
email: s.larsen@astro.ru.nl

Abstract. Throughout most of the Local Group, globular clusters (GCs) remain recognisable as extended objects in ground-based images taken in good seeing conditions. However, studying the full extent of the GC systems is challenging because of the large sky area that needs to be surveyed and recent years have seen dramatic progress in our knowledge of GC populations in nearby galaxies, thanks to large imaging surveys. At the same time, techniques for deriving detailed abundances from integrated-light spectra of GCs are maturing so that detailed comparisons of the chemical composition for GCs in different galaxies can now be made. Such comparisons may shed important light on the properties of proto-galactic fragments that were accreted onto galaxy halos. Nevertheless, our census of Local Group GCs probably remains far from complete, in particular at low luminosities and for very extended clusters.

Keywords. Globular Clusters; Galaxies

---

1. Introduction

Globular clusters (GCs) are classical tracers of galaxy halos. Although it now possible to image diffuse stellar halo light to impressively low levels of surface brightness (e.g., Mihos and McConnachie, these proceedings), GCs can be identified and properties such as ages, chemical composition, and kinematics can be studied in detail at much greater distances than for individual stars. In the context of galaxy halos, it is of particular relevance to note that, although metal-poor halos typically account for only a small percentage of the total stellar mass in a galaxy, the fractions of the total GC populations that are associated with halos can be large. This means that GCs can be efficiently employed to study these, otherwise difficult to access, components of galaxies.

While some of the brighter stellar clusters in the Magellanic Clouds were already catalogued by Dunlop (1828) and included in Herschel's *Catalogue of Nebulae and Clusters of Stars* (Herschel 1864), the study of extragalactic globular clusters started in earnest with Hubble's identification of 140 GC candidates in M31 (Hubble 1932). Hubble found a mean magnitude of $\langle V \rangle = 16.7$ for the M31 GCs, corresponding to $\langle M_V \rangle \approx -7.85$ when using the modern values of the distance and extinction towards M31 (Riess *et al.* 2012; Schlafly & Finkbeiner 2011). This is already quite close to modern estimates of the turn-over of the globular cluster luminosity function (GCLF) in M31 (Huxor *et al.* 2014) and other galaxies (e.g. Larsen *et al.* 2001). More detailed photometric work (Kron & Mayall 1960) showed the integrated colours of M31 GCs to be somewhat redder than those of their Milky Way counterparts, although uncertain corrections for interstellar reddening made it difficult to conclude whether this difference was intrinsic to the clusters (e.g., due to differences in age and/or metallicity) or could be caused by different amounts of extinction. From spectroscopic and photometric observations, van den Bergh (1969) found that GCs in M31 are indeed more metal-rich on average than those in the Milky Way, a result that has since been confirmed by many other studies (Huchra *et al.* 1991; Barmby *et al.* 2000; Perrett *et al.* 2002; Beasley *et al.* 2005; Caldwell *et al.* 2011). In the same paper, van den Bergh (1969) also noted that the GCs in the Fornax dwarf spheroidal galaxy appeared to have very low metallicities compared to those in the Milky Way and M31. It was thus clear already from these early studies that the properties of GC systems in different galaxies can differ substantially, and that such differences may provide important hints to the formation and chemical enrichment histories of their parent galaxies.

2. GCs in the Local Group: overview

Table 1 lists the Local Group member galaxies with known GC populations, based primarily on the catalogue by Harris *et al.* (2013). It is worth noting that this catalogue contains data for more than 400 extragalactic GC systems, the most distant of which are located well beyond the Coma galaxy cluster. Clearly, within the Local Group the GC system of M31 is the most populous by a large margin in terms of absolute numbers. However, when normalising the numbers of GCs to the host galaxy luminosities, expressed by the GC specific frequency ($S_N \equiv N_{\rm GC} \times 10^{0.4(M_V+15)}$; Harris & van den Bergh 1981), the well-known trend for S_N to increase for lower luminosity galaxies becomes apparent. Data for larger samples of galaxies show that the behaviour of S_N versus host galaxy absolute magnitude is actually U-shaped with a minimum between $M_V \sim -18$ and $M_V \sim -20$ (Harris *et al.* 1991; Miller & Lotz 2007; Peng *et al.* 2008; Georgiev *et al.* 2010; Harris *et al.* 2013; Mieske *et al.* 2014).

Although the galaxies in Table 1 are generally well studied, the numbers of known GCs have increased significantly in recent years for many of them. The PAndAS survey has revealed about 100 previously uncatalogued GCs in the outer parts of M31 (Huxor *et al.* 2008; 2014). Another interesting case is NGC 6822 ("Barnard's galaxy") which, until a few years ago, was thought to host only a single old GC, whereas 7 additional clusters have recently been identified in this galaxy (Hwang *et al.* 2011; Huxor *et al.* 2012). Recent additions to the census of Local Group GCs also include three clusters in NGC 147 and one in NGC 185 (Veljanoski *et al.* 2013).

For completeness, Table 1 also includes Local Group members (according to Mateo 1998) brighter than $M_V = -13$ (i.e., corresponding to the Fornax dSph) that do *not* host known GC populations. It should be clear from the preceding remarks that absence of evidence is, especially in these cases, not necessarily evidence of absence, particularly since some of these systems are quite distant and might merit further study. In the case of M32, however, the absence of a significant GC population does appear to be real, and may be attributable to dynamical erosion processes and/or stripping (Brockamp *et al.* 2014).

3. GC systems and the accretion histories of galaxy halos

3.1. *Metallicity distributions*

As already noted, the GC system of M31 is by far the richest in the Local Group. Based on spectroscopy of 150 GCs in M31, Huchra *et al.* (1991) found the metallicity distribution to be broad, ranging between [Fe/H] ≈ -2 and [Fe/H] ≈ 0, i.e., a range comparable to that seen in the Milky Way, with no obvious trend with luminosity (i.e., no evidence for significant mass-dependent self-enrichment within GCs) and only a weak radial gradient.

A recurrent theme in the discussion of the M31 GC metallicity distribution is the question of *bimodality*. In the Milky Way there is a fairly clear separation into a metal-poor group ($\langle$[Fe/H]$\rangle \approx -1.5$) with halo-like kinematics and spatial distribution, and a

Table 1. Globular cluster systems in the Local Group.

Galaxy	M_V	N_{GC}	S_N
M31	−21.8	∼ 450	0.86
Milky Way	−21.3	∼ 160	0.48
M33	−19.0	∼ 50	∼ 1.3
LMC	−18.4	16	0.70
SMC	−16.8	1	0.19
NGC 205	−16.7	11	2.3
NGC 6822	−15.5	8	5.0
NGC 147	−15.5	10	6.3
NGC 185	−15.4	8	5.5
WLM	−14.8	1	1.2
Sagittarius	−13.9	8	22
Fornax	−13.0	5	32
Galaxies with no known GCs:			
M32	−16.7		
NGC 3109	−15.7		
IC 10	−15.7		
IC 1613	−14.7		
Sext A	−14.6		
Sext B	−14.2		

more metal-rich population ($\langle[\mathrm{Fe/H}]\rangle \approx -0.5$) that is more naturally associated with the bulge and/or thick disc (Zinn 1985; Minniti 1996). This distinction is much less clear in M31; the metallicity distribution found by Huchra *et al.* (1991) does not display clearly distinct peaks, although a stronger hint of bimodality is present in the larger sample (229 GCs) of Perrett *et al.* (2002). Barmby *et al.* (2000) found evidence for two peaks at $[\mathrm{Fe/H}] = -1.4$ and $[\mathrm{Fe/H}] = -0.6$, i.e., quite similar to the peaks in the MW metallicity distribution. More recently, however, Caldwell *et al.* (2011) analysed the metallicity distribution of 322 GCs with spectroscopic observations and found no evidence of bimodality.

Bimodal *colour* distributions are commonly observed in extragalactic globular cluster systems (Elson & Santiago 1996; Kundu & Whitmore 2001; Larsen *et al.* 2001; Peng *et al.* 2006), the usual interpretation being that they reflect underlying bimodal *metallicity* distributions. However, non-linearities in the colour-metallicity relations may cause significant distortion of metallicity distributions when mapped to colour space. In particular, it has been argued that the rapid change in horizontal branch morphology of old stellar populations at intermediate metallicities can cause an inflection point in the colour-metallicity relation, so that clusters will tend to avoid intermediate colours. This effect can potentially produce bimodal colour distributions even if the underlying metallicity distributions are unimodal (Yoon *et al.* 2006; Cantiello & Blakeslee 2007). While the GC metallicity distributions for M31 discussed above are generally based on spectroscopic measurements, the relations between spectroscopic line indices and metallicities may be subject to similar effects (Kim *et al.* 2013) and indeed Caldwell *et al.* (2011) argued that non-linear transformations were required in their analysis.

Despite these complications, it is clear that the metallicity distributions of GC systems can differ substantially (e.g., Larsen *et al.* 2005), and at least in the Milky Way (where metallicities can be measured directly via high-dispersion spectroscopy of individual stars), the evidence for bimodality is strong. Presumably, these differences reflect differences in the corresponding formation- and assembly histories of the GC systems. Historically, bimodal GC metallicity distributions were predicted as a consequence of the "major merger" formation scenario for elliptical galaxies, in which the metal-poor GCs would represent the original (halo) GCs in gas-rich disc galaxies, and the metal-rich clusters were formed in the starburst accompanying the merger (Schweizer 1987; Ashman

& Zepf 1992). Other scenarios included accretion of metal-poor GCs from dwarf galaxies (Côté *et al.* 1998) or an *in-situ* "multi-phase" collapse (Forbes *et al.* 1997). Modern theoretical work now seeks to reproduce GC metallicity distributions in the context of hierarchical galaxy formation models and incorporates elements of all of the older ideas (Muratov & Gnedin 2010; Tonini 2013; Kruijssen 2015). By coupling cosmological merger trees with plausible assumptions about chemical evolution, GC formation efficiencies, and cluster disruption, such models are starting to provide more detailed insight into some of the mechanisms that may shape GC metallicity distributions. For example, galaxies that have accreted a larger fraction of their mass from small satellites may be expected to have a more prominent metal-poor GC population (Tonini 2013). As both models and observations continue to improve, other properties of GC sub-populations such as kinematics, detailed abundances, and age distributions, may provide important constraints on the accretion- and merger histories of galaxies.

3.2. *GCs in halos vs. dwarf galaxies*

Because of the long dynamical time scales in the outer parts of galaxy halos, this is where the signatures of accretion events are expected to be most readily visible. However, the low surface brightness and large extent on the sky (particularly for Local Group galaxies) represent significant observational challenges. Within the Local Group, the full extent of the M31 GC system has only recently become clear, thanks in large part to the PAndAS survey which has now mapped the M31 GC population to distances beyond 100 kpc from the centre of the galaxy (Huxor *et al.* 2008; 2014). Additional GC candidates beyond 100 kpc have also been identified in SDSS imaging (di Tullio Zinn & Zinn 2013). It is now clear that the M31 GC system is significantly more extended than that of the Milky Way; currently 91 GCs are known with (projected) galactocentric distances of $R_{\rm proj} > 25$ kpc and 12 with $R_{\rm proj} > 100$ kpc in M31. In the Milky Way the corresponding numbers are ~ 13 and ~ 1, respectively (Huxor *et al.* 2014), so the difference remains quite significant even after accounting for the overall greater number of GCs in M31. The spatial distribution of the GCs in M31 appears to correlate well with the stellar overdensities observed in the halo, from which it has been suggested that up to $\sim 80\%$ of the outer halo GCs in M31 may have been accreted (Mackey *et al.* 2010). Interestingly, searches for GCs at large distances from the centre of the third spiral in the Local Group, M33, have revealed only a handful of objects with $R_{\rm proj} > 10$ kpc (Cockcroft *et al.* 2010).

3.2.1. *Luminosity functions*

It was noted by van den Bergh (1998) that the GCs in the outer part of the Galactic halo (beyond $R = 80$ kpc) have a luminosity function which differs significantly from the LF seen in the inner part of the GC system, which is peaked at $M_V \approx -7.5$. The outer halo GCs are mostly fainter than $M_V = -6$, but one cluster (NGC 2419) is brighter than $M_V = -9$. Van den Bergh (1998) thus suggested that the GCLF in the outer Galactic halo may be bimodal, and noted that the Sagittarius dwarf galaxy appears to display a similarly bimodal GCLF, suggesting that the Searle-Zinn fragments that formed the halo may have resembled the Sagittarius dwarf. It has further been suggested that an accretion origin is especially likely for the "young halo" clusters in the Milky Way (Mackey & van den Bergh 2005; Forbes & Bridges 2010).

Drawing definitive conclusions from the small number of clusters in the outer Milky Way halo is difficult, but better statistics are available in M31. Huxor *et al.* (2014) found a similarly bimodal GCLF in the outer halo of M31, with peaks at $M_V \sim -7.5$ and at $M_V \sim -5.5$, although the exact location of the fainter peak is uncertain because of completeness effects. Again, this resembles the GCLF of the Sagittarius dwarf, and would

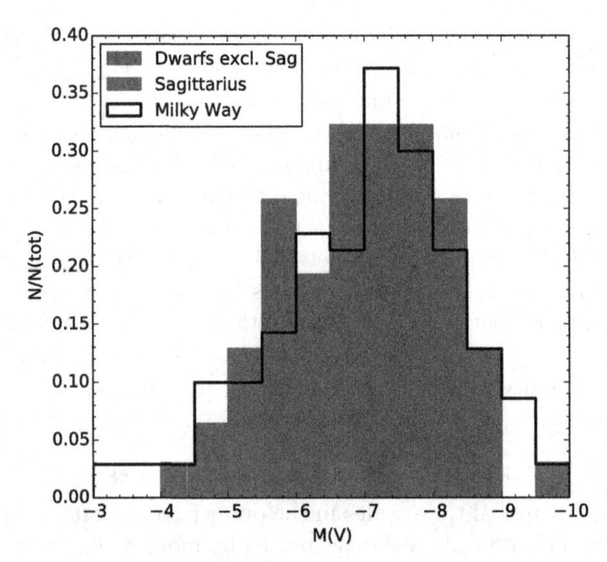

Figure 1. Luminosity functions for GCs in the Milky Way, Local Group dwarf galaxies (excl. Sagittarius) and the Sagittarius dwarf.

appear to be consistent with the idea that many of the outer halo GCs in M31 have been accreted from Sagittarius-like fragments.

Given that the Sagittarius dwarf is currently in the process of being accreted by the Milky Way, along with its ~ 8 GCs, it is perhaps not surprising that it has been used as a benchmark for comparison with halo GCs. Nevertheless, it may be worth asking how representative the GC system of the Sagittarius dwarf is of GCs in dwarf galaxies in general. Figure 1 shows the GC luminosity functions for the Milky Way, the Sagittarius dwarf, and other dwarf galaxies (with $-16 < M_V < -13$) in the Local Group. It is clear that the GCLF in Sagittarius is indeed quite different from the global GCLF in the Milky Way; however, it also differs from the combined GCLF of the remaining dwarfs. Indeed, the GCLFs of the remaining dwarfs (individually or combined) are consistent with being drawn from that of the Milky Way, with a K-S test yielding a p-value of 0.69 when comparing the Milky Way and combined dwarf galaxy GCLFs. Instead, the comparison of the Sagittarius vs. Milky Way GCLFs yields $p = 0.03$. It would be interesting to investigate in more detail to what extent these differences can be attributed to effects of dynamical evolution and the special circumstances of Sagittarius, in particular.

3.2.2. *Metallicities*

A comparison of the chemical composition of GCs in the halos of large galaxies with those in dwarf galaxies may provide additional clues to the properties of the fragments that built up halos. In this section we comment on the overall metallicities; the detailed chemical composition will be considered in Sect. 3.2.3.

As noted in the introduction, the GCs in the Fornax dSph are much more metal-poor on average than those in the Milky Way halo. More generally, there is a correlation between the metallicities of GC (sub)-populations and host galaxy luminosity/mass (e.g. Larsen *et al.* 2001; Peng *et al.* 2006). One difficulty associated with measuring accurate metallicities at the extremes of the distribution is that traditional integrated-light methods (broad-band colours, spectroscopic line indices) rely on calibrations that are less well established at low metallicities. However, most studies agree that the four most metal-poor clusters in Fornax (Fornax 1, 2, 3, and 5) have metallicities of [Fe/H] ≈ -2 or

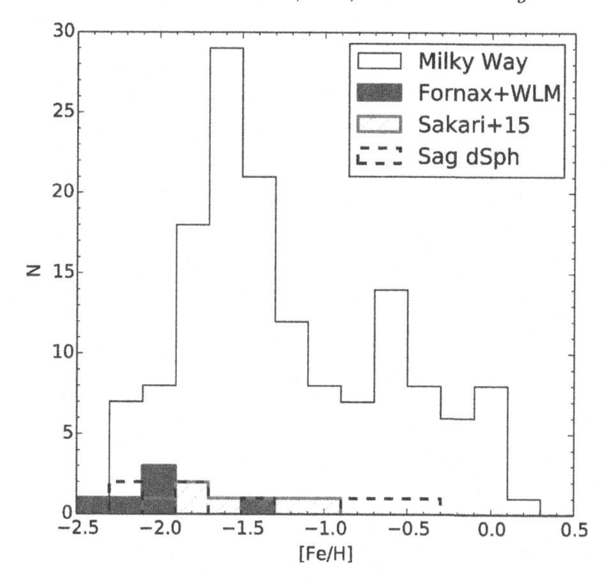

Figure 2. GC metallicity distributions for the Milky Way, Fornax dSph, Sagittarius dwarf, and outer halo GCs in M31.

below (Strader *et al.* 2003), which is significantly lower than the typical metallicities of halo GCs in the Milky Way or M31. Abundance measurements that do not rely on intermediate calibration steps are now available from high-dispersion spectroscopy, either for individual stars (Fornax 1, 2, and 3; Letarte *et al.* 2006) or from integrated light (Fornax 3, 4, and 5; Larsen *et al.* 2012). These measurements confirm that Fornax 1, 2, 3, and 5 all have $[Fe/H] < -2$, whereas Fornax 4 has $[Fe/H] \approx -1.4$.

Figure 2 shows the metallicity distributions for GCs in the Milky Way and Sagittarius (Harris 1996), the Fornax and WLM dwarf galaxies (Larsen *et al.* 2012; 2014), and outer M31 halo GCs from PAndAS (Sakari *et al.* 2015). The M31 outer halo GCs have similar metallicities to halo GCs in the Milky Way, whereas the Fornax and WLM GCs are evidently much more metal-poor. A K-S test yields a probability of only 0.003 that the Fornax+WLM GCs are drawn from the same metallicity distribution as the Milky Way GCs (restricting the comparison to $[Fe/H] < -1$). For Sagittarius, the corresponding comparison yields $p = 0.15$, i.e., no significant difference.

From these comparisons, it appears that the LFs and metallicity distributions of GCs in the *outer halos* of M31 and the Milky Way are similar to those of the Sagittarius dwarf galaxy and that the GCs in the outskirts of these large spirals might indeed have originated in fragments resembling Sagittarius. For the halo GC populations as a whole, the situation is less clear. While the metallicity distributions remain consistent with those in Sagittarius, the GCLFs are quite different. Clearly, a direct comparison of GCLFs is complicated by the possible role of dynamical evolution, which may have affected the GCLFs in different environments differently. It appears unlikely, however, that a significant fraction of the GC population in the large spirals originated in fragments resembling the Fornax dSph, as the metallicities of the Fornax GCs are too low.

3.2.3. *Detailed chemical composition*

While GCs lend themselves to spectroscopic studies at relatively high spectral resolution because of their modest velocity dispersions (typically 5–10 km s^{-1}), most spectroscopic work on GCs has, until recently, been based on methods developed primarily for analysis of galaxies at relatively low spectral resolution. However, in recent years several groups have developed analysis techniques that can take advantage of the large amount of information that is potentially available in an integrated-light, high-dispersion GC spectrum (McWilliam & Bernstein 2008; Larsen *et al.* 2012; Sakari *et al.* 2013; Colucci *et al.* 2014). While the approaches adopted by the various authors differ in detail, they may in general be seen as extensions of classical simple stellar population models to high spectral resolution, in which abundances of individual elements can be varied and the effect on the integrated spectra compared with observations.

It now appears within reach to apply "chemical tagging" (Freeman & Bland-Hawthorn 2002) to identify groups of GCs that may have a common origin. For example, one may exploit the differences in elemental abundance ratios as a function of metallicity in dwarf galaxies when compared with larger galaxies, such as the shift in the location of the "knee" in the [α/Fe] vs. [Fe/H] relation as a function of host galaxy mass (Tolstoy *et al.* 2009). This shift is well established for field stars, and is also seen in the GC system of the Fornax dSph, where Fornax 4, the most metal-rich of the clusters, has a noticeably lower [α/Fe] ratio than GCs of comparable metallicity in the Milky Way, but following the trend seen for field stars in Fornax (Larsen *et al.* 2012; Hendricks *et al.* 2014). Combining abundance information with other diagnostics, such as kinematics and spatial location, may provide a promising avenue towards identifying groups of GCs that once belonged to a common progenitor. Using this approach, Sakari *et al.* (2015) have identified several GCs that might be associated with stellar streams in the outer M31 halo.

A second potential application is to study the phenomenon of multiple stellar populations in GCs using integrated-light observations. In Milky Way GCs, it is now well established that the abundances of light elements (e.g., C, N, O, Na, Mg, Al) display substantial star-to-star variations within a given cluster (e.g., Carretta, Piotto, these proceedings). Integrated-light observations of extragalactic GCs have, in several cases, revealed depleted [Mg/Fe] ratios in clusters that otherwise appear to have normal [α/Fe] ratios, as well as enhanced [Na/Fe] ratios (Colucci *et al.* 2009; Larsen *et al.* 2012,2014; Sakari *et al.* 2015). These observations may be indicative of the Mg/Al and Na/O anti-correlations, whereby a fraction of the stars would have depleted Mg and enhanced Na abundances, thus driving the mean abundances of these elements down and up, respectively. A particularly exciting prospect is that such methods might be used to search for abundance anomalies in young massive clusters, such as those in the "Antennae" galaxies, which are generally too distant for individual stars within the clusters to be studied in detail spectroscopically.

4. Acknowledgements

I am grateful to the Leids Kerkhoven-Bosscha Fonds (LKBF) and the IAU for travel grants that made it possible for me to attend this symposium.

References

Ashman, K. M. & Zepf, S. E. 1992, *ApJ*, 384, 50
Barmby, P., Huchra, J. P., Brodie, J. P., *et al.* 2000, *AJ*, 119, 727
Beasley, M. A., Brodie, J. P., Strader, J., *et al.* 2005, *AJ*, 129, 1412
Brockamp, M., Küpper, A. H., Thies, I., *et al.* 2014, *MNRAS*, 441, 150
Caldwell, N., Schiavon, R., Morrison, H., *et al.* 2011, *AJ*, 141, 61
Cantiello, M. & Blakeslee, J. P. 2007, *ApJ*, 669, 982
Cockcroft, R., Harris, W. E., Ferguson, A. M. N., *et al.* 2011, *ApJ*, 730, 112
Colucci, J. E., Bernstein, R. A., Cameron, S., *et al.* 2009, *ApJ*, 704, 385
Colucci, J. E., Bernstein, R. A., & Cohen, J. G. 2014, *ApJ*, 797, 116

Côté, P., Marzke, R. O., & West, M. J. 1998, *ApJ*, 501, 554
di Tullio Zinn, G. & Zinn, R. 2013, *AJ*, 145, 50
Dunlop, J. 1828, *Phil. Trans. Royal Astron. Soc.*, 118, 113
Elson, R. A. W. & Santiago, B. X. 1996, *MNRAS*, 278, 617
Forbes, D. A., Brodie, J. P., & Grillmair, C. J. 1997, *ApJ*, 113, 1652
Forbes, D. A. & Bridges, T. 2010, *MNRAS*, 404, 1203
Freeman, K. & Bland-Hawthorn, J. 2002, *ARA&A*, 40, 487
Georgiev, I. Y., Puzia, T. H., Goudfrooij, P., & Hilker, M. 2010, *MNRAS*, 406, 1967
Harris, W. E. & van den Bergh, S. 1981, *AJ*, 86, 1627
Harris, W. E. 1991, *ARA&A*, 29, 543
Harris, W. E. 1996, *AJ*, 112, 1487
Harris, W. E., Harris, G. L. H., & Alessi, M. 2013, *ApJ*, 772, 82
Hendricks, B., Koch, A., Lanfranchi, G. A., *et al.* 2014, *ApJ*, 785, 102
Herschel, J. 1864, *Phil. Trans. Royal Astron. Soc.*, 154, 1
Hubble, E. 1932, *ApJ*, 76, 44
Huchra, J. P., Kent, S. M., & Brodie, J. P. 1991, *ApJ*, 370, 495
Huxor, A. P., Tanvir, N. R., Ferguson, A. M. N., *et al.* 2008, *MNRAS*, 385, 1989
Huxor. A. P., Ferguson, A. M. N. Veljanoski, J., *et al.* 2012, *MNRAS*, 429, 1039
Huxor, A. P., Mackey, A. D., Ferguson, A. M. N., *et al.* 2014, *MNRAS*, 442, 2165
Hwang, N., Lee, M. G., Lee, J. C., *et al.* 2011, *ApJ*, 738, 58
Kim, S., Yoon, S.-J., Chung, C., *et al.* 2013, *ApJ*, 768, 138
Kinman, T. D. 1963, *AJ*, 137, 213
Kron, G. E. & Mayall, N. U. 1960, *AJ*, 65, 581
Kruijssen, J. M. D. 2015, *MNRAS*, in press
Kundu, A. & Whitmore, B. C. 2001, *AJ*, 121, 2950
Larsen, S. S., Brodie, J. P., Huchra, J. P., *et al.* 2001, *AJ*, 121, 2974
Larsen, S. S., Brodie, J. P., & Strader, J. 2005, *A&A*, 443, 413
Larsen, S. S., Brodie, J. P., & Strader, J. 2012, *A&A*, 546, A53
Larsen, S. S., Brodie, J. P., Forbes, D. A., & Strader, J. 2014, *A&A*, 565, A98
Letarte, B., Hill, V., Jablonka, P., *et al.* 2006, *A&A*, 453, 547
Mackey, A. D. & van den Bergh, S. 2005, *MNRAS*, 360, 631
Mackey, A. D., Huxor, A. P., Ferguson, A. M. N., *et al.* 2010, *ApJ*, 717, L11
Mateo, M. 1998, *ARA&A*, 36, 435
McWilliam, A. & Bernstein, R. A. 2008, *ApJ*, 684, 326
Mieske, S., Küpper, A. H. W., & Brockamp, M. 2014, *A&A*, 565, L6
Miller, B. W. & Lotz, J. M. 2007, *ApJ*, 670, 1074
Minniti, D. 1996, *ApJ*, 459, 175
Muratov, A. L. & Gnedin, O. Y. 2010, *ApJ*, 718, 1266
Peng, E. W., Côté, P., Jordán, A., *et al.* 2006, *ApJ*, 639, 838
Peng, E. W., Jordán, A., Côté, P., *et al.* 2008, *ApJ*, 681, 197
Perrett, K. M., Bridges, T. J., Hanes, D. A., *et al.* 2002, *AJ*, 123, 2490
Riess, A. G., Fliri, J., & Valls-Gabaud, D. 2012, *ApJ*, 745, 156
Sakari, C. M., Shetrone, M., Venn, K., *et al.* 2013, *MNRAS*, 434, 358
Sakari, C. M., Venn, K. A., Mackey, A. D., *et al.* 2015, *MNRAS*, 448, 1314
Schlafly, E. F. & Finkbeiner, D. P. 2011, *ApJ*, 737, 103
Schweizer, F. 1987, in: Nearly Normal Galaxies, New York, Springer, p. 18, ed: S. M. Faber
Strader, J., Brodie, J. P., Forbes, D. A., *et al.* 2003, *AJ*, 125, 1291
Tolstoy, E., Hill, V., & Tosi, M. 2009, *ARA&A*, 47, 371
Tonini, C. 2013, *ApJ*, 762, 39
van den Bergh, S. 1969, *ApJS*, 19, 145
van den Bergh, S. 1998, *ApJ*, 505, L127
Veljanoski, J., Ferguson, A. M. N., Huxor, A. P., *et al.* 2013, *MNRAS*, 435, 3654
Zinn, R. 1985, *ApJ*, 293, 424
Yoon, S.-J., Yi, S. K., & Lee, Y.-W. 2006, *Science*, 311, 1129

The General Assembly of Galaxy Halos: Structure, Origin and Evolution
Proceedings IAU Symposium No. 317, 2015
A. Bragaglia, M. Arnaboldi, M. Rejkuba & D. Romano, eds.

doi:10.1017/S174392131500842X

Globular clusters as tracers of the halo assembly of nearby central cluster galaxies

Michael Hilker[1] **and Tom Richtler**[2]

[1]European Southern Observatory, Karl-Schwarzschild-Str. 2,
D-85748, Garching bei München, Germany
email: mhilker@eso.org

[2]Departamento de Astronomía, Universidad de Concepción, Concepción, Chile
email: tom@astroudec.cl

Abstract. The properties of globular cluster systems (GCSs) in the core of the nearby galaxy clusters Fornax and Hydra I are presented. In the Fornax cluster we have gathered the largest radial velocity sample of a GCS system so far, which enables us to identify photometric and kinematic sub-populations around the central galaxy NGC 1399. Moreover, ages, metallicities and [α/Fe] abundances of a sub-sample of 60 bright globular clusters (GCs) with high S/N spectroscopy show a multi-modal distribution in the correlation space of these three parameters, confirming heterogeneous stellar populations in the halo of NGC 1399. In the Hydra I cluster very blue GCs were identified. They are not uniformly distributed around the central galaxies. 3-color photometry including the U-band reveals that some of them are of intermediate age. Their location coincides with a group of dwarf galaxies under disruption. This is evidence of a structurally young stellar halo "still in formation", which is also supported by kinematic measurements of the halo light that point to a kinematically disturbed system. The most massive GCs divide into generally more extended ultra-compact dwarf galaxies (UCDs) and genuine compact GCs. In both clusters, the spatial distribution and kinematics of UCDs are different from those of genuine GCs. Assuming that some UCDs represent nuclei of stripped galaxies, the properties of those UCDs can be used to trace the assembly of nucleated dwarf galaxies into the halos of central cluster galaxies. We show via semi-analytical approaches within a cosmological simulation that only the most massive UCDs in Fornax-like clusters can be explained by stripped nuclei, whereas the majority of lower mass UCDs belong to the star cluster family.

Keywords. galaxies: halos, galaxies: kinematics and dynamics, galaxies: star clusters, galaxies: clusters: individual (Fornax, Hydra I)

---

1. Introduction

Central cluster galaxies host systems of thousands of globular clusters (GCs) which populate their halos out to several tens of effective radii. They are good probes to trace the assembly history of the diffuse and extended stellar halos residing in the cores of galaxy clusters. The colors and spectral line indices of GCs can be used to identify and characterize sub-populations of metal-poor and metal-rich as well as young GCs. Together with kinematic information from large radial velocity samples of GCs one can reconstruct the assembly history of different halo components.

The most nearby galaxy clusters, for which their central GC systems (GCSs) have been photometrically and/or spectroscopically studied in detail, are Virgo (e.g., Durrell *et al.* 2014, Romanowsky *et al.* 2012), Fornax (e.g., Bassino *et al.* 2006, Schuberth *et al.* 2010), Hydra I (e.g. Hilker 2002, Misgeld *et al.* 2011, Richtler *et al.* 2011), and Centaurus (e.g., Hilker 2003, Mieske *et al.* 2009). The number of radial velocity confirmed GCs around the central galaxies reaches 1000 GCs for M87 in Virgo and NGC 1399 in Fornax. The

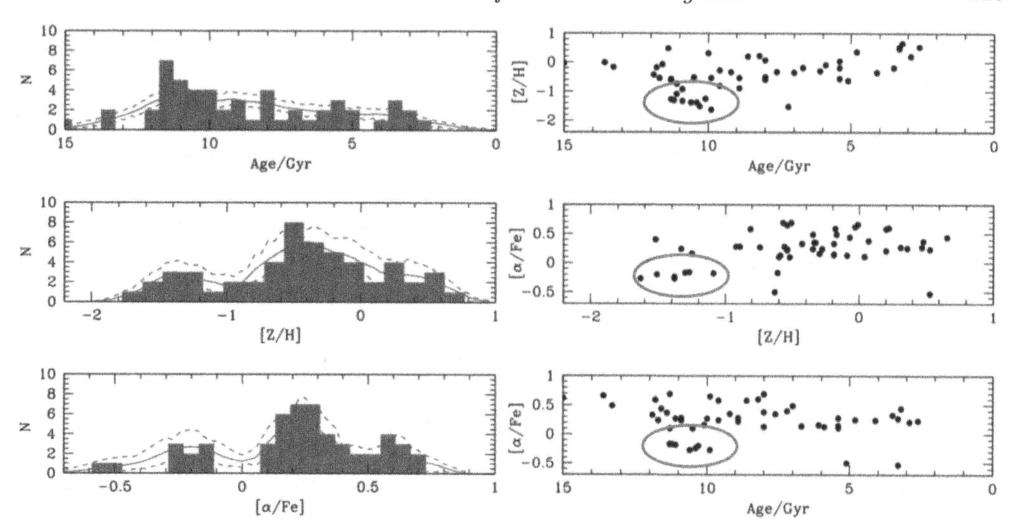

Figure 1. Left panels: distribution of ages, metallicities and [α/Fe] abundances for $\sim$ 50 Fornax GCs with high S/N spectroscopy. Right panels: correlations of the three parameters shown in the left. A distinct 'chemical' sub-group is highlighted by a red ellipse.

central GCSs are characterized by bimodal color distributions. Blue GCs are commonly interpreted to be metal-poor, and due to their extended spatial distribution are regarded as good tracers of the old, metal-poor halo of the central galaxies.

The high mass end of the GC mass function is dominated by so-called ultra-compact dwarf galaxies (UCDs), which first have been discovered in the core of the Fornax cluster (Minniti *et al.* 1998, Hilker *et al.* 1999, Drinkwater *et al.* 2000). It is under debate, which fraction of UCDs originate from stripped nucleated (dwarf) galaxies. The kinematic and stellar population properties of stripped nuclei-UCDs can be used to constrain the contribution of disrupted satellite galaxies to the build-up of central cluster galaxy halos. The most convincing case of a stripped nuclei origin of a UCD is the discovery of a supermassive black hole in M60-UCD1, one of the most massive and densest UCDs in the Virgo cluster (Seth *et al.* 2014). The supermassive black hole comprises 15% of the UCD's total mass.

2. The globular cluster system of the Fornax cluster

The Fornax cluster has a very well studied GC population. GC counts from photometric surveys revealed that there exist $\sim 6,450 \pm 700$ GCs within 83 kpc projected distance around the central cluster galaxy NGC 1399 (Dirsch *et al.* 2003). Within 300 kpc of NGC 1399 the GC number counts increase to $\sim 11,100 \pm 2,400$ GCs (Gregg *et al.* 2009, derived from the data of Bassino *et al.* 2006). The color distribution of GCs in Fornax is bimodal. The spatial distribution of red (mostly metal-rich) GCs follows the light of the central galaxy, whereas blue (mostly metal-poor GCs) are more widely distributed, suggesting that they trace the more extended halo population of NGC 1399.

Schuberth *et al.* 2010 analysed the kinematics of $\sim$ 700 GCs around NGC 1399 out to 80 kpc. They found that bimodality also exists in the kinematic properties of red and blue GCs. Red GCs are 'well-behaved', showing a gently decreasing velocity dispersion profile with increasing galactocentric distance. In contrast, blue GCs have a a generally higher velocity dispersion at all distances with apparent substructures in the velocity vs. distance diagram. This might be an imprint of recent galaxy interaction or merging events that leave accreted (mostly blue) GCs unmixed in phase space. Possible donor galaxies of accreted halo GCs are the nearby giant early-type galaxies NGC 1404 (see simulations by Bekki *et al.* 2003) and NGC 1387, but also dwarf galaxies that got entirely disrupted.

New spectroscopic surveys of GCs in the core region of the Fornax cluster are being analysed. A large VLT/VIMOS multi-object survey (PI: Capaccioli) covering the central square degree of the cluster will more than double the sample of radial velocity members. This will allow us to search for kinematic substructures in the halo of NGC 1399.

Deep VLT/FORS2 spectroscopy on ~ 50 bright GCs ($M_V < -9$ mag) allowed us to measure Lick indices and thus derive their ages, metallicities and $[\alpha/\mathrm{Fe}]$ abundances (Hilker, Puzia, *et al.*, in preparation). In Fig. 1 we show the distributions and correlations of these three quantities. Besides very old GCs there also exist a sizable number of metal-rich intermediate-age GCs (2-7 Gyr). And a striking feature in the correlation plots is a distinct group of seven old metal-poor GCs with sub-solar $[\alpha/\mathrm{Fe}]$ abundances. These GCs are restricted in projected galactocentric distance to NGC 1399 to a range of 18-31 kpc, five of them even to a range of 21-26 kpc, and thus might represent a 'chemo-dynamical sub-group' pointing to a common progenitor galaxy. Further kinematic analysis and correlations in phase space have to show whether this statement holds true.

3. Blue and *blue* globular clusters in the Hydra I cluster

The central galaxy NGC 3311 of the Hydra I cluster possesses a very rich GCS ($\sim 16,000$ GCs, Wehner *et al.* 2008) and a large population of UCDs (Misgeld *et al.* 2011). Dynamical analysis of 118 bright GCs and the light around NGC 3311 revealed a steeply rising velocity dispersion profile, reaching values of 800 km s^{-1} at 100 kpc galactocentric distance (Richtler *et al.* 2011), comparable to the velocity dispersion of the cluster galaxies. This might either point to a massive dark halo around NGC 3311 or indicate kinematic substructure in the halo that mimics a dark halo.

Hilker (2002, 2003) noticed that there exist very blue GC candidates around NGC 3311, with $0.70 < (V - I) = 0.85$ mag, much bluer than the blue peak of metal-poor GCs ($(V - I)_{\mathrm{blue}} = 0.9$). Those GCs are not centred on the galaxy but show an offset towards the north-east. Their blue $(V - I)$ color can be interpreted in two ways. Either these GCs are old but very metal-poor, with $[\mathrm{Fe/H}] \simeq -2.5$ dex, or they are metal-rich but rather young (1-5 Gyr). In order to break the age-metallicity degeneracy in the $(V - I)$ color, and thus uncover the nature of those blue GCs, we took U-band images in the cluster core with FORS at the VLT. In Fig. 2 we show the distribution of GCs in the $(U - V)$-$(V - I)$ 2-color diagram (left panels). With help of the PARSEC model grid for single stellar populations (Bressan *et al.* 2012) one can select GCs of different ages and/or metallicities and study their spatial distributions (right panels in Fig. 2). Whereas old GCs are mostly centred on NGC 3311, the distribution of young GCs (< 2 Gyr) is totally different. Most of them are displaced towards the North, coinciding with the location of a group of dwarf galaxies (Misgeld *et al.* 2008). Others are located south-east of NGC 3311 in the wake of the spiral galaxy NGC 3312, which is cruising through the Hydra I cluster at high speed, as evidenced by its compressed eastern edge.

The offset distribution of young, blue GCs coincides with the location of an offset faint stellar envelope around NGC 3311 (Arnaboldi *et al.* 2012), an offset X-ray halo (Hayakawa *et al.* 2004) and a region of high velocity dispersion in the halo light (Hilker *et al.*, in prep.). Taken all together, this points to a non-equlibrium situation in the core of Hydra I. The central galaxy seems not to be at rest with the cluster potential well, and infalling substructure is building up the central halo and intra-cluster light. Thus we are witnessing 'ongoing formation' of a central cluster halo.

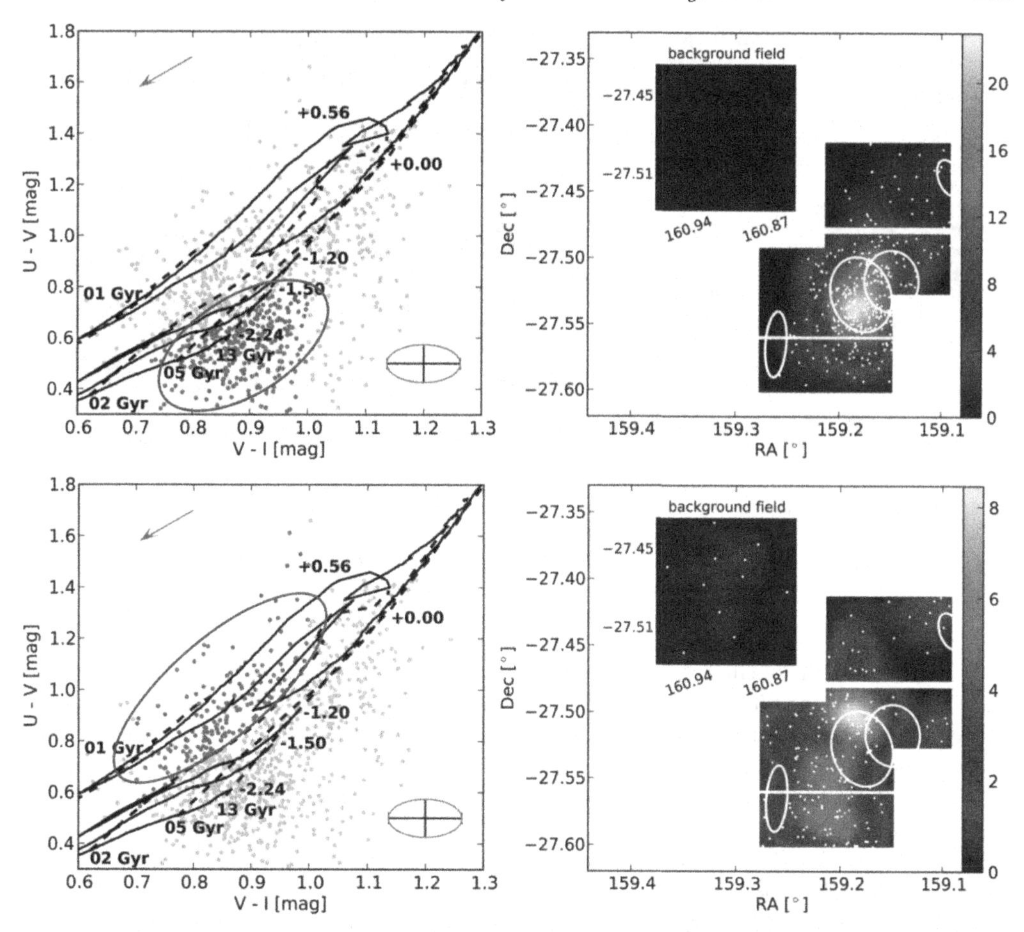

Figure 2. *Left panels*: GCs around NGC 3311 in Hydra I in the $(U-V)$-$(V-I)$ color-color space. Single stellar population tracks from PARSEC models (Bressan *et al.* 2012) for various ages (dashed lines) and metallicities (solid lines) are overlaid. In the upper panel, the region of old GCs (dark grey dots) is highlighted by a blue ellipse, in the lower panel the region for young (<2 Gyr) GCs. *Right panels*: spatial distribution of the selected old, metal-poor GCs (upper panel) and the young GCs (lower panel). White ellipses indicate the central major galaxies in Hydra I. The scale on the right are numbers per square arcmin. Note that the comparison background field is located several degrees east of the cluster.

4. The most massive GCs=UCDs: two formation channels

As mentioned in the introduction, very massive GCs cannot easily be distinguished from so-called ultra-compact dwarf galaxies (UCDs). One should rather think of the different formation channels that can lead to rather compact ($r_{\rm eff}$ = 3-100 pc) objects in the mass range $10^6 < M < 10^8 M_\odot$. One viable channel is the disruption of nucleated (dwarf) galaxies on radial orbits that pass the central cluster galaxies at small perigalactic radii and leave a 'naked' UCD-like stripped nucleus behind. However, cosmological simulations combined with a semi-analytic description to identify disrupted satellite galaxies suggest that this only explains the observed number of UCDs more massive than $M > 10^{7.3} M_\odot$ (Pfeffer *et al.* 2014). The observed number of lower mass UCDs is much larger than that of predicted stripped nuclei. They should, thus, be of star cluster origin, either formed as very massive genuine globular cluster (e.g., Murray 2009) or being the result of merged star cluster complexes (Fellhauer & Kroupa 2002).

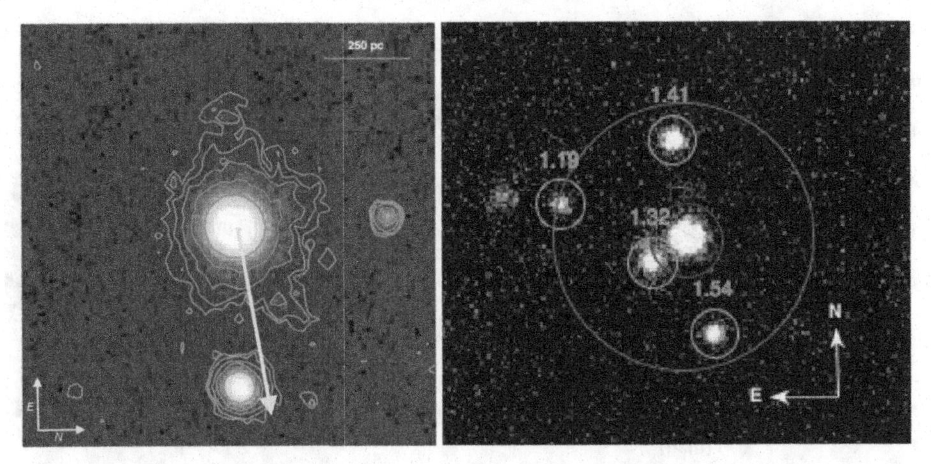

Figure 3. Left: UCD in the Fornax cluster that shows prominent tidal tail-like structures that extend out to 350 pc. The arrow indicates the direction towards the central Fornax galaxy NGC 1399. Right: Remote UCD (blue circle) at 85 kpc distance to NGC 1399 that harbours four GC candidates (green circles) within 1 kpc (red circle). The objects are labeled with their Washington $C - T_1$ colors (from Dirsch *et al.* 2003). The figures are taken from Voggel, Hilker & Richtler (2016).

In order to constrain the formation of UCDs we have studied the structural composition and clustering properties of 97 UCDs in the halo of NGC 1399, the central Fornax galaxy (see contribution by Voggel, Hilker & Richtler, this volume). We found evidence for faint stellar envelopes around several UCDs with effective radii of up 90 pc. One particularly extended UCD shows clear signs of tidal tails extending out to ~350 pc (see Fig. 3, left panel). This is the first time that a tidal tail has been detected around a UCD. But the most striking result is that we detect, in a statistical sense, a local overdensity of GCs on scales of $\leqslant 1$ kpc around UCDs. In particular blue (likely metal-poor) GCs are clustered around UCDs. These could either be remnant GCs of a formerly rich GCS around a disrupted nucleated dwarf galaxy, or surviving star clusters of a merged super star cluster complex (e.g., Brüns *et al.* 2009). A remote UCD, 85 kpc south of NGC 1399, possesses four GC candidates within 1 kpc radius, well within its tidal radius of 1.36 kpc, but shows no signs of a faint envelope in the same radius (see Fig. 3). The nature of this configuration is intriguing, pointing to a progenitor object that had a very rich substructure, maybe a complex star cluster system formed in- or outside a former host galaxy. Radial velocity measurements have to show whether the companion GCs are indeed bound to the host UCD.

5. Summary and outlook

Our general conclusions from the findings in the Fornax and Hydra I galaxy clusters shown in this contribution can be summarized as follows:

• In general, old globular clusters are good tracers of spheroid (red GCs) and halo (blue GCs) populations of ellipticals.

• The predominant GC population in the outer halo regions are the blue GCs. They trace the halo assembly history.

• Kinematics together with metal abundances of GCs is a powerful tool to find substructures and trace recent accretion events.

• In an appropriate 3-color space, sub-populations of blue GCs can be identified. Blue is not the same as *blue*!

• Ultra-compact dwarf galaxies (=the most massive star clusters) are a mixed bag of objects: most of them (>80%) are of star cluster origin.

• Extended stellar envelopes and overdensities of star clusters around them might hint to the accretion of nucleated dwarf galaxies or evolved super star cluster complexes.

In the coming years we will see great progress in extra-galactic globular cluster science. The Virgo and Fornax clusters are being scrutinized by photometric multi-wavelength wide-field surveys, covering the U- to the K-band. GCs and UCDs serve as one of the main tracers of the spatial distribution of baryonic structure in these clusters. Massive spectroscopic follow-up surveys are or will be launched to collect radial velocities and element abundances of thousands of GCs and UCDs around the central cluster galaxies in order to find chemo-dynamical substructures that constrain their halo assembly history.

References

Arnaboldi, M., Ventimiglia, G., Iodice, E., Gerhard, O., & Coccato, L. 2102, *A&A*, 545, 37

Bassino, L. P., Faifer, F. R., Forte, J. C., Dirsch, B., Richtler, T., Geisler, D., & Schuberth, Y. 2006, *A&A*, 451, 789

Bekki, K., Forbes, D. A., Beasley, M. A., & Couch, W. J. 2003, *MNRAS*, 344, 1344

Bressan, A., Marigo, P., Girardi, L., Salasnich, B., Dal Cero, C., Rubele, S., & Nanni, A. 2012, *MNRAS*, 427, 127

Brüns, R. C., Kroupa, P., & Fellhauer, M. 2009, *ApJ*, 702, 1268

Dirsch, B., Richtler, T., Geisler, D., Forte, J. C., Bassino, L. P., & Gieren, W. P. 2003, *AJ*, 125, 1908

Drinkwater, M. J., Jones, J. B., Gregg, M. D., & Phillipps, S. 2000, *PASA*, 17, 227

Durrell, P. R., Côté, P., Peng, E. W., Blakeslee, J. P., Ferrarese, L., *et al.* 2014, *ApJ*, 794, 103

Fellhauer, M. & Kroupa, P. 2002, *MNRAS*, 330, 642

Gregg, M.D., Drinkwater, M.J., Evstigneeva, E., Jurek, R., Karick, A.M., Phillipps, S., Bridges, T., Jones, J.B., Bekki, K., & Couch, W.J. 2009 *AJ*, 137, 498

Hayakawa, A., Furusho, T., Yamasaki, N. Y., Ishida, M., & Ohashi, T. 2004, *PASJ*, 56, 743

Hilker, M., Infante, L., Vieira, G., Kissler-Patig, M., & Richtler, T. 1999, *A&AS*, 134, 75

Hilker, M. 2002, in: D. Geisler, E.K. Grebel, & D. Minniti (eds.), *Extragalactic Star Clusters, IAU Symp. 207* (San Francisco: Astronomical Society of the Pacific), p. 281

Hilker, M. 2003, in: M. Kissler-Patig (ed.), *Extragalactic Globular Cluster Systems: Proceedings of the ESO Workshop, Garching* (Springer-Verlag), p. 173

Mieske, S., Hilker, M., Misgeld, I., Jordán, A., Infante, L., & Kissler-Patig, M. 2009, *A&A*, 498, 705

Minniti, D., Kissler-Patig, M., Goudfrooij, P., & Meylan, G. 1998, *AJ*, 115, 121

Misgeld, I., Mieske, S., & Hilker, M. 2008 *A&A*, 486, 697

Misgeld, I., Mieske, S., Hilker, M., Richtler, T., Georgiev, I. Y., & Schuberth, Y. 2011, *A&A*, 531, 4

Murray, N. 2009, *ApJ*, 691, 946

Pfeffer, J., Griffen, B. F., Baumgardt, H., & Hilker, M. 2014, *MNRAS*, 444, 3670

Richtler, T., Salinas, R., Misgeld, I., Hilker, M., Hau, G. K. T., Romanowsky, A. J., Schuberth, Y., & Spolaor, M. 2011, *A&A*, 531, 119

Romanowsky, A.J., Strader, J., Brodie, J.P., Mihos, J.C., Spitler, L.R., Forbes, D.A., Foster, C., & Arnold, J.A. 2012 *ApJ*, 748, 29

Schuberth, Y., Richtler, T., Hilker, M., Dirsch, B., Bassino, L.P., Romanowsky, A.J., & Infante, L. 2010 *A&A*, 513, 52

Seth, A. C., van den Bosch, R., Mieske, S., Baumgardt, H., & Brok, M. Den, *et al.* 2014, *Nature*, 513, 398

Voggel, K., Hilker, M., & Richtler, T. 2016 *A&A*, 586, A102

Wehner, E. M. H., Harris, W. E., Whitmore, B. C., Rothberg, B., & Woodley, K. A. 2008, *ApJ*, 681, 1233

The General Assembly of Galaxy Halos: Structure, Origin and Evolution
Proceedings IAU Symposium No. 317, 2015
A. Bragaglia, M. Arnaboldi, M. Rejkuba & D. Romano, eds.

doi:10.1017/S1743921315008479

Recent Results from SPLASH: Chemical Abundances and Kinematics of Andromeda's Stellar Halo

Karoline M. Gilbert[1], Rachael Beaton[2], Claire Dorman[3] and the SPLASH collaboration

[1]Space Telescope Science Institute,
3700 San Martin Dr., Baltimore, MD 21218, USA
email: kgilbert@stsci.edu

[2]The Observatories of the Carnegie Institution of Washington
813 Santa Barbara St., Pasadena, CA 91101, USA
email: rbeaton@obs.carnegiescience.edu

[3]UCO/Lick Observatory, University of California at Santa Cruz,
1156 High Street, Santa Cruz, CA 95064, USA
email: cdorman@ucolick.org

Abstract. Large scale surveys of Andromeda's resolved stellar populations have revolutionized our view of this galaxy over the past decade. The combination of large-scale, contiguous photometric surveys and pointed spectroscopic surveys has been particularly powerful for discovering substructure and disentangling the structural components of Andromeda. The SPLASH (Spectroscopic and Photometric Landscape of Andromeda's Stellar Halo) survey consists of broad- and narrow-band imaging and spectroscopy of red giant branch stars in lines of sight ranging in distance from 2 kpc to more than 200 kpc from Andromeda's center. The SPLASH data reveal a power-law surface brightness profile extending to at least two-thirds of Andromeda's virial radius (Gilbert *et al.* 2012), a metallicity gradient extending to at least 100 kpc from Andromeda's center (Gilbert *et al.* 2014), and evidence of a significant population of heated disk stars in Andromeda's inner halo (Dorman *et al.* 2013). We are also using the velocity distribution of halo stars to measure the tangential motion of Andromeda (Beaton *et al.*, in prep).

Keywords. galaxies: halos, galaxies: abundances, galaxies: individual (M31), galaxies: structure.

---

1. Introduction

The Milky Way (MW) and Andromeda (M31) provide the two best opportunities for in depth studies of stellar halos. Moreover, they provide complementary perspectives. While our internal vantage point provides us with exquisite detail of the inner halo of the MW, it is difficult to study the outer halo due to the large distance uncertainties, low stellar densities and vast survey areas required for even a partial view of the halo. Conversely, our external vantage point provides a global view of the halo of Andromeda. While Andromeda's distance has historically precluded measurements that are commonplace for the MWs halo, the last decade has seen tremendous progress. This has largely been due to the photometric and spectroscopic observations obtained by the PAndAS (Pan-Andromeda Archaeological Survey; McConnachie *et al.* 2009) and SPLASH (Spectroscopic and Photometric Landscape of Andromeda's Stellar Halo) collaborations.

The SPLASH collaboration has utilized the Mosaic camera on the Kitt Peak 4 m Mayall Telescope to obtain broad-band (M and T_2) and narrow-band (DDO51) imaging and the DEIMOS spectrograph on the Keck II 10 m telescope to obtain spectroscopy of

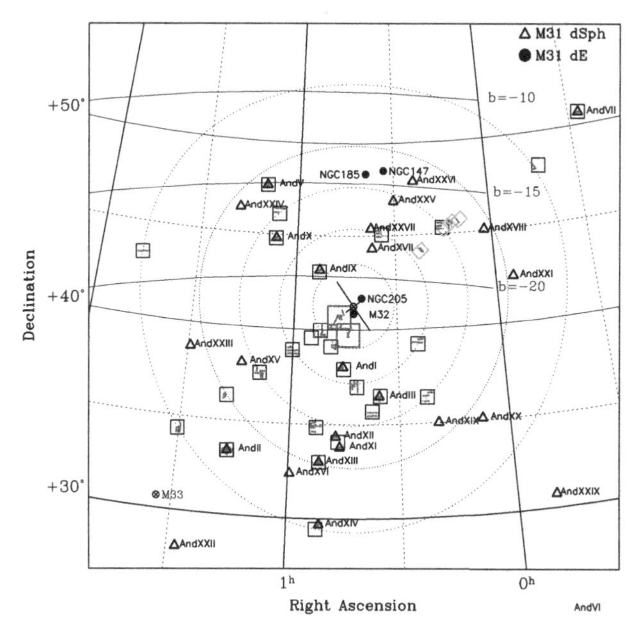

Figure 1. Location of the SPLASH photometric (larger rectangles) and spectroscopic (small, narrow rectangles) observations used in the measurement of the surface brightness and metallicity profiles of Andromeda's stellar halo. Figure from Gilbert *et al.* (2012).

individual M31 stars. The surface-gravity sensitive DDO51 imaging enables the selection of stars that are likely to be red giants at the distance of Andromeda, greatly increasing the efficiency of the spectroscopic observations. The SPLASH collaboration has imaged 78 fields and obtained $>20{,}000$ stellar spectra in Andromeda's disk, dwarf galaxies, and halo, in fields ranging from $2-230$ kpc in projected distance from Andromeda's center.

This dataset has led to the discovery and characterization of Andromeda's extended, metal-poor stellar halo (Guhathakurta *et al.* 2005, Gilbert *et al.* 2006, Kalirai *et al.* 2006a, & Courteau *et al.* 2011), and has been used to measure the global properties of the halo (Gilbert *et al.* 2012 & 2014) and to characterize the inner stellar halo and disk (Dorman *et al.* 2012, 2013, & 2015). It has been used to study the properties of Andromeda's dwarf satellites (Majewski *et al.* 2007, Kalirai *et al.* 2009, Howley *et al.* 2008, Kalirai *et al.* 2010, Tollerud *et al.* 2012, Ho *et al.* 2012 & Howley *et al.* 2013), to identify and characterize tidal debris features (Guhathakurta *et al.* 2006, Kalirai *et al.* 2006b, Gilbert *et al.* 2007, 2009a, 2009b), and led to the discovery of the continuation of Andromeda's giant southern stream (Gilbert *et al.* 2007, Fardal *et al.* 2008, 2012).

2. Surface Brightness Profile and Metallicity Gradient

SPLASH spectroscopic and photometric observations in 38 halo fields (Fig. 1) have been combined to measure the radial surface brightness and metallicity profile of Andromeda's stellar halo. These lines-of-sight span all quadrants of the halo and range from $9-230$ kpc in projected distance from Andromeda's center. The stellar spectra were used to identify secure samples of Andromeda red giant branch stars (removing Milky Way dwarf star contaminants; Gilbert *et al.* 2006) and to identify stars associated with kinematically cold tidal debris features (Gilbert *et al.* 2007, 2009b, & 2012).

The surface brightness profile of Andromeda's stellar halo is consistent with a single power-law extending to a projected distance of more than 175 kpc from Andromeda's

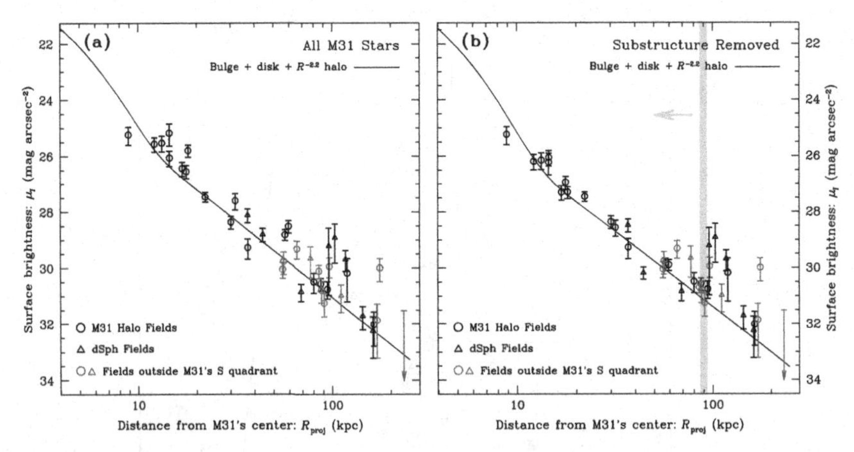

Figure 2. Surface brightness profile of Andromeda's stellar halo, with (left) and without (right) stars associated with kinematically cold tidal debris features. Andromeda's stellar halo is consistent with a single power-law with an index of -2.2 over a radial range of 10 – 175 kpc in projected distance from Andromeda's center. Tidal debris features have only been identified in fields less than 90 kpc from Andromeda's center: the low number of Andromeda stars in the outer halo fields ($R_{\rm proj} > 90$ kpc) prevent identification of multiple kinematical components. Figures are from Gilbert *et al.* (2012).

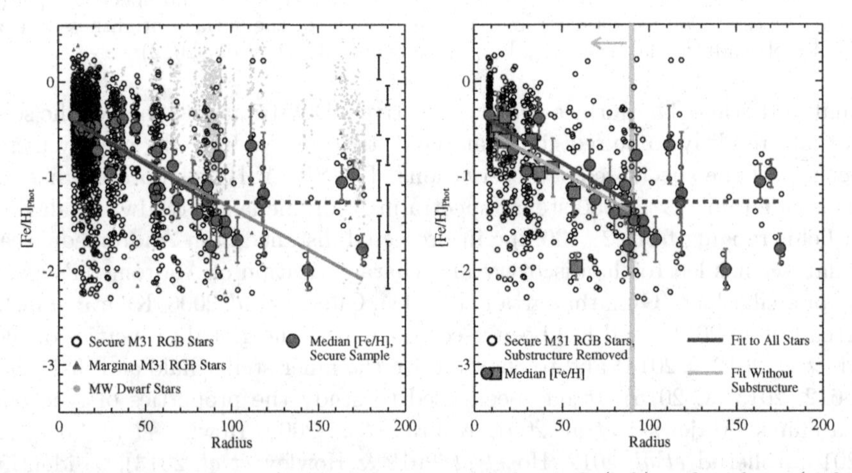

Figure 3. Metallicity profile of Andromeda's stellar halo, with (left) and without (right) stars associated with kinematically cold tidal debris features. As in Fig. 2, tidal debris features have only been identified and removed in fields less than 90 kpc from Andromeda's center. In the right panel, large squares denote the median [Fe/H] in the fields where tidal debris features have been identified and removed. Figures are from Gilbert *et al.* 2014.

center ($\sim 2/3$ the virial radius; Fig. 2). The surface brightness profile is derived from the ratio of counts of spectroscopically confirmed Andromeda red giant branch stars to Milky Way dwarf stars (Gilbert *et al.* 2012). The right panel shows the surface brightness profile once tidal debris features have been removed, using the results of maximum-likelihood, multi-Gaussian fits to each field's velocity distribution. Tidal debris features have been identified in half of the SPLASH spectroscopic fields within 90 kpc of Andromeda's center.

Andromeda's stellar halo shows clear evidence of a metallicity gradient, extending to $\sim$100 kpc (Fig. 3), with a total decrease of $\sim$1 dex (Gilbert *et al.* 2014). The [Fe/H] estimates are based on a comparison of the star's position in the color-magnitude diagram with stellar isochrones (assuming an age of 10 Gyr and the solar value of [α/Fe]). In the

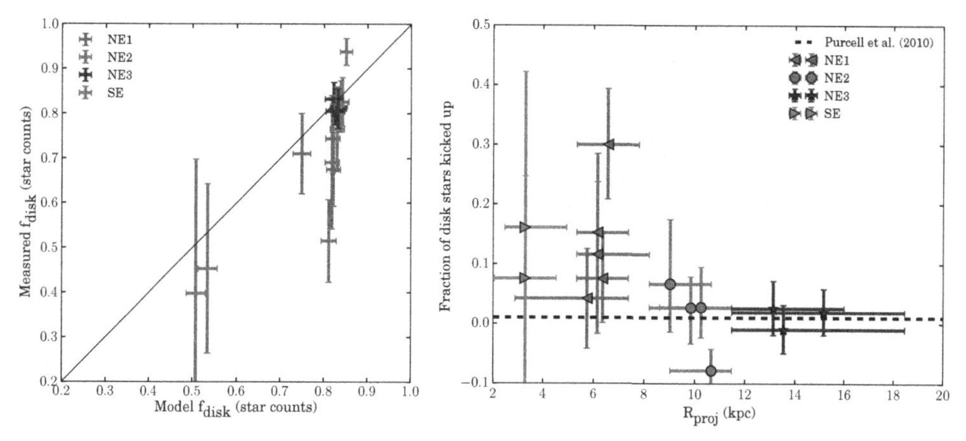

Figure 4. Evidence for a kicked-up disk component in Andromeda's inner halo. *Left:* Comparison of the disk fraction from a kinematical analysis of the velocity distribution of stars (measured $f_{\rm disk}$) with the disk fraction favored by a model that includes unresolved surface brightness data and the stellar luminosity function, as well as the stellar velocity distribution (model $f_{\rm disk}$). The disk fraction favored by the model is systematically higher than the measured dynamically cold disk fraction, suggesting there is a population of dynamically hot ("kicked-up") disk stars. *Right:* In order to reconcile the observed stellar luminosity function and the stellar kinematics, a fraction of the stars with a disk-like luminosity function must be dynamically hot. This panel shows the dynamically hot fraction of disk stars as a function of radius. Vertical error bars denote 1σ uncertainties while horizontal error bars show the full radial range of each spatial region shown in the figure. Figures are from Dorman *et al.* (2013).

right panel, stars that may be associated with tidal debris features have been removed, and a gradient of -1 dex over 100 kpc in radius remains. This shows that the metallicity gradient in Andromeda's stellar halo is *not* driven by the tidal debris features included in the SPLASH dataset. For a subset of spectra with the highest S/N, spectroscopic estimates of [Fe/H] based on the equivalent width of the Calcium Triplet were compared with the photometric estimates of [Fe/H]. On average the spectroscopic and photometric estimates agree, and the same gradient in [Fe/H] with radius is found.

Broad inferences about Andromeda's merger history can be made by comparing the surface brightness and metallicity profiles to simulations of stellar halo formation. The lack of a downward break in the surface brightness profile may indicate that Andromeda has undergone a fairly large number of recent, low-mass accretion events (Gilbert *et al.* 2012). The large-scale metallicity gradient extending over 100 kpc may indicate that the majority of the stars in the halo were contributed by one to a few early, relatively massive ($\sim 10^9\,M_\odot$) accretion events (see discussion in Gilbert *et al.* 2014). However, to make more concrete statements about the luminosity function and time of accretion of the satellites that formed Andromeda's stellar halo, we will need to measure the α-element abundances of halo stars. Vargas *et al.* (2014a,b) published the first [α/Fe] measurements of Andromeda stars (including 4 halo stars and measurements in 9 dwarf galaxies). The SPLASH dataset provides a rich archive for measuring [α/Fe] in many more halo fields.

3. Kinematics of M31's Inner Spheroid

Dorman *et al.* (2013) combined stellar kinematics from SPLASH (>5000 spectra), the stellar luminosity function derived from resolved HST imaging (from the PHAT survey; Dalcanton *et al.* 2012), and unresolved surface photometry in Andromeda's disk to model the relative strength of each of the structural components (bulge, disk, and halo) as a

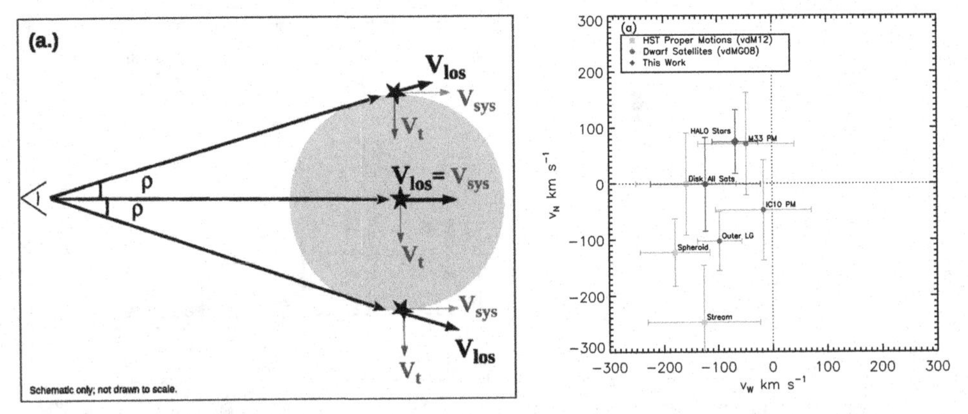

Figure 5. *Left:* Schematic illustration of the center-of-mass motion of an object in the sky plane along two dimensions — v_{sys} along the line-of-sight to the center of mass and v_t perpendicular to v_{sys} — and the effect of observing tracers at lines-of-sight at widely separated angles, ρ. For a tracer at $\rho \sim 0$, the line-of-sight of the center of mass and of the tracer are aligned, whereas for $\rho > 0$, this is no longer the case (i.e., $v_{los} \neq v_{sys}$) and the magnitude of the difference depends on both the radial (ρ) and azimuthal (ϕ) location of the tracer. *Right:* Measurements of Andromeda's tangential motion; v_W and v_N are the plane of sky motions to the West and North. Measurements from different sources are compared: direct HST proper motions (square) and statistical tests using the line-of-sight velocities of dwarf satellite galaxies (circles) and halo stars (diamond). In the on-line proceedings, each point is color-coded based on the source of the measurement: van der Marel *et al.* (2012; yellow), van der Marel & Guhathakurta (2008; red), or Beaton *et al.* (in prep; blue).

function of radius. The best-fitting models favor a disk fraction that is systematically higher than the disk fraction measured solely from an analysis of the stellar velocity dispersion (Fig. 4). The number of stars with a disk-like luminosity function is 5.2%±2.1% larger than the number of stars with dynamically cold, disk-like kinematics. This implies Andromeda's inner regions contain a population of stars with a disk-like luminosity function but spheroid-like kinematics. This is the first direct evidence that there are stars that were born in Andromeda's disk and subsequently dynamically heated into the halo; they are now kinematically indistinguishable from Andromeda's inner stellar halo. The number of stars that have been dynamically heated are consistent with that expected from simulations (see discussion in Dorman *et al.* 2013).

4. Tangential Motion of M31 From Halo Stars

Knowledge of the orbit of Andromeda is central to understanding the past, current, and future dynamical state of the Local Group. This requires determining Andromeda's three dimensional motion, an extremely challenging measurement. Recently, Sohn *et al.* (2012) used HST imaging obtained over a long time baseline to measure Andromeda's tangential motion via the observed proper motion of Andromeda's stars.

Indepedent confirmation of Andromeda's transverse velocity can be made using the kinematics of halo populations. This technique relies on the projection of the three dimensional center-of-mass motion on the line-of-sight motions of bound objects (Fig. 5, *left*). This technique is most powerful when applied over the large spatial baseline provided by tracer populations in Andromeda's halo: dwarf satellites, globular clusters (van der Marel & Guhathakurta 2008, van der Marel *et al.* 2012) and halo stars.

Beaton *et al.* (in prep) are employing the SPLASH dataset to make the first measurement of Andromeda's tangential motion using line-of-sight velocities of individual halo stars. We also incorporate the kinematics of Andromeda's dwarf satellites and halo globular clusters. Our model's only assumption is that the tracer stars come from a single virialized component (van der Marel & Guhathakurta 2008). Following Gilbert *et al.* (2012), we exclude stars likely to belong to dwarf satellites or tidal debris features. The greatest potential source of bias is from unidentified substructure. However, the magnitude of this effect can be estimated empirically from the data and is added to the total uncertainty. We find broad agreement between our measurement of Andromeda's tangential motion and previous measurements (Fig. 5, *right*). The error-weighted mean of all independent measurements is (v_W, v_N) = (-96.4 ± 22.4 km s^{-1}, -44.7 ± 25.4).

5. Summary

The SPLASH collaboration has obtained a rich dataset of photometric and spectroscopic observations spanning the Andromeda system. Among the many results from the SPLASH survey, four have been highlighted here. This dataset has been used to measure the surface brightness and metallicity profile to two-thirds of Andromeda's virial radius (Gilbert *et al.* 2012 & 2014). In conjunction with the PHAT HST imaging of Andromeda's disk, it has been used to show that there is a significant population of kicked up disk stars in Andromeda's inner stellar halo (Dorman *et al.* 2013). Finally, it is providing an independent measurement of Andromeda's tangential motion (Beaton *et al.*, in prep).

References

Courteau, S., Widrow, L. M., McDonald, M., *et al.* 2011, *ApJ*, 739, 20
Dalcanton, J. J., Williams, B. F., Lang, D., *et al.* 2012, *ApJ*, 200, 18
Dorman, C. E., Guhathakurta, P., & Fardal, M. A. *et al.* 2012, *ApJ*, 752, 147
Dorman, C. E., Guhathakurta, P., & Seth, A. C. *et al.* 2015, *ApJ*, 803, 24
Dorman, C. E., Widrow, L. M., Guhathakurta, P., *et al.* 2013, *ApJ*, 779, 103
Fardal, M. A., Babul, A., Guhathakurta, P., Gilbert, K. M., & Dodge, C. 2008, *ApJL*, 682, L33
Fardal, M. A., Guhathakurta, P., Gilbert, K. M., *et al.* 2012, *MNRAS*, 423, 3134
Gilbert, K. M., Fardal, M., Kalirai, J. S., *et al.* 2007, *ApJ*, 668, 245
Gilbert, K. M., Font, A. S., Johnston, K. V., & Guhathakurta, P. 2009, *ApJ*, 701, 776
Gilbert, K. M., Guhathakurta, P., Beaton, R. L., *et al.* 2012, *ApJ*, 760, 76
Gilbert, K. M., Guhathakurta, P., Kalirai, J. S., *et al.* 2006, *ApJ*, 652, 1188
Gilbert, K. M., Guhathakurta, P., Kollipara, P., *et al.* 2009b, *ApJ*, 705, 1275
Guhathakurta, P., Ostheimer, J. C., Gilbert, K. M., *et al.* 2005, ArXiv (arXiv:astro-ph/0502366)
Guhathakurta, P., Rich, R. M., Reitzel, D. B., *et al.* 2006, *AJ*, 131, 2497
Ho, N., Geha, M., Munoz, R. R., *et al.* 2012, *ApJ*, 758, 124
Howley, K. M., Geha, M., Guhathakurta, P., *et al.* 2008, *ApJ*, 683, 722
Howley, K. M., Guhathakurta, P., van der Marel, R., Geha, M., *et al.* 2013, *ApJ*, 765, 65
Kalirai, J. S., Beaton, R. L., Geha, M. C., *et al.* 2010, *ApJ*, 711, 671
Kalirai, J. S., Gilbert, K. M., Guhathakurta, P., *et al.* 2006a, *ApJ*, 648, 389
Kalirai, J. S., Guhathakurta, P., Gilbert, K. M., *et al.* 2006b, *ApJ*, 641, 268
Kalirai, J. S., Zucker, D. B., Guhathakurta, P., *et al.* 2009, *ApJ*, 705, 1043
Majewski, S. R., Beaton, R. L., Patterson, R. J., Kalirai, J. S., *et al.* 2007, *ApJL*, 670, L9
McConnachie, A. W., Irwin, M. J., Ibata, R. A., *et al.* 2009, *Nature*, 461, 66
Sohn, S. T., Anderson, J., & van der Marel, R. P. 2012, *ApJ*, 753, 7
Tollerud, E. J., Beaton, R. L., Geha, M. C., *et al.* 2012, *ApJ*, 752, 45
van der Marel, R. P. & Guhathakurta, P. 2008, *ApJ*, 678, 187
van der Marel, R. P., Fardal, M., Besla, G., *et al.* 2012, *ApJ*, 753, 8
Vargas, L. C., Geha, M. C., & Tollerud, E. J. 2014a, *ApJ*, 790, 73
Vargas, L. C., Gilbert, K. M., Geha, M., *et al.* 2014b, *ApJL*, 797, L2

The General Assembly of Galaxy Halos: Structure, Origin and Evolution
Proceedings IAU Symposium No. 317, 2015
A. Bragaglia, M. Arnaboldi, M. Rejkuba & D. Romano, eds.

doi:10.1017/S1743921315006742

Globular Cluster Streams as Galactic High-Precision Scales

Andreas H.W. Küpper[1]†, Eduardo Balbinot[2], Ana Bonaca[3], Kathryn V. Johnston[1], David W. Hogg[4], Pavel Kroupa[5] and Basilio X. Santiago[6,7]

[1]Department of Astronomy, Columbia University,
550 West 120th Street, New York, NY 10027, USA
email: akuepper@astro.columbia.edu
[2]Department of Physics, University of Surrey,
Guildford GU2 7XH, UK
[3]Department of Astronomy, Yale University,
New Haven, CT 06511, USA
[4]Center for Cosmology and Particle Physics, Department of Physics, New York University,
4 Washington Place, New York, NY 10003, USA
[5]Helmholtz-Institut für Strahlen- und Kernphysik (HISKP), University of Bonn,
Nussallee 14-16, 53115 Bonn, Germany
[6]Departamento de Astronomia, Universidade Federal do Rio Grande do Sul,
Av. Bento Gonçalves 9500, Porto Alegre 91501-970, RS, Brasil
[7]Laboratório Interinstitucional de e-Astronomia - LIneA,
Rua Gal. José Cristino 77, Rio de Janeiro, RJ - 20921-400, Brasil

Abstract. Tidal streams of globular clusters are ideal tracers of the Galactic gravitational potential. Compared to the few known, complex and diffuse dwarf-galaxy streams, they are kinematically cold, have thin morphologies and are abundant in the halo of the Milky Way. Their coldness and thinness in combination with potential epicyclic substructure in the vicinity of the stream progenitor turns them into high-precision scales. With the example of Palomar 5, we demonstrate how modeling of a globular cluster stream allows us to simultaneously measure the properties of the disrupting globular cluster, its orbital motion, and the gravitational potential of the Milky Way.

Keywords. methods: numerical, Galaxy: halo, Galaxy: structure, globular clusters: individual (Palomar 5), dark matter

---

1. Why thin globular cluster streams are so valuable

Within the past decade, the number of wide-field imaging surveys and spectroscopic campaigns has grown exponentially. In the vast amount of deep, high-quality data that has become available, a multitude of thin and cold stellar streams has been discovered. Due to their faintness, all of these thin streams were found in the halo of the Milky Way, and most probably originate from disrupting or disrupted globular clusters (e.g., Bonaca, Geha & Kallivayalil 2012, Grillmair *et al.* 2013, Bernard *et al.* 2014, Koposov *et al.* 2014).

Similar to the longer, but more diffuse, dwarf galaxy streams like the Sagittarius stream (e.g., Johnston *et al.* 2005, Law & Majewski 2010), thin globular cluster streams (GCS) are valuable tracers of the Galactic gravitational potential (Bonaca *et al.* 2014). Their coherence in phase space makes them ideal instruments for measuring the mass and shape of the otherwise invisible dark halo of the Galaxy, as has been demonstrated by Koposov *et al.* (2009) with the example of GD-1.

† Hubble Fellow

But the pure existence of thin streams in the halo of the Milky Way tells us even more: Pearson *et al.* (2015) found that GCS are very sensitive to the non-sphericity of the gravitational potential of their host galaxy. The authors showed that triaxial halo configurations can cause stream fanning – a broadening and diffusion of the stream perpendicular to the orbital motion of the satellite. Stream fanning makes the already faint and cold GCS harder to detect in imaging surveys, as their surface density is pushed beyond the detection limit. In a systematic study of orbits within a triaxial galaxy potential, Price-Whelan *et al.* (2015) demonstrated that thin streams can only occupy specific (regular) regions of orbital space. The existence of the many observed thin streams will, therefore, tell us something about the shape of the Milky Way's gravitational potential.

Furthermore, kinematically cold GCS are powerful potential probes as they inhibit dynamical substructure caused by apparent epicyclic motion of stars escaping the gravitational potential of the cluster (Küpper, MacLeod & Heggie 2008, Just *et al.* 2009). This substructure can be well understood, as it solely depends on the mass of the globular cluster, its orbital motion, and the shape of the galactic gravitational potential (Küpper, Lane & Heggie 2012). Long, thin streams of globular clusters that exhibit substructure are therefore high-precision scales of the host galaxy potential. The high achievable precision is due to the unique properties of GCS that let us accurately constrain their progenitor's stellar mass and orbit within the Galaxy:

(*a*) since globular clusters have simple compositions compared to dwarf galaxies, the progenitors of GCS can be well characterized. Mass estimates are accurate up to the uncertainties of globular cluster mass-to-light ratios, i.e., about a factor of two,

(*b*) the simple stellar compositions of globular clusters, furthermore, allow for a clearer separation of GCS stars from fore- and background contaminations,

(*c*) the relatively small widths of GCS enables a precise location of the streams, and

(*d*) their cold compositions (i.e., velocity spread among stream members) allows for accurate velocity information along the GCS.

GCS with epicyclic overdensities therefore contain a lot of different information (cluster mass, orbital motion, host gravitational potential), which has to be decoded and disentangled via modeling. The more information is available on the mass of the respective globular cluster, its orbital motion or the host gravitational potential, the better we can constrain the other components. We demonstrated this for the Milky Way globular cluster Palomar 5.

2. Modeling Palomar 5 and its tidal stream

The Milky Way globular cluster Palomar 5 (Pal 5) shows a thin, > 20 deg long tidal stream, which was first discovered by Odenkirchen *et al.* (2001) in commissioning data of the Sloan Digital Sky Survey (SDSS). A detailed review of the available observational data on Pal 5 and its stream can be found in Küpper *et al.* (2015).

In this publication, we extensively modeled Pal 5 and its tidal stream. We used a difference-of-Gaussian procedure to detect the densest and most significant regions within the Pal 5 stream. This ansatz also allowed us to locate potential epicyclic overdensities within the prominent tidal stream. We combined this surface density information with radial velocity measurements along the stream from Odenkirchen *et al.* (2009). Both over-dense regions and radial velocity measurements are shown in Fig. 1.

Similar to the FAST FORWARD method developed in Bonaca *et al.* (2014), we used streakline models of the Pal 5 stream to evaluate the likelihood of a given set of model parameters. Our stream models encompassed 10 free parameters, describing the progenitor globular cluster (mass, mass-loss rate, distance, proper motion), the position and

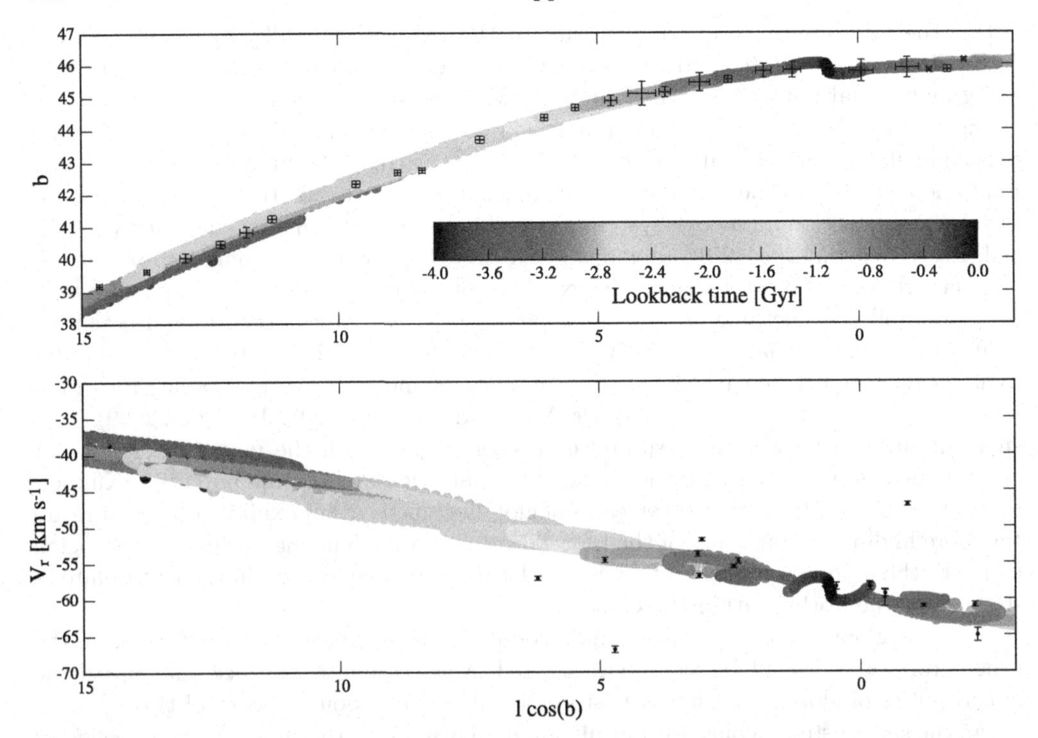

Figure 1. Streakline model of Pal 5 consisting of 2 × 4000 test particles that were released from the progenitor in intervals of 1 Myr. The color coding in both panels shows their release times. Upper panel: projection of the stream on the sky. Black data points are locally over-dense surface density regions of Pal 5-like stars in SDSS data. Lower panel: radial velocity gradient along the stream with measured velocities of red giants lying in projection within the stream from Odenkirchen *et al.* (2009).

motion of the Sun, and the properties of the Galactic gravitational potential (mass, size, and q_z – the flattening perpendicular to the Galactic disk). Posterior probability distributions of the parameters were obtained through Markov-chain Monte Carlo sampling using the freely available code *emcee* (Foreman *et al.* 2013).

We were able to tightly constrain all 10 model parameters and give uncertainties for each one, which demonstrates the power of substructured GCS when modeled with streaklines in a Bayesian framework. For example, we found the shape of the Galactic halo potential within the inner 19 kpc with a value of $q_z = 0.95^{+0.16}_{-0.12}$ close to being spherical. Koposov *et al.* (2009) came to a similar conclusion fitting orbits to the significantly longer, but fainter GD-1 stream. With three free model parameters, they were able to rule out a halo flattening smaller than $q_z = 0.89$ with 90% confidence, but could not give an upper limit on its possible prolateness. Similarly, results from modeling of the long and diffuse Sagittarius dwarf galaxy stream vary between strongly prolate (Helmi 2004) and oblate (Johnston *et al.* 2005), both without uncertainty estimates. Our modeling of Pal 5 is therefore a significant improvement over previous investigations.

3. Outlook

New observational data on Pal 5 and its stream will help to inform and improve further modeling of the stream. For example, Kuzma *et al.* (2014) published 39 additional radial velocity measurements of red giants along the Pal 5 stream. Moreover, Fritz & Kallivayalil

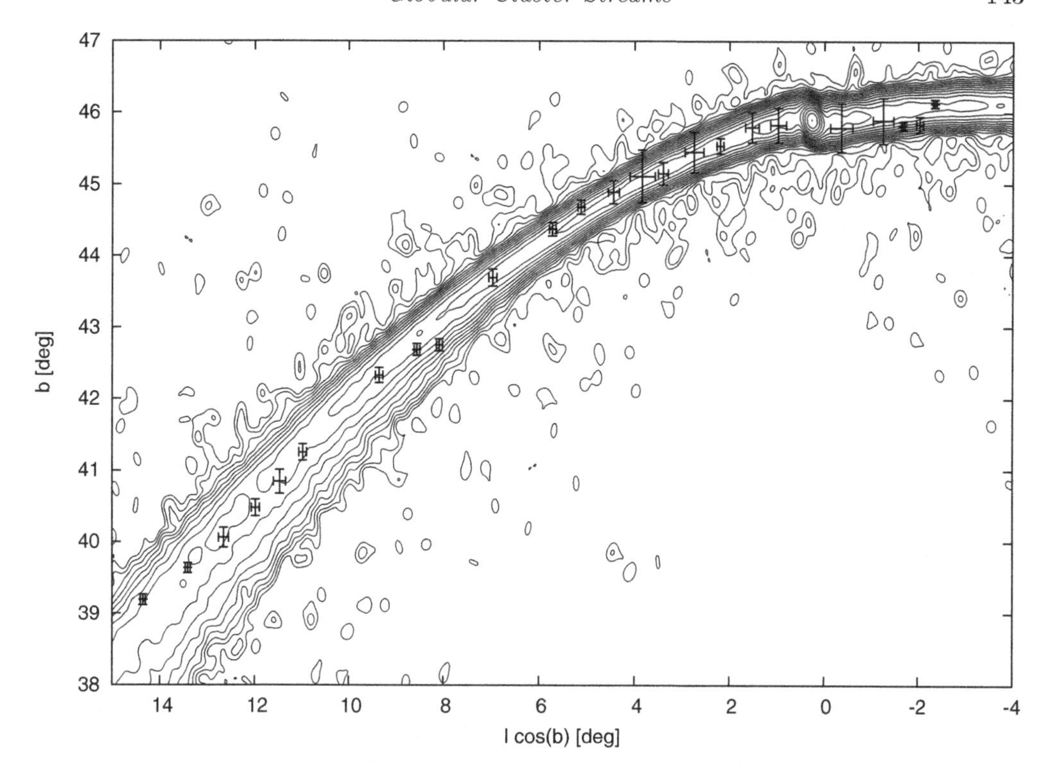

Figure 2. Contour density map of one of our first N-body models of Pal 5. Contour levels are equally spaced in log density. Black data points are the same as in Fig. 1. Like the streakline model, the N-body stream is thin along the whole extent, and is most collimated at about $l\cos(b) = 4$ deg, where the stream shows its most pronounced overdensities in SDSS data.

(2015) obtained 15 year baseline proper motions of Pal 5 from SDSS data in combination with LBT imaging data. Further data from, e.g., HST, Spitzer, Keck and CTIO will give us an even deeper and more detailed look on the Pal 5 stream and its kinematics in the near future.

But we can already use the constraints from our current streakline modeling on Pal 5's orbit within the Galactic tidal field to verify our understanding of the destruction of the cluster. Using the parameters of our best-fit models, we are currently following up our simplified stream modeling with accurate N-body models of the disrupting Pal 5. We aim at extending previous investigations on Pal 5's evolution by Dehnen *et al.* (2004) and Mastrobuono-Battisti *et al.* (2012), to link the internal evolution of the cluster to the appearance of its tidal stream.

Our first N-body models with the GPU-enabled, direct N-body code NBODY6 (Aarseth 2003) show good agreement with the overall properties of the streakline models (Fig. 2), but also give us new exciting challenges. Internal properties of Pal 5, such as its half-light radius, are observationally well determined and will give us further constraints on the evolution of Pal 5 (cf., Zonoozi *et al.* 2011; Zonoozi *et al.* 2014). Ultimately, full N-body modeling of Pal 5 and its stream will give us the unique opportunity to understand the birth, life and death of a globular cluster in unprecedented detail.

References

Aarseth, S. J. 2003, *Gravitational N-Body Simulations* (Cambridge University Press)

Bernard, E. J., Ferguson, A. M. N., Schlafly, E. F., *et al.* 2014, *MNRAS*, 443, L84
Bonaca, A., Geha, M., & Kallivayalil, N. 2012, *ApJ* (Letters), 760, L6
Bonaca, A., Geha, M., Küpper, A. H. W., Diemand, J., Johnston, K. V., & Hogg, D. W. 2014, *ApJ*, 795, 94
Dehnen, W., Odenkirchen, M., Grebel, E. K., & Rix, H.-W. 2004, *AJ*, 127, 2753
Foreman-Mackey, D., Hogg, D. W., Lang, D., & Goodman, J. 2013, *PASP*, 125, 306
Fritz, T. K. & Kallivayalil, N. 2015, *ApJ*, 811, 123
Grillmair, C. J., Cutri, R., Masci, F. J., Conrow, T., Sesar, B., Eisenhardt, P. R. M., & Wright, E. L. 2013, *ApJ* (Letters), 769, L23
Helmi, A. 2004, *ApJ*, 610, L97
Johnston, K. V., Law, D. R., & Majewski, S. R. 2005, *ApJ*, 619, 800
Just, A., Berczik, P., Petrov, M. I., & Ernst, A. 2009, *MNRAS*, 392, 969
Koposov, S. E., Rix, H.-W., & Hogg, D. W. 2010, *ApJ*, 712, 260
Koposov, S. E., Irwin, M., Belokurov, V., *et al.* 2014, *MNRAS*, 442, L85
Küpper, A. H. W., MacLeod, A., & Heggie, D. C. 2008, *MNRAS*, 387, 1248
Küpper, A. H. W., Lane, R. R., & Heggie, D. C. 2012, *MNRAS*, 420, 2700
Küpper, A. H. W., Balbinot, E., Bonaca, A., Johnston, K. V., Hogg, D. W., Kroupa, P., & Santiago, B. X. 2015, *ApJ*, 803, 80
Kuzma, P. B., Da Costa, G. S., Keller, S. C., & Maunder, E. 2015, *MNRAS*, 446, 3297
Law, D. R. & Majewski, S. R. 2010, *ApJ*, 714, 229
Mastrobuono-Battisti, A., Di Matteo, P., Montuori, M., & Haywood, M. 2012, *A&A*, 546, L7
Odenkirchen, M., Grebel, E. K., Rockosi, C. M., *et al.* 2001, *ApJ* (Letters), 548, L165
Odenkirchen, M., Grebel, E. K., Kayser, A., Rix, H.-W., & Dehnen, W. 2009, *AJ*, 137, 3378
Pearson, S., Küpper, A. H. W., Johnston, K. V., & Price-Whelan, A. M. 2015, *ApJ*, 799, 28
Price-Whelan, A. M., Johnston, K. V., Valluri, M., Pearson, S., Küpper, A. H. W., & Hogg, D. W. 2016, *MNRAS*, 455, 1079
Zonoozi, A. H., Küpper, A. H. W., Baumgardt, H., Haghi, H., Kroupa, P., & Hilker, M. 2011, *MNRAS*, 411, 1989
Zonoozi, A. H., Haghi, H., Küpper, A. H. W., Baumgardt, H., Frank, M. J., & Kroupa, P. 2014, *MNRAS*, 440, 3172

The General Assembly of Galaxy Halos: Structure, Origin and Evolution
Proceedings IAU Symposium No. 317, 2015
A. Bragaglia, M. Arnaboldi, M. Rejkuba & D. Romano, eds.

doi:10.1017/S1743921315008364

Stellar kinematics and dark matter in dwarf galaxies

Giuseppina Battaglia[1,2]

[1]Instituto de Astrofisica de Canarias,
38205 La Laguna, Tenerife, Spain
email: gbattaglia@iac.es

[2]Universidad de La Laguna, Dpto. Astrofisica,
38206 La Laguna, Tenerife, Spain

Abstract. In this review I will discuss the current status on determinations of the dark matter content and distribution in Milky Way dwarf spheroidals, for which the available data-sets allow the application of sophisticated mass modeling techniques.

Keywords. stars: kinematics; galaxies: dwarf; galaxies: halos; Local Group; dark matter

1. Background

The surroundings of the Milky Way (MW) are populated by a wealth of small passively-evolving satellite galaxies: the "classical" dwarf spheroidal galaxies (dSphs), whose existence has been known since a very long time (e.g. the Sculptor dSph was discovered by Shapley 1938), and the so-called "ultra-faint dwarfs" (UFDs), whose discovery we mainly owe to the SDSS.

If in dynamical equilibrium, "classical" dSphs have the highest mass-to-light ratios known to date, with $M/L \sim$ 100s $M_{\odot}/L_{\odot}$. At face value, UFDs exhibit even larger dynamical mass-to-light ratios, although these values are subject to numerous important caveats. Hence, in terms of their internal dynamics, these systems might well be key in constraining the nature of dark matter (DM).

Aim of several studies in the literature has been to compare the DM content and distribution inferred from the internal kinematics of this population of small galactic systems to the predictions from cosmological theories of structure formation in a DM context. In particular, ΛCold Dark Matter pure DM N-body simulations show that the density profile of the haloes formed follow very specific functional forms, which are rather steep near the centre (such as the Navarro, Frenk & White profile, NFW, Navarro *et al.* 1995, 1996, and more recently the Einasto form: Navarro *et al.* 2010 and references therein). These high-resolution simulations give robust predictions also for the sub-haloes mass function, which in the case of MW-sized main halos is established down to a simulations' resolution limit smaller than the mass estimates for the faintest dSphs (see e.g. Springel *et al.* 2008).

However, making a link between a luminous satellite to what should be its corresponding sub-halo in a pure DM N-body simulation is not a straightforward matter. It has in fact become increasingly clear that the inclusion of baryonic effects in such simulations can have a profound impact on the properties of the sub-haloes that will become luminous satellites, such as reducing the central dark matter density, enhancing tidal stripping etc. (e.g. Zolotov *et al.* 2012). On the other hand, this sensitivity to baryonic effects might be useful for constraining model ingredients, such as for example the efficiency of supernovae feedback (e.g. Breddels, Vera-Ciro & Helmi, 2015).

Independently of the comparison with cosmological simulations, determinations of the DM content and distribution of dSphs are central also to interpret results from particle physics experiments of indirect dark matter detection. For example, the γ-ray differential flux from DM decay (annihilation) over a solid-angle can be thought of as the multiplication of two factors, one dependent on the characteristics of the DM particle and the other one being the so-called "astrophysical factor" (Bergström *et al.* 1998). The latter is simply the integral along the line-of-sight (LOS) and over the solid angle of the (squared, in the case of annihilation) DM density. Hence robust determinations of the DM content and distribution of the targeted dSph result in a well constrained astrophysical contribution to the measured signal (see e.g. Bonnivard *et al.* 2015 for the impact of sample sizes, presence of Galactic contaminants in the sample etc. on knowledge of the dSphs' "astrophysical factor").

In the following I will focus only on the MW "classical dSphs"† as these are the objects for which the largest spectroscopic samples of LOS velocities for individual probable member stars are available (e.g. ~2000 and 3000 in Sculptor and Fornax, respectively), which - beside a more robust determination of the "observables" - also allows the application of sophisticated mass modeling techniques. For thorough discussions on the observed internal kinematics of these galaxies, the techniques used to extract information on their gravitational potential, implication of the findings and related uncertainties we refer the reader to recent review articles by Battaglia, Helmi & Breddels (2013), Walker (2013), Strigari (2013). Here I will briefly summarize some of the salient findings in studies of the DM content and distribution of MW "classical" dSphs, with a focus on the assumptions that have been made and their potential impact on the results, closely following the article by Battaglia, Helmi & Breddels (2013).

1.1. *Some notes*

Determining the mass content of a system requires observations of the kinematics of suitable tracers. Since dSphs are devoid of a neutral interstellar medium, the only tracers available are stars. The heliocentric distances to MW dSphs have made it unfeasible to obtain accurate proper motions of individual stars in these galaxies with current facilities. Hence to-date the analyses of their internal kinematics are based on their LOS velocity distributions (LOSVD) and their moments.

Since the internal kinematics of dSphs is found to be dominated by random motions, they are commonly treated as non-rotating systems. Another common assumption, less justified from an observational point of view (see Sect. 3), is to regard both the stellar and DM component of dSphs as spherical.

The above, together with the lack of knowledge on the other two components of the LOS velocity vector of the individual stars, causes the so-called "mass-anisotropy" degeneracy, a well-known and cumbersome issue in the dynamical modeling of these objects. It is important to remember that the "mass-anisotropy" degeneracy is a degeneracy between the mass *profile*, the anisotropy *profile* $\beta(r)$ and the tracer density distribution.

Finally, the majority of mass modeling studies of dSphs assumes these galaxies are in dynamical equilibrium, hence that the internal kinematics of the tracer population can be safely used to infer the underlying gravitational potential.

In Sect. 2 I will describe results from mass modeling studies carried out under the above assumptions, and mainly comment on the efforts that have been made to relieve or circumvent the "mass-anisotropy" degeneracy. In Sect. 3 I will comment on the possible effect of relaxing the assumption of sphericity and on the possible impact of tidal effects

† Hereafter I will use the terms "classical dSphs" and "dSphs" as interchangeable

onto the internal kinematics of dSphs's. Newtonian dynamics will be assumed throughout this article.

2. Results from dynamical equilibrium modeling analyses

The first attempt in determining the LOS velocity dispersion profile of dSphs was made by Mateo *et al.* (1991) from a sample of ~30stars. Nowadays all the MW classical dSphs have LOS velocity dispersion profiles based on 100s of member stars, where this overwhelming progress has been largely due to wide-area spectrographs with a high multiplex power mounted on 4-8m class telescopes. The LOS velocity dispersion profiles for these galaxies are observed to be approximately flat out to the last measured point, that is often out to and beyond their nominal King tidal radius. This is considered as the best evidence that in dSphs the DM has a different spatial distribution than the stars, so that mass-follows-light models can be excluded in the hypothesis of dynamical equilibrium (see e.g. Evans, An & Walker 2009 for the implications on nonphysical values of the velocity anisotropy in the mass-follows-light case; but see e.g. Łokas 2009 for a different take on how stars should be considered as bound to the dSphs and the effect it has on the shape of the LOS velocity dispersion profile).

It is however still debated whether dSphs inhabit cored or cuspy DM haloes. A Jeans analysis of the LOS velocity dispersion profile of a non-rotating spherical object cannot distinguish between different functional forms of the density profile for extended DM haloes, as exemplified in Fig. 1: there we can see that the LOS velocity dispersion profile of the Sculptor dSph is very well fitted by a cored DM profile traced by stars with an isotropic velocity distribution in the center becoming slightly radial at larger radii; this fit is essentially indistinguishable from the best-fitting cuspy model in the case of the stars having a slightly tangential anisotropy. The kind of conclusions that can be inferred by this type of analysis though is that dSphs display very large dynamical mass-to-light ratios (e.g. Kleyna *et al.* 2001; Koch *et al.* 2007; Walker *et al.* 2007; Battaglia *et al.* 2008)

Much effort has gone in the direction of including higher moments of the LOSVD in the analysis to constrain the DM density profile (e.g. Łokas 2009; Breddels *et al.* 2013; Mamon, Biviano & Boué 2013; Richardson & Fairbairn 2014), in particular the fourth moment because of the information it carries about the velocity anisotropy of the stars. Observationally, the fourth moment of the LOSVD has been derived for most classical MW dSphs and, although the details differ among works, current measurements suggest that the LOSVDs of stars in dSphs are not dramatically different from Gaussians in most cases, so that one can conclude that the velocity ellipsoid is neither strongly radial nor strongly tangential, hence in the regime where both cored and cuspy profile would be consistent with the data.

An alternative way for relieving the "mass-anisotropy" degeneracy was proposed by Battaglia *et al.* (2008), where the idea was to use the multiple "chemo-dynamical" stellar components observed in dSphs (e.g. Tolstoy *et al.* 2004; Battaglia *et al.* 2006) as independent tracers of the same gravitational potential. They carried out a spherical Jeans analysis of the Sculptor dSph, modeling it as a metal-rich centrally concentrated component, and a metal-poor hot and extended one, both embedded in a dark matter halo. This analysis shows that models in which the velocity anisotropy is constant as a function of radius for both the metal-rich and a metal-poor component can be excluded (see Battaglia 2007). While the metal-poor component is better fit with a nearly flat anisotropy profile, the metal-rich one, because of its rapidly falling velocity dispersion profile, requires a radially anisotropic ellipsoid. In this case, the cored models provided better fits but that NFW models could not be ruled out.

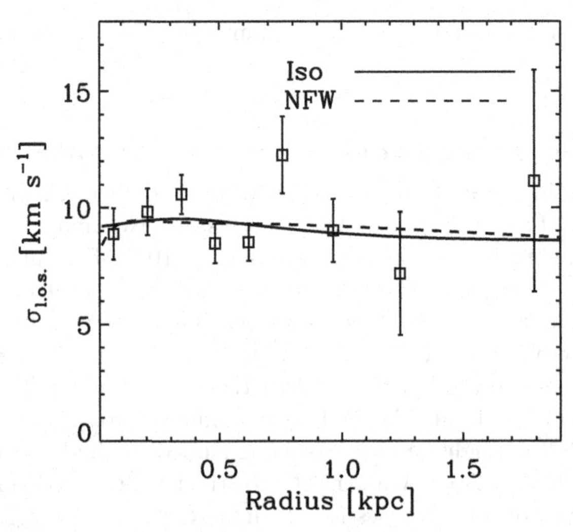

Figure 1. Example of the "mass-anisotropy" degeneracy. The squares with error-bars show the observed Sculptor dSph LOS velocity dispersion profile overlaid to the best-fitting NFW model with constant β (dashed line; c=35, virial mass = 6.1×10^8 $M_\odot$and $\beta = -0.5$) and pseudo-isothermal sphere with an Osipkov-Merrit anisotropy profile (solid line; core radius = 0.5kpc, M(<1.8kpc)= 3.2×10^8 $M_\odot$and anisotropy radius = 0.4kpc. Both models share the same χ^2 value. Data and analysis by Battaglia *et al.* (2008).

Another very interesting result of the recent literature is the finding that for dispersion-supported systems there exists one radius where the enclosed mass inferred from the LOS velocity dispersion is largely insensitive to the velocity anisotropy (see Strigari *et al.* 2007, Walker *et al.* 2009, Wolf *et al.* 2010 for an analytic proof using the spherical Jeans equation) for a system with a flat projected velocity dispersion profile. The enclosed mass at the radius can be expressed in a very simple way, just in terms of the (luminosity-weighted) LOS velocity dispersion and this characteristic radius. Walker & Peñarrubia (2011) have analyzed a set of dSphs and applied this type of mass estimator to the dSph's metal-rich and metal-poor component independently. This yields the enclosed mass at two different radii, and hence to a measurement of the slope. Application of this method to the Fornax and Sculptor dSph rules out NFW-like profiles at high significance levels, $\gtrsim$96% and $\gtrsim$99%, respectively.

The large number of LOS velocities available for some of the dSphs has also allowed the application of more sophisticated and non-parametric mass modeling techniques, such as e.g. Schwarzschild modeling, which is widely used in integrated light spectroscopic studies of more distant galaxies. So far this has been done treating the stellar component of dSphs as a single component. Some of the advantages of the Schwarzschild modeling are that the distribution function is guaranteed to be positive - unlike for the Jeans modeling - and that the velocity anisotropy is an outcome of the analysis. Breddels *et al.* (2013), Breddels & Helmi (2013), Jardel *et al.* (2013) have carried out Schwarzschild modeling of MW dSphs in the spherical case, while Jardel & Gebhardt (2012) have considered a non-spherical light distribution. Even though neither this method can make a distinction between cored and cuspy profiles for the individual dSphs, a striking result from the work of Breddels & Helmi 2013 is that the mass distribution of each of the analyzed dSphs is the same for all the DM models used (NFW, Einasto, cored models) from about the half-mass radius to the last measured kinematic data point. This means that the mass distribution is robustly determined over a large range of radii, and that a slope of the density profile can be determined at some intermediate point.

All in all, there is still debate as to whether the spherical dynamical modeling of dSphs favors cored or cuspy DM density distributions. In general, using two (or multiple) components disfavor NFW/cuspy profiles for dSphs, at least for Fnx and Scl, as also shown by Amorisco & Evans (2012), Agnello & Evans (2012) - but see Strigari, Frenk & White (2014). Efforts are under way to use multiple stellar populations also in Schwarzschild modeling (G. van de Ven talk, EWASS 2015). At the same time it would also be desirable to understand the extent down to which these systems's properties are better described using a few independent components, rather than to assume that the properties of the stars change gradually throughout the system, and understand how this might affect the modeling.

3. The possible impact of assumptions

3.1. *Sphericity*

The stellar component of dSphs appears flattened on the sky, with a mean projected ellipticity ∼0.3 and values ranging from ∼0.1 to ∼0.6. For an axisymmetric system this means that the ratio between the shortest and longest *projected* axis is between 0.4 & 0.9. Since we lack information on the inclination of the dSphs's stellar component with respect to the plane of the sky, we can only say that, unless we are seeing the dSph edge-on, most likely its 3D stellar density distribution is flatter than what we see in projection. It is then clear that the density distribution of the dSphs stellar component might deviate considerably from the assumption of sphericity. As for the DM halo, the expectations differ according to the characteristics of the DM particle. In ΛCold Dark Matter (pure N-body + semi-analytics) simulations it appears that those sub-haloes that are expected to become luminous satellites tend to be closer to spherical with respect to field sub-haloes as a consequence of tidal stripping (e.g. Barber *et al.* 2015).

The effect of moving away from the assumption of sphericity has started to be investigated by several groups (e.g. Jardel & Gebhardt 2012, Hayashi & Chiba 2012, Kowalczyk *et al.* 2013, Laporte *et al.* 2013, Bonnivard *et al.* 2015). Hayashi & Chiba (2012) explore the variation of the LOS velocity dispersion profile along the projected major and minor axes for different flattenings and density distributions of the DM and stellar components. The application of these axisymmetric Jeans models to the MW dSphs results into significantly flattened DM haloes for several systems, and in values of the mass within a spheroid of major-axis length of 300pc that can differ up to an order of magnitude with respect to the values from a spherical analysis, depending on the flattening retrieved for the system. It should be noticed that also this analysis might suffer from the assumptions made on the shape of the velocity ellipsoid, specifically that the velocity dispersion in the radial R and vertical z direction are identical, as there is a degeneracy between the flattening of the velocity ellipsoid and the flattening of the halo. However it is clearly worth to understand how relaxing the assumption of sphericity may affect the results from different methods.

3.2. *Dynamical equilibrium*

The dSphs are small galaxies orbiting a much larger system, hence one of the questions related to their dynamical modeling is to what extent they can be considered in dynamical equilibrium.

Clear signs of the tidal disturbance exerted by the MW and M31 over some of their small companions are observed in the form of tidal tails or isophote twists around surviving dSphs (e.g. NGC 205, NGC 147, Carina, Hercules: Choi *et al.* 2002; Crnojević et 2014; Battaglia *et al.* 2012; McMonigal *et al.* 2014; Roderick *et al.* 2015), and stellar streams of heavily or entirely disrupted objects (e.g. the spectacular cases of the Sagittarius stream around the MW, Ibata *et al.* 1994; the Giant Stream around M31, McConnachie *et al.* 2010). Other systems show outer breaks in their stellar number density profiles and/or increasing LOS velocity dispersion profiles (e.g. Leo I, Sohn *et al.* 2007, Mateo *et al.* 2008), features that might be caused by tidal disturbance, although this need not be the only explanation.

N-body simulations of dwarf-like systems orbiting a MW-like potential agree in that the central velocity dispersion (or the dispersion at the half-light radius) is a good indicator of the present maximum circular velocity and bound mass, as long as the objects retain a bound core (e.g. Muñoz *et al.*, 2008; Peñarrubia *et al.*, 2008b; Klimentowski *et al.*, 2009; Kazantzidis *et al.*, 2011). However, at large radii, the presence of unidentified unbound stars can significantly inflate the measured LOS velocity dispersion (e.g. Read *et al.* 2006, Muñoz *et al.* 2008, Klimentowski *et al.* 2009).

The impact of tidal disturbance is expected - and it has been shown - to depend on several (mainly unknown) factors, such as the degree of embedding of the stellar component in the DM halo, the DM density profile, the orbit of the object around the host galaxy. However, even if an object has suffered a large degree of mass loss, this does not necessarily imply lack of current dynamical equilibrium. Tidally perturbed stars progressively become unbound and are eventually dispersed, with the object eventually settling in a new equilibrium configuration (Peñarrubia *et al.*, 2009). Also, the degree of contamination of kinematic data-sets from stars originating in tidal tails can vary along the orbit, as a consequence of the varying orientation of the inner regions of the tidal tails with respect to the LOS (e.g. Klimentowski *et al.* 2009). Given the difficulty in unambiguously identifying unbound stars lingering close to the main body of the MW dSphs, it is quite possible that, depending on the specific orbital history and current location along the orbit, the observed kinematics of stars is a faithful tracer of their underlying potential in some of the MW dSphs and not in others.

N-body simulations tailored to reproduce the observed structural and internal kinematic properties of *specific* MW dSphs appear particularly well suited to explore this issue and so far confirm the expectation that not all dSphs are in the same situation. For example, it has been shown that, for *observationally motivated orbits*, the observed Carina's structural and internal kinematic properties are well matched also by modeling this galaxy as a tidally disturbed system initially embedded in "mass-follows-light" DM halo, rather than in an extended halo (Muñoz *et al.* 2008), while on the other hand this is not the case for the Fornax dSph, for which the effects produced by the Galactic tidal field are relatively small, and the final system is close to equilibrium in its own gravitational potential and appears to require an extended DM halo (Battaglia, Sollima & Nipoti 2015).

It is remarkable that also the internal kinematics of tidal streams can potentially be used to infer information on the DM density distribution of the tidally disturbed dwarf galaxy (Errani, Peñarrubia, Tormen 2015).

4. A few closing remarks

Much progress has been made in the study of the internal kinematics of dSphs, with impressive spectroscopic data-sets being already in place. The near future will see other major advances: the Gaia mission is likely to provide more accurate systemic proper motions for the MW dSphs and hence a better understanding of their orbital history around the MW; several fiber spectrographs with much larger field-of-view, multiplex power and simultaneous wavelength coverage at intermediate spectral resolution than the existing ones are planned to be mounted on 4m and 8m class telescopes in the next years (e.g. WHT/WEAVE, VISTA/4MOST, VLT/MOONS, Subaru/PFS). This appears as the right time for modelers to explore in depth what the need for future data-sets is, so that the full information on discrete LOS velocities can be exploited at the same time allowing to take into account other observed properties such as non-spherical density distributions and "multiple stellar components".

References

Agnello, A. & Evans, N. W., Aug. 2012, *ApJ*(Letters) 754, L39.
Amorisco, N. C. & Evans, N. W., Jan. 2012, *MNRAS* 419, 184–196.
Barber, C., Starkenburg, E., Navarro, J. F., & McConnachie, A. W., 2015, *MNRAS*, 447, 1112
Battaglia, G., *et al.*, 2006, *A&A*, 459, 423
Battaglia, G. 2007, Ph.D. Thesis
Battaglia, G., Helmi, A., Tolstoy, E., Irwin, M., Hill, V., & Jablonka, P., 2008, *ApJ*, 681, L13
Battaglia, G., Tolstoy, E., Helmi, A., Irwin, M., Parisi, P., Hill, V., & Jablonka, P., 2011, *MNRAS*, 411, 1013
Battaglia, G., Irwin, M., Tolstoy, E., de Boer, T., & Mateo, M., 2012, *ApJ*, 761, L31
Battaglia, G., Helmi, A., & Breddels, M., 2013, *New Astron. Revs*, 57, 52
Battaglia, G., Sollima, A., & Nipoti, C. 2015, arXiv:1509.02368
Bergström, L., Ullio, P., & Buckley, J. H. 1998, *Astroparticle Physics*, 9, 137
Binney, J. & Mamon, G. A., 1982, *MNRAS*, 200, 361
Binney, J. & Tremaine, S., 2008, *Galactic Dynamics* 2nd Ed., Princeton University Press, Princeton
Bonnivard, V., Combet, C., Daniel, M., *et al.* 2015, *MNRAS*, 453, 849
Breddels, M. A. & Helmi, A., 2013, *A&A*, 558, A35 (BH13)
Breddels, M. A., Helmi, A., van den Bosch, R. C. E., van de Ven, G., & Battaglia, G., 2013, *MNRAS*, 433, 3173
Breddels, M. A., Vera-Ciro, C., & Helmi, A. 2015, arXiv:1507.03995
Choi, P. I., Guhathakurta, P., & Johnston, K. V., 2002, *AJ*, 124, 310
Crnojević, D., *et al.*, 2014, *MNRAS*, 445, 3862
Di Cintio, A., Brook, C. B., Macciò, A. V., Stinson, G. S., Knebe, A., Dutton, A. A., & Wadsley, J., 2014, *MNRAS*, 437, 415
Dinescu, D. I., Keeney, B. A., Majewski. S. R., & Girard, T. M., 2004, *AJ*, 128, 687
Errani, R., Peñarrubia, J., & Tormen, G. 2015, *MNRAS*, 449, L46
Evans, N. W., An, J., & Walker, M. G. 2009, *MNRAS*, 393, L50
Hayashi, K. & Chiba, M., 2012, *ApJ*, 755, 145
Ibata, R. A., Gilmore, G., & Irwin, M. J., 1994, *Nature*, 370, 194
Jardel, J. R. & Gebhardt, K., 2012, *ApJ*, 746, 89
Jardel, J. R., Gebhardt, K., Fabricius, M. H., Drory, N., & Williams, M. J., 2013, *Apj* 763, 91.
Kazantzidis, S., Łokas, E. L., Callegari, S., Mayer, L., & Moustakas, L. A., 2011, *ApJ*, 726, 98
Kleyna, J. T., Wilkinson, M. I., Evans, N. W., & Gilmore, G., 2001, *ApJ*, 563, L115
Klimentowski, J., Łokas, E. L., Kazantzidis, S., Prada, F., Mayer, L., & Mamon, G. A., 2007, *MNRAS*, 378, 353
Klimentowski, J., Łokas, E. L., Kazantzidis, S., Mayer, L., Mamon, G. A., & Prada, F., 2009, *MNRAS*, 400, 2162
Koch, A., Kleyna, J. T., Wilkinson, M. I., Grebel, E. K., Gilmore, G. F., Evans, N. W., Wyse, R. F. G., & Harbeck, D. R., 2007, *AJ*, 134, 566
Kowalczyk, K., Łokas, E. L., Kazantzidis, S., & Mayer, L. 2013, *MNRAS*, 431, 2796
Laporte, C. F. P., Walker, M. G., & Peñarrubia, J. 2013, *MNRAS*, 433, L54
Łokas, E. L. 2009, *MNRAS*, 394, L102
Mamon, G. A., Biviano, A., & Boué, G. 2013, *MNRAS*, 429, 3079
Mateo, M., Olszewski, E., Welch, D. L., Fischer, P., & Kunkel, W., 1991, *AJ*, 102, 914
Mateo, M., Olszewski, E. W., & Walker, M. G., 2008, *ApJ*, 675, 201

Mayer, L., Governato, F., Colpi, M., Moore, B., Quinn, T., Wadsley, J., Stadel, J., & Lake, G., 2001, *ApJ*, 559, 754

McConnachie, A. W., Ferguson, A. M. N., Irwin, M. J., Dubinski, J., Widrow, L. M., Dotter, A., Ibata, R., & Lewis, G. F., 2010, *ApJ*, 723, 1038

McMonigal, B., *et al.*, 2014, *MNRAS*, 444, 3139

Méndez, R. A., Costa, E., Gallart, C., Pedreros, M. H., Moyano, M., & Altmann, M., 2011, *AJ*, 142, 93

Muñoz, R. R., *et al.*, 2006, *ApJ*, 649, 201

Muñoz, R. R., Majewski, S. R., & Johnston, K. V., 2008, *ApJ*, 679, 346

Navarro, J. F., Frenk, C. S., & White, S. D. M., 1995, *MNRAS*, 275, 720 (NFW)

Navarro, J. F., Eke, V. R., & Frenk,. C. S., 1996, *MNRAS*, 283, L72

Navarro, J. F., Ludlow, A., Springel, V., *et al.* 2010, *MNRAS*, 402, 21

Peñarrubia, J., McConnachie, A. W., & Navarro, J. F. 2008, *Apj*, 672, 904

Peñarrubia, J., Navarro, J. F., & McConnachie, A. W., 2008, *ApJ*, 673, 226

Peñarrubia, J., Navarro, J. F., McConnachie, A. W., & Martin, N. F., 2009, *ApJ*, 698, 222

Peñarrubia, J., Navarro, J. F., McConnachie, A. W., & Martin, N. F. 2009, *Apj*, 698, 222

Peñarrubia, J., Benson, A. J., Walker, M. G., Gilmore, G., McConnachie, A. W., & Mayer, L., 2010, *MNRAS*, 406, 1290

Piatek, S., Pryor, C., Bristow, P., Olszewski, E. W., Harris, H. C., Mateo, M., Minniti, D., & Tinney, C. G., 2007, *AJ*, 133, 818

Read, J. I., Wilkinson, M. I., Evans, N. W., Gilmore, G., & Kleyna, J. T., 2006, *MNRAS*, 367, 387

Richardson, T. & Fairbairn, M., 2014, *MNRAS*, 441, 1584

Roderick, T. A., Jerjen, H., Mackey, A. D., & Da Costa, G. S., 2015, *ApJ*, 804, 134

Shapley, H. 1938, *Harvard College Observatory Bulletin*, 908, 1

Sohn, S. T., *et al.*, 2007, *ApJ*, 663, 960

Sohn, S. T., Besla, G., van der Marel, R. P., Boylan-Kolchin, M., Majewski, S. R., & Bullock, J. S., 2013, *ApJ*, 768, 139

Springel, V., Wang, J., Vogelsberger, M., *et al.* 2008, *MNRAS*, 391, 1685

Strigari, L. E., Bullock, J. S., Kaplinghat, M., *et al.* 2007, *ApJ*, 669, 676

Strigari, L. E., 2013, *PhR*, 531, 1

Strigari, L. E., Frenk, C. S., & White, S. D. M. 2014, arXiv:1406.6079

Tolstoy, E., Irwin, M. J., Helmi, A., *et al.* 2004, *ApJ*(Letters), 617, L119

Vera-Ciro, C. A., Sales, L. V., Helmi, A., & Navarro, J. F. 2014, *MNRAS*, 439, 2863

Walker, M. G., Mateo, M., Olszewski, E. W., Gnedin, O. Y., Wang, X., Sen, B., & Woodroofe, M., 2007, *ApJ*, 667, L53

Walker, M. G., Mateo, M., & Olszewski, E. W. 2008, *ApJ*(Letters), 688, L75

Walker, M. G., Mateo, M., & Olszewski, E. W., 2009, *AJ*, 137, 3100

Walker, M. G. & Peñarrubia, J., 2011, *ApJ*, 742, 20

Walker, M., 2013, in *Planets Stars, and Stellar Systems*, Vol. 5, ed. T. Oswalt, & G. Gilmore, Springer, Berlin, 1039

Wolf, J., Martinez, G. D., Bullock, J. S., *et al.* 2010, *MNRAS*, 406, 1220

Zolotov, A., *et al.*, 2012, *ApJ*, 761, 71

The General Assembly of Galaxy Halos: Structure, Origin and Evolution
Proceedings IAU Symposium No. 317, 2015
A. Bragaglia, M. Arnaboldi, M. Rejkuba & D. Romano, eds.

doi:10.1017/S1743921315009837

Globular Clusters, Dwarf Galaxies, and the Assembly of the M87 Halo

Eric W. Peng[1,2], Hong-Xin Zhang[3,4], Chengze Liu[5] and Yiqing Liu[1]

[1]Department of Astronomy, Peking University, Beijing, China
email: peng@pku.edu.cn

[2]Kavli Institute for Astronomy and Astrophysics, Peking University, Beijing, China

[3]National Astronomical Observatories, Chinese Academy of Sciences, Beijing 100012, China

[4]Departamento de Astronomía y Astrofísica, Pontificia Universidad Católica de Chile, 7820436 Macul, Santiago, Chile

[5]Center for Astronomy and Astrophysics, Department of Physics and Astronomy, Shanghai Jiao Tong University, Shanghai 200240, China

Abstract. At the center of the nearest galaxy cluster, the Virgo cluster, lies the massive cD galaxy, M87 (NGC 4486). Using data from the Next Generation Virgo Cluster Survey, we investigate the relationship between M87, its globular clusters (GCs), and satellite dwarf galaxies. We find that the kinematics of GCs and ultra-compact dwarfs (UCDs) are different, indicating that UCDs are not simply massive GCs. We also identify a morphological sequence of envelope fraction around UCDs correlated with cluster-centric distance that suggest UCDs are the result of tidal stripping. Lastly, we find that the [α/Fe] abundance ratios of low-mass early-type galaxies in Virgo exhibit a strong negative gradient within $\sim$ 400 kpc of M87, where the galaxies closest to M87 have the highest values. These satellite galaxies are likely the surviving counterparts of accreted dwarfs that contribute stars to the metal-poor, α-rich stellar halos of massive galaxies. Together, these results describe a dense environment that has had a strong and continuing impact on the evolution of its low-mass neighbors.

Keywords. galaxies: abundances; galaxies: evolution; galaxies: clusters : individual (Virgo); galaxies: halo; galaxies: dwarf; galaxies: individual (M87/NGC4486); galaxies: kinematics and dynamics; galaxies: stellar content

---

1. Introduction

Massive galaxies are a product of a continuous interaction with their environment. These galaxies grow by accreting gas and stars from nearby satellites. The evolution of satellite galaxies is, in turn, altered by their proximity to their host's deep gravitational potential. Low-mass galaxies, with their large numbers and their high susceptibility to environmental effects (e.g., ram pressure, tidal stripping and harassment), are some of the best targets for studying the connection between environment and the evolution of both low and high mass galaxies.

Ultra-compact dwarf galaxies (UCDs) are a particularly interesting class of object as they are thought to be the remnant nuclear star cluster of what was previously a nucleated dwarf galaxy (e.g., Bekki *et al.* 2003; Côté *et al.* 2006). If so, they are unique signposts of disrupted galaxies that have long ago donated their stars to the parent halo. Some evidence, however, has indicated that UCDs may simply be the most massive globular clusters (GCs) (Mieske *et al.* 2012), in which case, they may have less significance for the assembly history of stellar halos.

To address these and other questions, we have embarked on the Next Generation Virgo Cluster Survey (NGVS, Ferrarese *et al.* 2012), a Large Program on the Canada-France-Hawaii Telescope (CFHT) to perform deep *u* g' i' z'* imaging of 104 deg^2 of the Virgo cluster out to the virial radii of its two main subclusters. In this contribution, we focus on the central $2° \times 2°$ around the cD galaxy, M87, a region where we have also obtained deep K_s-band photometry (Muñoz *et al.* 2014). One of the key advantage of NGVS imaging in the study of UCDs is that our image quality is excellent (median i'-band seeing of 0.55"). This allows us to resolve and measure sizes for objects in Virgo that are larger than 10 pc, a value typically used as a lower bound for objects considered UCDs.

2. Kinematics of ultra-compact dwarfs around M87

Using the NGVS imaging to identify candidate UCDs and GCs, we have carried out a systematic spectroscopic survey of compact stellar systems around M87 with the 2dF/AAOmega multi-fiber spectrograph on the 3.9-meter AAT and the Hectospec multi-fiber spectrograph on the 6.5-meter MMT. Combined with data from the literature (e.g., Hanes *et al.* 2001; Strader *et al.* 2011), we analyzed a sample of 97 UCDs and 911 GCs, all spectroscopically confirmed, within 1.5 deg (430 kpc) of M87 (Zhang *et al.* 2015). We separated UCDs from GCs based on their size (r_h, or half-light radius) and luminosity. UCDs in our sample all have $r_h > 11$ pc and $g' \leqslant 21.5$ mag ($M_g \leqslant -9.6$).

Our comparison of UCD and GC kinematics revealed two significant differences. We analyzed the rotation of the UCD system compared to the systems of blue and red GCs. We found that while the blue GCs show little or no systematic rotation about M87, the UCDs have a significant rotation amplitide (~ 150 km/s) around an axis in-between the major and minor photometric axes. This difference suggests that the UCDs in M87 are not simply the most massive and largest GCs, for if their formation mechanism was the same, we would expect the populations to have similar kinematics.

We also inferred the orbital anisotropies of the UCDs and GCs based on the spherically symmetric Jeans equation. The anisotropy parameter, $\beta_r \equiv 1 - \sigma_t^2/2\sigma_r^2$, measures the relative amount of tangential and radial velocity dispersion. We estimated the profile of β_r as a function of distance from M87 for UCDs, blue GCs, and red GCs, shown in Figure 1. Once again, the blue GCs, with which UCDs are most often associated because of their similar colors, have kinematics different from the UCDs. The blue GCs have tangential orbits in both the inner and outer regions. The UCDs follow the blue GCs, having tangential orbits within ~ 30 kpc, but then diverge from the blue GCs at larger radii, where the UCDs have predominantly radial orbits.

It is not surprising that the inner UCDs have more circular orbits, as our size-defined UCDs have at least an order of magnitude lower average density than GCs of similar luminosity, making them more vulnerable to tidal disruption at small pericenters. At large radii, however, the transition to radial orbits is suggestive of an origin that involves some tidal stripping, as would be the case if UCDs were the remnants of nucleated dwarf galaxies. Although we cannot say for certain what the origins of UCDs are, the differing kinematic profiles in the outer region between UCDs and blue GCs again suggests that they are not a single population of objects. The full description of the analysis and interpretation are in Zhang *et al.* (2015).

3. A morphological sequence of ultra-compact dwarfs

If UCDs really are the remnant nuclear star clusters of stripped galaxies, we might expect to see UCDs in different stages of being tidally stripped. Using the deep NGVS

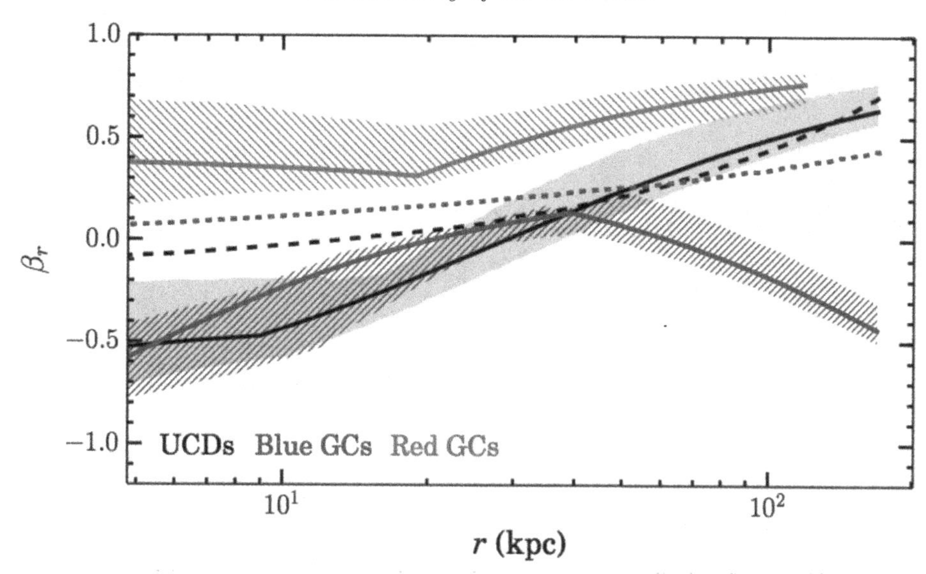

Figure 1. Variation of the orbital anisotropy as a function of the 3D radius from M87, from Zhang *et al.* (2015). The profiles for UCDs, blue GCs, and red GCs are represented as black, blue, and red solid curves, respectively. Following the same color code, the hatched regions of different styles mark the 68% confidence intervals for blue GCs, and red GCs. The gray shaded region marks the 68% confidence interval for the UCDs. The short dashed curves (black for UCDs, blue for blue GCs) represent the anisotropy profiles predicted by a universal relation between the number density slope and β for relic high-σ density peaks as found in cosmological simulations by Diemand *et al.* (2005).

imaging, we have performed a photometric study of the UCDs around M87 (Liu, C. *et al.* 2015). The NGVS g'-band images reach a limiting surface brightness of 29 mag arcsec^{-2}, giving us a new ability to detect faint envelopes around UCDs.

When we examined our UCD sample, we found that while many showed no sign of any extended envelope, a number did possess diffuse halos of varying prominence. Figure 2a shows a mosaic of nucleated, early-type dwarfs with UCDs. We find that there appears to be a continuous morphological sequence of following the visual prominence of the diffuse halo, going from a typical nucleated dwarf galaxy to a "naked" UCD.

We quantified the fraction of light contained in the envelope, and divided the UCDs into two categories, those with a visible envelope, and those with none. Figure 2b shows the cumulative fraction of these two groups, as well as that of the dwarf galaxy nuclei, as a function of galactocentric distance. The UCDs without any visible envelope are the most centrally concentrated, followed by those with visible envelopes, and then by the dwarf galaxy nuclei. This result suggests that the location of the UCD or nucleus within the M87 potential well is a determining factor in the strength of any surrounding diffuse halo. Similar to the kinematic evidence described in the previous section, the radial dependence of the existence of diffuse stellar envelopes around UCDs in M87 is consistent with a tidal stripping origin for UCDs.

4. [α/Fe] in early-type Virgo dwarfs

Even without the aforementioned example of the tidally disrupted progenitors of UCDs, it has already been well-established that low-mass galaxies contribute their stars to the stellar halos of more massive galaxies. It is also known that low-mass galaxies residing in

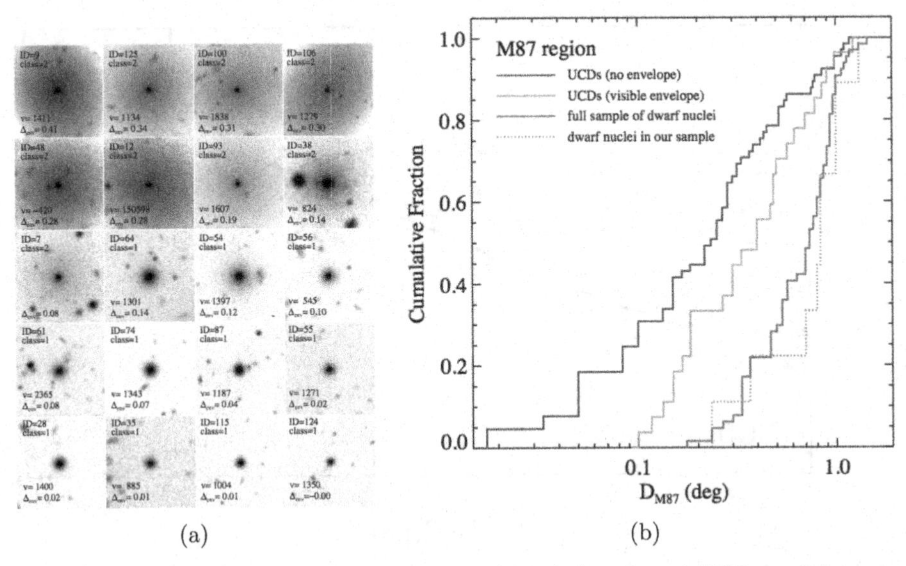

Figure 2. *Left (a):* Mosaic of images showing 20 objectively-selected UCD candidates around M87, from Liu, C. *et al.* (2015). The first nine panels show known nucleated dwarf galaxies from Binggeli *et al.* (1985) or new discoveries from the NGVS. The objects have been loosely arranged according to the visual prominence of an underlying diffuse halo, which is unmistakable in the first objects and non-existent in last ones. *Right (b):* Cumulative distributions for D_{M87} (the projected distance from M87) for UCD candidates in the M87 region (blue and cyan lines), the full sample of 64 dwarf nuclei (red solid line), and the nine additional candidates subse- quently classified as dwarf nuclei (red dotted line). The cleaned sample of 92 UCD candidates has been divided into two classes by measuring of the prominence of an outer envelope.

dense environments, or near a more massive neighbor, can have their evolution strongly altered. As a consequence, the types of stars accreted onto stellar halos are influenced by the host galaxies themselves. Greene *et al.* (2012, 2013) measured the mean ages, [Fe/H], and [α/Fe] for the outer halos of 25 massive ETGs, finding that these stellar populations had old ages, relatively low metallicity ([Fe/H]~ -0.5), and high [α/Fe]$\sim +0.3$, a combination that had not been found in the low mass galaxies that could plausibly form an accreted population. [α/Fe], an indicator of early star formation timescales, is typically low for low-mass galaxies, unlike the values found in the stellar halos of massive galaxies. To further explore this connection, we have investigated the timescales of star formation in low-mass Virgo early-type galaxies (ETGs) as a function of their local environment.

Peng *et al.* (2008) reported indirect evidence that star formation in low-mass ETGs at the center of Virgo must have proceeded rapidly, finding that galaxies closer to M87 had higher GC specific frequencies (S_N). Assuming that massive star clusters favor intense bursts of star formation, then two galaxies with the same stellar mass but different S_N must have different timescales over which the bulk of their stars were formed. Here, and in Liu, Y. *et al.* (2015), we report the results of a study of [α/Fe] in 11 low-mass Virgo ETGs.

Figure 3 shows [α/Fe] plotted versus projected distance from M87 for these 11 Virgo dwarf ETGs. We found a strong gradient where the galaxies closest to M87 have the highest values of [α/Fe]. This trend is only obvious, however, when looking at galaxies within 0.4 Mpc (0.26 times the virial radius of Virgo A), all of which have super-solar [α/Fe]. Beyond this projected radius, the sample galaxies have [α/Fe] values around

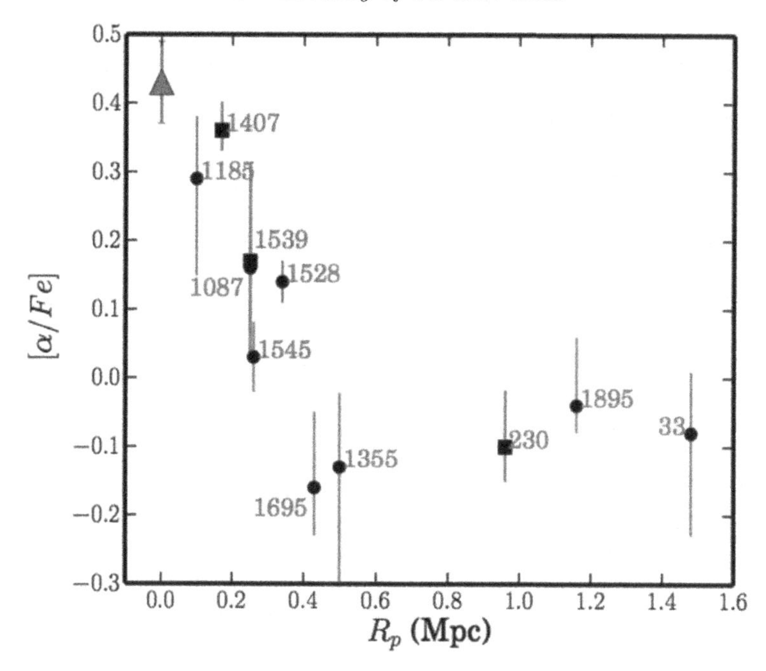

Figure 3. SSP-equivalent [α/Fe] for low mass Virgo early-type galaxies as a function of projected distance from M87, from Liu, Y. *et al.* (2015). Black circles are the low mass ETGs in our sample, labeled by their VCC number. The error bars represent 68% confidence levels. [α/Fe] decreases with cluster-centric distance in the inner region, and then has a flat distribution in the outer regions, at values close to solar. The red triangle represents M87 within $R_e/2$ with 1-σ error bar, from the work of McDermid *et al.* (2015).

or below solar. The innermost galaxies have [α/Fe] values comparable to that of M87 itself suggesting that: 1) The low-mass ETGs we observe are plausible "survivors" of a parent population from which many galaxies have already accreted onto M87, and 2) Environment is the controlling factor for star formation timescales in dense regions, perhaps through quenching processes such as ram pressure stripping and strangulation.

5. Conclusions

In this contribution, we present results from three different investigations of the satellite population (GCs, UCDs, and dwarfs) around the Virgo cluster cD galaxy, M87. Both the kinematics and structural properties of UCDs support the idea that they are not simply massive GCs, but perhaps the remnant nuclei of tidally stripped dwarf galaxies. Our measurement of [α/Fe] in low-mass ETGs around M87 shows that the dense environment in which the innermost galaxies reside has led to shorter star formation timescales within the central 0.4 Mpc. Taken together, these studies show that the assembly of M87's stellar halo is a result of an interplay between its deep gravitational potential and its many low-mass satellites. Future studies of the detailed chemical abundances and the internal kinematics of these satellite systems, and of the M87 halo itself, will be important for establishing the details of these links.

References

Bekki, K., Couch, W. J., Drinkwater, M. J., & Shioya, Y. 2003, *MNRAS*, 344, 399
Binggeli, B., Sandage, A., & Tammann, G. A. 1985, *AJ*, 90, 1681

Côté, P., Piatek, S., Ferrarese, L., *et al.* 2006, *ApJS*, 165, 57
Ferrarese, L., Côté, P., Cuillandre, J.-C., *et al.* 2012, *ApJS*, 200, 4
Diemand, J., Madau, P., & Moore, B. 2005, *MNRAS*, 364, 367
Greene, J. E., Murphy, J. D., Comerford, J. M., Gebhardt, K., & Adams, J. J. 2012, *ApJ*, 750, 32
Greene, J. E., Murphy, J. D., Graves, G. J., Gunn, J. E., Raskutti, S., Comerford, J. M., & Gebhardt, K. 2013, *ApJ*, 776, 64
Hanes, D. A., Côté, P., Bridges, T. J., *et al.* 2001, *ApJ*, 559, 812
Liu, C., Peng, E. W., Côté, P., *et al.* 2015, *ApJ*, 812, 34
Liu, Y., Peng, E. W., Blakeslee, J. P., *et al.* 2015, *ApJ*, submitted
McDermid, R. M., Alatalo, K., Blitz, L., *et al.* 2015, *MNRAS*, 448, 3484
Mieske, S., Hilker, M., & Misgeld, I. 2012, *A&A*, 537, A3
Muñoz, R. P., Puzia, T. H., Lançon, A., *et al.* 2014, *ApJS*, 210, 4
Peng, E. W., Jordán, A., Côté, P., *et al.* 2008, *ApJ*, 681, 197
Strader, J., Romanowsky, A. J., Brodie, J. P., *et al.* 2011, *ApJS*, 197, 33
Zhang, H.-X., Peng, E. W., Côté, P., *et al.* 2015, *ApJ*, 802, 30

The General Assembly of Galaxy Halos: Structure, Origin and Evolution
Proceedings IAU Symposium No. 317, 2015
A. Bragaglia, M. Arnaboldi, M. Rejkuba & D. Romano, eds.

doi:10.1017/S1743921316000107

Chemical Abundances of Metal-poor stars in Dwarf Galaxies

Kim A. Venn[1], Pascale Jablonka[2], Vanessa Hill[3], Else Starkenburg[4], Bertrand Lemasle[5], Matthew Shetrone[6], Mike Irwin[7], John Norris[8], David Yong[8], Gerry Gilmore[7], Stephania Salvadori[9], Asa Skuladottir[9] and Eline Tolstoy[9]

[1]Dept. of Physics & Astronomy, University of Victoria, Victoria, BC, V8P 5C2, Canada
email: kvenn@uvic.ca

[2]Laboratoire d'astrophysique, EPFL, Observatoire de Sauverny, 1290 Versoix, Switzerland

[3]Obs. de la Cote d'Azur, 06304 Nice Cedex 4, France

[4]Leibniz-Institute fur Astrophysik Potsdam, 14482 Potsdam, Germany

[5]Anton Pannekoek Inst. for Astronomy, Univ. of Amsterdam, 1090 GE Amsterdam, Netherlands

[6] McDonald Observatory, Univ. of Texas at Austin, Austin TX 78712, USA

[7] Institute of Astronomy, Univ. of Cambridge, Madingley Road, Cambridge, CB3 0HA, UK

[8] RSAA, Australian National Univ., Mount Stromlo Observatory, Weston ACT 2611, Australia

[9] Kapteyn Astronomical Inst., Univ. of Groningen, 9747 AD Groningen, Netherlands

Abstract. Stars in low-mass dwarf galaxies show a larger range in their chemical properties than those in the Milky Way halo. The slower star formation efficiency make dwarf galaxies ideal systems for testing nucleosynthetic yields. Not only are alpha-poor stars found at lower metallicities, and a higher fraction of carbon-enhanced stars, but we are also finding stars in dwarf galaxies that appear to be iron-rich. These are compared with yields from a variety of supernova predictions.

Keywords. stars: abundances, galaxies: dwarf, evolution, Local Group, stellar content.

---

1. Introduction

It has been known for more than a decade now that "the overwhelming majority of Milky Way stars, those in the Galactic thick disk and thin disk, seem to have nothing at all to do with dwarf galaxy origins" (Gilmore 2012). As summarized by Tolstoy *et al.* (2009), the ratio of alpha-elements to iron are lower in the stars in dwarf galaxies for metallicities above [Fe/H]~ -2. This has been found for stars in the Sculptor, Carina, Fornax, and LMC dwarf galaxies, and the Sagittarius dwarf galaxy remnant. The early results have been confirmed by larger data samples (e.g., Lemasle *et al.* 2012, 2014, Hendricks *et al.* 2014), and extended to lower mass dwarfs, such as Bootes I and Segue I (e.g., Feltzing *et al.* 2009, Norris *et al.* 2010, Frebel *et al.* 2014).

The picture is less clear when comparing the chemistry of the stars in the Galactic halo with those in dwarf galaxies with metallicities [Fe/H] < -2.0. An examination of the stellar abundances of [Ca/Fe] and [Mg/Fe] in stars below metallicity [Fe/H] $= -2$ is shown in Fig. 1. The uncertainties in each individual stars are typically < 0.2 dex. While the general trend of high [alpha/Fe] ratios (above the solar ratio) appears in all systems, the dispersion is larger in the dwarf galaxies, and includes several stars with very low [alpha/Fe] abundances (below the solar ratio).

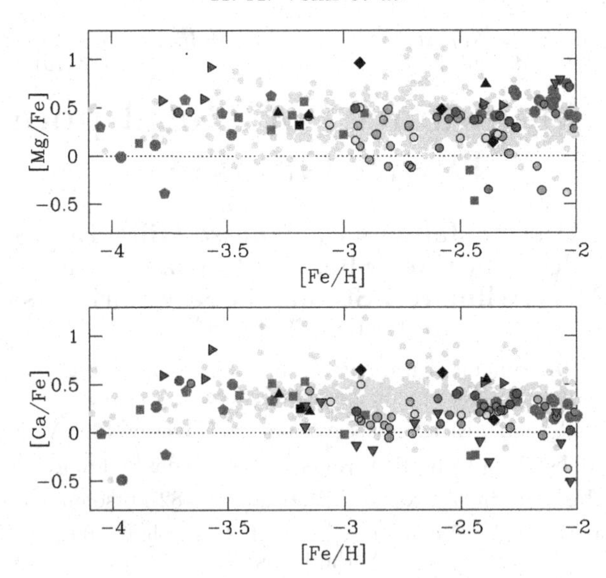

Figure 1. [Mg/Fe] and [Ca/Fe] for stars in the Milky Way (grey) and dwarf galaxies. These include abundances summarized by Frebel 2010, and additional new references for Sculptor (blue; Jablonka *et al.* 2015, Simon *et al.* 2015, Starkenburg *et al.* 2013, Tafelmeyer *et al.* 2010), Sextans (green; Tafelmeyer *et al.* 2010) Carina (orange; Norris *et al.* 2016, Venn *et al.* 2012, Lemasle *et al.* 2012) Draco and UMi (yellow), and the ultra faint dwarf galaxies (black; Frebel 2014 for Segue 1 and Aden *et al.* 2011 for Hercules).

2. Element Abundances in Metal-Poor Stars

At the lowest metallicities, the detailed chemical abundances have been modelled as enrichment by a variety of core collapse SN, including a range of progenitor masses, SN explosion energies, and mixing and fallback prescriptions. Iwamoto *et al.* (2005) have found that "faint supernovae" with extensive mixing and fallback during the SN II explosion produce decreasing yields with increasing atomic number (mitigated by the odd-even effect); and that these yields reproduce the element distribution for the ultra metal-poor stars HE1327-2326 and HE0107-5240 better than predictions from the metal-free massive models by Heger & Woosley (2002, 2008). Neither of these models produce the heavy neutron capture elements.

The core collapse supernova models by Wanajo (2013) can reproduce the solar-system r-process distribution by tuning the initial core masses. These models are also able to reproduce the heavy element distribution in metal-poor Galactic stars, such as CS 31082-001. However, Wanajo *et al.* (2014) have also been able to make these same heavy element distributions with a variety of compact binary mergers. This latter scenario is an interesting alternative for the site of the r-process (also see Shen *et al.* 2015), though rely on timescale arguments for making and merging two neutron stars.

In Fig. 2, the [Ba/Fe] ratios for metal-poor stars in the Galaxy and dwarf galaxies are shown. The general distribution and dispersions are quite similar for the majority of stars, implying the sites and yields for the r-process are similar in these systems. Again, the uncertainties in the abundances ratios are estimated as < 0.2 dex. Only the upper limits on [Ba/Fe] for stars in the dwarf galaxies show an unexpected result, as highlighted by Frebel *et al.* (2014). The intermediate metallicity stars (with [Fe/H] > -2.5) in the ultra faint dwarf galaxies, Segue 1, Com Ber, and Hercules, have extremely low [Ba/Fe] values or upper limits. This suggests these stars formed from gas that was not enriched in r-process elements, at all, yet they have high iron abundances. This is not seen in the

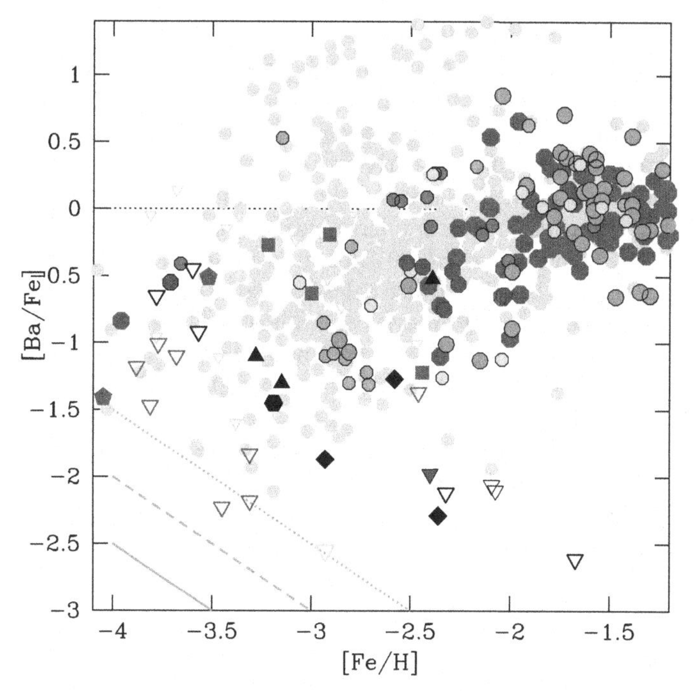

Figure 2. [Ba/Fe] for stars in the Milky Way (grey) and dwarf galaxies (colours as in 1). Upper limits for stars in dwarf galaxies (only) are noted as open downward-pointing open triangles. Detection limits for the Ba II 4554 A spectral line are shown as green lines (10 mA as dotted, 3 mA as dashed, 1 mA as solid), for a red giant with T = 4500 K, following Roederer (2013).

Galaxy nor the classical dwarf galaxies. This observation has been interpreted by Frebel & Norris (2015) as the result of a single (or few) high-mass metal-free supernova event that also removed the gas in these lowest mass galaxies.

3. Iron-rich Stars in Dwarf Galaxies

At intermediate metallicities, [Fe/H] > -2.5, another phenomenon is found in dwarf galaxies, where some stars appear to be enriched in iron-group elements. This is seen as an unusually low [X/Fe] ratios for nearly all elements, except those in the iron-group. One example is Star 612 in the Carina dwarf galaxy (Venn *et al.* 2012), though others have now been found in Carina (Norris *et al.* 2016) and Sculptor (Jablonka *et al.* 2015). Similar stars have been found in the (outer) halo of the Galaxy by Ivans *et al.* (2003) and Jacobson *et al.* (2015).

One possibility is that the iron-enrichment comes from inhomogeneous mixing of the interstellar medium at later times. Pockets of SN Ia material may occur if the gas is poorly mixed. Some chemical evolution models (e.g., Revaz & Jablonka 2012, Wise *et al.* 2012, Romano *et al.* 2015) support inhomogeneous mixing on sufficiently long timescales only in the lower mass dwarf galaxies.

Another option has been suggested by Kobayashi *et al.* (2015) where SN Iax have a more significant role in the chemical evolution of a dwarf galaxy. SN Iax are those from a higher mass (hybrid, C+O+Ne) white dwarf, formed at lower metallicities (even as low as [Fe/H] $= -2.5$).

4. Carbon-rich Stars in Dwarf Galaxies

Carbon-enhancements ([C/Fe] > 0.7) are commonly found in metal-poor stars in the Galaxy and typically associated with binary systems (e.g., Starkenburg *et al.* 2014). However, the CEMP-no stars may be a different class of these stars, formed from primordial faint supernovae and dominating the early metal enrichment in dwarf galaxies. While CEMP-no stars are found in the Galaxy and the ultra faint dwarf galaxies (e.g., Beers & Christlieb 2005, Norris *et al.* 2013), they have only recently been found also in the classical dwarf galaxies (Skuladottir *et al.* 2015). Initial interpretations were that the early chemical evolution of the classical dwarf galaxies may have differed from those of the ultra faint dwarfs, however a recent analysis of cosmological models of dwarf galaxies has shown that CEMP-no stars would have formed in all systems, and with equal fractions in the populations with [Fe/H] < -4 (Salvadori *et al.* 2015). The difficulty is in finding these stars once the peak in the metallicity distribution function moves from [Fe/H] = -3 to -2, as in the higher mass classical dwarf galaxies. Also the classical dwarf galaxies are further away, meaning that only the brightest stars can be observed. Comparisons of the metallicities of the stars from CaT surveys versus those from high resolution studies shows that more metal-poor stars are detected at lower magnitudes (e.g., Lemasle *et al.* 2012).

5. Conclusions

The element abundance ratios in the most metal-poor stars ([Fe/H] < -3) in dwarf galaxies are excellent tests for the variety of SN II yields and other supernova models. Most elements show larger dispersions in [X/Fe] in dwarf galaxies, than similar stars in the Galactic halo. Spectral lines of some of the more interested elements (such as barium) can be so weak that we are near the detection limits of our high resolution spectrographs, even on the 8-10 meter telescopes. Also, the CEMP stars are likely in all dwarf galaxies, though can be harder to find in the more distant and higher mass (higher mean metallicity) classical dwarf galaxies.

The intermediate-metallicity stars ([Fe/H] ~ -2) can show the largest deviations from similar metallicity stars in the Galactic halo, due to differences in the later chemical evolution stages of the dwarf galaxies. This can include effects due to inhomogeneous mixing of the interstellar medium, including SN Ia pockets of iron-enrichment, in the lower mass dwarf galaxies.

References

Aden, D. *et al.* 2011, *A&A*, 525, 153
Feltzing, S. *et al.* 2009, *A&A*, 508, 1
Frebel, A. 2010, in Astron. Nachr. 331, 474
Frebel, A., Simon, J. D., Kirby, E. N. 2014, *ApJ*, 786, 74
Frebel, A., Norris, J. E. 2015, *ARAA*, 53, 631
Gilmore, G. 2012, in Assembling the Puzzle of the Milky Way, Le Grand-Bournand, France.
Heger, A., Woosley, S. E. 2003, *ApJ*, 567, 532
Heger, A., Woosley, S. E. 2010, *ApJ*, 724, 341
Hendricks, B. *et al.* 2014, *ApJ*, 785, 102
Ivans I., *et al.* 2003, *ApJ*, 592, 906
Iwamoto, N., *et al.* 2005, *Sci*, 309, 451
Jablonka, P., *et al.* 2015, *A&A*, 583, 67
Jablonka, P., *et al.* 2016, in prep
Kobayashi, C., *et al.* 2015, *ApJ*, 804, 24

Lemasle, B., *et al.* 2012, *A&A*, 538, 100
Lemasle, B., *et al.* 2014, *A&A*, 572, 88
Norris, J. E., *et al.* 2010, *ApJ*, 723, 1632
Norris, J. E., *et al.* 2013, *ApJ*, 762, 28
Norris, J. E., *et al.* 2016, in prep
Revaz, Y., Jablonka, P 2012, *A&A*, 538, 82
Roederer, I.U. 2013 *AJ*, 145, 26
Romano, D., Bellazzini, M., Starkenburg, E., Leaman, R. 2015, *MNRAS*, 446, 4220
Salvadori, S., Skuladottir, A., Tolstoy, E. 2015 *MNRAS*, 454, 1320
Shen, S., Cooke, R. J., Ramirez-Ruiz, E., *et al.* 2015, *ApJ*, 807, 115
Simon, J. D. *et al.* 2015, *ApJ*, 802, 93
Skuladottir, A., *et al.* 2015, *A&A*, 574, 129
Starkenburg, E., *et al.* 2014, *MNRAS*, 441, 1217
Starkenburg, E., *et al.* 2013, *MNRAS*, 429, 725
Tafelmeyer, M., *et al.* 2010, *A&A*, 524, A58
Tolstoy, E., *et al.* 2009, *ARAA*, 47, 371
Venn, K. A., *et al.* 2012, *ApJ*, 751, 102
Wanajo, S. 2013, *ApJ*, 770, 22
Wanajo, S. *et al.* 2014, *ApJ*, 789, 39
Wise, J. *et al.* 2012, *ApJ*, 745, 50

The General Assembly of Galaxy Halos: Structure, Origin and Evolution
Proceedings IAU Symposium No. 317, 2015
A. Bragaglia, M. Arnaboldi, M. Rejkuba & D. Romano, eds.

doi:10.1017/S1743921315006754

Chemical enrichment in Ultra-Faint Dwarf galaxies

Donatella Romano

INAF, Osservatorio Astronomico di Bologna,
Via Ranzani 1, I-40127, Bologna, Italy
email: donatella.romano@oabo.inaf.it

Abstract. Our view of the Milky Way's satellite population has radically changed after the discovery, ten years ago, of the first Ultra-Faint Dwarf galaxies (UFDs). These extremely faint, dark-matter dominated, scarcely evolved stellar systems are found in ever-increasing number in our cosmic neighbourhood and constitute a gold-mine for studies of early star formation conditions and early chemical enrichment pathways. Here we show what can be learned from the measurements of chemical abundances in UFD stars read through the lens of chemical evolution studies, point out the limitations of the classic approach, and discuss the way to go to improve the models.

Keywords. Galaxies: dwarf, galaxies: evolution, galaxies: individual: Boötes I, Segue 1, stars: abundances

1. Introduction

The past two decades have seen the advent of the era of large-scale digital sky surveys, which has had a dramatic impact on studies of the local universe (Ivezić *et al.* 2012). In particular, since 2005 the Sloan Digital Sky Survey (SDSS; York *et al.* 2000) has more than doubled the number of known Milky Way satellites, alleviating —if not providing a way out of— the 'missing satellite' problem (Klypin *et al.* 1999; Moore *et al.* 1999; see Tollerud *et al.* 2008). Nowadays deeper surveys are being conducted (the SkyMapper Southern Sky Survey, the Dark Energy Survey, Pan-STARRS...), that will greatly improve the areal coverage and eventually deliver a complete census of UFDs out to the Milky Way virial radius. This will set more stringent constraints to large-scale structure formation theories. In the meanwhile, the brightest giant star members of the newly-discovered satellites are promptly targeted for spectroscopy in order to shed light on a number of questions: did these small, sparse stellar systems suffer 'one-shot' or extended star formation? what is the level and significance of chemical inhomogeneity inside them? was stellar feedback effective in removing all the gas left over from the star formation process, or did the interaction with the environment play a major role? are the chemical properties of UFDs consistent with the idea that they are the surviving relics of the accretion processes that shaped the diffuse Milky Way halo in hierarchical theories of galaxy formation?

As for the last question, it has been noted that the cumulative metallicity distribution function (MDF) for the metal-poor tails of a number of UFDs well compares to that of the Milky Way halo (Kirby *et al.* 2008). Moreover, a significant number of UFD stars with [Fe/H]< -2.5 have [α/Fe] ratios consistent with those of halo stars of similar metallicity, within the uncertainties (Vargas *et al.* 2013). This would indicate that the UFDs are good candidates for the basic building blocks from which the halo was assembled. However, the abundances of neutron-capture elements in UFDs do not follow the typical halo

Table 1. Overview of currently available high-resolution spectroscopic data for UFD giants.

Object name	Instrument@Telescope	# of analysed member stars	Reference
Boötes I	HIRES@Keck	7	Feltzing *et al.* (2009)
	UVES@VLT	8	Gilmore *et al.* (2013)
	HDS@Subaru	6	Ishigaki *et al.* (2014)
Ursa Major II	HIRES@Keck	3	Frebel *et al.* (2010)
Coma Berenices	HIRES@Keck	3	Frebel *et al.* (2010)
Hercules	MIKE@Magellan	2	Koch *et al.* (2008)
Leo IV	MIKE@Magellan	1	Simon *et al.* (2010)
Segue 1	MIKE@Magellan+HIRES@Keck	6	Frebel *et al.* (2014)
Boötes II	HIRES@Keck	1	Koch & Rich (2014)
Segue 2	MIKE@Magellan	1	Roederer & Kirby (2014)

Note: last update July 31, 2015.

abundance pattern in the metallicity range $-3.0 \leqslant$[Fe/H]$\leqslant -2.5$ (Koch *et al.* 2008; Frebel *et al.* 2010), making the identification with the halo building blocks less straightforward (see also complementary arguments by Giuliana Fiorentino, these proceedings).

It is clear that, in order to unravel the UFDs' chemical enrichment history and, thus, (at least a piece of) the halo assembly history, one has to fully exploit the information encoded in the abundances of *different chemical elements, produced by different stellar factories on different timescales.* To do this, high-resolution spectroscopic observations in UFDs are badly needed. Unfortunately, high-resolution spectroscopy in these objects is very challenging and only a few stars per galaxy have been observed so far (see Table 1). In the following, I will mostly focus on Boötes I. Located at 66 $\pm$ 2 kpc distance from the Sun and with an absolute visual magnitude of $M_V = -6.3 \pm 0.2$ mag (McConnachie 2012), it is one of the brightest UFDs orbiting the Milky Way and it has the largest number of high-resolution abundance determinations, as well as a well-defined MDF (Lai *et al.* 2011).

2. Chemical evolution models

In Romano *et al.* (2015) we present one-zone chemical evolution models for Boötes I. The usual set of integro-differential equations (Talbot & Arnett 1971; Tinsley 1980; Matteucci 2001) is solved numerically and the evolution of the average abundances of several chemical elements in the interstellar medium (ISM) is followed by taking into account in detail the stellar lifetimes (no instantaneous recycling approximation). The adopted stellar yields are tested against the Milky Way data and shown to reproduce reasonably well the abundance patterns of several elements (Romano *et al.* 2010).

We run *classic models*, resting on a number of assumptions: *(i)* smooth infall of gas of primordial chemical composition provides the raw material for star formation; *(ii)* the stellar ejecta mix instantaneously with the ISM; *(iii)* galactic outflows develop when the thermal energy of the gas heated by supernova (SN) explosions exceeds its binding energy; *(iv)* the stellar initial mass function (IMF; Kroupa 2001) is constant in space and time. We also run *cosmologically-motivated models*, where we adopt mass assembly histories of candidate Boötes I-like galaxies selected from the mock satellite catalogue of Starkenburg *et al.* (2013). These were obtained by coupling the Munich semi-analytical model of galaxy formation (Li *et al.* 2010, and references therein) to the Aquarius dark matter simulations (Springel *et al.* 2008). According to Li *et al.* (2010) and Starkenburg *et al.* (2013), only part of the stellar ejecta is directly mixed to the cold ISM, while 95 per cent of the newly-produced metals are recycled through the hot galaxy's halo. Furthermore, the matter expelled through the outflow is stored in an *'ejected gas'* component and may

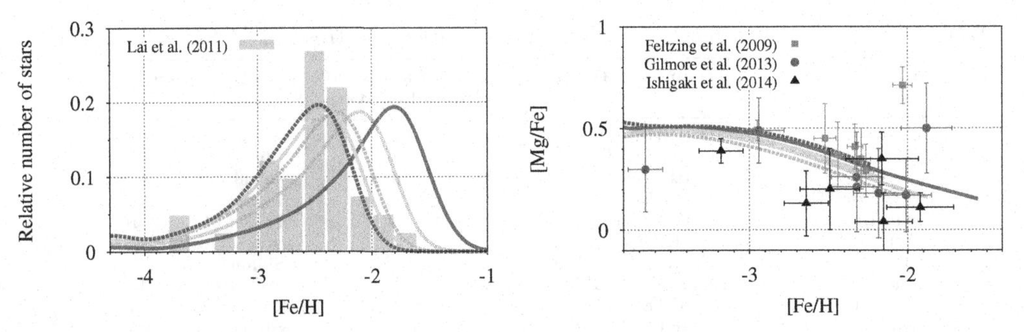

Figure 1. *Left panel:* theoretical *(curves)* and observed *(histogram)* MDFs of Boötes I. *Right panel:* theoretical *(curves)* and observed *(symbols)* [Mg/Fe] versus [Fe/H] relations for Boötes I. The high [Mg/Fe] = 0.71 measured by Feltzing *et al.* (2009) for the star Boo-127 has been revised downwards by both Gilmore *et al.* (2013) and Ishigaki *et al.* (2014).

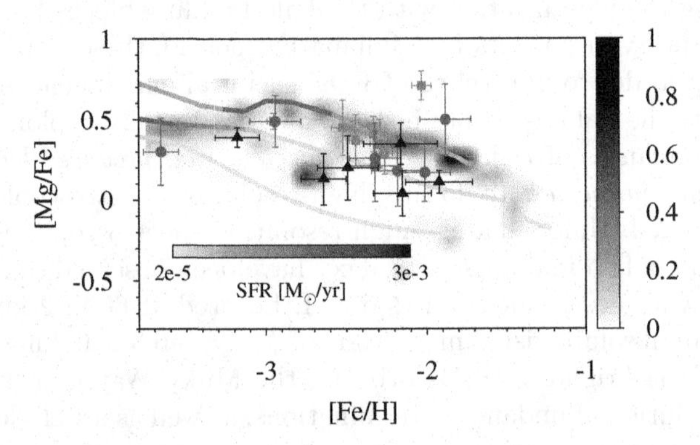

Figure 2. [Mg/Fe] versus [Fe/H] relations for the models run in a fully cosmological context. The density map shows the distribution of synthetic long-lived stars for 11 Boötes I candidates in the mock Milky Way satellite catalogue of Starkenburg *et al.* (2013) normalised to its maximum value. The curves show the predictions of three representative models and are colour-coded according to the strength of the star formation rate (see legend on the bottom of the plot).

eventually be re-accreted by the galaxy. We include all these gas flows in our models. Last but not least, the processes of stripping are included in the cosmologically-motivated models, while they are neglected in the classical models.

2.1. *A small fish in a big pond?*

In Fig. 1 we show the MDF *(left panel)* and [Mg/Fe] versus [Fe/H] relation of Boötes I stars *(right panel)* predicted by our classic chemical evolution model *(lines)* with different assumptions about the duration and efficiency of star formation, as well as about the dark and baryonic matter content of the galaxy (see Romano *et al.* 2015 for details on the model parameters). The model predictions are compared to the observations (legend on the top-left corner of each plot). Similarly, Fig. 2 shows the predictions of the fully cosmological models. In general, a good agreement between model predictions and data is found by assuming a very low efficiency of conversion of gas into stars, which confirms previous results for UFDs by Salvadori & Ferrara (2009) and Vincenzo *et al.* (2014). Successful models, in fact, require that $\sim 10^7$ $M_\odot$ of gas are present at the epoch of star formation to dilute the metals produced by SN explosions; out of these, only $\sim 6\times10^4$ $M_\odot$ end up as long-lived stars that still inhabit the galaxy (value of present-day stellar mass from McConnachie 2012). We do not find any clearcut evidence that galactic winds

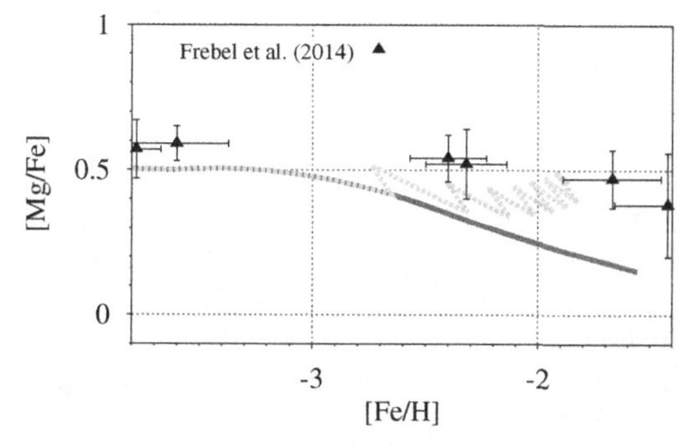

Figure 3. [Mg/Fe] versus [Fe/H] behaviour in Segue 1. The triangles are high-resolution spectroscopic observations of giant stars by Frebel *et al.* (2014). The lines are the predictions of two chemical evolution models. The duration of the star formation is set to 1 Gyr for both models, but in one case *(dashed line)* it is assumed that the ejecta from SNeIa is vented out of the galaxy much more efficiently than that from SNeII.

deprive the galaxy of all its residual gas; since no neutral gas is associated to Boötes I today, we conclude that stripping must have played a fundamental role.

The need for a substantially more massive progenitor can be easily understood. If each type II SN (SNII) produces 0.07 $M_\odot$ of Fe that fully mix within the considered volume, in the absence of metal-rich outflows <2 SNII explosions suffice to increase the metal content of a 10^5 $M_\odot$ gas cloud from [Fe/H] $\simeq -4$ to $\simeq -3$, while a metallicity [Fe/H] $\simeq -2$ is reached after $\sim$20 such events. Since up to 700 SNeII are expected to have exploded in Boötes I (when considering a present-day mass in stars of $\sim 6\times10^4$ $M_\odot$ and a canonical IMF), we end up with the request that some 10^6 $M_\odot$ of gas must have been present to dilute the freshly-produced metals and not to shift the peak of the MDF towards values higher than observed (Fig. 1, *left panel*). Since some tens of SNeIa are also expected to have polluted the ISM while Boötes I was still forming stars (see Romano *et al.* 2015, their figure 6), even more diluting gas is needed. It is worth stressing that our conclusion that Boötes I must have been much more massive in the past is based mainly on the assumption of a canonical IMF (Kroupa 2001): the steeper the IMF slope in the high-mass regime, the lower the number of core-collapse SNe and, hence, the lower the gas amount necessary to dilute the ejecta.

2.2. *Star formation histories and chemical enrichment*

Recently, Brown *et al.* (2014) combined photometric and spectroscopic data for five UFDs and found that these galaxies formed the bulk of their stars long ago, by $z \sim 6$ (12.8 Gyr ago). They did not experience any significant star formation afterwards, most likely because of the quenching effects of reionization. In principle, one might try to put more stringent constraints on the duration of the star formation in UFDs by using chemical evolution arguments. For instance, the occurrence of a knee in the [α/Fe] versus [Fe/H] relation can be interpreted as an indication for a significant contribution to the chemical evolution from SNeIa and the [Fe/H] value where the knee occurs related to the system's formation timescale (Matteucci & Brocato 1990, their figure 4). In practice, we analyse models for Boötes I where the bulk of the stellar population forms in $\sim$100 Myr or in $\sim$1 Gyr and find that all can fit the data reasonably well after a proper fine-tuning of the free parameters; the degeneracy is mainly due to the approximations introduced to treat the galactic outflows (see Romano *et al.* 2015, for a discussion).

Fig. 3 further illustrates this problem, by showing the predictions of two different models for the UFD Segue 1. This galaxy displays an almost constant [Mg/Fe] ratio in the

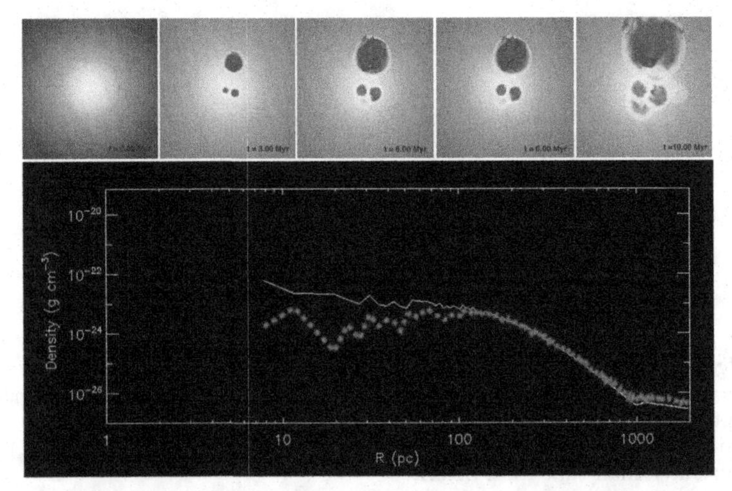

Figure 4. *Top:* gas density maps in the XY plane for the simulated Boötes I UFD at different times from 0 to 10 Myr, as labeled in the bottom right corner of each snapshot. The simulation uses a box size of 2 kpc. The maximal refinement level is set by $l_{\max} = 11$, which corresponds to a sub-parsec (0.9 pc) maximal spatial resolution. *Bottom:* gas density profiles at $t = 0$ Myr *(solid line)* and $t = 10$ Myr *(dashed line)*. After 10 Myr from the beginning of the simulation, the gas starts to be pushed towards the external regions, because of the energy release by multiple SNe exploding in associations.

metallicity range $-4 <$[Fe/H]< -1.5 (Frebel *et al.* 2014) —a clear signature of chemical enrichment from core-collapse SNe alone. In the framework of classic chemical evolution models, this is interpreted as an indication for a short-lasting star formation, of the order of 30 Myr or so. In fact, if the star formation lasts longer, a knee appears, due to Fe production from SNeIa (Fig. 3, *solid line*). However, if the products of SNIa nucleosynthesis are expelled from the galaxy before mixing with the ISM and more efficiently than SNII ejecta, the [Mg/Fe] ratio in the galaxy may be kept flat on timescales as long as ~1 Gyr (Fig. 3, *dashed line*). Such an assumption may be justified on the basis of 2D hydrodynamical simulations by Recchi *et al.* (2001), who found that the ejecta of SNeIa, exploding in a medium already rarefied by previous SNII explosions, find an easier way out of the galaxy.

3. Future work

Chemical evolution models have succeeded in explaining many observed properties of galaxies. However, when moving to stellar systems with lower and lower dynamical masses, severe uncertainties affect the model predictions: the treatment of mechanical feedback from stars, the conditions set for the development of large-scale outflows, the assumed mass loading factors of the ISM cause a high degeneracy of the proposed solutions. In order to make a quantum leap forward in our understanding of the formation and evolution of systems with very shallow potential wells, detailed chemical evolution models must be coupled self-consistently to 3D hydrocodes. To this aim, we have recently started to study the evolution of a galaxy resembling Boötes I by means of a customised version of the adaptive mesh refinement code RAMSES (Teyssier 2002). In Fig. 4 we show some preliminary results of our simulation, to be presented in a forthcoming paper.

Acknowledgements

I thank my collaborators M. Bellazzini, F. Calura, C. G. Few, A. D'Ercole, R. Leaman and E. Starkenburg. I acknowledge financial support from the IAU and from the Italian Ministero dell'Istruzione, dell'Università e della Ricerca (MIUR) through the grant *'The Chemical and Dynamical Evolution of the Milky Way and Local Group Galaxies'*, prot. 2010LY5N2T. The simulations discussed in this contribution are performed on the Galileo supercomputer at CINECA (Bologna, Italy), within projects supported by the Italian SuperComputing Resource Allocation (ISCRA).

References

Brown, T. M., *et al.* 2014, *ApJ*, 796, 91
Feltzing, S., Eriksson, K., Kleyna, J., & Wilkinson, M. I. 2009, *A&A* (Letters), 508, L1
Frebel, A., Simon, J. D., Geha, M., & Willman, B. 2010, *ApJ*, 708, 560
Frebel, A., Simon, J. D., & Kirby, E. N. 2014 *ApJ*, 786, 74
Gilmore, G., *et al.* 2013, *ApJ*, 763, 61
Ishigaki, M. N., Aoki, W., Arimoto, N., & Okamoto, S. 2014, *A&A*, 562, A146
Ivezić, Ž., Beers, T. C., & Jurić, M. 2012, *ARAA*, 50, 251
Kirby, E. N., *et al.* 2008, *ApJ* (Letters), 685, L43
Klypin, A., Kravtsov, A. V., Valenzuela, O., & Prada, F. 1999, *ApJ*, 522, 82
Koch, A., *et al.* 2008, *ApJ* (Letters), 688, L13
Koch, A. & Rich, R. M. 2014, *ApJ*, 794, 89
Kroupa, P. 2001, *MNRAS*, 322, 231
Lai, D. K., *et al.* 2011, *ApJ*, 738, 51
Li, Y.-S., De Lucia, G., & Helmi, A. 2010, *MNRAS*, 401, 2036
Matteucci, M. 2001, *The Chemical Evolution of the Galaxy* (Dordrecht: Kluwer)
Matteucci, M. & Brocato, E. 1990, *ApJ*, 365, 539
McConnachie, A. W. 2012, *AJ*, 144, 4
Moore, B., *et al.* 1999, *ApJ* (Letters), 524, L19
Recchi, S., Matteucci, F., & D'Ercole, A. 2001, *MNRAS*, 322, 800
Roederer, I. U. & Kirby, E. N. 2014, *MNRAS*, 440, 2665
Romano, D., Bellazzini, M., Starkenburg, E., & Leaman, R. 2015, *MNRAS*, 446, 4220
Romano, D., Karakas, A. I., Tosi, M., & Matteucci, F. 2010, *A&A*, 522, A32
Salvadori, S. & Ferrara, A. 2009, *MNRAS* (Letters), 395, L6
Simon, J. D., Frebel, A., McWilliam, A., Kirby, E. N., & Thompson, I. B. 2010, *ApJ*, 716, 446
Springel, V., *et al.* 2008, *MNRAS*, 391, 1685
Starkenburg, E., *et al.* 2013, *A&A*, 549, A88
Talbot, R. J., Jr. & Arnett, W. D. 1971, *ApJ*, 170, 409
Teyssier, R. 2002, *A&A*, 385, 337
Tinsley, B. M. 1980, *Fundam. Cosm. Phys.*, 5, 287
Tollerud, E. J., Bullock, J. S., Strigari, L. E., & Willman, B. 2008, *ApJ*, 688, 277
Vargas, L. C., Geha, M., Kirby, E. N., & Simon, J. D. 2013, *ApJ*, 767, 134
Vincenzo, F., Matteucci, F., Vattakunnel, S., & Lanfranchi, G. A. 2014, *MNRAS*, 441, 2815
York, D. G., Adelman, J., Anderson, J. E., *et al.* 2000, *AJ*, 120, 1579

The General Assembly of Galaxy Halos: Structure, Origin and Evolution
Proceedings IAU Symposium No. 317, 2015
A. Bragaglia, M. Arnaboldi, M. Rejkuba & D. Romano, eds.

doi:10.1017/S1743921315008467

Multiple populations in the Sagittarius nuclear cluster M 54 and in other anomalous globular clusters

A. P. Milone

Research School of Astronomy & Astrophysics, Australian National University, Mt Stromlo Observatory, via Cotter Rd, Weston, ACT 2611, Australia. email: milone@mso.anu.edu.au

Abstract. M 54 is the central cluster of the Sagittarius dwarf galaxy. This stellar system is now in process of being disrupted by the tidal interaction with the Milky Way and represents one of the building blocks of the Galactic Halo.
Recent discoveries, based on the synergy of photometry and spectroscopy have revealed that the color-magnitude diagram (CMD) of some massive, anomalous, Globular Clusters (GCs) host stellar populations with different content of heavy elements. In this paper, I use multi-wavelength *Hubble Space Telescope* (*HST*) photometry to detect and characterize multiple stellar populations in M 54. I provide empirical evidence that this GC shares photometric and spectroscopic similarities with the class of anomalous GCs. These findings make it tempting to speculate that, similarly to Sagittarius nuclear cluster M 54, other anomalous GCs were born in an extra-Galactic environment.

1. Introduction

Several studies, based on high-precision photometry, have established that the CMD of nearly all the GCs is made of multiple stellar sequences, that have been identified among all the evolutionary stages (e.g. Milone *et al.* 2012; Piotto *et al.* 2015). The multiple sequences of the majority of GCs correspond to stellar populations with different content of those light elements involved in H-burning reactions like C, N, O, Na, Mg, and Al (e.g. Gratton *et al.* 2004, 2012; Marino *et al.* 2008; Yong *et al.* 2008) and different helium abundance (e.g. D'Antona *et al.* 2002, 2005; Piotto *et al.* 2007; Milone *et al.* 2014). Noticeable, while star-to-star variations in light elements and helium have been observed in nearly all the clusters, most of them have homogeneous iron content (e.g. Carretta *et al.* 2009). In this context, one of the most-intriguing discoveries of the last years is that a small, but still increasing, number of massive GCs host stellar populations with different metallicity (Marino *et al.* 2009, 2011a, 2015; Da Costa *et al.* 2009; Carretta *et al.* 2010a,b; Yong *et al.* 2014; Johnson *et al.* 2015).

In this paper, I will investigate M 54, which is a very massive GCs ($M \sim 2 \times 10^6$ $M_{\odot}$, McLaughlin & Van der Marel 2005) located in the nuclear region of the Sagittarius dwarf spheroidal galaxy (Sgr). Due to its present-day position, it has been speculated that M 54 is the nucleus of the Sgr dwarf and that the stellar system including M 54 and the Sgr represents the local counterpart of nucleated dwarf ellipticals (e.g. Sarajedini & Layden 1995). As an alternative, M 54 formed in an external region of the dwarf and has been then dragged into the bottom of the Sgr potential well because of decay of the orbit due to dynamical friction (e.g. Monaco *et al.* 2005; Bellazzini *et al.* 2008). In any case, it is widely accepted that M 54 is associated with the Sgr and that it formed and evolved in the environment of a dwarf spheroidal galaxy.

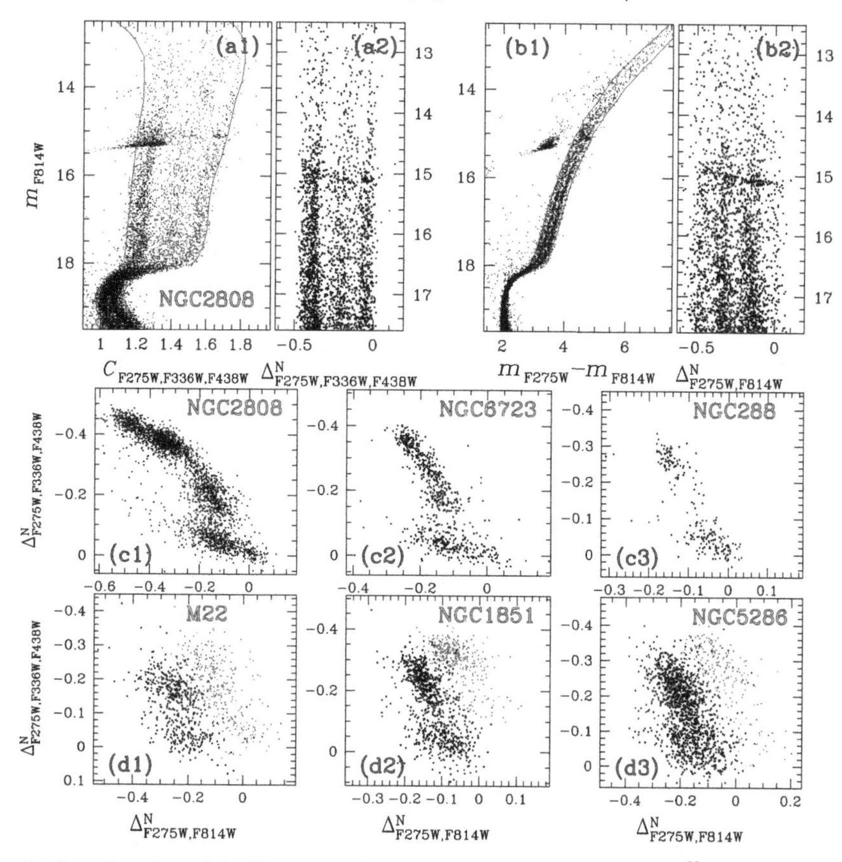

Figure 1. Panels a1 to b2 illustrate the procedure to derive the $\Delta^{N}_{F275W,F336W,F438W}$ vs. $\Delta^{N}_{F275W,F814W}$ diagram of RGB stars in NGC 2808 plotted in the panel c1. Panels c2, c3, d1, d2, and d3 show the same diagram but for the monometallic GCs NGC 6723 and NGC 288 and for the anomalous GCs M 22, NGC 1851, and NGC 5286. Stars in the anomalous RGB are colored red (see text for details).

In the following, I will present multi-wavelength *HST* photometry of M 54 and of the surrounding Sgr dwarf from the *UV legacy survey of Galactic GCs* (Piotto *et al.* 2015) and identify the multiple stellar populations within the M 54+Sgr stellar system. Moreover, I will use both spectroscopy and photometry to investigate possible connections between M 54 and the other anomalous GCs exhibiting metallicity variations.

2. Mapping multiple stellar populations in GCs

Recent papers have revealed that any CMD or two-color diagram made with appropriate combination of filters is very efficient to identify multiple stellar populations in GCs. In particular, the $m_{F275W} - m_{F814W}$ color and the pseudo-color $C_{F275W,F336W,F438W} = (m_{F275W} - m_{F336W}) - (m_{F336W} - m_{F438W})$ introduced by Milone *et al.* (2013) are very sensitive to stellar populations with different helium and light-element abundance and have been recently used by Piotto *et al.* (2015) in their survey of GCs to identify stellar populations in a large number of clusters. As an example, panels a1 and b1 of Fig. 1 show m_{F814W} vs. $C_{F275W,F336W,F438W}$ and m_{F814W} vs. $m_{F275W} - m_{F814W}$ for NGC 2808 and clearly reveal its multiple RGB.

In order to combine information from both diagrams and identify stellar populations within NGC 2808, Milone *et al.* (2015a) have introduced the method illustrated in the upper panels of Fig. 1, where the blue and the red fiducial superimposed on each diagram mark the bluest and the reddest envelope of the RGB. These fiducials are used to verticalize the two CMDs in a way that the blue and the red fiducials translate into vertical lines. The verticalized, $m_{\rm F814W}$ vs. $\Delta^{\rm N}_{\rm F275W,F336W,F438W}$ and $m_{\rm F814W}$ vs. $\Delta^{\rm N}_{\rm F275W,F814W}$ diagrams are plotted in panels a2 and b2 for RGB stars in NGC 2808† The resulting $\Delta^{\rm N}_{\rm F275W,F336W,F438W}$ vs. $\Delta^{\rm N}_{\rm F275W,F814W}$ plot is shown in panel (c1) for NGC 2808 and reveals its five distinct populations (see Milone *et al.* 2015a for details). Alvio Renzini has named this diagram 'chromosome map' and I will keep this nickname in the following.

Panels c2–d3 of Fig. 1 show the chromosome map for five additional GCs. Among them NGC 6723 and NGC 288 have homogeneous abundance of iron and neutron-capture elements, similarly to NGC 2808. On the contrary, M 22, NGC 1851, and NGC 5286 are anomalous GCs and host two distinct groups of stars with different metallicity and s−process elements-abundance (Marino *et al.* 2009, 2012, 2015; Yong *et al.* 2008, 2009, 2015; Villanova *et al.* 2010; Lee 2015; Lim *et al.* 2015; Carretta *et al.* 2010a).

Noticeably, RGB stars in anomalous GCs distribute along two parallel sequences, with the anomalous stars enhanced in heavy elements (red points in Fig. 1) having, on average, larger $\Delta^{\rm N}_{\rm F275W,F814W}$ values than stars with standard chemical composition. In contrast, the sequence of anomalous stars is not present in monometallic GCs. This finding makes the chromosome map a powerfull tool to identify anomalous GCs from photometry of RGB stars.

3. Multiple stellar populations in M 54

In the last two decades several papers on the photometry of stellar populations in the system including M 54 and the Sgr dwarf have been published by different groups (see Siegel *et al.* 2007 and references therein). In contrast, very little attention has been dedicated to stellar populations within M 54.

A strong evidence that the CMD of M 54 is not consistent with a simple population has been provided by Piotto *et al.* (2012) who have discovered that the sub-giant branch (SGB) of this cluster is bimodal in the optical $m_{\rm F606W}$ vs. $m_{\rm F606W} - m_{\rm F814W}$ CMD. This is an unusual feature for Galactic GCs, indeed while the SGB of clusters without metallicity variations is narrow and well defined when observed in visual filters, a split or broadened SGB is a distinctive feature of anomalous GCs with variation in [Fe/H] and [(C+N+O)/Fe] (Milone *et al.* 2008; Cassisi *et al.* 2008; Ventura *et al.* 2009; Sbordone *et al.* 2011; Marino *et al.* 2009, 2012).

More recently, multi-wavelength photometry from the *HST* UV survey of Galactic GCs (Piotto *et al.* 2015) has revealed that M 54 hosts a complex system of multiple stellar populations that are clearly visible along the main sequence, the SGB, the RGB, and horizontal branch in the $m_{\rm F275W}$ vs. $m_{\rm F275W} - m_{\rm F814W}$ CMD of Fig. 2a. The chromosome map of RGB stars plotted in the panel b of Fig. 2 clearly separates the M 54 members (black points) from the Sgr stars (aqua points). The Hess diagram shown in the panel c for M 54 reveals that the chromosome map of this cluster consists of two sequences of stars in close analogy with what observed in anomalous GCs (see Fig. 1). Moreover, the fact that each sequence exhibits two or more distinct clumps demonstrates that both

† Specifically, Milone *et al.* (2015a) have defined:
$\Delta^{\rm N}_{\rm X} = W[(X - X_{\rm blue\ fiducial})/(X_{\rm red\ fiducial} - X_{\rm blue\ fiducial})] - 1$,
where X=($m_{\rm F275W} - m_{\rm F814W}$) or ($C_{\rm F336W,F438W,F814W}$) and $X_{\rm blue\ fiducial}$ and $X_{\rm red\ fiducial}$ are obtained by subtracting the color of the fiducial at the corresponding F814W magnitude from the color of each star. The constant, W, is chosen as the distance between the red and the blue fiducial two F814W magnitudes above the turn off.

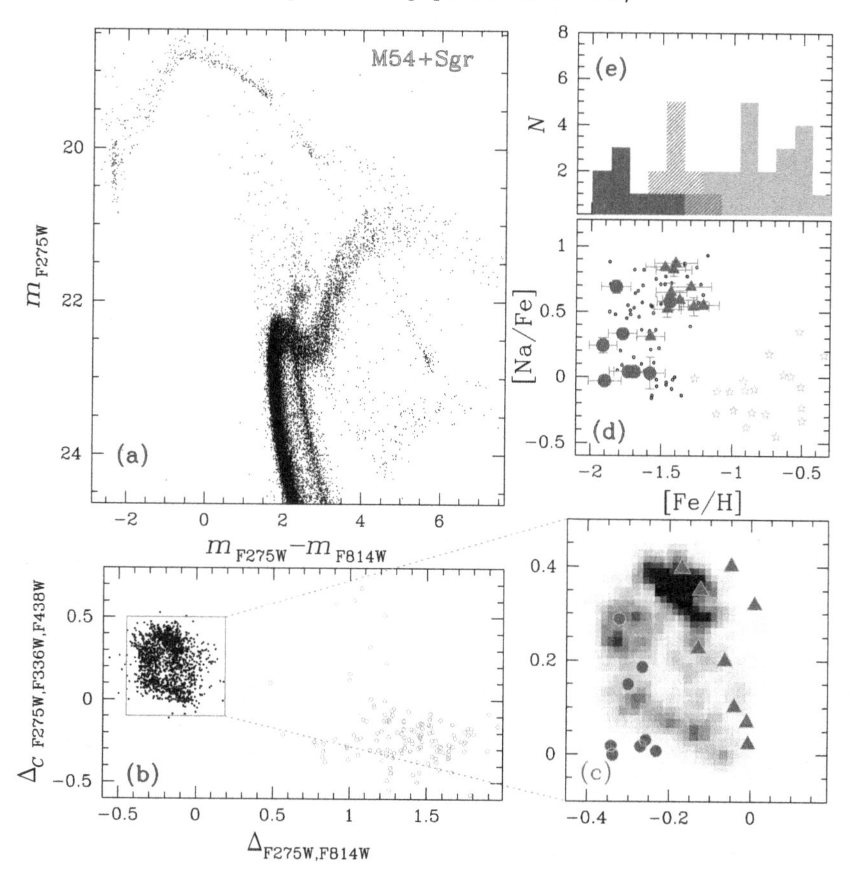

Figure 2. m_{F275W} vs. $m_{F275W} - m_{F814W}$ CMD for stars in a ~3×3 region around the center of M 54 (panel a). Panel b shows the chromosome map for M 54 RGB stars (black points) and for Sgr stars (aqua circles), while the Hess diagram of M 54 stars is plotted in panel c. Panel d shows [Na/Fe] vs. [Fe/H] from Carretta *et al.* (2010b), and the histograms of the iron distribution is provided in panel e for stars in the standard and anomalous population of M 54 and for stars in the Sgr dwarf spheroidal galaxy. Blue and red colors in panels c, d, and e indicate normal and anomalous stars in M 54, respectively, while Sgr stars are represented with aqua color codes.

the normal and the anomalous group host sub-populations with different helium and light-element abundance.

Carretta *et al.* (2010b) have derived homogeneous abundances of Fe, O, Na from high-resolution spectroscopy for more than 100 stars in M 54 and in the surrounding nucleus of the Sgr dwarf galaxy. They have concluded that M 54 exhibits intrinsic iron dispersion of about 0.2 dex and found a very-extended Na-O anticorrelation among cluster stars. Panel (d) of Fig. 2 shows [Na/Fe] vs. [Fe/H] from Carretta and collaborators for M 54 members (black symbols) and for Sgr stars (aqua symbols).

For 18 stars both spectroscopy and *HST* photometry is available. Among them, the eight stars represented with large blue dots in Fig. 2 belong to the normal RGB, and the ten stars in the anomalous RGB are plotted with red triangles. Stars in the anomalous RGB are enhanced in [Fe/H], and have, on average, higher [Na/Fe] values than the standard RGB, in close analogy with what observed in M 22 and in the other anomalous GCs. As shown in panel d and e of Fig. 2, Sgr stars have solar sodium abundance relative to iron and define a tail at high metallicity in the histogram of the [Fe/H] distribution.

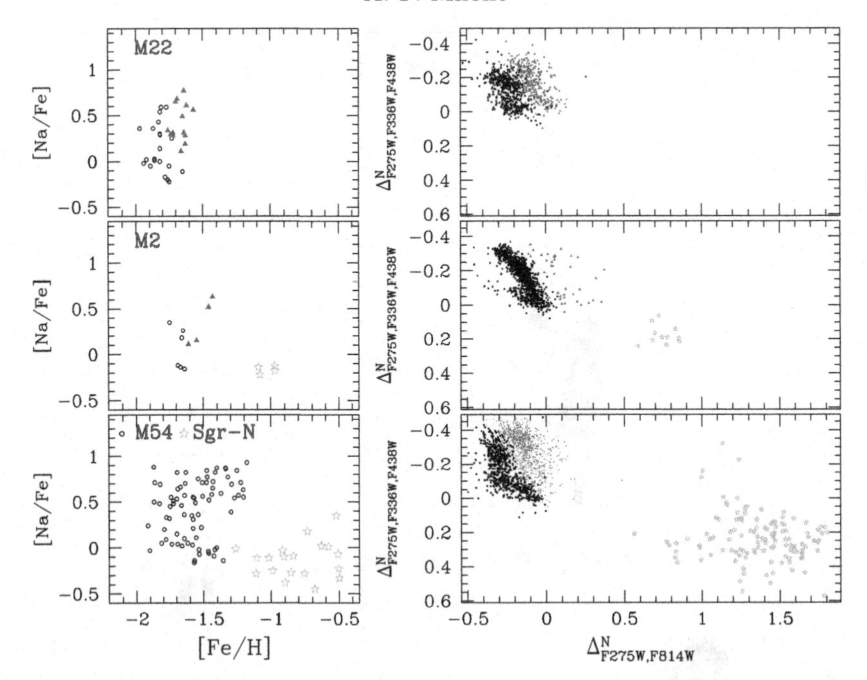

Figure 3. *Left Panels:* [Na/Fe] as a function of [Fe/H] for M 22 (from Marino *et al.* 2009, 2011a, top) M 2 (from Yong *et al.* 2014, middle) and M 54+Sgr (from Carretta at el. 2010b, bottom). *Right Panels:* Chromosome map of M 22, M 2, and M 54+Sgr (from Milone *et al.* 2015b and in preparation). Stars in the anomalous RGB are colored red while I used aqua starred symbols to represent stars in the Sgr dwarf and in the extreme population of M 2.

4. Comparison with anomalous GCs

Further information on multiple stellar population in M 54 comes from the comparison with anomalous GCs. In Fig. 3 I show [Na/Fe] vs. [Fe/H] from high-resolution spectroscopy (left panels) and the chromosome maps derived from the *HST* photometry presented by Piotto *et al.* (2015) for two anomalous GCs, namely M 22 and M 2 (upper and middle panel) and for the stellar system formed by M 54 and the Sgr (lower panel).

M 22 is the prototype of anomalous GCs and hosts two distinct groups of stars with different metallicity and different abundance of s-process elements and C+N+O (Marino *et al.* 2009; 2011a). When compared with M 22, M 2 exhibits a more-extreme chemical composition. Indeed it hosts three main stellar components, composed of metal-poor, metal-intermediate, and metal-rich stars (Yong *et al.* 2014; Milone *et al.* 2015b). While the metal-poor, metal-intermediate population resemble M 22, the metal-rich population is present in M 2 only.

The stellar system including M 54 and the Sgr shares strong similarities with M 2. M 54 seems to include populations similar to the iron-poor and iron-intermediate population of M 2, while the chemistry of Sgr stars is similar to the abundance pattern of the extreme population of M 2. Unlike the case of ω Cen, however, where most of the metal-rich stars are strongly enhanced in [Na/Fe] (Marino *et al.* 2011b), both the Sgr stars and the population of M 2 exhibit low [Na/Fe]. In addition, the extreme population of M 2 has lower α-element abundance than the bulk of M 2 stars in close analogy with what observed in the Sgr nucleus, where Sgr stars have lower [α/Fe] than M 54. Similar conclusions come from the comparison of the chromosome maps plotted in the right panels of Fig. 3.

In summary, I have provided evidence that the chromosome maps of GCs are efficient tools to identify anomalous GCs with heavy-element variations. Both spectroscopy and photometry show that the Sgr nuclear cluster, M 54 shares similarities with anomalous GCs like M 2, M 22, NGC 5286, NGC 1851, and ω Cen. These findings make it tempting to speculate that, similarly to M 54, the other anomalous GCs are the remnants of dwarf galaxies tidally disrupted by the interaction with the Milky Way.

Acknowledgements

I thank G. Piotto, L. Bedin, J. Anderson, I. King, A. Marino, D. Nardiello and the other collaborators involved in 'The *HST* UV Legacy Survey of Galactic Globular Clusters' and A. Bragaglia who has revised this manuscript. I acknowledge support by the Australian Research Council through Discovery Early Career Researcher Award DE150101816.

References

Bellazzini, M., Ibata, R. A., Chapman, S. C., *et al.* 2008, *AJ*, 136, 1147
Carretta, E., Bragaglia, A., Gratton, R., D'Orazi, V., & Lucatello, S. 2009, *A&A*, 508, 695
Carretta, E., Gratton, R. G., Lucatello, S., *et al.* 2010, *ApJ*, 722, L1
Carretta, E., Bragaglia, A., Gratton, R. G., *et al.* 2010, *ApJ*, 714, L7
Cassisi, S., Salaris, M., Pietrinferni, A., *et al.* 2008, *ApJ*, 672, L115
D'Antona, F., Caloi, V., Montalbán, J., Ventura, P., & Gratton, R. 2002, *A&A*, 395, 69
D'Antona, F., Bellazzini, M., Caloi, V., *et al.* 2005, *ApJ*, 631, 868
Da Costa, G. S., Held, E. V., Saviane, I., & Gullieuszik, M. 2009, *ApJ*, 705, 1481
Gratton, R., Sneden, C., & Carretta, E. 2004, *AA&A*, 42, 385
Gratton, R. G., Carretta, E., & Bragaglia, A. 2012, *A&AR*, 20, 50
Johnson, C. I., Rich, R. M., Pilachowski, C. A., *et al.* 2015, *AJ*, 150, 63
Lee, J.-W. 2015, *ApJS*, 219, 7
Lim, D., Han, S.-I., Lee, Y.-W., *et al.* 2015, *ApJS*, 216, 19
Marino, A. F., Villanova, S., Piotto, G., *et al.* 2008, *A&A*, 490, 625
Marino, A. F., Milone, A. P., Piotto, G., *et al.* 2009, *A&A*, 505, 1099
Marino, A. F., Sneden, C., Kraft, R. P., *et al.* 2011, *A&A*, 532, A8
Marino, A. F., Milone, A. P., Piotto, G., *et al.* 2011, *ApJ*, 731, 64
Marino, A. F., Milone, A. P., Sneden, C., *et al.* 2012, *A&A*, 541, A15
Marino, A. F., Milone, A. P., Karakas, A. I., *et al.* 2015, *MNRAS*, 450, 815
McLaughlin, D. E. & van der Marel, R. P. 2005, *ApJS*, 161, 304
Milone, A. P., Bedin, L. R., Piotto, G., *et al.* 2008, *ApJ*, 673, 241
Milone, A. P., Piotto, G., Bedin, L. R., *et al.* 2012, *ApJ*, 744, 58
Milone, A. P., Marino, A. F., Piotto, G., *et al.* 2013, *ApJ*, 767, 120
Milone, A. P., Marino, A. F., Dotter, A., *et al.* 2014, *ApJ*, 785, 21
Milone, A. P., Marino, A. F., Piotto, G., *et al.* 2015, *ApJ*, 808, 51
Milone, A. P., Marino, A. F., Piotto, G., *et al.* 2015, *MNRAS*, 447, 927
Monaco, L., Bellazzini, M., Bonifacio, P., *et al.* 2005, *A&A*, 441, 141
Piotto, G., Bedin, L. R., Anderson, J., *et al.* 2007, *ApJ*, 661, L53
Piotto, G., Milone, A. P., Anderson, J., *et al.* 2012, *ApJ*, 760, 39
Piotto, G., Milone, A. P., Bedin, L. R., *et al.* 2015, *AJ*, 149, 91
Sarajedini, A. & Layden, A. C. 1995, *AJ*, 109, 1086
Sbordone, L., Salaris, M., Weiss, A., & Cassisi, S. 2011, *A&A*, 534, A9
Siegel, M. H., Dotter, A., Majewski, S. R., *et al.* 2007, *ApJ*, 667, L57
Ventura, P., Caloi, V., D'Antona, F., *et al.* 2009, *MNRAS*, 399, 934
Villanova, S., Geisler, D., & Piotto, G. 2010, *ApJ*, 722, L18
Yong, D., Grundahl, F., Johnson, J. A., & Asplund, M. 2008, *ApJ*, 684, 1159
Yong, D. & Grundahl, F. 2008, *ApJ*, 672, L29
Yong, D., Grundahl, F., D'Antona, F., *et al.* 2009, *ApJ*, 695, L62
Yong, D., Roederer, I. U., Grundahl, F., *et al.* 2014, *MNRAS*, 441, 3396
Yong, D., Grundahl, F., & Norris, J. E. 2015, *MNRAS*, 446, 3319

The General Assembly of Galaxy Halos: Structure, Origin and Evolution
Proceedings IAU Symposium No. 317, 2015
A. Bragaglia, M. Arnaboldi, M. Rejkuba & D. Romano, eds.

doi:10.1017/S1743921315008820

Investigating the earliest epochs of the Milky Way halo

Else Starkenburg[1] and the Pristine Team†

[1]Leibniz Institute for Astrophysics (AIP)
An der Sternwarte 16, 18842 Potsdam, Germany
email: estarkenburg@aip.de

Abstract. Resolved stellar spectroscopy can obtain knowledge about chemical enrichment processes back to the earliest times, when the oldest stars were formed. In this contribution I will review the early (chemical) evolution of the Milky Way halo from an observational perspective. In particular, I will discuss our understanding of the origin of the peculiar abundance patterns in various subclasses of extremely metal-poor stars, taking into account new data from our abundance and radial velocity monitoring programs, and their implications for our understanding of the formation and early evolution of both the Milky Way halo and the satellite dwarf galaxies therein. I conclude by presenting the "Pristine" survey, a program on the Canada-France-Hawaii Telescope to study this intriguing epoch much more efficiently.

Keywords. stars: abundances, Galaxy: evolution, galaxies: abundances, early universe

---

1. Introduction

One field of study that has been very confined to the Milky Way galaxy and some surrounding galaxies is that of individual metal-poor stars. Stars keep a chemical imprint from their birth cloud during their lives and also preserve kinematical information of their infall orbits (if they were born outside of the potential they live in today) for long times. Therefore, studying the present-day stars can teach us about the past and provides a unique opportunity to study galaxy formation in general. Lessons learned from local well-studied stars and stellar systems can be applied to our understanding of the physics of star formation and the assembly of galaxies in the Universe.

The first generations of stars are particularly interesting. As we believe the metallicity of the galaxy is built up over generations, the lowest metallicity stars that still exist today probably represent our closest approach to studying the very first stars. These stars carry the imprint of very few supernovae and it is possible that they have been formed even before the epoch of reionization. At the present day a very limited number of stars are studied - particularly in high-resolution - with an iron-to-hydrogen ratio of less than a thousandth of that of the Sun, i.e., [Fe/H]<-3 or "extremely metal-poor" (see for some recent overviews and results Aoki *et al.* 2013; Cohen *et al.* 2013; Spite *et al.* 2013; Yong *et al.* 2013a; Placco *et al.* 2014). Going even lower, the numbers get more and more limited. At the present day, we know just eight stars with [Fe/H]<-4.5 (Christlieb *et al.* 2002; Frebel *et al.* 2005; Norris *et al.* 2007; Caffau *et al.* 2011; Hansen *et al.* 2014; Keller *et al.* 2014; Bonifacio *et al.* 2015), of which four show [Fe/H]<-5 (hyper

† Thanks to the full Pristine Team: PIs Else Starkenburg & Nicolas Martin, Co-Is: Piercarlo Bonifacio, Elisabetta Caffau, Raymond Carlberg, Patrick Côté, Patrick François, Stephen Gwyn, Vanessa Hill, Rodrigo Ibata, Julio Navarro, Alan McConnachie, Ruben Sanchez-Janssen & Kim Venn. Additional thanks to Matthew Shetrone, Alan McConnachie and Kim Venn for their role in the CEMP velocity monitoring survey.

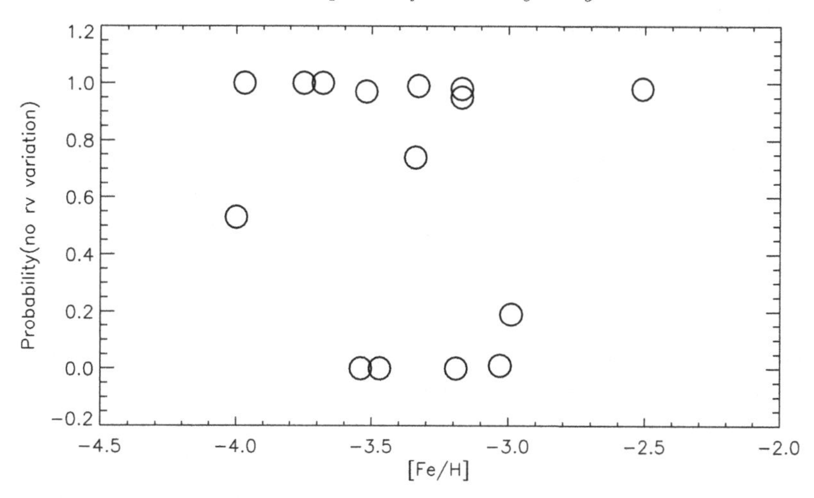

Figure 1. Data taken from Starkenburg *et al.* (2014, see also references therein) Table 1. For their sample of studied CEMP-no stars, here we plot the probability that the radial velocity observations can be explained by the uncertainty in the radial velocity determination solely. For this purpose, 3σ error bars are used (see for a discussion Section 3.2 of Starkenburg *et al.* 2014). It is clear that although some stars exhibit radial velocity variations, for a majority of the sample the data does not require a binary companion to explain the observations.

metal-poor). These likely belong to the oldest objects we know in the Universe. That we know so few of these very pristine stars that provide such important clues on the earliest stages of star formation is explained by the fact that they are generally vastly outnumbered by stars of later generations in any component of the Galaxy.

2. Carbon-enhanced metal-poor (CEMP) stars

Reviewing the sample of extremely metal-poor stars in the Milky Way, it immediately stands out that about a third of these stars show a puzzling high enhancement in carbon (e.g., Yong *et al.* 2013b). As these stars span a range in stellar parameters and populate both the giant branch and main-sequence (turn-off) branch, it is clear that this carbon-enhancement is not produced in the stars themselves. Do these chemical anomalies tell us something important about the very first stages of star formation?

A substantial fraction of CEMP stars also show an overabundance in heavy s-process elements (measured as an enhancement in barium) and are called CEMP-s (Beers & Christlieb 2005). Stars with low or normal barium values are on the other hand called CEMP-no. A generally accepted explanation across the community is that the CEMP-s stars obtain their overabundant carbon from a companion star that was initially more massive and previously reached the stage of carbon fusion in its interior, dredged this up and has deposited it through winds - along with s-process elements - onto the surface of the extremely metal-poor star. Strong evidence in favour of this scenario was found from repeated radial velocity measurements. The fraction of confirmed binaries in the CEMP-s class is so high that all such stars might indeed be in binary systems (Lucatello *et al.* 2005, and references therein); thus, the nucleosynthetic signatures of these stars are likely attributable to their binarity.

From a theoretical standpoint, it is insufficiently understood if carbon could be transferred from a companion without s-process elements (Suda *et al.* 2004), making also CEMP-no stars potentially AGB polluted. To address these issues we have undertaken to monitor the velocities of all CEMP-no stars below [Fe/H]=-3 from the compilation in Norris *et al.* (2013) in order to understand whether they - or a subsample of them - owe their peculiar chemical composition as the result of being in a binary system. From the first analysis of this (partial) dataset observed with the Hobby-Eberly Telescope and presented in Starkenburg *et al.* (2014) we can already firmly conclude that although there are some clear binaries in this population, most of them show no signature of variations. This is illustrated here in Figure 1. A more statistical analysis, including literature data, shows that the two different classes of CEMP stars - those with high barium and those without - have different binary properties. Through comparison with simulations, we demonstrate in Starkenburg *et al.* (2014) that all barium-enhanced populations are best represented by a ~100% binary fraction with a shorter period distribution of at maximum ~20,000 days. This result greatly strengthens the hypothesis that a similar binary mass transfer origin is responsible for their chemical patterns. The complete CEMP-no dataset is however clearly inconsistent with the binary properties of the CEMP-s class, thereby strongly indicating a different physical origin of their carbon enhancements.

3. CEMP stars: what do they tell us?

For the origin of the CEMP-no class an alternative explanation for their chemical pattern - if they indeed are not in binary systems - is that these stars are truly the oldest stars we see and formed from gas clouds already imprinted with a large overabundance of carbon and other light elements by the first stars (see for an overview e.g., Norris *et al.* 2013; Karlsson, Bromm & Bland-Hawthorn 2013). Various models predict that gas could be enriched in carbon-rich material by the supernova and/or wind ejecta of fast rotating Pop III stars (e.g., Fryer, Woosley & Heger 2001; Meynet, Ekström & Maeder 2006; Cescutti *et al.* 2013), or supernovae explosions with fall-back, thereby expelling mainly lighter elements (e.g., Umeda & Nomoto 2003; Limongi, Chieffi & Bonifacio 2003). The fact that almost all of the most metal-poor stars known are of the CEMP-no class (note however the exception at [Fe/H]=-4.7 from Caffau *et al.* (2011)) seems to favor such an explanation. On the other hand, it has been suggested that the most iron-poor stars were in fact born with [Fe/H] ~-2, more in line with their carbon abundance pattern, and that their iron abundance has been dust-depleted since (Venn & Lambert 2008). A follow-up study shows that many of the ultra metal-poor stars do not show any infrared excess, making this interpretation less likely. While for some stars such an interpretation can be ruled out based on their precise abundance patterns, this is not yet the case for all of them (Venn *et al.* 2014).

A large majority of the extremely metal-poor stars studied today live in a volume fairly close to the Sun (see Figure 2). There are however some indications for a dependence on environment of the number of carbon-enhanced metal-poor stars (and of their type) with either disk height (Frebel *et al.* 2006); in the inner versus outer halo (Carollo *et al.* 2014); and in the Galactic halo versus classical dwarf galaxies (Starkenburg *et al.* 2013; Skúladóttir *et al.* 2015). However, statistics are lacking and a larger sample of these stars is needed to understand the chemical signatures of the first epoch of star formation and the - perhaps very exotic - phenomena that are proposed to have existed at this early epoch.

4. Finding the needle in the haystack

A high fraction of the known extremely metal-poor stars were discovered in the HK and Hamburg/ESO (HES) surveys (Beers, Preston & Shectman 1985; Christlieb *et al.* 2002). Both these surveys were centered around the very strong Ca H&K features in

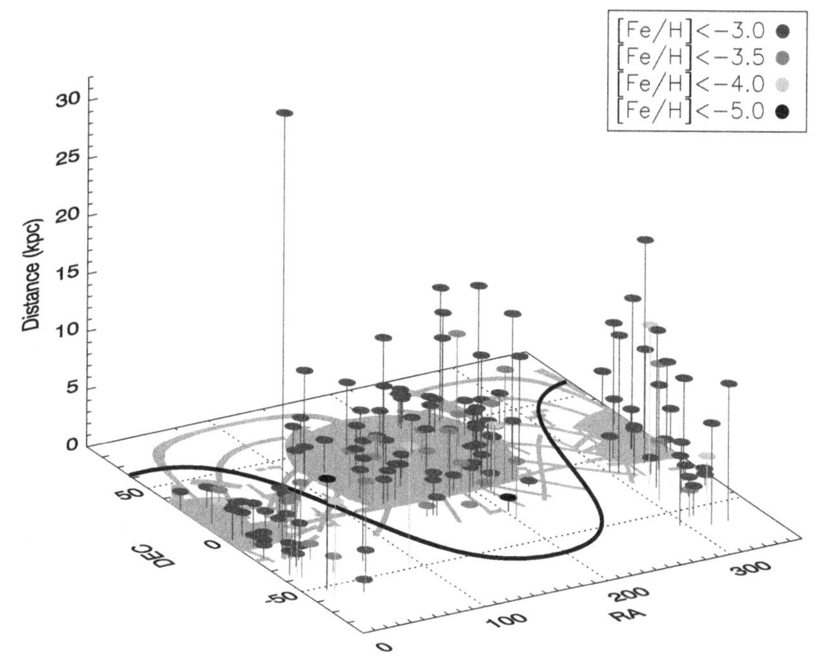

Figure 2. A figure showing the distance from the Sun versus the RA and DEC coordinates for a large sample of extremely metal-poor stars. Distances are taken from the work of Carollo *et al.* (2014) who have analyzed both the sample from Aoki *et al.* (2013) and the dataset presented in Yong *et al.* (2013a). This figure is meant to illustrate how relatively local our knowledge of metal-poor stars still is. The footprint of the Sloan Digital Sky Survey (SDSS) is shown (grey contours) as well as the location of the Milky Way thin disk (black solid line).

the spectrum, using an objective grism technique. Any very metal-poor stars will show hardly any loss of flux in the regions of the broad Ca H and K features (i.e. the lines will be much weaker, the effect is illustrated in Figure 3) setting them apart from more metal-rich stars of the same colour. The target lists from these surveys have been dominating the field of research into pristine stars for many years (e.g., see for a recent compilation of results from the HES survey targets Cohen *et al.* 2013).

Furthermore, the Sloan Digital Sky Survey (SDSS) has significantly contributed to the number of metal-poor stars known in its spectroscopic follow-up mainly carried out as part of the Sloan Extension for Galactic Understanding and Exploration (SEGUE, Yanny *et al.* 2009). Yet the number of stars with spectroscopic follow-up is a very small fraction of the total area surveyed and its selection function is complex, focussing on many different types of stars. So far ∼70 extremely metal-poor stars from SDSS/SEGUE have been followed with HDS on Subaru (Aoki *et al.* 2013), a comparable number is currently being followed up with X-shooter on the VLT (Caffau *et al.* 2013).

5. CFHT-Pristine

Utilizing the new narrow-band Ca H&K filter for MegaCam on the 4-meter Canada-France-Hawaii telescope on Mauna Kea the *Pristine* survey aims to mine a significant area in the footprint of the SDSS in the Northern Hemisphere. The filter is centered around these same Ca H&K lines illustrated in Figure 3 to allow a very efficient search for metal-poor star candidates. In the first semester the filter has been available, already over ∼600 square degrees of area have been photometrically surveyed. The survey area

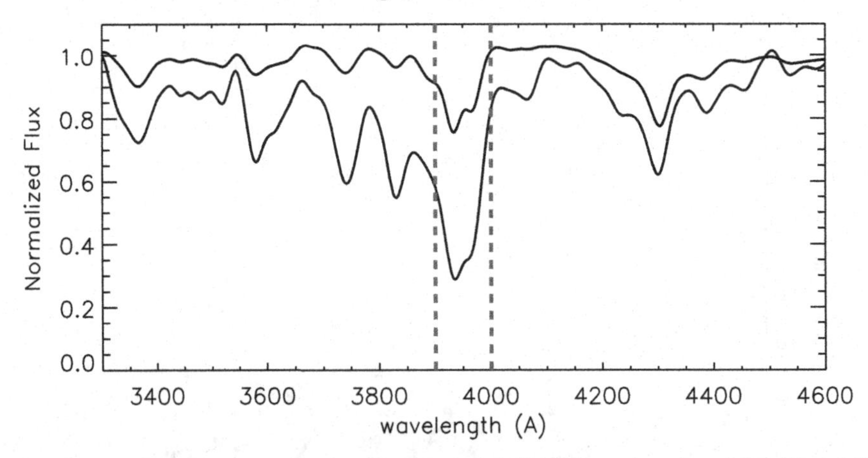

Figure 3. A synthetic spectrum of two red giants with [Fe/H]=-1 (grey) and [Fe/H]=-3 (black), made using MARCS models and the Turbospectrum code Alvarez & Plez (1998). Dashed vertical lines indicate the approximate placing of the boundaries for the Ca H&K filter. Even in these spectra that are convolved to lower resolution (R $\sim$ 1500) the Ca H&K lines clearly stand out as the strongest features in this part of the spectrum and are very sensitive to the stars metallicity. By comparing the flux in such a narrow band filter to that in broader band filters (sensitive to the overall color of the star), stars of different metallicities can be split up easily.

deliberately includes in its footprint a range of galactic latitudes (30$<$ b $<$73) and some exciting Galactic substructure features. It includes for instance part of the Sagittarius stellar stream (without being overwhelmed by it). From the first analysis of *Pristine* photometry, combined with SDSS broad-band colours and results from the overlapping SEGUE spectroscopy (Lee *et al.* 2008), we can confirm that indeed this new survey is very well equipped to distinguish the sparse extremely metal-poor targets from the more metal-rich bulk of stars in the Milky Way.

We note that this strategy (using a narrow-band Ca H&K filter in conjunction with broad-band colors) is very similar to the pending SkyMapper survey. SkyMapper is run on an automated wide field 1.35 meter survey telescope at Siding Spring Observatory (e.g., Keller *et al.*, 2007) and is designed to map all of the 20.000 deg^2 of the Southern Hemisphere in the Sloan-like ugriz filters and additionally a (broader version of a) Ca H&K filter to find extremely metal- poor stars. From their preparatory investigations and first results overlapping with medium-resolution spectra the SkyMapper Team also finds this filter presents a very clean distinction between the sought for metal-poor stars and the more metal-rich contaminants (Keller, Skymapper Team & Aegis Team 2012). The *Pristine* program will be less comprehensive than SkyMapper as we have broad-band colours already available from SDSS. In terms of coverage, the two surveys are very much complementary, as they are directed towards the two different hemispheres.

These narrow-band efforts are expected to alleviate the current lack of statistical samples of extremely metal-poor stars. They will thereby contribute significantly to a new era in Galactic archaeology studies, in particular when combined with other future datasets from the Gaia satellite and upcoming large multi-object spectroscopic surveys.

6. Bibliography

References

Alvarez R. & Plez B., 1998, A&A, 330, 1109
Aoki W. *et al.*, 2013, AJ, 145, 13

Beers T. C. & Christlieb N., 2005, ARA&A, 43, 531
Beers T. C., Preston G. W., & Shectman S. A., 1985, AJ, 90, 2089
Bonifacio, P. *et al.*, 2015, A&A, 579, A28
Caffau, E. *et al.*, 2011, Nature, 477, 67
—, 2013, A&A, 560, A71
Carollo D., Freeman K., Beers T., Placco V., Tumlinson J., & Martell S., 2014, ApJ, submitted, (arXiv:1401.0574)
Cescutti G., Chiappini C., Hirschi R., Meynet G., & Frischknecht U., 2013, A&A, 553, A51
Christlieb, N. *et al.*, 2002, Nature, 419, 904
Cohen J. G., Christlieb N., Thompson I., McWilliam A., Shectman S., Reimers D., Wisotzki L., & Kirby E., 2013, ApJ, 778, 56
Frebel, A. *et al.*, 2005, Nature, 434, 871
—, 2006, ApJ, 652, 1585
Fryer C. L., Woosley S. E., & Heger A., 2001, ApJ, 550, 372
Hansen, T. *et al.*, 2014, ApJ, 787, 162
Karlsson, T., Bromm, V., & Bland-Hawthorn, J., 2013, Reviews of Modern Physics, 85, 809
Keller, S. C. *et al.*, 2014, Nature, 506, 463
Keller S. C., Skymapper Team, & Aegis Team, 2012, in Astronomical Society of the Pacific Conference Series, Vol. 458, Galactic Archaeology: Near-Field Cosmology and the Formation of the Milky Way, Aoki W., Ishigaki M., Suda T., Tsujimoto T., Arimoto N., eds., p. 409
Lee, Y. S. *et al.*, 2008, AJ, 136, 2022
Limongi, M., Chieffi, A., & Bonifacio, P., 2003, ApJ, 594, L123
Lucatello, S., Tsangarides, S., Beers, T. C., Carretta, E., Gratton, R. G., & Ryan S. G., 2005, ApJ, 625, 825
Meynet G., Ekström S., & Maeder A., 2006, A&A, 447, 623
Norris, J. E., Christlieb, N., Korn, A. J., Eriksson, K., Bessell, M. S., Beers, T. C., Wisotzki, L., & Reimers, D., 2007, ApJ, 670, 774
Norris, J. E. *et al.*, 2013, ApJ, 762, 28
Placco, V. M., Frebel, A., Beers, T. C., Christlieb, N., Lee, Y. S., Kennedy, C. R., Rossi, S., & Santucci, R. M., 2014, ApJ, 781, 40
Skúladóttir, Á., Tolstoy, E., Salvadori, S., Hill, V., Pettini, M., Shetrone, M. D., & Starkenburg, E., 2015, A&A, 574, A129
Spite, M., Caffau, E., Bonifacio, P., Spite, F., Ludwig, H.-G., Plez, B., & Christlieb, N., 2013, A&A, 552, A107
Starkenburg, E. *et al.*, 2013, A&A, 549, A88
Starkenburg, E., Shetrone, M. D., McConnachie, A. W., & Venn, K. A., 2014, MNRAS, 441, 1217
Suda, T., Aikawa M., Machida, M. N., Fujimoto, M. Y., & Iben, Jr. I., 2004, ApJ, 611, 476
Umeda H. & Nomoto K., 2003, Nature, 422, 871
Venn K. A. & Lambert D. L., 2008, ApJ, 677, 572
Venn, K. A., Puzia, T. H., Divell, M., Côté, S., Lambert, D. L., & Starkenburg, E., 2014, ApJ, 791, 98
Yanny, B. *et al.*, 2009, AJ, 137, 4377
Yong, D. *et al.*, 2013a, ApJ, 762, 26
—, 2013b, ApJ, 762, 27

The General Assembly of Galaxy Halos: Structure, Origin and Evolution
Proceedings IAU Symposium No. 317, 2015
A. Bragaglia, M. Arnaboldi, M. Rejkuba & D. Romano, eds.

doi:10.1017/S1743921315010479

Metallicity Gradients in the Halos of Elliptical Galaxies

Jenny E. Greene[1], Chung-Pei Ma[2], Andrew Goulding[1], Nicholas J. McConnell[3], John P. Blakeslee[3], Timothy Davis[4], Jens Thomas[5]

[1]Department of Astrophysics, Princeton University, Princeton, NJ 08544, USA
[2]Department of Astronomy, University of California, Berkeley, CA 94720, USA
[3]Dominion Astrophysical Observatory, NRC Herzberg Institute of Astrophysics, Victoria, BC V9E 2E7, Canada
[4]Centre for Astrophysics Research, University of Hertfordshire, Hatfield, Herts AL10 9AB, UK
[5]Max Planck-Institute for Extraterrestrial Physics, Giessenbachstr. 1, D-85741 Garching, Germany

Abstract. We discuss the stellar halos of massive elliptical galaxies, as revealed by our ambitious integral-field spectroscopic survey MASSIVE. We show that metallicity drops smoothly as a function of radius out to $\sim 2.5R_e$, while the $[\alpha/\mathrm{Fe}]$ abundance ratios stay flat. The stars in the outskirts likely formed rapidly (to explain the high ratio of alpha to Fe) but in a relatively shallow potential (to explain the low metallicities). This is consistent with expectations for a two-phase growth of massive galaxies, in which the second phase involves accretion of small satellites. We also show some preliminary study of the gas content of these most MASSIVE galaxies.

Keywords. galaxies: elliptical and lenticular, cD, galaxies: evolution, galaxies: kinematics and dynamics, galaxies: stellar content

1. Introduction

Although they comprise the most massive galaxies in the present-day universe, the assembly history of elliptical galaxies remains poorly understood. In this contribution, we discuss our ongoing ambitious integral-field spectroscopic (IFS) survey, MASSIVE (Ma et al. 2014). Our goal is to observe the $\sim$ 100 most massive early-type galaxies within $\sim$ 100 Mpc with the large-format IFS instrument, the George and Cynthia Mitchell Visible Integral Replicable Unit Prototype (hereafter the Mitchell spectrograph Hill et al. 2008).

Our survey is designed to complement existing IFS studies focused on early-type galaxies, and in particular ATLAS^{3D} (Cappellari et al. 2011). We complement their work in two important ways. First, the Mitchell spectrograph has a field of view of 2 arcmin, which allows us to investigate the stellar kinematics and stellar populations of our galaxies to beyond twice their effective radius. With this wide radial coverage, we are able to probe the dark matter halo potential with our stellar kinematics, as well as measure radial gradients in stellar populations out to large radius. Second, we survey galaxies to a volume limit of 107 Mpc (to include the Coma cluster) and thus probe the very massive end of the galaxy mass function. We show the difference in sample demographics between the two surveys in Figure 1. We show a representative sample of galaxy images from MASSIVE in Figure 2.

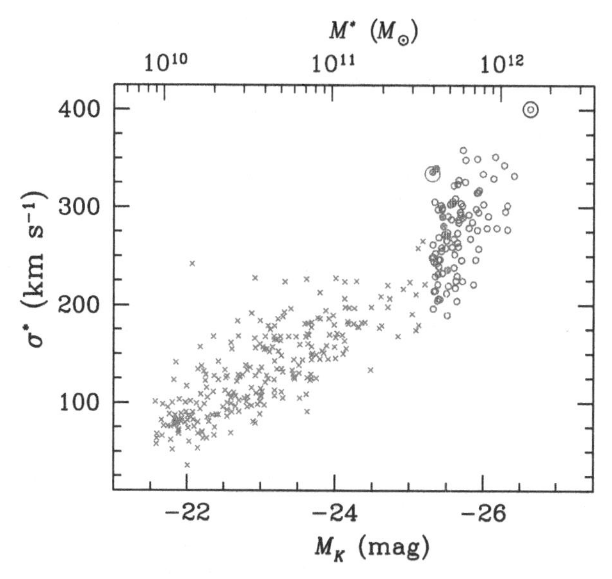

Figure 1. Adapted from Ma *et al.* 2014. We plot the stellar velocity dispersion versus the absolute magnitude (Faber & Jackson 1976). In blue crosses, we show the ATLAD^{3D} galaxies while the MASSIVE galaxies are shown as red open circles. For context, we highlight in large circles NGC 4889 (red) and M87 (blue).

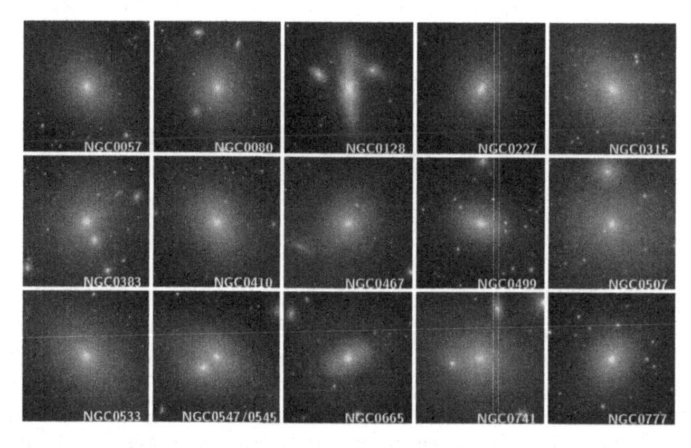

Figure 2. Three-color images from the SDSS, adapted from Ma *et al.* 2014.

2. The Sample

The MASSIVE sample contains 116 early-type galaxies in the northern sky with distance $D < 108$ Mpc and absolute K-band magnitude $M_K < -25.3$ mag (stellar mass $M^* > 10^{11.5} M_\odot$). The survey volume includes the Coma Cluster and is more than an order of magnitude larger than that probed by ATLAS^{3D}, enabling us to obtain a statistical sample of early-type galaxies at the highest end of the galaxy mass function. The survey galaxies are selected based on their total stellar mass via the K-band luminosities from 2MASS. Details of the distance determination, morphological cut, and other selection criteria are described in Ma et al. (2014).

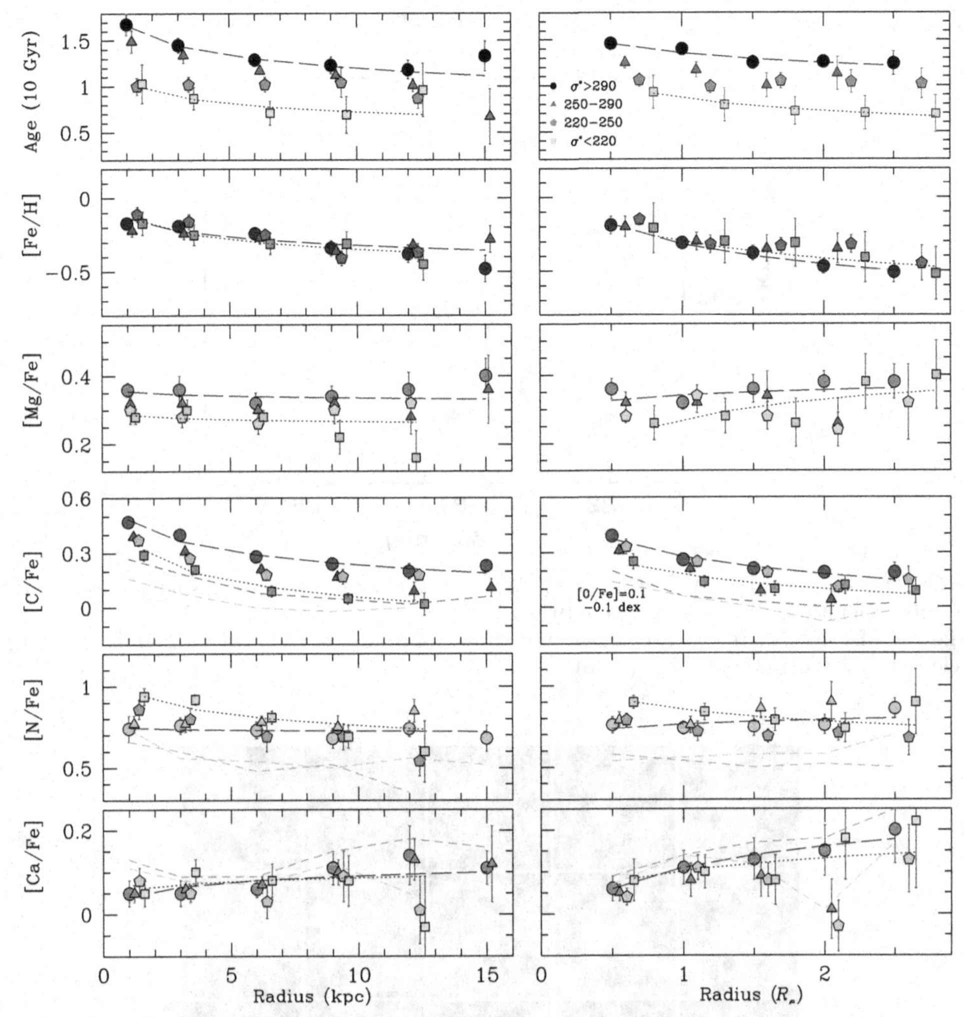

Figure 3. Radial gradients in age, [Fe/H], [Mg/Fe], [C/Fe], [N/Fe], and [Ca/Fe], adapted from Greene et al. (2015). The measurements are made on four sets of stacked spectra binned on stellar velocity dispersion (see figure key), and are shown as a function of R in kpc (left) or R/R_e. The lines are power-law fits (see Greene *et al.* 2015). Note the decline with radius in [Fe/H] and [C/Fe] in contrast with the radially constant age, [Mg/Fe], [N/Fe], and [Ca/Fe]. To indicate systematic errors in the light elements due to the unknown oxygen abundance, we also show the resulting models assuming [O/Fe]= 0.1 rather than the default [O/Fe]= 0.5 (keeping [O/Fe] constant with radius in both cases) ($\sigma_* > 290$ km s^{-1} coadd in dash and $220 < \sigma_* < 290$ km s^{-1} in dot-dash). The [C/Fe] lines with alternate oxygen abundance have been offset by -0.1 dex for presentation purposes.

3. Stellar Population Gradients

Radial gradients in stellar populations can elucidate when, how rapidly, and in what environment, the stars in galaxy outskirts were formed (e.g., White 1980; Kobayashi 2004; Greene et al. 2013; Hirschmann et al. 2015). At the same time, the kinematics of the stars (e.g., V/σ_*, the level of radial anisotropy, etc.) may provide insight into how these stars joined the galaxy (e.g., Wu et al. 2014; Arnold et al. 2014; Raskutti et al. 2014; Naab et al. 2014; Röttgers et al. 2014). We focus most on the former measurements here.

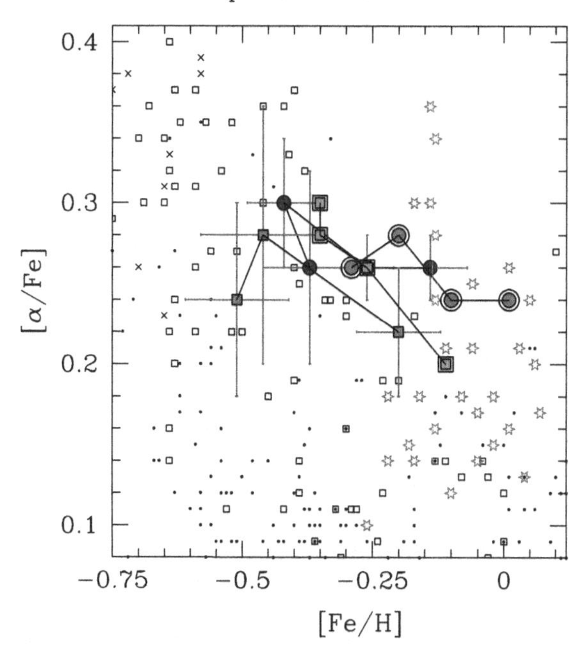

Figure 4. Adapted from Greene et al. (2013), we show galaxy centers from Graves et al. (2009) in red stars, well-aligned with the centers of our galaxies (large red circles and blue squares). As we follow the massive galaxy tracks outward from the locus of red stars, the radius grows. However, the [α/Fe] stays roughly constant (to marginally rising) while the [Fe/H] drops outward. For reference, we show Milky Way stars from Venn et al. (2004), including disk stars (small dots), thick disk stars (open squares), and halo stars (crosses).

Historically, measuring reliable stellar population parameters beyond the half-light radius of galaxies has been quite a challenge. Resolved stellar population studies, beautifully discussed at this meeting, are only able to reach a few of the nearest early-type galaxies (e.g., Kalirai et al. 2006; Harris et al. 1999; Rejkuba et al. 2005; Harris et al. 2007; Crnojević et al. 2013; Pastorello et al. 2014; Peacock et al. 2015; Williams et al. 2015), while only a handful of existing observations probe such a wide radial range (Carollo et al. 1993; Carollo & Danziger 1994; Mehlert et al. 2003; Kelson et al. 2006; Spolaor et al. 2010; Pu et al. 2010; Pu & Han 2011; Weijmans et al. 2009; Murphy et al. 2011). Imaging surveys have also made great strides recently using stacking techniques (e.g., D'Souza et al. 2014), which complement our approach.

The IFS data allows us to radially average our data in elliptical annuli and boost the signal-to-noise relative to classic long-slit techniques. In Figure 3 we show the average radial gradients in galaxies binned on different properties (here we show bins of stellar velocity dispersion Greene et al. 2015). We find no significant gradient in age, gently declining [Fe/H], and flat radial behavior in [α/Fe].

It is useful to look at the radial behavior of the galaxies in the oft-used [α/Fe] vs [Fe/H] plane (Figure 4). Here we see that the stars in the outskirts of massive elliptical galaxies share most in common with thin-disk stars in our own galaxy. This makes sense if the stars were formed rapidly (at high redshift) in a relatively shallow potential well.

3.1. *Dependence on Group Richness*

In Greene *et al.* (2015) we leverage the large MASSIVE sample to investigate the possible dependence of radial gradients on the richness of the large-scale galaxy environment. Specifically, despite being a massive galaxy survey, roughly 40% of the galaxies in MAS-

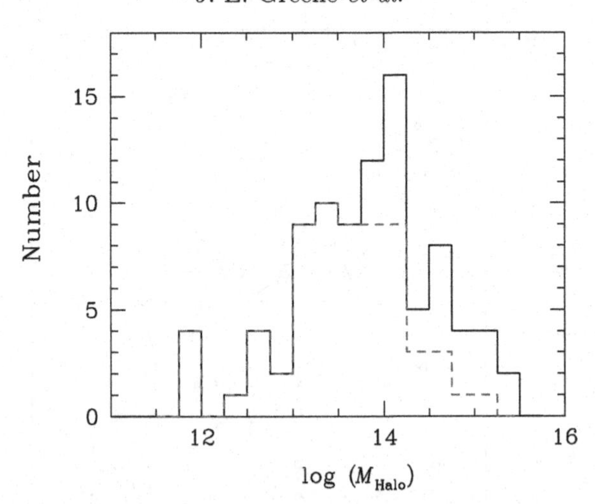

Figure 5. Distribution of dark matter halo masses for the 90 MASSIVE galaxies that reside in groups in the HDC catalog (black histogram), from Crook *et al.* Among the 90, 65 are the brightest group galaxies (BGGs) in their respective groups (red histogram).

SIVE live in isolated environments (Figure 5). Using the group catalog of Crook et al. (2007), we divide the galaxies into groups (with more than three members) and field galaxies. By forcing identical distributions in σ_*, we are able to investigate whether, at fixed σ_*, there are differences in radial gradients in the cluster and the field (Figure 6).

We see only slight differences in the radial gradients as a function of group richness. At higher density (and fixed σ_*) the galaxies are slightly older, slightly [Fe/H] poorer, and slightly alpha-enhanced. We might expect such trends if galaxies in richer environments form earlier, but cannot rule out the alternate possibility that at fixed σ_*, galaxies in richer environments may be smaller (Greene et al. 2015). Trying to focus only on the brightest galaxies in a group cut down our sample size considerably.

4. Gas

We have also been working to characterize the gas content of the MASSIVE galaxies. The mass and phase of the gas provides additional insight into the growth mechanisms of these galaxies, as well as providing complementary dynamical tracers.

The dominant phase of gas in massive elliptical galaxies is hot, X-ray emitting gas. Study of the distribution, temperature, and luminosity of the hot gas (A. Goulding *et al.* in preparation; Figure 7) both allows us to understand the origin and kinematics of the gas, and also study the dark matter halo mass. Disturbances in the X-ray emission are strong indicators of recent merger or nuclear activity. Combining our measurements with those from ATLAS^{3D} (e.g., Sarzi et al. 2013; Kim & Fabbiano 2015) will give us a very broad baseline in stellar mass, σ_*, rotation, and halo mass, with which to study the origin and kinematics of the hot gas (e.g., Negri et al. 2014).

Although elliptical galaxies are gas poor, they are not always gas free (e.g., Knapp et al. 1989). We have initiated a CO survey, and in our initial pilot survey detected six of the eleven galaxies that we targeted (Davis *et al.* submitted). We have a more comprehensive search ongoing with the IRAM-30m telescope (PI Davis) to map out the CO detection fraction as a function of mass and σ_*. From our pilot survey, we find intriguing evidence that the most massive galaxies rotate significantly faster than expected from the Tully-Fisher relation of less massive early-type galaxies (Davis *et al.* submitted).

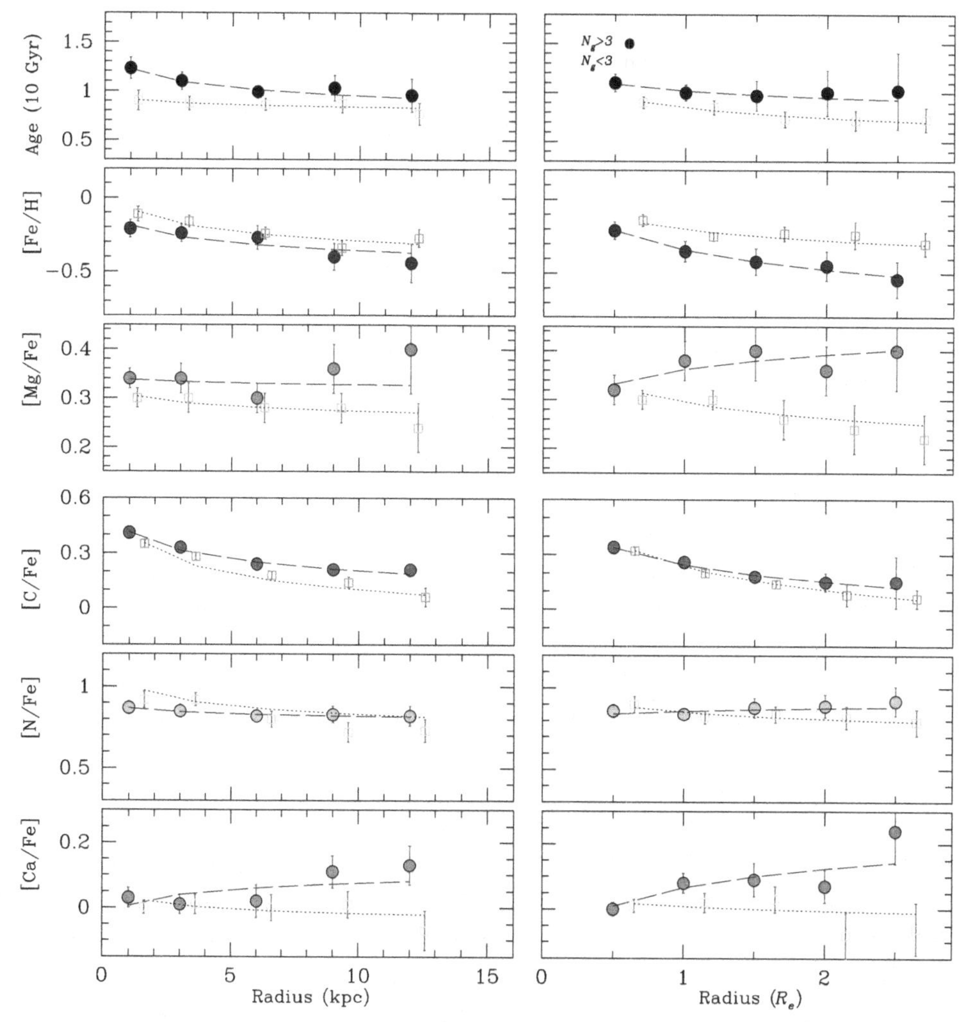

Figure 6. Radial gradients in age, [Fe/H], [Mg/Fe], [C/Fe], [N/Fe], and [Ca/Fe] as above, but now in bins of group richness, controlling for the distribution of σ_*. Low density (open squares) have three or fewer neighbors with $L > L^*$, while filled circles include everything else. The lines represent power-law fits to the gradients.

To complement the study of the molecular gas, we are also mapping the ionized gas using [O II] emission in our Mitchell cubes (Pandya *et al.* in prep). Thus far, we detect roughly one-fifth of the galaxies, with a range of morphologies ranging from organized rotating disks to patchy, disorganized regions with no ordered rotation. In a few cases, the gas extends to beyond four times the effective radius, potentially signaling cooling flows in these systems. We find that the detection fraction is a strong function of σ_*. Note that nearly all ATLAS^{3D} galaxies are detected in ionized gas (Sarzi et al. 2010) and that the angular momentum of the ionized and CO gas tend to align when they are both detected (Davis et al. 2011).

5. Future Work

To go further in our understanding of the assembly of elliptical galaxies, we must combine our stellar population measurements with the kinematic information that we can also extract from our spectra (M. Veale *et al.* in prep). We will ask whether there

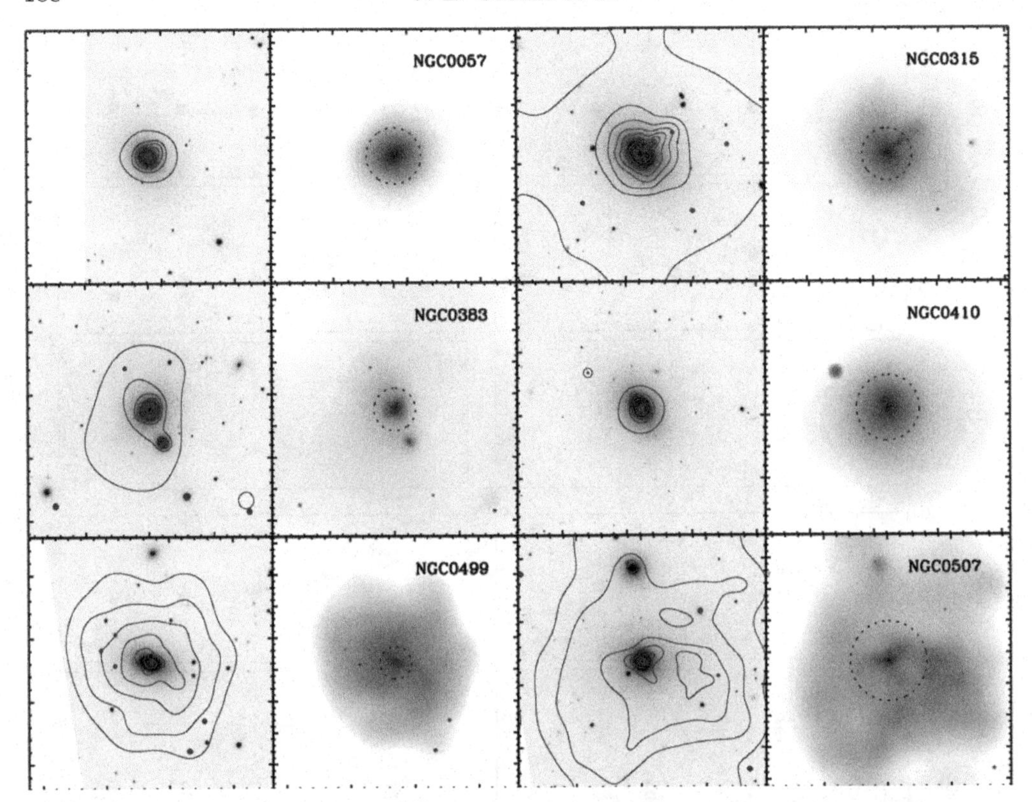

Figure 7. 4 × 4 arcminute cutouts of the MASSIVE galaxies with publicly available archival *Chandra* X-ray observations performed with the ACIS instrument, adapted from Goulding *et al.* in preparation. Left column shows the optical *i*-band SDSS DR12 Atlas image, logarithmic X-ray contours are overlaid in solid blue. The logarithm of the adaptively-smoothed and vignetting and exposure-corrected Chandra ACIS-S images are shown to the right of the optical images. Dashed circles represent one effective radius derived from optical imaging (see Ma *et al.* 2014 for details).

is any relationship between the angular momentum content of the stars at large radius and the stellar population gradients. We will use the molecular and ionized gas as an independent probe of the enclosed mass and investigate the initial mass function of these galaxies. Finally, we will use our upcoming *HST* observations (PI Blakeslee) to look for links between galaxy structure and stellar populations.

References

Arnold, J. A., Romanowsky, A. J., Brodie, J. P., Forbes, D. A., Strader, J., Spitler, L. R., Foster, C., Blom, C., Kartha, S. S., Pastorello, N., Pota, V., Usher, C., & Woodley, K. A. 2014, *ApJ*, 791, 80

Cappellari, M., Emsellem, E., Krajnović, D., McDermid, R. M., Serra, P., Alatalo, K., Blitz, L., Bois, M., Bournaud, F., Bureau, M., Davies, R. L., Davis, T. A., de Zeeuw, P. T., Khochfar, S., Kuntschner, H., Lablanche, P.-Y., Morganti, R., Naab, T., Oosterloo, T., Sarzi, M., Scott, N., Weijmans, A.-M., & Young, L. M. 2011, *MNRAS*, 416, 1680

Carollo, C. M., & Danziger, I. J. 1994, *MNRAS*, 270, 743

Carollo, C. M., Danziger, I. J., & Buson, L. 1993, *MNRAS*, 265, 553

Crnojević, D., Ferguson, A. M. N., Irwin, M. J., Bernard, E. J., Arimoto, N., Jablonka, P., & Kobayashi, C. 2013, *MNRAS*, 432, 832

Crook, A. C., Huchra, J. P., Martimbeau, N., Masters, K. L., Jarrett, T., & Macri, L. M. 2007, *ApJ*, 655, 790

Davis, T. A., et al. 2011, *MNRAS*, 417, 882

D'Souza, R., Kauffman, G., Wang, J., & Vegetti, S. 2014, *MNRAS*, 443, 1433

Faber, S. M., & Jackson, R. E. 1976, *ApJ*, 204, 668

Graves, G. J., Faber, S. M., & Schiavon, R. P. 2009, *ApJ*, 693, 486

Greene, J. E., Janish, R., Ma, C.-P., McConnell, N. J., Blakeslee, J. P., Thomas, J., & Murphy, J. D. 2015, *ApJ*, 807, 11

Greene, J. E., Murphy, J. D., Graves, G. J., Gunn, J. E., Raskutti, S., Comerford, J. M., & Gebhardt, K. 2013, *ApJ*, 776, 64

Harris, G. L. H., Harris, W. E., & Poole, G. B. 1999, *AJ*, 117, 855

Harris, W. E., Harris, G. L. H., Layden, A. C., & Wehner, E. M. H. 2007, *ApJ*, 666, 903

Hill, G. J., et al. 2008, in Society of Photo-Optical Instrumentation Engineers (SPIE) Conference Series, Vol. 7014, Society of Photo-Optical Instrumentation Engineers (SPIE) Conference Series

Hirschmann, M., Naab, T., Ostriker, J. P., Forbes, D. A., Duc, P.-A., Davé, R., Oser, L., & Karabal, E. 2015, *MNRAS*, 449, 528

Kalirai, J. S., et al. 2006, *ApJ*, 648, 389

Kelson, D. D., Illingworth, G. D., Franx, M., & van Dokkum, P. G. 2006, *ApJ*, 653, 159

Kim, D.-W., & Fabbiano, G. 2015, *ApJ*, accepted (arXiv:1504.00899)

Knapp, G. R., Guhathakurta, P., Kim, D.-W., & Jura, M. A. 1989, *ApJS*, 70, 329

Kobayashi, C. 2004, *MNRAS*, 347, 740

Ma, C.-P., Greene, J. E., McConnell, N., Janish, R., Blakeslee, J. P., Thomas, J., & Murphy, J. D. 2014, *ApJ*, 795, 158

Mehlert, D., Thomas, D., Saglia, R. P., Bender, R., & Wegner, G. 2003, *A&A*, 407, 423

Murphy, J. D., Gebhardt, K., & Adams, J. J. 2011, *ApJ*, 729, 129

Naab, T., Oser, L., Emsellem, E., Cappellari, M., Krajnović, D., McDermid, R. M., Alatalo, K., Bayet, E., Blitz, L., Bois, M., Bournaud, F., Bureau, M., Crocker, A., Davies, R. L., Davis, T. A., de Zeeuw, P. T., Duc, P.-A., Hirschmann, M., Johansson, P. H., Khochfar, S., Kuntschner, H., Morganti, R., Oosterloo, T., Sarzi, M., Scott, N., Serra, P., Ven, G. v. d., Weijmans, A., & Young, L. M. 2014, *MNRAS*, 444, 3357

Negri, A., Ciotti, L., & Pellegrini, S. 2014, *MNRAS*, 439, 823

Pastorello, N., Forbes, D. A., Foster, C., Brodie, J. P., Usher, C., Romanowsky, A. J., Strader, J., & Arnold, J. A. 2014, *MNRAS*, 442, 1003

Peacock, M. B., Strader, J., Romanowsky, A. J., & Brodie, J. P. 2015, *ApJ*, 800, 13

Pu, S.-B., & Han, Z.-W. 2011, Research in Astronomy and Astrophysics, 11, 909

Pu, S. B., Saglia, R. P., Fabricius, M. H., Thomas, J., Bender, R., & Han, Z. 2010, *A&A*, 516, A4

Raskutti, S., Greene, J. E., & Murphy, J. D. 2014, *ApJ*, 786, 23

Rejkuba, M., Greggio, L., Harris, W. E., Harris, G. L. H., & Peng, E. W. 2005, *ApJ*, 631, 262

Röttgers, B., Naab, T., & Oser, L. 2014, *MNRAS*, 445, 1065

Sarzi, M., Alatalo, K., Blitz, L., Bois, M., Bournaud, F., Bureau, M., Cappellari, M., Crocker, A., Davies, R. L., Davis, T. A., de Zeeuw, P. T., Duc, P.-A., Emsellem, E., Khochfar, S., Krajnović, D., Kuntschner, H., Lablanche, P.-Y., McDermid, R. M., Morganti, R., Naab, T., Oosterloo, T., Scott, N., Serra, P., Young, L. M., & Weijmans, A.-M. 2013, *MNRAS*, 432, 1845

Sarzi, M., et al. 2010, *MNRAS*, 402, 2187

Spolaor, M., Kobayashi, C., Forbes, D. A., Couch, W. J., & Hau, G. K. T. 2010, *MNRAS*, 408, 272

Venn, K. A., Irwin, M., Shetrone, M. D., Tout, C. A., Hill, V., & Tolstoy, E. 2004, *AJ*, 128, 1177

Weijmans, A.-M., et al. 2009, *MNRAS*, 398, 561

White, S. D. M. 1980, *MNRAS*, 191, 1P

Williams, B. F., Dalcanton, J. J., Gilbert, E. F. B. M., Guhathakurta, P., Dorman, C., Lauer, T. R., Seth, A. C., Kalirai, J. S., Rosenfield, P., & Girardi, L. 2015, *ApJ*, accepted (arXiv:1501.06631)

Wu, X., Gerhard, O., Naab, T., Oser, L., Martinez-Valpuesta, I., Hilz, M., Churazov, E., & Lyskova, N. 2014, *MNRAS*, 438, 2701

The General Assembly of Galaxy Halos: Structure, Origin and Evolution
Proceedings IAU Symposium No. 317, 2015
A. Bragaglia, M. Arnaboldi, M. Rejkuba & D. Romano, eds.

doi:10.1017/S1743921315010959

Kinematics and Angular Momentum in Early Type Galaxy Halos

Jean P. Brodie[1], **Aaron Romanowsky**[1,2] **and the SLUGGS team**[3]

[1]UC Observatories, University of California
1156 High St, Santa Cruz, CA 95064, USA
email: `jbrodie@ucsc.edu`

[2] San José State University, San Jose, CA
email: `aaron.romanowsky@sjsu.edu`

[3]http://sluggs.ucolick.org

Abstract. We use the kinematics of discrete tracers, primarily globular clusters (GCs) and planetary nebulae (PNe), along with measurements of the integrated starlight to explore the assembly histories of early type galaxies. Data for GCs and stars are taken from the SLUGGS wide field, 2-dimensional, chemo-dynamical survey (Brodie *et al.* 2014). Data for PNe are from the PN.S survey (see contributions by Gerhard and by Arnaboldi, this volume). We find widespread evidence for 2-phase galaxy assembly and intriguing constraints on hierarchical merging under a lambda CDM cosmology.

Keywords. galaxies: elliptical and lenticular, galaxies: star clusters, galaxies: formation, galaxies: abundances, galaxies: kinematics and dynamics

---

1. Introduction

It is now generally agreed that galaxies form in two phases. A widely accepted scenario involves an early phase, occurring at a redshift of 2 or earlier, that produces a relatively compact nugget that grows over time by continually accreting lower mass satellites in dry minor mergers (e.g., Oser *et al.* 2010, Naab *et al.* 2014). Alternatively, large star forming disks may evolve passively until quenched by process that relate to their densities or velocity dispersions, perhaps also increasing somewhat in size via dry minor mergers (van Dokkum *et al.* 2015). Given that more than 90% of the total mass and angular momentum of a galaxy lie beyond one effective radius (R_e), it stands to reason that testing models for the assembly of galaxy halos will require observations out to large galactocentric radius.

The SAGES Legacy Unifying Globular Clusters and GalaxieS (SLUGGS) survey (Brodie *et al.* 2014) uses SUBARU/SuprimeCam imaging and Keck/DEIMOS spectroscopy to generate 2-dimensional metallicity and kinematic data out to $\sim$3 R_e for galaxy starlight and out to $\sim$10 R_e for globular clusters (GCs) in 25 nearby early type galaxies. Here we report initial results from the SLUGGS survey, whose observational component is nearing completion. We also include some results from the Planetary Nebula Spectrograph Galaxy Survey (PN.S), which uses planetary nebulae to explore kinematics and dynamics in 33 nearby galaxies (Arnaboldi *et al.*, 2016, in preparation).

Underpinning the use of GCs to unravel the formation histories of galaxies is the fact that GC formation accompanies all the major star forming events in a galaxy's history. Typically containing 10^5 to 10^6 stars, GCs are bright enough to allow integrated spectroscopy out to distances in excess of 50 Mpc. The vast majority of GCs are as old

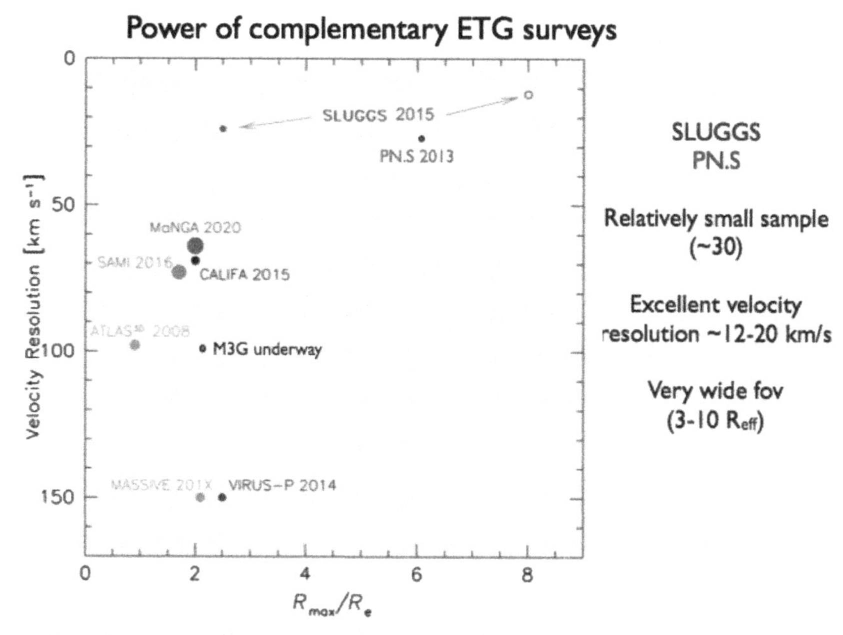

Figure 1. Complementary ETG surveys. Many other surveys are ongoing that target early type galaxies. While SLUGGS and PN.S provide excellent velocity resolution and wide field coverage, other surveys have much larger galaxy samples.

as we can measure them (>10 Gyr) and are bright beacons that were "along for the ride" during all the mergers and acquisitions that have built the galaxies we see today.

GC systems typically divide into two subpopulations, a blue, metal-poor population that appears to trace the build up of galaxy halos, and a red, metal-rich, population that is linked to bulge development. The subpopulations are distinct not only in metallicity (e.g., Brodie *et al.* 2012), but are also kinematically distinct (Pota *et al.* 2013). Like red GCs, planetary nebulae appear to be closely linked to the starlight in ETGs (Coccato *et al.* 2009).

Many surveys are underway that are targeting ETGs. Figure 1 shows a figure of merit for these surveys that is an update of the figure in Brodie *et al.* (2014). SLUGGS and PN.S were designed to offer wide field coverage with very high velocity resolution, but the galaxy sample is relatively small ($\sim$30). Other surveys complement this work by providing the large galaxy samples, albeit with poorer resolution and field coverage. Taken together, these surveys are offering unprecedented insight into the formation histories of ETGs.

2. Angular Momentum

Figure 2 shows specific angular momentum, λ_R versus radial extent for SLUGGS stars and PNe. The definitions of λ_R is different in the two panels. The SLUGGS version uses a local definition defined in successive annuli. The PN.S version is cumulative (as in Atlas^{3D} analyses (Emsellem *et al.* 2011). We definite a local version to preserve evidence of radial transitions. Evident from both tracers is the trend for galaxies that were defined as slow rotators based on observations in their central regions, to remain slow with increasing radius. Centrally defined fast rotators may continue to rise, plateau or decline with increasing radius. A similar result was obtained form the VIRUS-P survey of 33 massive galaxies (Raskutti *et al.* 2014).

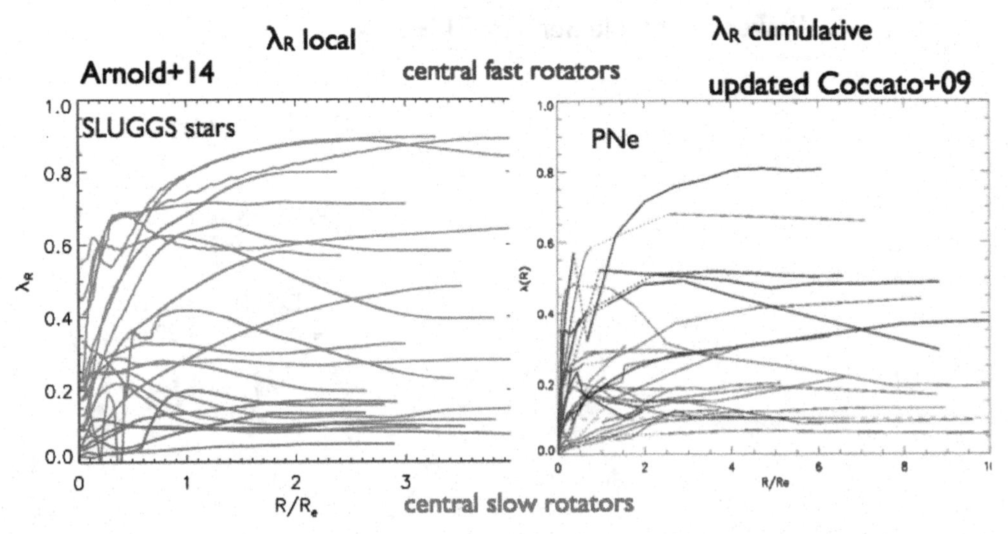

Figure 2. Specific Angular Momentum. Left panel: local specific angular momentum versus radial extent in units of effective radius for ETG stars from the SLUGGS survey. Right panel: cumulative specific angular momentum versus radial extent for ETGs from the P.NS survey (courtesy of L.Coccato and M.Arnaboldi). Both surveys reveal the same trends. Central slow rotators (red) remain slow. Central fast rotators (blue) may rise, plateau or fall with increasing radius.

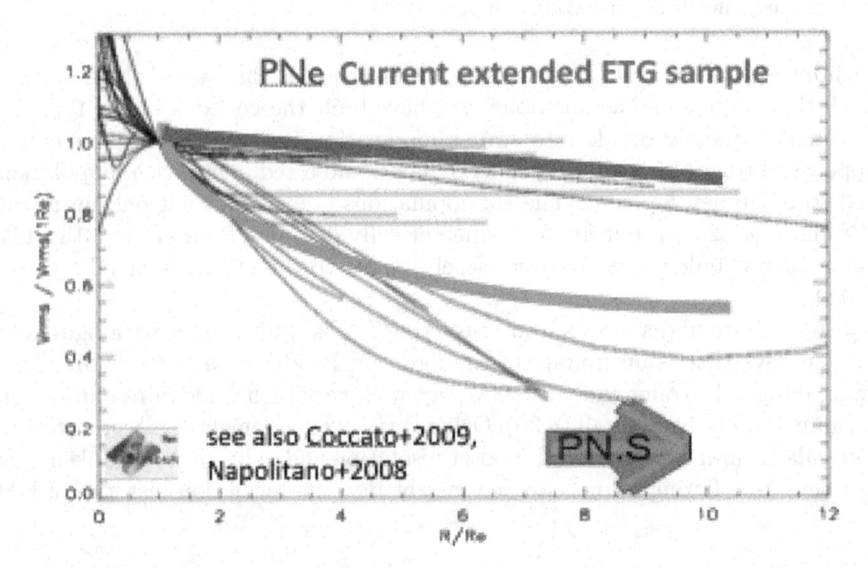

Figure 3. The radial distribution of V_{rms} for galaxies in the PN.S survey (courtesy of L.Coccato and M.Arnaboldi).

3. Velocity Dispersion

Figure 3 is a plot of root mean square velocity (V_{rms}, a proxy for velocity dispersion) against radial extent from PN.S observations of planetary nebulae. The PN.S team find a dichotomy between galaxies displaying nearly flat and steeply declining profiles. Possible explanations for such an effect include a dark matter dichotomy or anisotropy projection effects. See the contribution by Napolitano in this volume for further discussion of this point. In Figure 4 we show the rms velocity as a function of radius for GCs from the

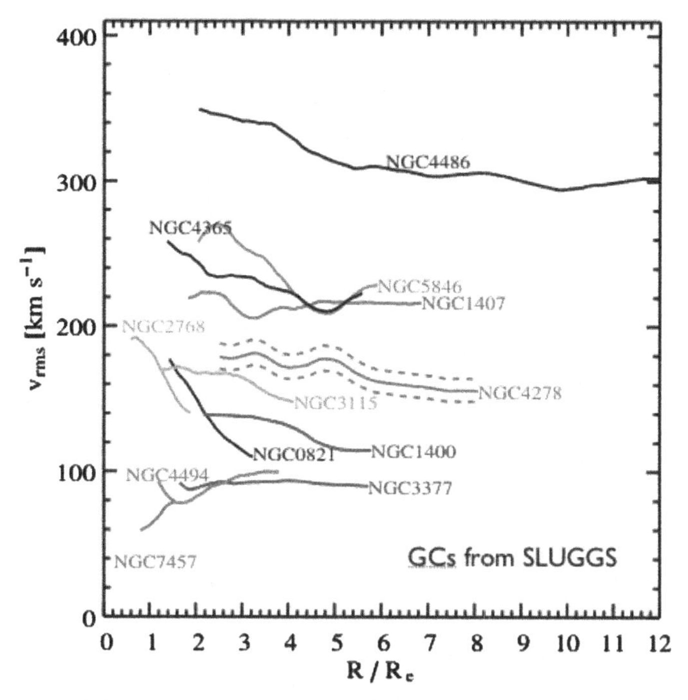

Figure 4. The radial distribution of V_{rms} of GCs for a subsample of the SLUGGS galaxies.

SLUGGS survey. Although we do not see a dichotomy in our data, not all of the SLUGGS galaxies have yet been included.

4. Mass and Dark Matter

Pota *et al.* (2015) carried out multi population dynamical modeling of NGC 1407 using the spherical Jeans equation and employing stars, metal-rich and metal-poor GCs as three independent tracers of the dark matter distribution. Using a Bayesian MCMC analysis, we determined that different anisotropies are needed to fit the profiles and that the metal-poor GCs have tangential anisotropy. This result for blue (metal-poor) GCs is inconsistent with expectations from hierarchical merging. Kurtosis measurements for a larger number of SLUGGS galaxies (Pota *et al.* 2013) reveal that the majority of blue GCs are on tangential orbits, while there is a mix of radial and tangential orbits for red (metal-rich) GCs and for PNe.

Cappellari *et al.* (2015) used Jeans axisymmetric models (JAM) on a combination of ATLAS^{3D} and SLUGGS data for 14 galaxies classified as fast rotators based on their central ($<1R_e$ kinematics). The JAM models allow spatially varying anisotropy and quite general profiles for the dark matter; no restriction on slope is imposed. This simple axisymmetric model fits the data for all 14 galaxies remarkably well and yields a power law density profile with exponent 2.19 ± 0.04 over the full range covered by the data ($0.1R_e > r > 4R_e$; see Figure 6). The scatter among the 14 galaxies is only 0.14. Since the a power law density profile is not a generic prediction of lambda CDM cosmology, this results offers tight constraints on the cosmological models. It has long been known, e.g., from observations of the gas, that the rotation curves of spiral galaxies flatten at large radius, reflecting the interplay between dark matter and baryons. This is the first

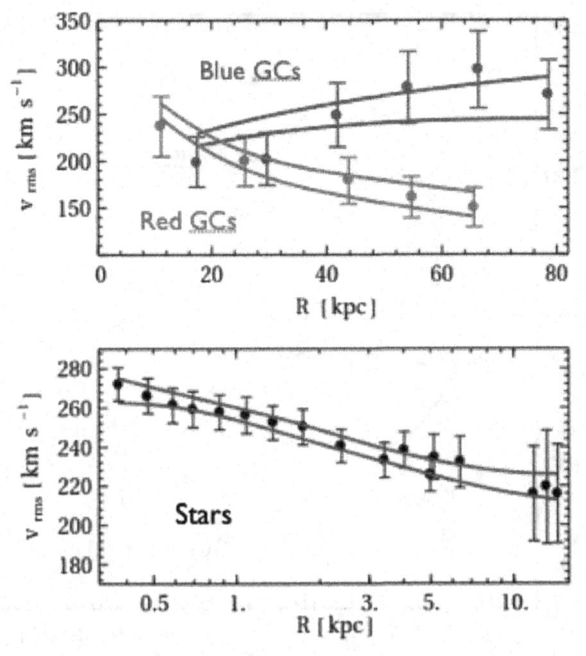

Figure 5. The best fit to generalized NFW profiles for metal rich and metal poor GCs (top panel) and stars (bottom panel). Based on figures in Pota *et al.* (2015). The solid lines are the 1σ boundaries of the fits.

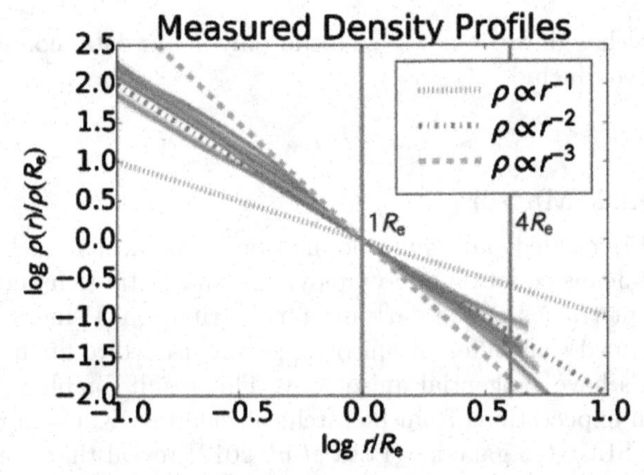

Figure 6. Measured total density profiles from Cappellari *et al.* (2015). Solid lines are the individual galaxy measurements which display very little scatter about the isothermal relation with exponent 2.19±0.04.

indication that the same flattening occurs in early type galaxies revealing a surprising "dark matter conspiracy" across markedly different galaxy types.

5. Velocity Position Phase Space

Simulations of satellite infall (e.g. Bullock & Johnston 2005) show that satellite galaxy accretion can set up a temporary set of nested chevrons in the velocity-position phase space of the accreted material, due to repeated passage near to the center of the more massive (accreting) galaxy. Direct evidence of this effect was reported by Romanowsky

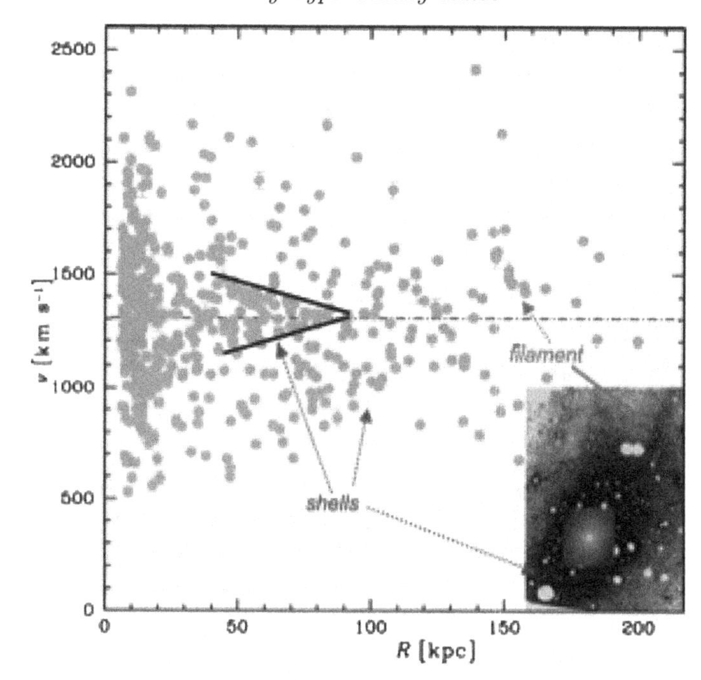

Figure 7. Velocity-position phase space for GCs around M87 reveals chevron structures that are characteristic of recent massive accretion events. The figure is adapted from Romanowsky *et al.* (2011). The image in the bottom right hand corner of the plot is from Mihos *et al.* (2005).

et al. (2011), based on high precision radial velocities of more than 500 GCs associated with M87, the massive elliptical galaxy at the center of the Virgo cluster (Figure 7). We inferred that M87 had acquired an L* galaxy, bringing in ∼1000 GCs within the last Gyr. Such characteristic chevrons are erased on a timescale of less than a Gyr, so their presence allows the timing of the accretion event to be estimated. Recently, the tally of GCs with high precision velocities has risen to more than 1700 and the chevron structure still persists. A chevron is also seen in independently in the phase space distribution of ∼300 PNe (Longobardi *et al.* 2015).

6. Summary

Wide field surveys of early type galaxies (such as SLUGGS and PN.S) are revealing widespread evidence in favor of the two-phase paradigm of galaxy assembly. In particular, we find a wide range of stellar and PNe radial profiles. Inner slow rotators remain slow; inner fast rotators can rise, plateau, or fall with increasing radius from the center of the galaxy. Multi-population dynamical modeling using stars, blue and red GCs as 3 independent probes of the gravitational potential provides estimates of DM density, the total enclosed mass, anisotropy, and M/L. We find many instances of unexplained tangential anisotropy for red GCs and PNe, while the majority of the blue GCs have tangential anisotropy. This result is contrary to expectations based on hierarchical merging models. Jeans axisymmetric modeling for stars from a combination of ATLAS^{3D} data for galaxy centers and SLUGGS data for the outer regions (to typically 4R_e) reveals a remarkable dark matter/baryon conspiracy to produce power law density profiles, with exponent 2.19 ± 0.04, with very little galaxy-to-galaxy scatter (<0.14). Substructure in velocity-position phase space in the distribution of GCs and PNe provides evidence for recent accretion in M87, including an estimate of the mass of the accreted galaxy and event timing.

7. Acknowledgements

This work was supported by NSF grant AST-1211995.

References

Brodie, J. P., Romanowsky, A. J., Strader, J., *et al.*, 2014, *ApJ*, 796, 52
Brodie, J. P., Usher, C., Conroy, C., *et al.*, 2012, *ApJ*, 759, L33
Bullock, J. S., & Johnston, K. V., 2005, *ApJ*, 635, 931
Cappellari, M., Romanowsky, A. J., Brodie, J. P., *et al.* 2015, *ApJ*, 804, L21
Coccato, L., Gerhard, O., Arnaboldi, M., *et al.*, 2009, *MNRAS*, 394, 1249
Emsellem, E., Cappellari, M., Krajnović, D., *et al.*, 2011, *MNRAS*, 414, 888
Longobardi, A., Arnaboldi, M., Gerhard, O., & Mihos, J. C., 2015, *A&A*, 579, L3
Mihos, J. C., Harding, P., Feldmeier, J., & Morrison, H., 2005, *ApJ*, 631, L41
Naab, T., Oser, L., Emsellem, E., *et al.*, 2014, *MNRAS* 444, 3357
Napolitano, G. M., Esposito, G., & Rosa, L. 2008, *arXiv* 0810.2952
Oser, L., Ostriker, J. P., Naab, T., Johansson, P. H., & Burkert, A., 2010, *ApJ*, 725, 2312
Pota, V., Romanowsky, A. J., Brodie, J. P., *et al.*, 2015, *MNRAS*, 450, 3345
Pota, V., Forbes, D. A., Romanowsky, A. J., Brodie, J. P., *et al.* 2013, *MNRAS*, 428, 389
Raskutti, S., Greene, J. E., & Murphy, J. D., 2014, *ApJ*, 786, 23
Romanowsky, A. J., Strader, J., Brodie, J. P., *et al.*, 2012, *ApJ*, 748, 29
van Dokkum, P. G., Nelson, E. J., Franx, M., *et al.*, 2015, *eprint arXiv:1506.03085*

The General Assembly of Galaxy Halos: Structure, Origin and Evolution
Proceedings IAU Symposium No. 317, 2015
A. Bragaglia, M. Arnaboldi, M. Rejkuba & D. Romano, eds.

doi:10.1017/S1743921315008443

Stellar populations of stellar halos: Results from the Illustris simulation

B. A. Cook, C. Conroy, A. Pillepich and L. Hernquist

Harvard-Smithsonian Center for Astrophysics
60 Garden St., Cambridge, MA 02138
Contact email: bcook@cfa.harvard.edu

Abstract. The influence of both major and minor mergers is expected to significantly affect gradients of stellar ages and metallicities in the outskirts of galaxies. Measurements of observed gradients are beginning to reach large radii in galaxies, but a theoretical framework for connecting the findings to a picture of galactic build-up is still in its infancy. We analyze stellar populations of a statistically representative sample of quiescent galaxies over a wide mass range from the Illustris simulation. We measure metallicity and age profiles in the stellar halos of quiescent Illustris galaxies ranging in stellar mass from 10^{10} to $10^{12} M_{\odot}$, accounting for observational projection and luminosity-weighting effects. We find wide variance in stellar population gradients between galaxies of similar mass, with typical gradients agreeing with observed galaxies. We show that, at fixed mass, the fraction of stars born in-situ within galaxies is correlated with the metallicity gradient in the halo, confirming that stellar halos contain unique information about the build-up and merger histories of galaxies.

1. Introduction

Stellar halos are diffuse regions of stars ubiquitously found surrounding galaxies (with a few notable exceptions, van Dokkum *et al.* 2014). They are observed to extend to many times a galaxy's effective radius (Martínez-Delgado *et al.* 2010), where dynamical timescales are very long compared to the ages of their host galaxies (Eggen *et al.* 1962). Detections of streams and tidal features in integrated light images of the halos in the Milky Way (Helmi *et al.* 1999), Andromeda (Ibata *et al.* 2001), and external galaxies (Martinez-Delgado *et al.* 2014) imply that stellar halos result from mergers and accretion as galaxies grow hierarchically, a picture anticipated by numerical simulations (e.g., Johnston *et al.* 1996; Bullock & Johnston 2005; Cooper *et al.* 2010; Pillepich *et al.* 2014). The long timescales at large radius can help preserve the information content of each galaxy's merger history, which has been used as motivation for observations of kinematics and populations of the stellar halo (Bell *et al.* 2008; Schlaufman *et al.* 2009).

Stellar populations – characterized by the metallicities and ages of stars – are clues to when and in what systems stars originally formed (Greene *et al.* 2015). The mass-metallicity relation in galaxies (Tremonti *et al.* 2004), combined with hierarchical accretion, implies that metallicity gradients should indicate the relative contributions of more-or-less massive progenitor systems to particular regions of a galaxy and its halo. Different merger histories should leave distinct imprints on observed stellar population gradients (Hirschmann *et al.* 2015), although stars formed *in-situ*, or within the galaxy in which they now reside, may still be significant in the inner regions of the halo (Font *et al.* 2011). For massive, *early-type* galaxies (usually red, quiescent, and elliptical), the two-phase formation scenario outlines the contributions of mergers and in-situ star formation to the formation of a metallicity gradient (Spolaor *et al.* 2010; Pastorello *et al.* 2014). The early phase of dissipative collapse leads to an steep, negative gradient from in-situ stars, while the later phase involving accretion of stars from smaller satellite galaxies tends to flatten these gradients (Kobayashi 2004).

Observations of stellar population gradients are beginning to accumulate for large samples of galaxies (Spolaor *et al.* 2010; Pastorello *et al.* 2014) and for individual giant ellitpicals (Coccato *et al.* 2010, 2011), and they are reaching continually larger radii through integral-field spectroscopy (Delgado *et al.* 2015; Greene *et al.* 2015). Observations find a wide variety of gradients between galaxies with similar masses and morphologies, which is to be expected if particular merger histories shape the gradients in a stochastic way. Unfortunately, the observational need for detailed predictions connecting gradients to merger histories have so far outpaced numerical simulations. Due to limitations in computing resources, hydrodynamical simulations have so far only resolved stellar halo populations around individual galaxies (Abadi *et al.* 2006; Hirschmann *et al.* 2015; Cooper *et al.* 2015), a method which lacks the statistical power required to replicate the observed halo-to-halo variations. Large samples of galaxy halos have been produced in N-body simulations, some of which include semi-analytic models and stellar-tagging techniques (Cooper *et al.* 2010), but these do not include baryonic physics effects, which may shape the dark matter distribution, and must rely on complicated fitting functions to generate "stellar particles" with realistic orbital properties (Bailin *et al.* 2014).

In this work, we measure stellar population gradients in a sample of quiescent galaxies from the hydrodynamical cosmological simulation Illustris. With the statistical power of Illustris' large cosmological volume and its self-consistent model for star and galaxy formation, we are able to match the observed galaxy-to-galaxy variance in gradients and show that these variations are indeed reflective of different galactic merger histories.

2. Methods

2.1. *The Simulations*

The Illustris simulations (Vogelsberger *et al.* 2014a; Genel *et al.* 2014; Nelson *et al.* 2015) are a suite of N-body+hydrodynamical cosmological simulations (106.5 Mpc on a side), run at multiple resolutions with the adaptive mesh code *AREPO* (Springel 2010; Vogelsberger *et al.* 2013). The simulations model key physical processes for the formation of galaxies, including stellar formation, evolution, and feedback, chemical enrichment, radiative cooling, supermassive black hole growth, and feedback from AGN. The highest resolution run (Illustris-1, hereafter simply Illustris) has a mass resolution of $m_{DM} = 6.26 \times 10^6 M_\odot$ and $m_{baryon} \sim 1.26 \times 10^6 M_\odot$ for the dark matter and baryonic components, respectively. At $z = 0$ gravitational forces for stellar particles are resolved to a softening length of 0.7 kpc. Illustris was run from $z = 127$ to $z = 0$ using ΛCDM cosmological paramters consistent with *WMAP9* ($\Omega_\Lambda = 0.7274, \Omega_m = 0.2726, h = 0.704$, Hinshaw *et al.* 2013).

At $z = 0$, the Illustris volume contains more than 4×10^4 well-resolved galaxies (Vogelsberger *et al.* 2014a), with a reasonable diversity of morphologies and colors, including early-type and late-type galaxies (Torrey *et al.* 2015). The most massive central galaxies (as identified by the *FOF* and *SUBFIND* algorithms, Springel *et al.* 2001; Dolag *et al.* 2009) reproduced in the simulation have stellar masses within their stellar half-mass radii of $M_* \sim 1 - 2 \times 10^{12} M_\odot$. The simulation reproduces the observed $z = 0$ mass-metallicity relation in galaxies (Vogelsberger *et al.* 2014b), and a reasonable relation between galaxy mass and stellar ages once luminosity weighting is properly taken into account.

2.2. *The Quiescent Galaxy Sample*

Our goal to reproduce measurements in the outer regions of early-type galaxies motivates the selection criteria of our sample. We select central galaxies (which are not satellites/subhalos of more massive parents) with stellar masses $M_* \geqslant 10^{10} M_\odot$†. Each galaxy in this mass range is resolved with at least a few hundred star particles beyond 8 effective radii, ensuring that we constrain gradients all the way throughout the low-density outer regions and up to 10 times the effective radius.

In this work, we examine the properties of simulated *quiescent* galaxies, in comparison to observed *early-type* galaxies. Of all Illustris galaxies with $M_* > 10^{10} M_\odot$, we select the 352 quiescent galaxies with specific star formation rates SSFR $\leqslant 10^{-11.5}$ yr^{-1} within twice the stellar half-mass radius. See Cook *et al.* (In Prep.) for details.

2.3. *Fitting Stellar Population Gradients*

We simulate observational projection effects by projecting the radii of star particles from the center of their host galaxies against a random line-of-sight. We also account for observational biases by weighting the contribution of each star particle relative to its V-band lumionsity‡. Star particles identified by *SUBFIND* as bound to smaller subhalos or satellites are not included in the analysis.

Using these observational considerations, we measure the 2-D azimuthally-averaged values of stellar metallicity and age in five logarithmically-spaced radial bins over a chosen projected radius range. We focus on three particular ranges in units of the V-band effective radius (R_e ¶), which we label: the inner galaxy (0.1 - 1 R_e), the outer galaxy (1 - 2 R_e), and the stellar halo (2 - 10 R_e). We then calculate a logarithmic gradient by fitting a line to these averages, equally weighting each radius bin: $f(r) = f(R_e) + \nabla_f \log_{10}(r/R_e)$, with f the mean log-metallicity ([Z/H]) or the mean age (in Gyrs) in each bin.

3. Results and Implications

3.1. *Comparisons to Observed Gradients*

The measured metallicity and age gradients in the inner and outer galaxy ranges ($< 2R_e$) are shown as a function of central velocity dispersion (σ_0, within $\frac{1}{8}R_e$) in Fig. 1. Observations can constrain gradients in these radius ranges for large samples of galaxies (Spolaor *et al.* 2010; Pastorello *et al.* 2014; Greene *et al.* 2015; Delgado *et al.* 2015), but so far there are only a few individual cases for comparison in the stellar halo ($> 2R_e$) range. We compare our measurements to observations of individual galaxies from Spolaor *et al.* (2010) (inner galaxy) and stacked measurements by Greene *et al.* (2015) (outer galaxy).

The typical values of our metallicity and age gradients agree with observations: metallicity gradients are negative – outer regions are more metal poor – while age gradients are roughly flat – inner and outer regions have similar average ages. Of particular importance is the fact that our measurements also reproduce the observed scatter in gradients between galaxies of similar masses. Illustris, with its large statistical sampling of galaxies, is uniquely able to replicate this galaxy-galaxy variance, whereas more individualized simulations cannot replicate the wide variety of observed gradients.

† Throughout the paper, M_* denotes the total mass in stars within the stellar half-mass radius (R_{half}) after removing the contributions of gravitationally-bound satellites identified by *SUBFIND*.

‡ Star particles in each galaxy are assigned luminosities in several common observational bands using single-age stellar population SED templates (Torrey *et al.* 2015).

¶ Each galaxy's effective radius is the projected radius within which one-half of its total V-band luminosity is located. This is calculated as an average over 100 random lines of sight.

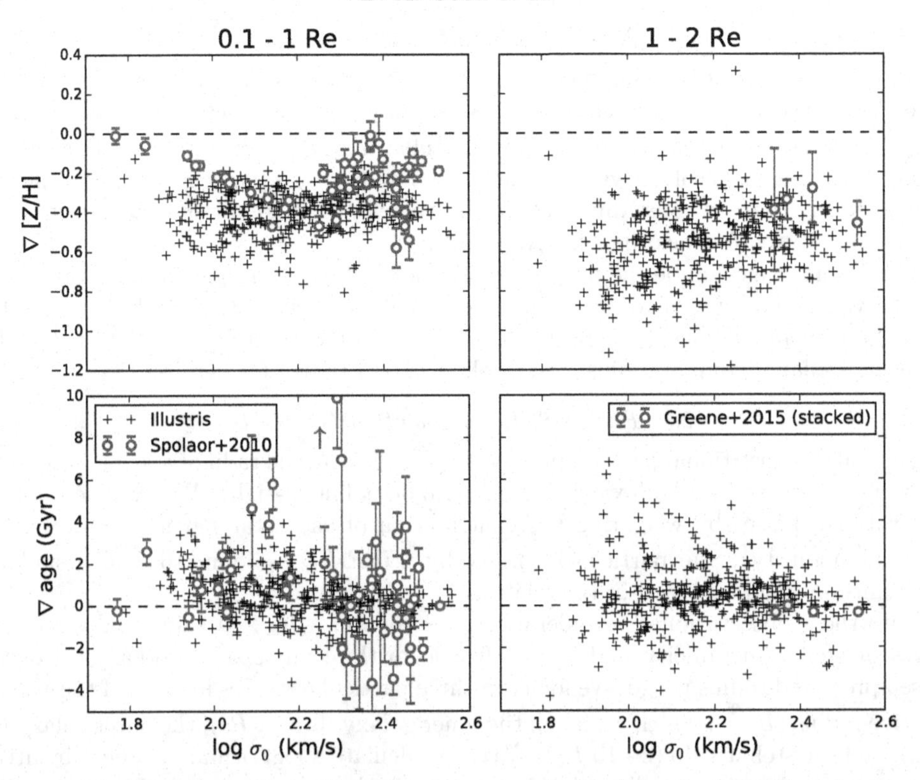

Figure 1. Measurements of metallicity (*top*) and age (*bottom*) gradients in Illustris quiescent galaxies (black cross marks). These are calculated in the inner (0.1 - 1 R_e, *left*) and outer (1 - 2 *Re*, *right*) galaxy ranges, and include projection along a random line of sight and luminosity-weighting. We compare to the measurements of Spolaor *et al.* (2010) and Greene *et al.* (2015) in the inner and outer galaxy, respectively.

Measured stellar populations in the halo of NGC 4889 (Coccato *et al.* 2010), the Brightest Central Galaxy (BCG) of the Coma cluster, show a similarly steep metallicity gradient to 1.2 R_e, but a shallower gradient from $1.2-4$ R_e (-0.1 ± 0.2) than typical for the most massive Illustris galaxies. A flat metallicity gradient is also found in the halo of NGC 3311, the BCG of the Hydra I cluster (Coccato *et al.* 2011), although there are signatures of significant substructure. Illustris has only a few galaxies in the regime of these most massive systems, and future simulations with larger volumes should provide improved statistics at this massive end.

3.2. *Relating Metallicity Gradients to Merger Properties*

Here, we study whether the scatter in stellar population gradients among Illustris galaxies at fixed mass – which matches the scatter observed in early-type galaxies – can be connected to the particular merger histories of each galaxy. We quantify the contribution of mergers to a galaxy's build-up in terms of the *in-situ fraction*: the mass fraction of a galaxy's stars which were formed within the galaxy or its main progenitor branch. Galaxies with low in-situ fractions were primarily built-up from mergers and accretion of smaller systems, while galaxies with high in-situ fractions have had little influence from mergers.

In the hierarchical model of galaxy formation, galaxies grow continually through mergers with increasingly massive neighbors. This effect, reproduced in many cosmological simulations and seen clearly in Illustris (Rodriguez-Gomez *et al.* 2015), results in a strong

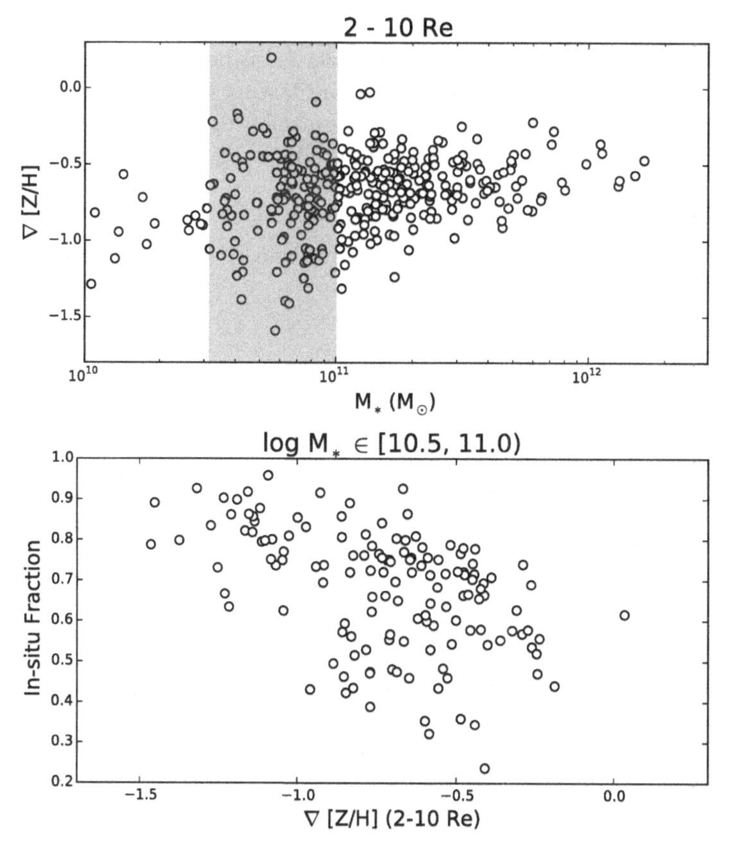

Figure 2. *Top*: The metallicity gradient in the stellar halo (2 - 10 R_e) of the quiescent Illustris sample, versus galaxy stellar mass. As at smaller radii, the metallicity gradient shows wide galaxy-galaxy variance within narrow mass ranges, such as between $10^{10.5}$ and $10^{11} M_\odot$ (shaded region). *Bottom*: The correlation between galaxy in-situ fraction and stellar halo metallicity gradient, for galaxies between in the mass range given above. Galaxies with lower in-situ fractions (mergers more significant in growth) have flatter metallicity gradients.

trend towards lower in-situ fractions in more massive galaxies. To disentangle this effect from the individual histories, we can focus our analysis to small ranges in stellar mass. Over a sufficiently narrow window, mass differences are small enough that the in-situ fraction should be driven primarily by the particular merger history of each galaxy.

In Fig. 2, we show the metallicity gradient in the stellar halo (2 - 10 R_e). Galaxies have overall negative metallicity gradients and there is a large scatter at fixed mass, just as in the inner and outer galaxy regions. When selecting galaxies in the narrow mass window from $10^{10.5}$ to $10^{11} M_\odot$ and comparing to the in-situ fraction, we see that there is a correlation. Low in-situ fractions are associated with flatter metallicity gradients in the stellar halo, while galaxies with higher in-situ fractions have steeper halo gradients.

This result suggests that much of the scatter in observed metallicity gradients can be explained through a variety of galactic merger histories. The process of building-up a galaxy through mergers appears to flatten the metallicity gradient. Measuring gradients into the outer stellar halo should provide clues about the relative importance of mergers in a particular early-type galaxy's evolution (see also, Pillepich *et al.* 2014).

We note that this correlation is significant only in the stellar halo regions (2 - 10 R_e) of our simulated galaxies. Metallicity gradients in the inner and outer galaxy regions

($< 2R_e$) show no association with the in-situ fraction. This emphasizes the importance of measuring gradients into the outer stellar halo (well beyond $2R_e$), where the significantly longer dynamical timescales result in more lasting imprints from mergers. Stellar population gradients in the halo retain the information content of merger histories; it is still unclear what leads to the wide scatter in gradients in the interior regions.

Acknowledgements

This material is based upon work supported by the NSF Graduate Research Fellowship Program under Grant No. DGE1144152.

References

Abadi, M. G., Navarro, J. F., & Steinmetz, M. 2006, http://dx.doi.org/10.1111/j.1365-2966.2005.09789.x *Mon. Not. R. Astron. Soc.*, 365, 747

Bailin, J., Bell, E. F., Valluri, M., *et al.* 2014, http://dx.doi.org/10.1088/0004-637X/783/2/95 *Astrophys. J.*, 783, 95

Bell, E. F., Zucker, D. B., Belokurov, V., *et al.* 2008, http://dx.doi.org/10.1086/588032 *Astrophys. J.*, 680, 295

Bullock, J. S. & Johnston, K. V. 2005, http://dx.doi.org/10.1086/497422 *Astrophys. J.*, 635, 931

Coccato, L., Gerhard, O., & Arnaboldi, M. 2010, http://dx.doi.org/10.1111/j.1745-3933.2010.00897.x *Mon. Not. R. Astron. Soc. Lett.*, 407, L26

Coccato, L., Gerhard, O., Arnaboldi, M., & Ventimiglia, G. 2011, http://dx.doi.org/10.1051/0004-6361/201117546 *Astron. Astrophys.*, 533, A138

Cooper, A. P., Parry, O. H., Lowing, B., Cole, S., & Frenk, C. 2015, http://arxiv.org/abs/1501.04630eprint, http://arxiv.org/abs/1501.04630arXiv:1501.04630

Cooper, A. P., Cole, S., Frenk, C. S., *et al.* 2010, http://dx.doi.org/10.1111/j.1365-2966.2010.16740.x *Mon. Not. R. Astron. Soc.*, 406, 744

Delgado, R. M. G., García-Benito, R., Pérez, E., *et al.* 2015, http://arxiv.org/abs/1506.04157eprint arXiv:1506.04157, 42

Dolag, K., Borgani, S., Murante, G., & Springel, V. 2009, http://dx.doi.org/10.1111/j.1365-2966.2009.15034.x *Mon. Not. R. Astron. Soc.*, 399, 497

Eggen, O. J., Lynden-Bell, D., & Sandage, A. R. 1962, http://dx.doi.org/10.1086/147433 *Astrophys. J.*, 136, 748

Font, A. S., McCarthy, I. G., Crain, R. A., *et al.* 2011, http://dx.doi.org/10.1111/j.1365-2966.2011.19227.x *Mon. Not. R. Astron. Soc.*, 416, 2802

Genel, S., Vogelsberger, M., Springel, V., *et al.* 2014, http://dx.doi.org/10.1093/mnras/stu1654 *Mon. Not. R. Astron. Soc.*, 445, 175

Greene, J. E., Janish, R., Ma, C.-P., *et al.* 2015, http://dx.doi.org/10.1088/0004-637X/807/1/11 *Astrophys. J.*, 807, 11

Helmi, A., White, S. D. M., de Zeeuw, P. T., & Zhao, H. 1999, http://dx.doi.org/10.1038/46980 *Nature*, Vol. 402, Issue 6757, pp. 53-55 (1999).

Hinshaw, G., Larson, D., Komatsu, E., *et al.* 2013, http://dx.doi.org/10.1088/0067-0049/208/2/19 *Astrophys. J. Suppl. Ser.*, 208, 19

Hirschmann, M., Naab, T., Ostriker, J. P., *et al.* 2015, http://dx.doi.org/10.1093/mnras/stv274 *Mon. Not. R. Astron. Soc.*, 449, 528

Ibata, R., Irwin, M., Lewis, G., Ferguson, A. M. N., & Tanvir, N. 2001, http://labs.adsabs.harvard.edu/adsabs/abs/2001Natur.412...49I/ *Nature*, Vol. 412, Issue 6842, pp. 49-52 (2001).

Johnston, K. V., Hernquist, L., & Bolte, M. 1996, http://dx.doi.org/10.1086/177418 *Astrophys. J.*, 465, 278

Kobayashi, C. 2004, http://dx.doi.org/10.1111/j.1365-2966.2004.07258.x *Mon. Not. R. Astron. Soc.*, 347, 740

Martinez-Delgado, D., D'Onghia, E., Chonis, T. S., *et al.* 2014, http://adsabs.harvard.edu/abs/2014arXiv1410.6368Meprint, http://arxiv.org/abs/1410.6368arXiv:1410.6368

Martínez-Delgado, D., Gabany, R. J., Crawford, K., *et al.* 2010, http://dx.doi.org/10.1088/0004-6256/140/4/962 *Astron. J.*, 140, 962

Nelson, D., Pillepich, A., Genel, S., *et al.* 2015, http://arxiv.org/abs/1504.00362eprint arXiv:1504.00362, http://arxiv.org/abs/1504.00362arXiv:1504.00362

Pastorello, N., Forbes, D. A., Foster, C., *et al.* 2014, http://dx.doi.org/10.1093/mnras/stu937 *Mon. Not. R. Astron. Soc.*, 442, 1003

Pillepich, A., Vogelsberger, M., Deason, A., *et al.* 2014, http://dx.doi.org/10.1093/mnras/stu1408 *Mon. Not. R. Astron. Soc.*, 444, 237

Rodriguez-Gomez, V., Genel, S., Vogelsberger, M., *et al.* 2015, http://dx.doi.org/10.1093/mnras/stv264 *Mon. Not. R. Astron. Soc.*, 449, 49

Schlaufman, K. C., Rockosi, C. M., Prieto, C. A., *et al.* 2009, http://dx.doi.org/10.1088/0004-637X/703/2/2177 *Astrophys. J.*, 703, 2177

Spolaor, M., Kobayashi, C., Forbes, D. A., Couch, W. J., & Hau, G. K. T. 2010, http://dx.doi.org/10.1111/j.1365-2966.2010.17080.x *Mon. Not. R. Astron. Soc.*, 408, 272

Springel, V. 2010, http://dx.doi.org/10.1111/j.1365-2966.2009.15715.x *Mon. Not. R. Astron. Soc.*, 401, 791

Springel, V., White, S. D. M., Tormen, G., & Kauffmann, G. 2001, http://dx.doi.org/10.1046/j.1365-8711.2001.04912.x *Mon. Not. R. Astron. Soc.*, 328, 726

Torrey, P., Snyder, G. F., Vogelsberger, M., *et al.* 2015, http://dx.doi.org/10.1093/mnras/stu2592 *Mon. Not. R. Astron. Soc.*, 447, 2753

Tremonti, C. A., Heckman, T. M., Kauffmann, G., *et al.* 2004, http://dx.doi.org/10.1086/423264 *Astrophys. J.*, 613, 898

van Dokkum, P. G., Abraham, R., & Merritt, A. 2014, http://dx.doi.org/10.1088/2041-8205/782/2/L24 *Astrophys. J.*, 782, L24

Vogelsberger, M., Genel, S., Sijacki, D., *et al.* 2013, http://dx.doi.org/10.1093/mnras/stt1789 *Mon. Not. R. Astron. Soc.*, 436, 3031

Vogelsberger, M., Genel, S., Springel, V., *et al.* 2014a, http://dx.doi.org/10.1093/mnras/stu1536 *Mon. Not. R. Astron. Soc.*, 444, 1518

—. 2014b, http://dx.doi.org/10.1038/nature13316 *Nature*, 509, 177

The General Assembly of Galaxy Halos: Structure, Origin and Evolution
Proceedings IAU Symposium No. 317, 2015
A. Bragaglia, M. Arnaboldi, M. Rejkuba & D. Romano, eds.

doi:10.1017/S1743921315006948

Gas accretion from halos to disks: observations, curiosities, and problems

Bruce G. Elmegreen[1]

[1]IBM T.J. Watson Research Center, 1101 Kitchawan Road, Yorktown Heights, NY 10598, USA
email: `bge@us.ibm.com`

Abstract. Accretion of gas from the cosmic web to galaxy halos and ultimately their disks is a prediction of modern cosmological models but is rarely observed directly or at the full rate expected from star formation. Here we illustrate possible large-scale cosmic HI accretion onto the nearby dwarf starburst galaxy IC10, observed with the VLA and GBT. We also suggest that cosmic accretion is the origin of sharp metallicity drops in the starburst regions of other dwarf galaxies, as observed with the 10-m GTC. Finally, we question the importance of cosmic accretion in normal dwarf irregulars, for which a recent study of their far-outer regions sees no need for, or evidence of, continuing gas buildup.

Keywords. accretion, galaxy formation, starbursts, dwarf irregulars

1. Introduction

Cosmic accretion onto galaxies is generally difficult to observe because the accreting halo gas is mostly ionized and the accretion rate can be irregular. Here we report fairly clear accretion via two streams of neutral gas onto the local starburst dwarf Irregular galaxy IC 10 observed with the Jansky Very Large Array and Green Bank Telescopes (Ashley *et al.* 2014). We also point out intriguing starburst hotspots in tadpole-shaped galaxies and extremely low metallicity dwarf galaxies that have metallicities lower than in the rest of the galaxy by a factor of ~ 5 (Sanchez Almeida *et al.* 2013; Sanchez Almeida *et al.* 2014a; Sanchez Almeida *et al.* 2015). These hotspots suggest fresh accretion and triggering of the starburst. On the other hand, observations of the far outer parts of 20 dwarf irregulars show slow star formation with a gas consumption time of 100 Gyr or more, suggesting no need for cosmic accretion to sustain their activity (Elmegreen & Hunter 2015). Whether normal galaxies accrete at significant rates or just starbursts do this is an open question. A review of accretion-fed star formation is in Sanchez Almeida *et al.* (2014b).

2. IC 10

IC 10 is a local group dwarf irregular starburst at a distance of 700 kpc. Neutral atomic hydrogen has been extensively studied, showing a large peripheral excess (e.g., Nidever *et al.* 2013). A recent analysis of the structure and motions of this HI suggest accretion along two streams, one from the far side in the north and another from the near side in the south (Ashley *et al.* 2014). Figure 1 shows JVLA and GBT maps of the HI velocities. Blue is approaching and red is receding. If we use the swirling spiral pattern seen in HI as an indication of the sense of rotation, i.e., so that the spirals are trailing, then the velocities indicate that the near side of the disk is in the south and the far side is in the north. Since the approaching velocities are also in the north, we presume they are on the far side. Thus we see that both the far side and the near side have streams of HI that are

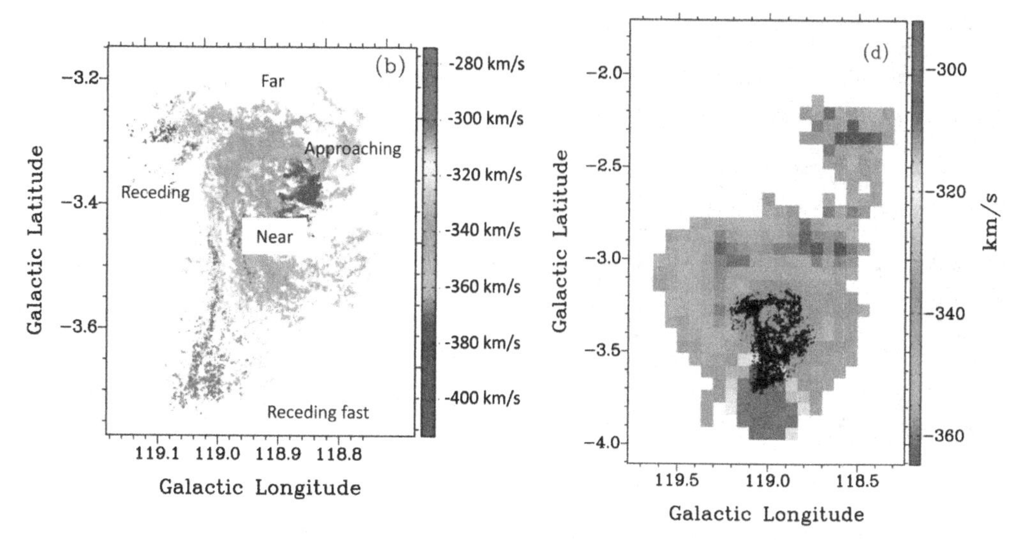

Figure 1. IC10 in HI with the VLA (left) and GBT (right), showing velocities as indicated by the color scale. Receding motions (red) in the south and approaching motions (blue) in the north indicate accretion toward the galaxy along two streams. The accretion rate is about equal to the star formation rate in this local group starburst. Images from Ashley *et al.* (2014) with labels added.

moving toward the center of the galaxy. This is accretion. There is also an indication of excessive turbulence in the northeast and southwest, as shown by the velocity dispersion maps (Ashley *et al.* 2014). These could be impact sites.

The accretion rates in the northern and southern streams can be estimated from the velocity gradients and the masses. They are 0.001 $M_\odot$ yr^{-1} in the north and 0.05 $M_\odot$ yr^{-1} in the south. The star formation rate is comparable, 0.08 $M_\odot$ yr^{-1}.

Clear examples of HI accretion like this are rare and the origin of the streaming gas is not known. It could be fall-back from a previous tidal interaction, perhaps with nearby M31, but it is fairly regular in structure and motion and fall-back would seem to be more cloudy and dispersed. There are no measurements of metallicity in these accretion streams. We predict it will be lower than in the main disk of IC 10.

3. Starburst hotspots with locally low metallicities

In a study of tadpole-shaped galaxies, which are common at high redshift (Elmegreen & Elmegreen 2010) and among extremely low metallicity dwarfs locally (Morales-Luis *et al.* 2011; Elmegreen *et al.* 2012), we determined metallicities in the HII regions using several techniques. In almost all cases (17 out of 23 studied) the metallicities were significantly lower in the brightest HII regions than elsewhere in the galaxies. This leads us to suspect that the metal-poor hotspots are accretion regions of nearly pristine gas.

Tadpole galaxies are elongated with one bright star-forming region at the "head" and a lower brightness tail that may be 5 times larger than the head. Many BCDs have tadpole or comet shapes (Loose & Thuan 1986; Noeske *et al.* 2000). Presumably most tadpoles are disks viewed edge-on because they have exponential profiles outside the hotspots and they show rotation (Sanchez Almeida *et al.* 2013).

Figure 2 shows an image of a tadpole with a metallicity and intensity trace to the side. The metallicity was calculated using the method of HII-CHI-mistry13 which involves a chi-squared fit to many spectral lines (Pérez-Montero 2014). The observations were taken

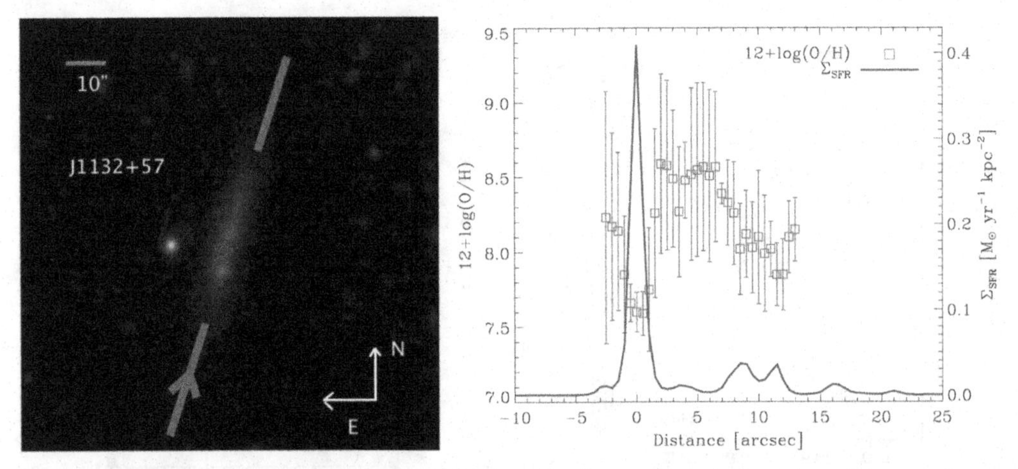

Figure 2. Tadpole galaxy SDSS image (left) with slit position indicated and scans of star formation rate (blue) and metallicity (red) on the right. The metallicity drops at the site of the peak star formation rate. Images from Sanchez Almeida *et al.* (2015).

with the OSIRIS instrument at the 10-m Gran Telescopio Canarias. The metallicity drop at the star formation hotspot is typical for this type of galaxy. The smallness of the spot suggests a short timescale for the accretion, significantly less than a disk orbit time, or $\sim$ 100 Myrs. The excess star formation rate corresponds to an excess gas mass in an accretion event which is consistent with the decrease in metallicity compared to the rest of the galaxy if the main disk is diluted with nearly pristine material (Sanchez Almeida *et al.* 2015). The actual accretion stream has not been seen in any of these galaxies yet, but it could be ionized and very faint.

A simulation of an accretion event that could make a tadpole galaxy with a metal-poor hotspot is in Verbeke *et al.* (2014). Metallicity drops associated with enhanced star-formation activity driven by gas accretion are also present in cosmological zoom-in simulations by Ceverino *et al.* (2015).

4. Outer Parts of Dwarf Irregulars

In a recent study of 20 local dwarf Irregular galaxies, Elmegreen & Hunter (2015) measured the properties of HI and star formation to very low intensity levels. The star formation rate was proportional to the HI gas column density to a power between 2 and 3, as is typical for dwarfs and the outer parts of spiral galaxies (Bigiel *et al.* 2008; Bigiel *et al.* 2011; Kennicutt & Evans 2012). The timescale for gas consumption in the furthest regions where star formation was detected in the FUV is around 100 Gyr, which is such a long time that no gas is needed to sustain star formation. The average radial profiles of FUV and HI are also fairly regular and exponential in shape, which suggests further that there is little active accretion in the outer parts.

Figure 3 shows the radial profiles of gas surface density (upper left), gas density (lower left), which is from the ratio of the column density to the scale height as fit from the stellar and gaseous surface densities; the gas velocity dispersion (upper right), and the free fall time in the midplane (lower right), which comes from the midplane density. In each panel, the left-hand axis is the quantity plotted in natural logs so the exponential scale length can be read from the figure (each drop by unity in the natural log corresponds to one exponential scale length on the abscissa). The right-hand axes are the physical units plotted in a base-10 log.

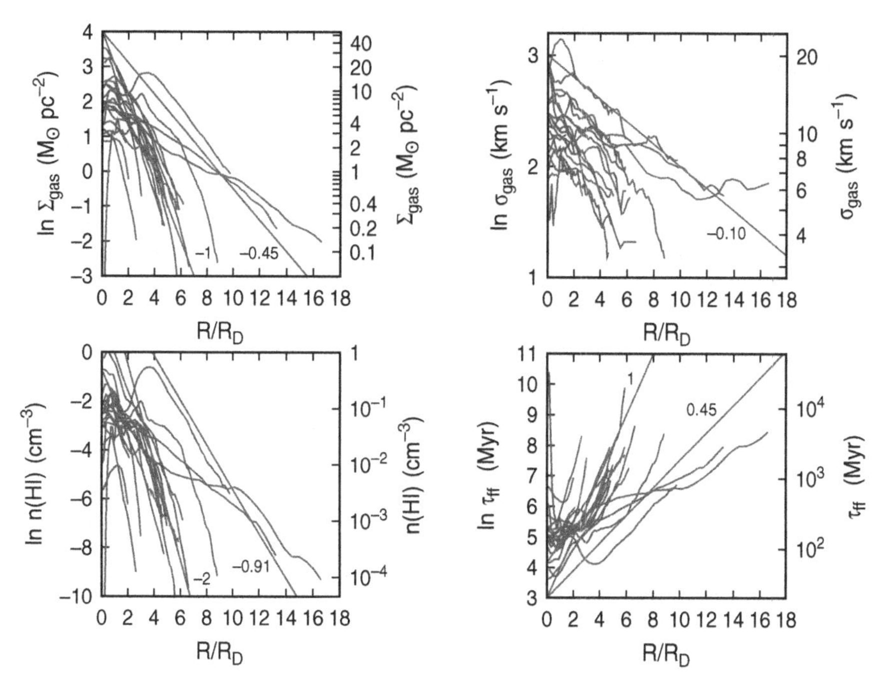

Figure 3. Radial profiles of HI gas surface density (upper left), density (lower left), velocity dispersion (upper right), and midplane free fall time (lower right) for 20 dIrr galaxies showing approximately exponential radial profiles to the far outer regions. The free fall time in the outer parts corresponds to a gas consumption time of $\sim$ 100 Gyr, which is so long that accretion is not needed to sustain star formation. Image from Elmegreen & Hunter (2015).

The figure shows HI surface densities corrected for projection down to $\sim 0.1\ M_{\odot}\ \mathrm{pc}^{-2}$, which is one-tenth that of a damped Lyman alpha cloud. The midplane density drops down to $\sim 10^{-3}\ \mathrm{cm}^{-3}$, at which point the Stromgren radius of an O5 type star is 30 kpc, much larger than the galaxy (suggesting escape of ionizing photons if there were such a star). The velocity dispersion continues to decrease with radius, but slowly, and the free fall time continues to increase with radius, getting to values larger than 1 Gyr. Considering the $\sim$ 1% efficiency of star formation elsewhere in our galaxies, this free fall time divided by the efficiency is $\sim$ 100 Gyr, and that is the gas consumption time, as mentioned above.

5. Summary

Three different types of observations about accretion onto galaxies have been shown. There is evidence for active accretion onto IC 10 in the form of two in-moving streams of neutral gas with velocity gradients of several km/s/kpc. These gradients correspond to accretion times on the order of a Gyr. The origin of this gas is unknown as is its metallicity, but its impact with the disk of this dwarf irregular could be what is triggering the current starburst, since the accretion rate equals about the star formation rate.

There is indirect evidence for low-metallicity accretion in a high fraction of extremely low metallicity galaxies or tadpole-shaped galaxies because the brightest spots of star formation have metallicities that are lower than in the rest of the disk by a factor of about 5. This accretion seems to be in the form of discrete metal-poor clouds that impact the disk and trigger an excess of star formation during a $\sim$ 100 Myr period.

Normal dwarf irregulars seem to have little accretion, however. Their HI is distributed somewhat uniformly with an exponential radial profile on average, and their midplane densities are so low that star formation can be sustained at the present rate for 100 Gyr. The star-forming disk in FUV is also fairly regular with an exponential profile.

References

Ashley, T., Elmegreen, B. G., Johnson, M., Nidever, D. L., Simpson, C. E., & Pokhrel, N. R. 2014, *AJ*, 148, 130

Bigiel F., Leroy A., Walter F., Brinks E., de Blok W. J. G., Madore B., Thornley M. D. 2008, *AJ*, 136, 2846

Bigiel F. *et al.* 2011, *ApJ*, 730, L13

Ceverino, D., Sánchez Almeida, J., Muñoz Tuñón, C., Dekel, A., Elmegreen, B. G., Elmegreen, D. M., & Primack, J. 2015, arXiv150902051C

Elmegreen, B.G. & Elmegreen, D.M. *ApJ*, 722, 1895

Elmegreen, D. M., Elmegreen, B. G., Sánchez Almeida, J., *et al.* 2012, *ApJ*, 750, 95

Elmegreen, B. G. & Hunter , D. A. 2015, *ApJ*, 805, 145

Kennicutt R. C. & Evans N. J. 2012, *ARA&A*, 50, 531

Loose, H.-H. & Thuan, T. X. 1986, in Star-Forming Dwarf Galaxies and Related Objects, ed. D. Kunth, T. X. Thuan, & J. T. T. Vân, Gif-sur-Yvette: Edition Frontières, p.73

Morales-Luis, A. B., Sánchez Almeida, J., Aguerri, J. A. L., & Muñoz-Tuñón, C. 2011, *ApJ*, 743, 77

Nidever, D. L., Ashley, T., & Slater, C. T., *et al.* 2013, *ApJL*, 779, L15

Noeske, K., Guseva, N., Frick, K., *et al.* 2000, *A&A*, 361, 33

Pérez-Montero, E. 2014, *MNRAS*, 441, 2663

Sánchez Almeida, J., Muñoz-Tuñón, C., Elmegreen, D. M., Elmegreen, B. G., & Méndez-Abreu, J. 2013, *ApJ*, 767, 74

Sánchez Almeida, J., Morales-Luis, A. B., Muñoz-Tuñón, C., Elmegreen, D. M., Elmegreen, B. G., & Méndez-Abreu, J. 2014, *ApJ*, 783, 45

Sánchez Almeida, J., Elmegreen, B. G., Muñoz-Tuñón, C., & Elmegreen, D. M. 2014, *A&ARev*, 22, 71

Sánchez Almeida, J., Elmegreen, B. G., Muñoz-Tuñón, C., Elmegreen, D. M., Pérez-Montero, E., Amorn, R., Filho, M. E., Ascasibar, Y., Papaderos, P., & Vlchez, J. M. 2015, *ApJL*, 810, L15

Verbeke, R., De Rijcke, S., Koleva, M., *et al.* 2014, *MNRAS*, 442, 1830

The General Assembly of Galaxy Halos: Structure, Origin and Evolution
Proceedings IAU Symposium No. 317, 2015
A. Bragaglia, M. Arnaboldi, M. Rejkuba & D. Romano, eds.

doi:10.1017/S1743921315007024

Studying stellar halos with future facilities

Laura Greggio[1], Renato Falomo[1] and Michela Uslenghi[2]

[1]INAF, Osservatorio Astronomico di Padova,
Vicolo dell'Osservatorio 5, 35122 Padova, Italy
emai:laura.greggio@oapd.inaf.it, renato.falomo@oapd.inaf.it

[2]INAF, Istituto di Astrofisica Spaziale e Fisica Cosmica,
Via Bassini 15, 20133 Milano, Italy

Abstract. Stellar halos around galaxies retain fundamental evidence of the processes which lead to their build up. Sophisticated models of galaxy formation in a cosmological context yield quantitative predictions about various observable characteristics, including the amount of substructure, the slope of radial mass profiles and three dimensional shapes, and the properties of the stellar populations in the halos. The comparison of such models with the observations provides constraints on the general picture of galaxy formation in the hierarchical Universe, as well as on the physical processes taking place in the halos formation. With the current observing facilities, stellar halos can be effectively probed only for a limited number of nearby galaxies. In this paper we illustrate the progress that we expect in this field with the future ground based large aperture telescopes (E-ELT) and with space based facilities as JWST.

Keywords. instrumentation: adaptive optics, galaxies: halos, galaxies: formation.

1. Introduction

Nowadays there is plenty of evidence for the presence of streams and substructures in the halo of galaxies, as is expected in the hierarchical models of galaxy formation (e.g. Atkinson, Abraham, & Ferguson 2013). These features, detected down to very low surface brightness ($\mu_V \sim 29$ mag/arcsec^2), are hard to measure against the sky background. Even more difficult is characterizing their stellar populations, estimating total magnitude and color, which trace respectively the mass and the age/metallicity of their stars. In recent years galaxy formation models in a cosmological context have become very sophisticated, yielding detailed predictions about the structure of stellar halos, the amount of substructure, the shapes of shells and streams, the characteristics of the stellar populations in the different components, and more (see Johnston *et al.* 2008; Font *et al.* 2011; Cooper *et al.* 2013; Pillepich *et al.* 2014). The predictions depend on the techniques adopted to construct the models (e.g. whether full hydrodynamical simulations, or n-body models with particle tagging), as well as on parameters describing the physical processes which occur in the troubled galaxy life (e.g. star formation (SF), the initial mass function, the feedback). Although the ubiquitous presence of substructures in galaxy halos qualitatively supports these models, we need to perform a quantitative comparison between predictions and observations, especially to constrain the models' parameters. This can be done in different ways, among which (i) analyzing the demographics of substructures of different types and the properties of their stellar populations; (ii) comparing the stellar density profile, checking for the presence and extent of an *in situ* component; (iii) measuring the metallicity gradient of stellar halos. For example, halos completely built with stars shed by accreted companions will hardly show a metallicity gradient, while according to full hydrodynamical simulations, a sizable metallicity difference between the outer and the inner halo should exist, since the latter hosts heated disk stars. These

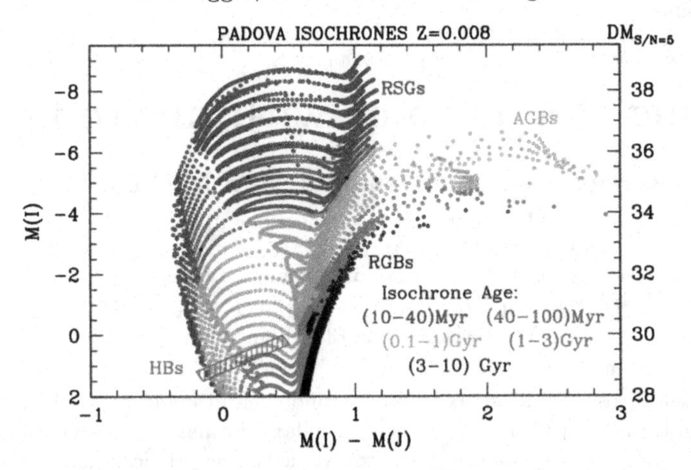

Figure 1. Isochrones computed with the CMD tool (stev.oapd.inaf.it/cgi-bin/cmd) for the indicated ages and metallicity.The right axis is labelled with the distance modulus at which sources of magnitude given on the left axis are measured with S/N=5 in a 3 hrs exposure wit MICADO. The locations of Red Supegiants (RSGs), Asymptotic Giant Branch (AGBs), Horizontal Branch (HBs) and Red Giant Branch (RGBs) stars are indicated.

kind of studies have been done only for nearby galaxies (Deason, Belorukov & Evans 2011; Greggio *et al.* 2014; Ibata *et al.* 2014; Crnojevic, this volume); with future large aperture telescopes we expect to pursue these issues on more distant objects, enriching the samples with a greater number and more types of galaxies. In this paper we illustrate some examples of these future possibilities, concentrating on photometry of individual sources, stars and Globular Clusters (GC), which trace the halo formation history.

2. Instrumental set up and stellar probes

Our results are based on simulations of images computed with the AETC † (v3.0) tool, adopting two instrumental set ups: the NIRcam camera on board of JWST, and the MICADO ‡ camera as baseline for the ELTCAM for the E-ELT telescope. The latter configuration exemplifies the more general case of a $\sim$ 30m-class ground based telescope, working close to the diffraction limit thanks to an efficient adaptive optics module.

The most important features of the MICADO and the NIRcam set ups include the field of view (FoV), respectively of $\simeq$ 0.78 and $\simeq$ 9.33 arcmin^2, the pixel size, of $\sim$ 3 and $\sim$ 32 mas , and the magnitude limits for an isolated point source, which, in 3 hrs of integration and at S/N=5, are respectively $I = 30.3$ and 28.8, $J = 29.5$ and 28.3. Thus, the NIRcam FoV is $\gtrsim$ 10 times wider than the MICADO FoV, but the latter yields a factor of $\sim$ 10 better resolution. This is important for the photometry in crowded fields, but becomes of little relevance in the low density, halo environment. On the other hand, the much larger collecting area of E-ELT allows deeper photometry with MICADO compared to NIRcam, by more than 1 mag. With these constraints how far can we detect the stellar probes of the halo formation history? Fig.1 shows theoretical isochrones for a wide range of ages. The right axis, labelled with M(I) + 30.3, indicates the distance modulus up to which the various stellar probes can be measured with good accuracy on a MICADO image. It appears that bright RSGs will be detected up to very large distances (hundreds of Mpc), but they trace only the most recent SF. Using AGB stars we will probe the SF

† http://aetc.oapd.inaf.it
‡ http://www.mpe.mpg.de/ir/micado

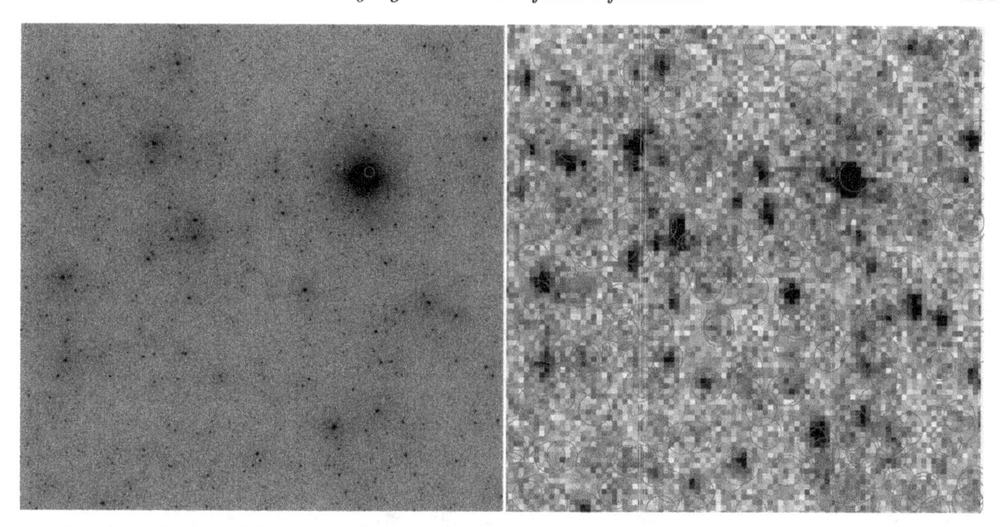

Figure 2. MICADO (left) and NIRcam (right) 3×3 arcsec^2 I-band simulated images of a stellar population typical for a stream in a galaxy halo at 8 Mpc distance (see text). A total integration time of 3 hrs has been adopted. Circles, with a diameter of 10 (left) and 6 (right) pixels, show the position of red clump stars ($28.5 \leqslant I \leqslant 29.5$).

of the last few Gyr, but to access the whole SF history we need to sample the bright RGB stars, which can be measured in the galaxy halos up $\sim$ 40 Mpc away. The RGB colors are almost insensitive to age, so these stars yield information on the integrated mass of the stellar population, but not on the detailed SF history. On the other hand, they can can be used to evaluate a photometric metallicity. Finally, red HB stars can be measured up to $\sim$ 10 Mpc, but blue HB stars can be detected well only within $\lesssim$ 5 Mpc.

In summary, we will be able to catch currently forming stars up to distances of hundreds of Mpcs, which will be very interesting for interacting galaxies. However, in order to derive the fundamental properties of galaxy halos we need to probe ages of several Gyrs, and the brightest objects we can use to achieve this goal are RGB stars.

3. Resolved stars in streams and smooth component in galaxy halos

The characterization of the SF history in the halo substructures and streams yields fundamental information on their progenitors. To illustrate this case we produced simulated images of a synthetic stellar population obtained with a constant SF rate over 10 Gyr, and located at a distance of 8 Mpc, with an average surface brightness of $\mu_V = 24.5$ mag/arcsec^2 (Fig. 2). The luminosity sampled in the field is only of 8×10^4 $L_{V,\odot}$, and there are 450 stars brighter than $I = 30$, but the total MICADO FoV is more than 300 times wider, providing large statistics. The impressive resolution of the MICADO image is readily appreciated, with faint stars clearly visible through the halo of the bright objects. The core helium burning stars (marked with red circles) are too faint and crowded to be measured well on the NIRcam image, while they are very well detected on the MICADO's. On the other hand, NIRcam will sample well the bright RGB stars, allowing us to measure the density profile, the shape, and the metallicity of the stellar halo. To better explore this matter we have simulated a 3 hrs integration full NIRcam I-band image containing one stellar population with an age of 6 Gyr, metallicity [Fe/H] = -1, and total luminosity of 10^7 $L_{V,\odot}$, located at a distance of 8 Mpc, for an average surface brightnes of $\mu_V \simeq 28$, typical of a galaxy halo. Photometry of the $\simeq$ 3500 stars brighter

Figure 3. The lenticular galaxy NGC4452 member of the Virgo Cluster, with superimposed the FoV of NIRcam (cyan) and MICADO (yellow).

than I = 27.5 ($M_I \leqslant$ -2) shows that they are measured with a mean error of $\lesssim$ 0.04 mag. Font *et al.* (2011) models for Milky Way type galaxies feature a metallicity gradient with [Fe/H] $\sim$ -1.2 and -0.5 respectively in the outer and inner halo regions, which implies a colour difference of $\Delta(I - J) \gtrsim 0.15$ mag between the inner and outer halo bright RGB stars. Comparing this color difference to the photometric accuracy mentioned above it appears that it will be possible to test these models, and in general derive fundamental information on the metallicity of stellar halos of galaxies at this distance.

The VIRGO cluster of galaxies is the closest rich aggregate of all morphological types. It is then interesting to examine to which extent we will be able to study the halos of its members. Fig. 3 shows a SDSS image of a VIRGO member galaxy, with superimposed the MICADO and the NIRcam FoV. With 3 hrs integration, the two set ups will allow sampling the upper 3 and 1.5 mag on the RGB: both limits are quite adequate to derive the photometric metallicity, though MICADO will yield more accurate results. On the other hand, due to its wide FoV, NIRcam will be more efficient to map the widespread stellar halo, while MICADO will allow exquisite photometry in the inner halo regions, at the interface with the disk, at high surface brightness.

Finally, we can ask how far we can detect substructures in galaxy halos using star counts, arguably the most robust way to trace overdensities at very low surface brightness. Due to the combination of RGB stars color and limiting magnitude, the J band is more efficient than the I band to this end. At half solar metallicity, the tip of the RGB is located at M(J) $\simeq$ -5.5; at the magnitude limits mentioned above for MICADO, we will

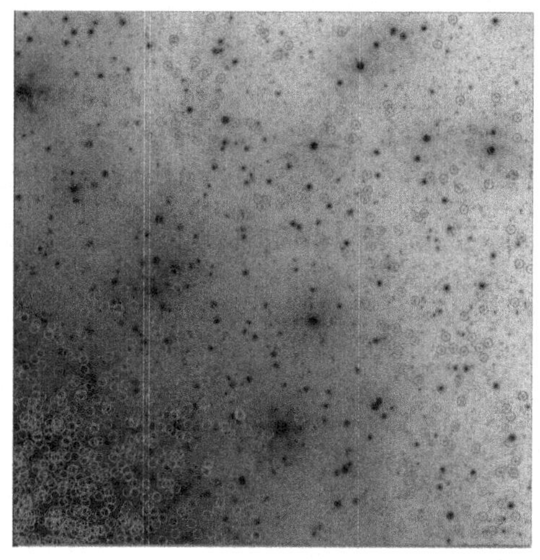

Figure 4. MICADO simulated image in the J-band of a 12 Gyr old stellar population, with [Fe/H]=-1.3, and M(V)=-9.7, adopting a 3 hrs exposure and a distance of 1 Mpc. Stars are distributed following a King profile with core and tidal radii of 1 and 30 pc, respectively. On the right panel we show a section of the image, extending from 1.5 pc to 8.7 pc from the cluster center (lower left and upper right corners), with turn-off stars ($J \simeq 28$) circled in red.

sample the upper magnitude on the RGB up to $m-M \sim 34$. At such distance, MICADO maps a field of $\sim$ 15 kpc on a side, so we will still need some mosaicking to sample the $\gtrsim$ 100 kpc extension of the halo, but we will be able to detect structures at very low surface brightness. A stellar population of 10^4 $L_{V,\odot}$ has $\sim$ 1.4 stars in this magnitude range, if its age is older than a few Gyr. Thus the surface brightness scales with the surface density σ of such stars according to $\mu_V = -4.8 - 2.5\log\sigma + m - M$. At $(m-M) = 34$, $\mu_V = 33$ corresponds to σ=0.03 stars/arcsec^2 which is almost 100 times higher than the expected density of foregorund stars in this magnitude range, estimated from the TRILEGAL † simulator. Therefore, we expect to be able to trace structures to very faint surface brightness; this is particularly important, since galaxy formation models predict that most substructures are found at $\mu_V \gtrsim 30$ (Johnston *et al.* 2008).

4. Globular Clusters and Globular Cluster systems in Galaxies

One important application for a large aperture telescope working at the diffraction limit deals with photometry of resolved stars in distant GCs for, e.g., their age dating, the detection of multiple populations and their characterization. To assess the capabilities in this field a large grid of simulations should be computed and analyzed with sophisticated photometric tools which allow a full exploitation of the data. Here we just remark that it will be possible to construct Color-Magnitude diagrams down to the turn offs of old clusters in the whole Local Group. Fig. 4 shows an example of a simulated image for a GC at 1 Mpc from us. The stars visible on the left panel are mostly bright giants, but the circled sources on the right panel are turn off stars, with $27.8 \leqslant J \leqslant 28.2$. These stars, very crowded inside $\sim$ 2 core radii, become well separated in the outer region.

Viewed as point sources, GCs hold fundamental information on the formation history of the parent galaxy, as implied by, e.g., the relation between the total halo mass and total mass of the GC system (Hudson, Harris & Harris 2014), the bimodal color distribution,

† stev.oapd.inaf.it/cgi-bin/trilegal

the spatial coincidence of clusters and streams (e.g. Blom *et al.* 2014). Assuming as typical the magnitude distribution of GCs in M31, which peaks at $M(J) = -9.2$ (Nantais *et al.* 2006), and using the above mentioned limiting magnitudes for MICADO, we will derive accurate photometry of the GC system in galaxies up to a distance of ~ 450 Mpc, and, thanks to the exquisite resolution, we will detect them even on top of the high surface brightness, central regions. In addition, the large volume accessible with future instrumentation will allow us to characterize the GC systems in hosts of Active Galactic Nuclei (AGNs) of various types and luminosity, and to study the possible connection between the GCs in the host galaxies and the formation of AGNs in their centers, using a sample of hundreds of objects.

5. Summary and Conclusions

Based on the results of simulations, we conclude that future telescopes will allow ample and detailed studies of stellar halos with sufficent accuracy to constrain models of galaxy formation. In particular it will be possible to:

- catch young stars up to distances of hundreds of Mpcs and witness current stars formation in large samples of interacting galaxies;
- costruct the color-magnitude diagram of stars in tidal features down to the red clump for galaxies up to ~ 10 Mpc away, and derive the metallicity distribution and its gradient in the halo of galaxies within 20 Mpc;
- map the stellar density in galaxy halos up to a distance of 50 Mpc; overdensities will be traceable at very faint limits of surface brightness ($\mu_V \sim 33$ mag/arcsec^2);
- analyze the color-magnitude diagram of resolved stars in globular clusters down to the old turn-off region in the whole Local Group;
- measure the luminosity, mass and color distribution of the globular cluster members of galaxies up to a redshift $z \simeq 0.1$.

Acknowledgements

We acknowledge support from INAF and MIUR through the *Progetto premiale T-REX*.

References

Atkinson, A. M., Abraham, R. G., & Ferguson, A. M. N. 2013, *ApJ*, 765, 28
Blom, C., Forber, D. A., Romanowsky, A. J., *et al.* 2014, *MNRAS*, 439, 2420
Cooper, A. P., D'Souza, R. Kauffmann, G., *et al.* 2013, *MNRAS*, 434, 3348
Deason, A. J., Belorukov, W., & Evans, N. W. 2011, *MNRAS*, 416, 2903
Font, A. S., McCarthy, I. G., Crain, R. A., *et al.* 2013, *MNRAS*, 416, 2802
Greggio, L., Rejkuba, M., Gonzales, O. A., *et al.* 2014, *AA*, 562, A73
Hudson, M. J., Harris, G. L., & Harris, W. E. 2014, *ApJ* (Letters), 787, L5
Ibata, R. A., Lewis, G. F., McConnachie, A. W., *et al.* 2014, *ApJ*, 780, 128
Johnston, K. V., Bullock, J. S., Sharma, S., *et al.* 2008, *ApJ*, 689, 936
Nantais, J. B., Huchra, J. P., Barmby, P., *et al.* 2006, *AJ*, 131, 1416
Pillepich, A., Vogelsberger, M., Deason, A. J., *et al.* 2014, *MNRAS*, 444, 237

The General Assembly of Galaxy Halos: Structure, Origin and Evolution
Proceedings IAU Symposium No. 317, 2015
A. Bragaglia, M. Arnaboldi, M. Rejkuba & D. Romano, eds.

doi:10.1017/S1743921315010935

The formation of the smooth halo component

Jorge Peñarrubia

Institute for Astronomy, University of Edinburgh, Royal Observatory, Blackford Hill, Edinburgh EH9 3HJ, UK
email: jorpega@roe.ac.uk

Abstract. The detection and characterization of debris in the integral-of-motion space is a promising avenue to uncover the hierarchical formation of the Milky Way. Yet, the fact that the integrals do not remain constant during the assembly process adds considerable complexity to this approach. Indeed, in time-dependent potentials tidal substructures tend to be effaced from the integral-of-motion space through an orbital diffusion process, which naturally leads to the formation of a 'smooth' stellar halo. In this talk I will introduce a new probability theory that describes the evolution of collisionless systems subject to a time-dependent potential. The new theory can be used to reconstruct the hierarchical assembly of our Galaxy through modelling the observed distribution of accreted stars in the integral-of-motion space.

Keywords. Galaxy: kinematics and dynamics

---

1. Introduction

Astrometric surveys of our Galaxy offer a unique opportunity to test key aspects of the current cosmological paradigm, wherein galaxies form throug a hierarchical accretion process. In principle, remnants of the Milky Way assembly will appear as stellar clumps in the integral-of-motion space (e.g. energy – angular momentum) that cluster around the integrals of the orbits of the progenitor systems from which they were tidally stripped. Hence, detecting and characterizing individual subtructures can be used to *uncover the accretion history of our Galaxy* (e.g. Helmi & White 1999; Helmi & de Zeeuw 2000). In practice, however, the fact that cosmological potentials evolve with time implies that integrals do not remain constant. Peñarrubia (2013) shows that substructures in the integral-of-motion space undergo *collisionless diffusion*, and that the inexorable cycle of deposition, and progressive dissolution, of tidal clumps naturally leads to the formation of a 'smooth' stellar halo.

There are several mechanisms that introduce a time dependence in the potential of our Galaxy. For example, cooling and infall of gas deepen the potential well of the host (e.g. adiabatic contraction), whereas supernova feedback may alter the density profile of the dark matter halo (e.g. Ponzten & Governato 2013). Accretion of massive satellites leads to the growth of the dark matter halo and break existing spatial symmetries, which in turn induces diffusion both in the energy and angular momentum dimensions. In order to uncover the accretion history of the Milky Way from observations of the integral-of-motion space one needs to *model* such diffusion processes.

Unfortunately, the relaxation of collisionless systems is still poorly understood. The most successful theory dates back to the violent relaxation theory of Lynden-Bell (1967), which shows that self-gravitating systems evolve toward a unique equilibrium state which can be accurate described by a combination of Fermi-Dirac distribution functions. The central difficulty in accepting the distribution function derived from Lynden-Bell theory is that it predicts infinite mass for the system. In other words, the variational problem that determines the most probable distribution function possesses no solution for any finite total mass. This shortcoming may be due to the short life of the process that drives relaxation, i.e. fluctuations of the gravitational field, which vanish on the time scale $(G\rho)^{-1/2}$, well before the thermodynamical equilibrium is attained. As a result, in most gravitating systems the evolution will be frozen in a subdomain of the available phase space (a.k.a. 'incomplete relaxation'). In addition, in the current cosmological paradigm galaxies can be rarely found in an equilibrium state. Indeed, the detection of tidal streams in the stellar halo of the Milky Way provides a clear-cut evidence that our Galaxy has not yet reached dynamical equilibrium.

2. A new probability theory

Recently, Peñarrubia (2015) proposes a new probability theory that describes the non-equilibrium state of collisionless gravitating systems subject to time-dependent gravitational forces. This theory can be used, for example, to calculate the distribution function of stellar tracers orbiting in a growing potential. In a spherical potential the non-equilibrium distribution function, $f(E,t) = N(E,t)/\omega(E,t)$, where N is the probability to find a particle in the energy interval $E, E + \mathrm{d}E$ at the time t and ω is the so-called density of states (e.g. Binney & Tremaine 1986), is found by convolving the initial DF with a transition probability (also called 'propagator') p_c

$$N(E,t) = \int \mathrm{d}E_0 N(E_0,t_0) p_c(E,t|E_0,t_0), \tag{2.1}$$

where $p_c(E,t|E_0,t_0)$ is the probability that a particle with energy E_0 at the time t_0 has an energy E at the time t.

Equation (2.1) is a remarkable result, as it suggests that one can describe the dynamical state of a gravitating system at the time t from the state at the time t_0 through a single probability convolution. This is akin to the "jumps" we are familiar with in Quantum Mechanics! As in Stochastic Calculus, one can also construct a serial sequence of κ-jumps, $t_0 < t_1 <, ..., < t_{\kappa-1} \leqslant t_\kappa$, with $t_\kappa = t_0 + \tau'$, where the end product of a jump is used as the initial condition for the next one. The result is a Markov chain, where

$$p_{c,\kappa}(E,t|E_0,t_0) = \int \mathrm{d}E_1 p_c(E_1,t_1|E_0,t_0) \int \mathrm{d}E_2 p_c(E_2,t_2|E_1,t_1) ... \\ \times \int \mathrm{d}E_{\kappa-1} p_c(E,t|E_{\kappa-1},t_{\kappa-1}) p_c(E_{\kappa-1},t_{\kappa-1}|E_{\kappa-2},t_{\kappa-2}). \tag{2.2}$$

We shall say that a Markov chain is *transitive* when the transition probability between the states t_0 and $t = t_0 + \tau'$ is independent of the number of intermediate steps, that is if $p_c = p_{c,1} = ... = p_{c,\kappa}$ for any value of κ. Peñarrubia (2015) shows that p_c is always transitive on time scales $\tau' \ll (\dot{\Phi}/\Phi_0)^{-1}$, where Φ is a time-dependent gravitational potential.

3. Transition Probabilities

The probability function $p_c(E,t|E_0,t_0)$ is the result of the convolution of two Gaussians

$$p_c(E,t|E_0,t_0) = \int p(E,t|IR^{-2},t_0) p(I|E_0,t_0) \mathrm{d}I. \tag{3.1}$$

In potentials that evolve in the linear regime p is the solution to Einstein's equation for freely diffusing particles

$$p(E,t|E_a,t_0) = \frac{1}{\sqrt{4\pi\tilde{D}(E_a,t)}} \exp\left[-\frac{[E - E_a + \tilde{C}(E_a,t)]^2}{4\tilde{D}(E_a,t)}\right], \tag{3.2}$$

Here E_a denotes the adiabatic energy, which varies with time, and $\tilde{C}$ and $\tilde{D}$ are the so-called drift and diffusion coefficients, respectively. The mean and the dispersion of the distribution are $\overline{E - E_a} = -\tilde{C}$, and $\overline{(E - E_a)^2} - \overline{E - E_a}^2 = 2\tilde{D}$, respectively. If the potential evolves adiabatically both coefficients $\tilde{C}$ and $\tilde{D}$ approach zero, and the probability function (3.2) becomes sharply peaked about $E = E_a$, which recovers the adiabatic solution exactly.

The second Gaussian is

$$p(I|E_0,t_0) = \frac{1}{\sqrt{4\pi\tilde{D}(E_0,t_0)R_0^4}} \exp\left[-\frac{[I - E_0R_0^2 - \tilde{C}(E_0,t_0)R_0^2]^2}{4\tilde{D}(E_0,t_0)R_0^4}\right]; \tag{3.3}$$

where $I = E_aR^2(t)$ is a dynamical invariant (constant of motion), and $R(t)$ is a scaling factor (see §4).

Hence, to integrate (2.1) and (3.1) we only need to calculate the following functions: $\tilde{C}(E,t)$, $\tilde{C}(E_0,t_0)$, $\tilde{D}(E,t)$ and $\tilde{D}(E_0,t_0)$. This implies that the problem of collisionless relaxation in the linear regime reduces to the derivation of four coefficients!

4. Diffusion coefficients

Peñarrubia (2013, 2015) uses dynamical invariants to derive *analytically* the diffusion coefficients for systems subject to power-law forces

$$F(r,t) = -\mu(t)r^n. \tag{4.1}$$

In such systems the drift and diffusion coefficients can be simply written as

$$\tilde{C}(E,t) = -(\dot{R}/R)\dot{\mathcal{I}}, \tag{4.2}$$

and

$$\tilde{D}(E,t) = B_n(\dot{R}/R)^2\mathcal{I}T, \tag{4.3}$$

where B_n is a positive constant that only depends on the power-law index of the force (4.1). The quantities $\mathcal{I}$ and T are, respectively, the moment of inertia and the mean kinetic energy of a particle ensamble with energy E, that is

$$\begin{aligned}\mathcal{I}(E,t) &= \langle r^2/2\rangle = \frac{1}{2\omega}\int r^2\delta(E-H)\mathrm{d}^3\mathbf{r}\mathrm{d}^3\mathbf{v} \\ &= \frac{(4\pi)^2}{2\omega}\int_0^{r_m} r^4[2(E-\Phi)]^{1/2}\mathrm{d}r,\end{aligned} \tag{4.4}$$

and

$$\begin{aligned}T(E,t) &= \langle v^2/2\rangle = \frac{1}{\omega}\int \frac{\mathbf{v}^2}{2}\delta(E-H)\mathrm{d}^3\mathbf{r}\mathrm{d}^3\mathbf{v} \\ &= \frac{(4\pi)^2}{6\omega}\int_0^{r_m} r^2[2(E-\Phi)]^{3/2}\mathrm{d}r,\end{aligned} \tag{4.5}$$

where $r_m(E,t)$ is the maximum radius that particles with energy E can reach, that is $\Phi(r_m,t) = E$.

The quantity $R(t)$ is a scaling factor, which for power-law forces is also a power law

$$R(t) = \left[\frac{\mu(t)}{\mu_0}\right]^{-1/(n+3)} . \tag{4.6}$$

Note that $\dot{R}/R = -(\dot{\mu}/\mu)/(n+3)$, and that the diffusion coefficient (4.3) is proportional to the variation of the potential *squared.* This implies that the diffusion of substructures in the integral-of-motion space is most effective during periods in which the Milky Way potential changes quickly (e.g. during the infall of massive satellites).

5. *N*-body tests

As an illustration, Peñarrubia (2015) considers logarithmic potentials that evolve linearly with time as

$$\frac{\mu(t)}{\mu_0} = 1 + \epsilon\frac{t - t_0}{P_0}, \tag{5.1}$$

where $P_0 = P(E, t_0) = 2\int_0^{r_m(E)} \mathrm{d}r/\sqrt{2(E - \Phi[r, t_0])}$ is the radial period of an orbit with $r_m = \mu_0 = 1$ at $t = t_0$.

N-body equilibrium realizations of a lowered Maxwellian distribution are generated in a logarithmic potential

$$f(E_0, t_0) = \begin{cases} A\left[e^{\frac{-E_0+\Phi_{\rm lim}}{\sigma^2}} - 1\right] & , E_0 < \Phi_{\rm lim}, \\ 0 & , E_0 \geqslant \Phi_{\rm lim}, \end{cases} \tag{5.2}$$

where A is a normalization factor, and $\Phi_{\rm lim} = \mu_0 \ln(r_{\rm lim})$ is an energy truncation. To simplify our models we set $\sigma^2 = \mu_0/2 = 3T$, where T is the kinetic energy associated with the potential (see Table 1).

Next, velocities of all particles are multiplied by a factor q. To highlight non-equilibrium features we choose a small value, $q = 0.2$, which leads to $|2T/W| = q^2 = 0.04$. Such a low virial ratio guarantees the collapse of the system on a time-scale comparable to its free-fall time. Fig. 1 shows ten snap-shots of a model orbiting in a time-dependent potential that evolves at a rate $\epsilon = -0.1$. Cold collapse happens early on ($\tau' \approx 0.2P_0$) and leads to the formation of shell structures that move progressively towards larger radii with time. By the end of the simulation, $\tau' \approx 2.0P_0$, the model has not yet reached dynamical equilibrium. By definition the system is in a state of 'incomplete relaxation'†.

Fig. 2 shows the drift coefficient $\tilde{C} = -(\dot{R}/R)\langle \mathbf{r}\cdot\mathbf{v}\rangle$ as a function of energy for the snap-shots shown in Fig. 1. As the system begins to collapse the averaged radial velocity of the particle ensemble is negative at all radii and $\partial\tilde{C}/\partial t > 0$ at all energies. At slightly later times, $\tau' \gtrsim 0.2P_0$, the coefficient $\tilde{C}$ begins to exhibit coherent fluctuations in the innermost regions of the potential (left side of the panels), as particles with short orbital period go through pericentre and start moving toward larger radii. In the outskirts, however, particles are still falling in from large distances, which translates into positive values of $\tilde{C}$. The negative crests are associated with the shell features of Fig. 1. Given that in a potential with $n = -1$ the orbital period decreases toward the central regions of the potential as

$$P(E, t) = (2\pi/\mu)^{1/2} r_m = (2\pi/\mu)^{1/2} r_{\rm lim} \exp(E/\mu),$$

† These models provide a useful representation of the dynamical state of the outer regions of galactic haloes and galaxy clusters, where dynamical times are comparable to the age of the Universe and phase mixing becomes very inefficient.

fluctuations in $\tilde{C}$ damp out progressively from inside out. By $\tau' = 1.9P_0$ the mean radial velocity of particles with $E - \Phi_{\rm lim} \lesssim -2$ is $\langle \mathbf{r} \cdot \mathbf{v} \rangle \approx 0$, signalling that the inner regions of the system are phase-mixed and evolving in a state of quasi-dynamical equilibrium.

Fig. 3 illustrates the complexity involved in describing the state of systems undergoing violent relaxation. Green dots show the distribution function of the N-body models plotted in Fig. 1 at three different snap-shots. The decreasing potential ($\epsilon < 0$) shifts the orbital energies to higher values, which leads to a non-monotonic increase of $f(E,t)$ at fixed energies. Interestingly, the distribution function is not completely smooth. Non-equilibrium features arise in the inner-most regions of the potential and propagate toward high energies as time goes by. After constructing a larger suite of N-body models (not shown here) we find that the amplitude of these fluctuations increases for larger ϵ and smaller q values, which correspond to faster growth rates and higher radial anisotropies, respectively.

Equation (2.1) (red solid lines) is able to capture these complexities and provide an accurate statistical description of the non-equilibrium state of the system. For simplicity we assume that the transition probability $p_c(E,t|E_0,t_0)$ corresponds to a single "jump" between t_0 and $t_0+\tau'$. The Green convolution is solved by setting $\tilde{C}(E_0,t_0) = 0$ and computing $\tilde{D}(E_0,t_0)$ analytically from Equation (4.3) . The coefficients $\tilde{C}(E,t)$ and $\tilde{D}(E,t)$ are measured from the phase-space coordinates of the N-body particles at $t = t_0 + \tau'$ as $\tilde{C}(E,t) = -(\dot{R}/R)\langle \mathbf{r} \cdot \mathbf{v} \rangle$ and $\tilde{D}(E,t) = (\dot{R}/R)^2 \langle (\mathbf{r} \cdot \mathbf{v})^2 \rangle$.

The lower panels of Fig. 3 plots the difference between the N-body distribution function and the adiabatic solution at fixed energy values (green dots). From Equation (5.1) particles respond adiabatically to a time-varying force if their orbital periods obey $P(\dot{\Phi}/\Phi_0) = \epsilon(P/P_0) \ll 1$. Since P increases exponentially with the particle energy we find the distribution N evolves adiabatically ($\Delta N_a \approx 0$) at $E \ll \Phi_{\rm lim}$, while strong departures from the adiabatic solution ($|\Delta N_a| \sim N$) are visible at $E \gtrsim \Phi_{\rm lim}$, where the potential changes significantly during an orbital period, i.e. $P(\dot{\Phi}/\Phi_0) \sim 1$. As a result, the linear approximation (solid magenta lines) becomes less accurate in the outskirts of the system. Note that similar deviations are also visible in the upper panels at $\tau' = 1.8P_0$, albeit with a lesser magnitude.

The existence of internal macroscopic motions ($\tilde{C} \neq 0$) leads to fluctuations of the distribution function which travel toward high energies, as shown in Fig. 2. Comparison with the solid magenta lines shows that the ripples and troughs of the fluctuations are located at energies where the gradient $\nabla(\tilde{C}N) \equiv \partial(\tilde{C}N)/\partial E$ finds local maxima and minima, respectively. Notice that by the end of the simulation ($\tau' = 1.8P_0$) relaxation is still 'incomplete'.

6. Summary

Peñarrubia (2015) proposes a probability theory that describes the non-equilibrium evolution of large particle ensembles orbiting in a time-dependent gravitational field. This theory states that in the linear regime the non-equilibrium state of collisionless systems can be obtained by a single convolution of the initial distribution function with a transition probability which is uniquely defined by 4 coefficients, $\tilde{C}(E,t)$, $\tilde{C}(E_0,t_0)$, $\tilde{D}(E,t)$ and $\tilde{D}(E_0,t_0)$.

In principle these results provide a simple tool to determine the evolution of stellar tracers in time-dependent potentials. In particular, it offers a simple method for modelling the evolution of accreted clumps in the integral-of-motion space of the Milky Way as follows: given the *observed* energy distribution function of the i-th clump, say $N_i(E,t)$, we would like to constrain the initial energy distribution $N_i(E_{\rm acc},t_{\rm acc})$, where $t_{\rm acc}$ is

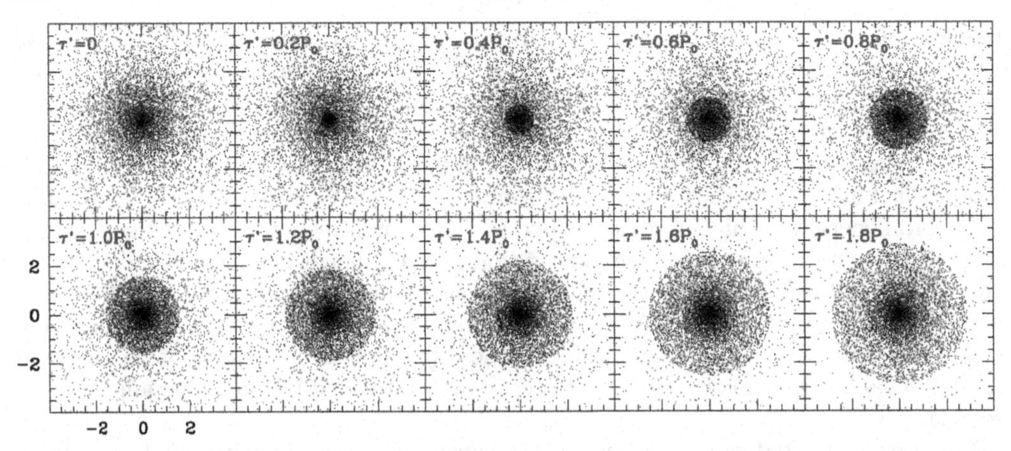

Figure 1. Cold collapse of a model with $|2T/W| = 0.04$ at $t = t_0$. Particles move in a time-dependent potential that varies at a constant rate $\epsilon = -0.1$. The system as a whole evolves toward an equilibrium configuration following a 'violent relaxation' process.

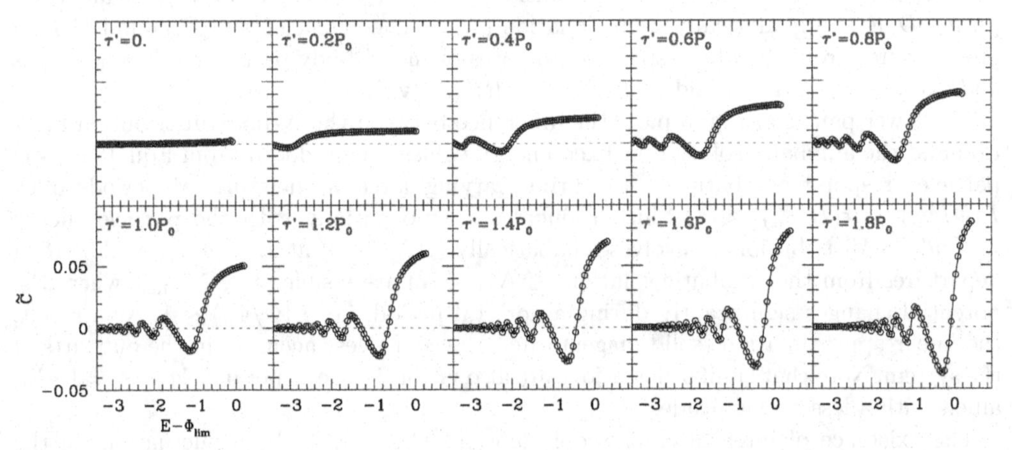

Figure 2. Drift coefficient $\tilde{C} = -(\dot{R}/R)\langle \mathbf{r} \cdot \mathbf{v} \rangle$ as a function of energy of the model shown in Fig. 1. Notice that at $t = t_0$ the system is phased mixed (i.e. $\langle \mathbf{r} \cdot \mathbf{v} \rangle = 0$) but out of virial equilibrium, $|2T/W| = 0.04$. By the end of the simulation only the internal regions of the potential have reached a state of quasi-dynamical equilibrium ($\tilde{C} \approx 0$).

the time of accretion, for a given potential evolution $\Phi(r, t)$. Both $N_i(E_{\rm acc}, t_{\rm acc})$ and $\Phi(r, t)$ are unknowns and need to be modelled. The former contains information on the properties of the i-th progenitor system, e.g. accretion time, orbit, mass, etc. The latter informs us how the Milky Way potential has changed from $t = t_{\rm acc}$ to the present and has the same functional form for all clumps detected in the integral-of-motion space.

In practice not all potentials admit analytical solutions to the diffusion equation. For example, in systems that are not initially virialized a derivation of the diffusion coefficients requires precise knowledge of the trajectories of individual particles in phase space, rendering the problem analytically intractable. In those cases the diffusion coefficients and the integrals (2.1) and (3.1) need to be solved numerically.

As a final remark it is worth mentioning that the new probability theory has further applications than those mentioned here. For example, we may get a deep insight into the evolution and stability of self-gravitating collisionless systems by studying the evolution

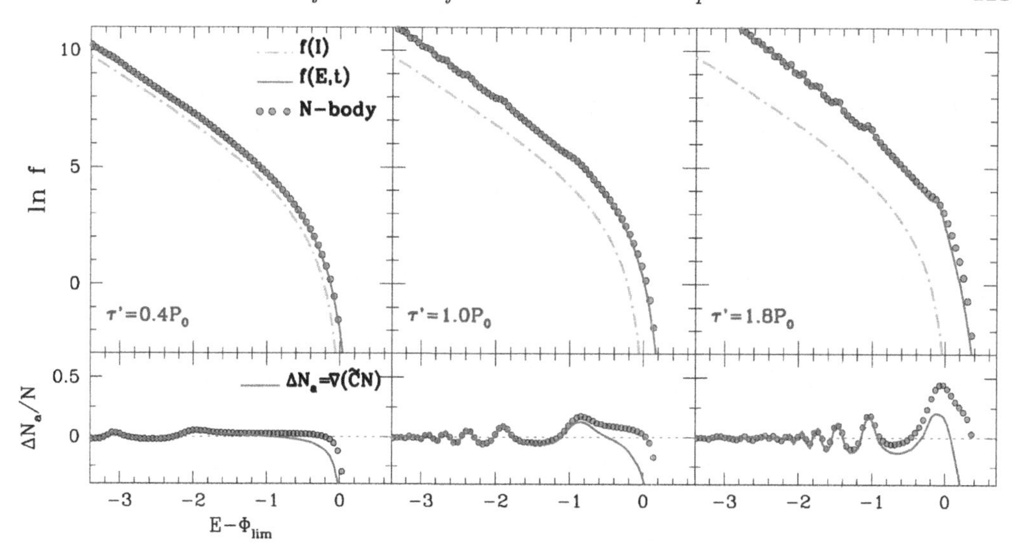

Figure 3. *Upper panels:* Distribution function $f = N/\omega$ of the models shown in Fig. 1 at three different snap-shots. Red solid lines correspond to the Green convolution given by Equation (2.1) with coefficients $\tilde{C}(E,t) = -(\dot{R}/R)\langle \mathbf{r}\cdot\mathbf{v}\rangle$ and $\tilde{D}(E,t) = (\dot{R}/R)^2\langle(\mathbf{r}\cdot\mathbf{v})^2\rangle$ measured from the phase-space locations of the N-body particles. *Lower panels:* Deviation of the N-body models from the adiabatic distribution (green dots). Magenta solid lines show deviations from the adiabatic distribution function calculated with our probability theory. Note that the ripples and troughs of the curves are located at energies where the gradient $\nabla(\tilde{C}N)$ finds local maxima and minima, respectively. Departures from the adiabatic solution are particularly strong in the outskirts of the system, $E \gtrsim \Phi_{\rm lim}$.

of particle ensembles driven by their own self-gravity, i.e.

$$\nabla^2\Phi(\mathbf{r},t) = 4\pi G\int f(\mathbf{r},\mathbf{v},t)\mathrm{d}^3v.$$

Also, our formalism can be straightforwardly extended to systems with a varying spatial symmetry, where both the angular momentum and the energy of individual particles are allowed to vary with time. In this case the Green function that solves the diffusion equation has now one extra dimension, i.e. $p(E,L,t|E',L',t_0)$, which defines the probability that a particle with an energy E' and angular momentum L' at the time $t = t_0$ has an energy in the interval $E, E+\mathrm{d}E$ and an angular momentum in the interval $L, L+\mathrm{d}L$ at the time $t = t_0 + \tau'$. Finally, the fact that the evolution of the microcanonical distribution has the same form as Einstein's equation for purely stochastic processes offers the tantalizing possibility to incorporate the effects of random particle-particle collisions into our probability theory in a natural way. Such an extension may be useful, for example, to study a range of dynamical processes taking place in planetary systems, dense stellar objects and/or self-interacting dark matter (SIDM) haloes.

References

Helmi, A. & de Zeeuw, P. T. 2000, *MNRAS*, 319, 657
Helmi, A. & White, S. D. M. 1999, *MNRAS*, 307, 495
Lynden-Bell, D. 1967, *MNRAS*, 136, 101
Peñarrubia, J. 2015, *MNRAS*, 451, 3537
Peñarrubia, J. 2013, *MNRAS*, 433, 2576
Pontzen, A. & Governato, F. 2013, *MNRAS*, 430, 121

The General Assembly of Galaxy Halos: Structure, Origin and Evolution
Proceedings IAU Symposium No. 317, 2015
A. Bragaglia, M. Arnaboldi, M. Rejkuba & D. Romano, eds.

doi:10.1017/S1743921315008558

Resolving the stellar halos of six massive disk galaxies beyond the Local Group

Antonela Monachesi[1], Eric F. Bell[2], David J. Radburn-Smith[3], Roelof S. de Jong[4], Jeremy Bailin[5,6], Benne Holwerda[7] and David Streich[4]

[1]Max Planck Institute for Astrophysics, Karl-Schwarzschild-Str. 1, Garching D-85748, Germany
email: antonela@mpa-garching.mpg.de
[2]Department of Astronomy, University of Michigan, 311 West Hall, 1085 South University Ave., Ann Arbor, MI 48109, USA
[3]Department of Astronomy, University of Washington, Seattle, WA 98195,USA
[4]Leibniz-Institut für Astrophysik Potsdam, D-14482 Potsdam, Germany
[5]Department of Physics and Astronomy, University of Alabama, Box 870324, Tuscaloosa, AL, 35487, USA
[6]National Radio Astronomy Observatory, P.O. Box 2, Green Bank, WV, 24944, USA
[7]University of Leiden, Sterrenwacht Leiden, Niels Bohrweg 2, NL-2333 CA Leiden, The Netherlands

Abstract. Models of galaxy formation in a hierarchical universe predict substantial scatter in the halo-to-halo stellar properties, owing to stochasticity in galaxies' merger histories. Currently, only few detailed observations of stellar halos are available, mainly for the Milky Way and M31. We present the stellar halo color/metallicity and density profiles of red giant branch stars out to ~60 kpc along the minor axis of six massive nearby Milky Way-like galaxies beyond the Local Group from the Galaxy Halos, Outer disks, Substructure, Thick disks and Star clusters (GHOSTS) HST survey. This enlargement of the sample of galaxies with observations of stellar halo properties is needed to understand the range of possible halo properties, i.e. not only the mean properties but also the halo-to-halo scatter, what a 'typical' halo looks like, and how similar the Milky Way halo is to other halos beyond the Local Group.

Keywords. galaxies: halos, galaxies: spiral, galaxies: individual (NGC 253, NGC 891, NGC 3031, NGC 4565, NGC 4945, NGC 7814), galaxies: photometry, galaxies: stellar content, galaxies: evolution

1. Introduction

Over the past decades different approaches have been used to observe stellar halos, a challenging task due to their faint surface brightnesses. Long exposure wide field imaging with photographic plates (Malin&Carter 1980) and with small telescopes and wide field CCDs (e.g., Zheng *et al.* 1999, Martínez-Delgado *et al.* 2010) has allowed panoramic mapping of the brightest overdensities in nearby galaxies, revealing numerous tidal streams. Evidence of stellar halo substructures (e.g., stellar streams, shells, etc) in the outer regions of galaxies was possible with these types of images, proving the importance of merging in the galaxy formation process. However, none of these techniques allow for a detailed study of the physical properties of individual halos predicted by models, such as their age and metallicity as a function of galactocentric distance.

One of the best approaches for characterizing the properties of nearby galactic stellar halos is to study their resolved stellar populations. It is possible to measure stellar densities of resolved stars reaching equivalent surface brightnesses as faint as $\mu_V \sim 33$ mag arcsec^{-2}, as well as measuring the stellar populations of those halos, which is crucial

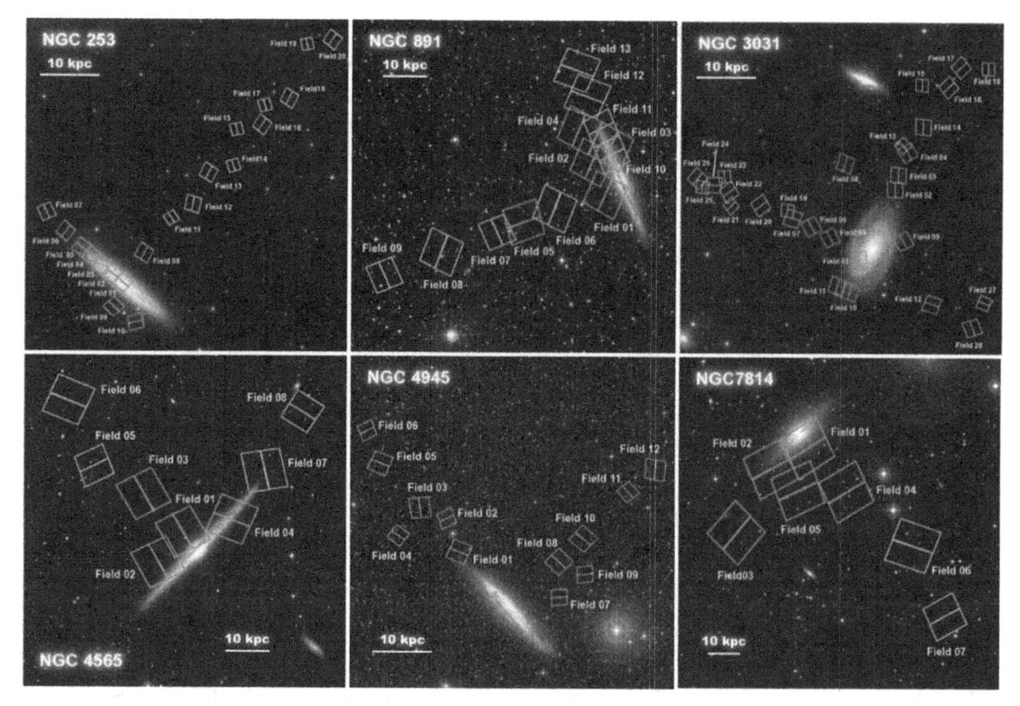

Figure 1. DSS colored images of the GHOSTS massive disk galaxies analyzed showing the locations of the *HST* ACS/WFC and WFC3/UVIS fields. North is up and east is to the left. ACS fields in green were introduced in R-S11 whereas ACS and WFC3 fields indicated in yellow are presented in Monachesi *et al.* (2016).

for testing model predictions (Monachesi *et al.* 2013). One such prediction is that there should be stellar population variations within a halo since the stellar population of halos should reflect the various satellites that form them. In particular, how a halo has formed and evolved is expected to leave strong imprint on its metallicity or abundance pattern (e.g., Font *et al.* 2006, Cooper *et al.* 2010, Gomez *et al.* 2012, Tissera *et al.* 2014).

The resolved stars of the Milky Way and M31 halos have been extensively studied. While both halos have similarities and qualitatively agree within the LCDM preferred paradigm of hierarchical formation, they exhibit clear differences. There is an order of magnitude difference in their stellar halo mass (e.g., Bell *et al.* 2008, Ibata *et al.* 2014) and a factor of 5 difference in metallicity and difference in gradient, which betray large differences in halo growth history (Deason *et al.* 2013, Gilbert *et al.* 2014). Cosmological models of galaxy formation predict that there should be large variations among the properties of individual halos in disk galaxies with similar mass (e.g., Bullock&Johnston 2005, hereafter BJ05, Cooper *et al.* 2010, Tissera *et al.* 2014). Given the stochasticity involved in the process of galaxy formation, predictions such as differences in metallicity profiles, fraction of stellar halo created in-situ and accreted, stellar halo morphology, etc., need to be compared against observations to differentiate between the models and quantify the predicted halo-to-halo scatter. This motivates the study of other stellar halos, which was one of the main goals of the GHOSTS survey.

2. The GHOSTS survey and sample of galaxies

The GHOSTS survey (Galaxy Halos, Outer disks, Substructure, Thick disks, and Star clusters survey, Radburn-Smith *et al.* 2011, hereafter R-S11) is an extensive HST program dedicated to resolve the stars in the outskirts of 16 nearby disk galaxies of different masses, luminosities, and inclinations. Various fields along the principal axes of each galaxy were observed reaching projected distances as far as $\sim$ 70 kpc from the galactic center. GHOSTS is the largest study of resolved stellar populations in the outer disk and halo of disk galaxies to date and its observations offer a direct test of model predictions. GHOSTS observations provide star counts and color-magnitude diagrams (CMDs) reaching typically 2–3 magnitudes below the tip of the red giant branch (TRGB). Using the RGB stars as tracers of the stellar halo population, we are able to study the size and shape of each stellar halo as well as the properties of their stellar populations such as age and metallicity.

In this work, we analyze the more massive disk Milky Way-like galaxies of our sample, namely NGC 253, NGC 891, NGC 3031, NGC 4565, NGC 4945, and NGC 7814. They have maximum rotation velocity $V_{max} \gtrsim 170$ km/s and their distances range from 3.5 to 15 Mpc. Two of these galaxies are edge-on and four are highly inclined. This facilitates studies of their halos, because we expect minimal to no disk contamination in projected distances beyond 5 or 10 kpc when they are observed along their minor axis. Fig. 1 shows the field locations of the GHOSTS galaxies studied here.

3. Results: Stellar halo density and color profiles

The data reduction steps and photometry were performed using the GHOSTS pipeline described in R-S11 and Monachesi *et al.* (2015). Stellar photometry on the images was performed using the ACS and WFC3 modules of DOLPHOT, a modified version of HSTPHOT (Dolphin 2000). The final output of DOLPHOT provides instrumental VEGA magnitudes, already corrected for CTE loss and with aperture corrections calculated using isolated stars. We show in Fig. 2 the CMDs of two fields, from NGC 3031 and NGC 253. The CMDs are dereddened for Galactic extinction and are free of contaminants. The unresolved background galaxies (the major source of contamination in our observations) were eliminated using photometric ‘culls’ determined from diagnostic parameters included in the photometric output of DOLPHOT. We can clearly see halo detections in the outer fields observed, with projected distances between 50 and 70 kpc from the galactic center. In general the CMDs of the fields shown in Fig. 1 are mostly populated by old RGB stars (older than 1 Gyr). There are however younger populations such as blue, extended main sequence stars ($<$ 500 Myr) or massive stars burning helium in their core (25–600 Myr old red and blue loop sequence stars). These appear primarily in the fields closer than $R \sim 15$ kpc to each galaxy, and especially along the major axis, which are dominated by disk stars.

The stellar density profiles and color profiles of two of the galaxies analyzed are also shown in Fig 2 for NGC 3031 (left panels) and NGC 253 (right panels). The results for all galaxies and details on how these results were obtained can be found in Monachesi *et al.* (2015) and Harmsen *et al.* (in prep.). Briefly, we selected RGB stars within a magnitude range that ensures all of the stars to be in a CMD region with completeness higher than 70%. This is typically from the TRGB down to $\sim$ 0.7 mag below it. We measured the projected stellar density in each of the fields observed and constructed their stellar density profiles (middle panels in Fig. 2). The equivalent $V-$band surface brightness values are indicated on the right hand $y-$axis. Errorbars indicate Poisson uncertainty. We fit power-law functions to the projected minor axis density profiles over 10 to 70 kpc and find slopes between $-2.2 > \alpha > -4$ for all our galaxies. Deviations from the power law fits that cannot be accounted for from Poisson uncertainty most likely indicate halo substructure. The scatter around the power law profiles is ~ 0.18 dex, which may suggest that there is abundant substructure (RMS/total $\sim 40\%$).

The minor axis stellar halo color profiles (bottom panels in Fig. 2) were obtained from the median $F606W - F814W$ colors of the selected RGB stars in each field as a function of projected radius. In order to have a clean stellar halo profile and to avoid the disk as much as possible we do not use the major axis fields to construct these profiles. Error bars indicate uncertainties in the median values calculated by bootstrapping our sample of stars as well as systematic uncertainties due to calibration which accounts for up to ~ 0.04 mag in colors. The red line in the color profiles shows a linear fit to the data. We find that half of the galaxies (NGC 4565, NGC 0891, and NGC 7814) show fits consistent with stellar halo color gradients whereas the remaining three galaxies (NGC 0253, NGC 3031, and NGC 4945) have rather flat color profiles. Because the colors of the RGB stars are more sensitive to metallicity than to age and because there is a direct relation between RGB colors and metallicities (Streich *et al.* 2014), one can assume that the color profiles reflect metallicity profiles. The right hand y-axes indicate the [Fe/H] values that the colors correspond to, calculated from the relation derived by Streich *et al.* (2014) and assuming $[\alpha/\mathrm{Fe}] = 0.3$. The metallicities [Fe/H] will be lower or higher for a given color in case of $[\alpha/\mathrm{Fe}]$ larger or lower than 0.3, respectively.

In Figure 3 we show the stellar halo color profiles of all the galaxies together, where we can see the diversity in the color profiles of massive disk galaxies. Since the galaxies studied in this work are all Milky Way-like galaxies, we find that there is a wide range in stellar halo colors and a diversity of color/metallicity profiles for galaxies of similar mass and luminosity. The black dashed line in this figure represents the average color profile of the 11 BJ05 stellar halo models, which were analyzed in the exact same way as the data (see Monachesi *et al.* 2013) and the shaded area represents the 1-σ model-to-model scatter from the average.

4. Conclusions and future work

We analyze the halo stellar populations of six massive highly inclined disk Milky Way-like galaxies from the GHOSTS survey. Several fields along the principal axes of each galaxies were observed from which we constructed CMDs showing halo populations out to projected distances between $\sim 50 - 70$ kpc along the minor axis. It is important to emphasize that *all* of these galaxies have halo stars out to at least 50 kpc *along the minor axis*, which is more than 50 scale heights of the Milky Way's thick disk. Thus, our observations show that highly inclined massive disk galaxies ($V_{max} \gtrapprox 170$ km/s) have clear extended stellar halos beyond the region where the disk dominates.

We use RGB stars along the minor axis of these galaxies as halo tracers and find that: 1) Half of the galaxies (NGC 0891, NGC 4565, NGC 7814) display a negative gradient in their color profiles, with bluer color in the outer regions. Three galaxies show flat color halo profiles (NGC 0253, NGC 3031, NGC 4945) reflecting negligible halo population variations as a function of galactocentric distances. 2) The projected minor axis stellar density profiles have power law slopes of $-4 < \alpha < -2$ over 10 to ~ 70 kpc and significant scatter around the power-law fit of ~ 0.18 dex, which likely suggests the existence of abundant substructure (RMS/total $\sim 40\%$).

The next step will be to quantitative compare our results against high resolution cosmological hydrodynamical simulations, such as EAGLE, (Schaye *et al.* 2015) that will help us to interpret the observations. We will generate simulated maps of RGB stars and analyze their properties in the exact same way as we did with the observations, highlighting areas of agreement with the data and areas where future improvement is required.

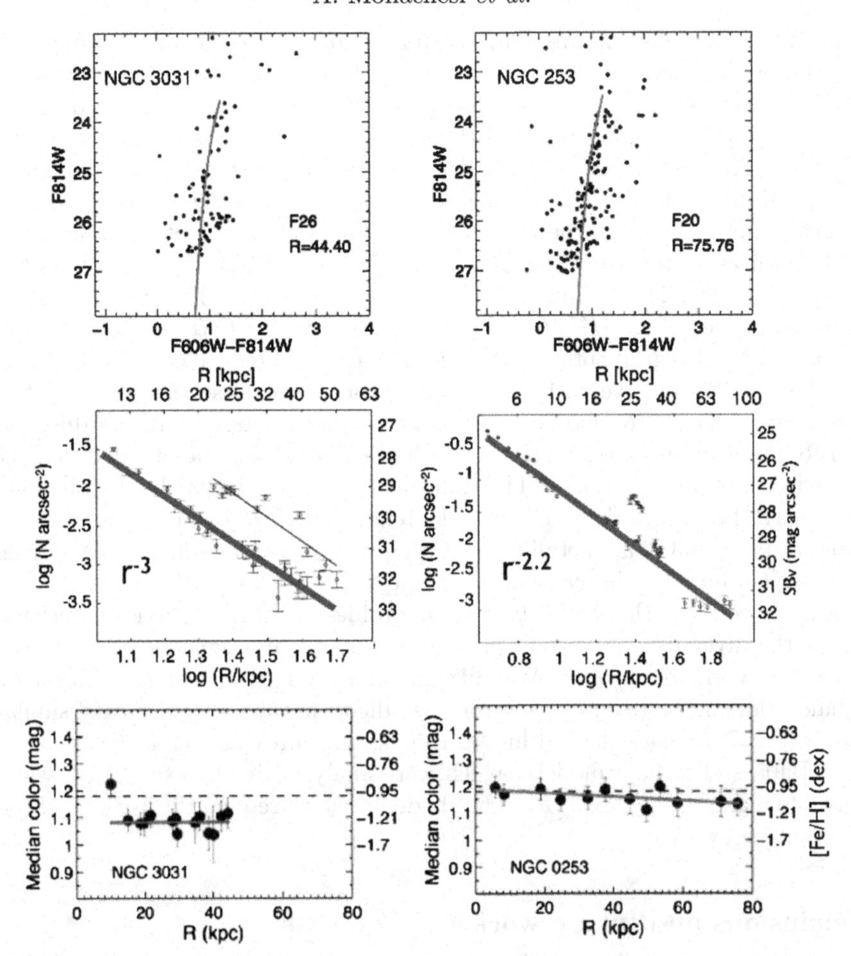

Figure 2. An example of our results for two of the GHOSTS galaxies analyzed: NGC 3031 (left panels) and NGC 253 (right panels). Results for all galaxies are presented in Monachesi *et al.* (2016) and Harmsen *et al.* (in prep.). Top panels: Color magnitude diagrams of stars, after unresolved background galaxies and other spurious detections have been removed, for two of the most distant fields. There are clear detections of RGB halo stars out to distances as far as ~ 70 kpc from the galactic center. Middle panels: Stellar density profiles along the minor axis (blue dots) and major axis (red dots) measured using the RGB stars between the TRGB and 0.7 mag below it. We fitted the minor axis stellar density profiles with power-law functions and find slopes between $-4 < \alpha < -2$ for all the GHOSTS massive galaxies. Deviations from these fits may indicate the presence of substructure in these halos. The equivalent $V-$band surface brightness is labeled on the right hand $y-$axis. We reach down to $\sim 32\mu_V$. Bottom panels: Minor axis median color profiles obtained using the RGB stars between the TRGB and 0.7 mag below. The dashed line represents the average color profile obtained using the 11 BJ05 realizations. Metallicities have been estimated using the relation derived in Streich *et al.* (2014) and are indicated on the right hand $y-$axis. These two galaxies have flat minor axis metallicity profiles.

Finally, we note that our results are obtained from pencil-beam HST observations which, even though we have several fields per galaxy such that our results are representative of a relatively large portion of the stellar halo, do not map the entire halos of these galaxies. It is therefore important to combine our results with panoramic views of these galaxies, such as those that are being obtained with the PISCeS survey (see Crnojević's contribution in this proceeding) and Hyper Suprime-Cam (Okamoto *et al.* 2015) in order to fully understand the assembling history of these galaxies.

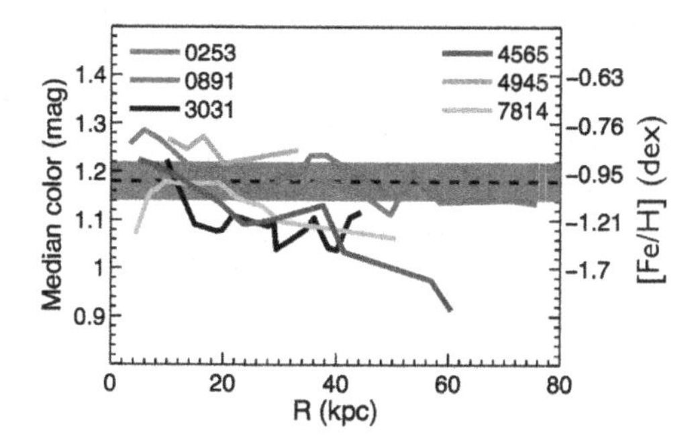

Figure 3. Minor axis color profiles of the stellar halos of the GHOSTS massive disk galaxies. Half of the galaxies display a negative gradient, with bluer colors in the outer regions. Three galaxies however show flat color halo profiles. The estimated metallicities are indicated on the right hand $y-$axis. The black dashed line indicates the average color profile of the 11 BJ05 stellar halo model realizations and the shaded area represents the 1-σ model-to-model scatter from the average.

References

Bell, E. F., Zucker, D. B., & Belokurov, V. *et al.* 2008, *ApJ*, 680, 295
Bullock, J. S. & Johnston, K. V. 2005, *ApJ*, 635, 931 (BJ05)
Cooper, A. P., Cole, S., & Frenk, C. S. *et al.* 2010, *MNRAS*, 406, 744
Deason, A. J., Belokurov, V., Evans, N. W., & Johnston, K. V. 2013, *ApJ*, 763, 113
Dolphin, A. E. 2000, *PASP*, 112, 1383
Font, A. S., Johnston, Kathryn V., Bullock, J. S., & Robertson, B. E. 2006, *ApJ*, 646, 886
Gilbert, K. M., Kalirai, J. S., & Guhathakurta, P. *et al.* 2014, *ApJ*, 796, 76
Gomez, F. A., Coleman-Smith, C. E., & O'Shea, B. W. *et al.* 2012, *ApJ*, 760, 112
Ibata, R. A., Lewis, G. F., & McConnachie, A. W. *et al.* 2014, *ApJ*, 780, 128
Malin, D. F. & Carter, D. 1980, *Nature*, 285, 643
Martínez-Delgado, D., Gabany, R. J., & Crawford, K. *et al.* 2010, *ApJ*, 140, 962
Monachesi, A., Bell, E. F., & Radburn-Smith, D. J. *et al.* 2013, *ApJ*, 766, 106
Monachesi, A., Bell, E. F., & Radburn-Smith, D. J. *et al.* 2016, *MNRAS*, 457, 1419
Okamoto, S., Arimoto, N., & Ferguson, A. M. N. *et al.* 2015, *ApJL*,809, 10
Radburn-Smith, D. J., de Jong, R. S., & Seth, A. C. *et al.* 2011, *ApJS*, 195, 18(R-S11)
Schaye, J., Crain, R. A., & Bower, R. G. *et al.* 2015, *MNRAS*, 446, 521
Streich, D., de Jong, R. S., & Bailin, J. *et al.* 2014, *A&A*, 563, A5
Tissera, P. B., Beers, T. C., Carollo, D., & Scannapieco, C. 2014, *MNRAS*, 439, 3128
Zheng, Z., Shang, Z., & Su, H. *et al.* 1999, *AJ*, 117, 2757

The General Assembly of Galaxy Halos: Structure, Origin and Evolution
Proceedings IAU Symposium No. 317, 2015
A. Bragaglia, M. Arnaboldi, M. Rejkuba & D. Romano, eds.

doi:10.1017/S1743921315009771

Stellar halos and the link to galaxy formation

Amina Helmi

Kapteyn Astronomical Institute
P.O. Box 800, 9700 AV Groningen, The Netherlands
email: ahelmi@astro.rug.nl

Abstract. I present a brief overview of how stellar halos may be used to constrain the process of galaxy formation. In particular, streams and substructure in stellar halos trace merger events but can also be used to determine the mass distribution of the host galaxy and hence put constraints on the nature of dark matter. Much of the focus of this contribution is on the Milky Way, but I also present an attempt to understand the kinematics of the globular cluster system of M31.

Keywords. Galaxy: halo; Galaxy: kinematics and dynamics; galaxies: formation; galaxies: halos; Local Group

---

1. Introduction

At first sight, it might appear perplexing to use stellar halos to understand the process of galaxy formation. For example, in a galaxy like the Milky Way more than 90% of the baryons (stars and gas) reside in the stellar disk and bulge/bar. The stellar halo contains likely less than 1% of the stellar mass (Helmi, 2008). How it can be claimed that such a component can be used to tackle galaxy formation?

While this argument is certainly valid and to address e.g. how most of the stars in the Milky Way formed one must study the Galactic disks and bulge, some of the physical processes associated to the build up of a galaxy as a whole leave also their imprints in the stellar halo. Let us therefore recall some specific properties of the halo that make it unique to probe certain aspects of galaxy formation.

First of all, stellar halos constitute a natural reservoir of merger debris. If mergers have been important in the build of the mass of the Galaxy (including dark matter), then the debris left behind is really the only way we have to access that history (Bullock & Johnston, 2005; Deason *et al.*, 2015).

Secondly, stellar halos are generally old, arguably contain the oldest stars we know of, not only in the field but also in globular clusters (Jofre & Weiss, 2011; Leaman *et al.* 2013). This implies that by studying their properties we have access to the physical conditions present in the early Universe.

Finally, for example, by recovering the building blocks of the Milky Way we have access to both the properties of galaxies at the earliest epochs (long gone, perhaps even the first disks formed), as well as to the streams that are sensitive probes of the mass distribution and its assembly (Sanders & Binney, 2013). These then help us pin down the nature of dark matter. Since dark matter is a fundamental ingredient of the current cosmological model, by studying its properties we also constrain galaxy formation in a broader sense.

2. Substructure in stellar halos: history

To study debris from merger events is very challenging, both observationally as well as theoretically. From the observational point of view one is limited in the case of external

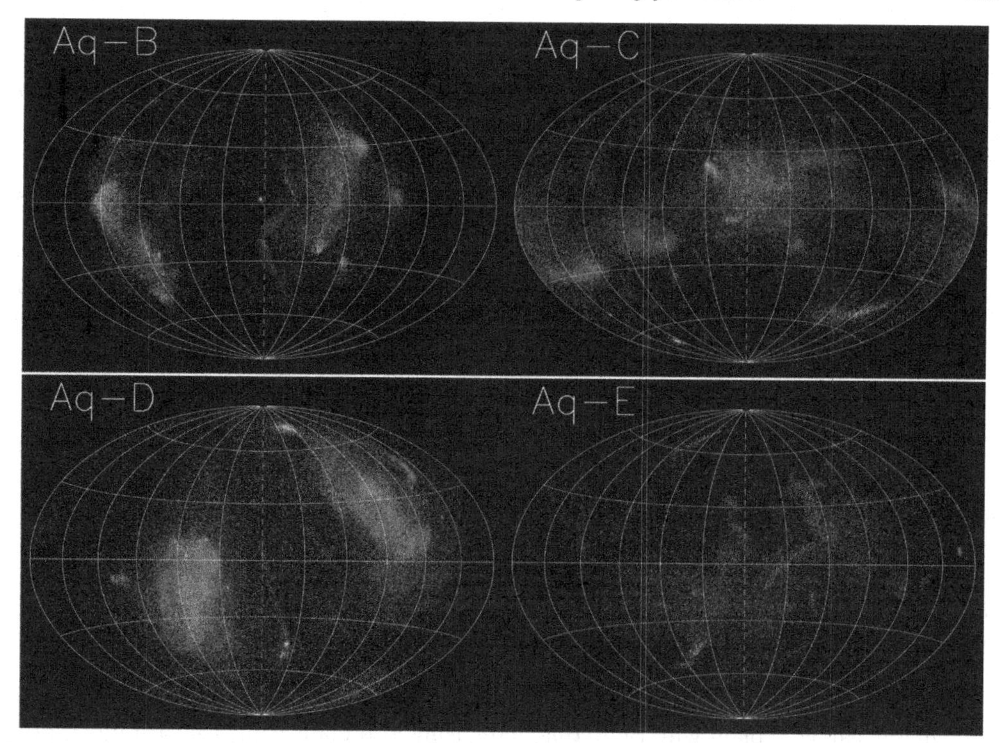

Figure 1. Sky distribution of stars in the Aquarius stellar halos, located at distances of 30-50 kpc from the "Sun" . The different colours correspond to stars originating in different progenitors. Even more distant substructures are distributed anisotropically, and this should become apparent in the next generation of photometric surveys (see Helmi *et al.* 2011).

galaxies by the low surface brightness of the tidal features, and hence the requirements on the photometric quality are very demanding. Nonetheless, significant progress is being made and astounding panoramic views of stellar halos and debris are becoming steadily available (Martinez-Delgado *et al.* 2010; Janowiecki *et al.* 2010; Duc *et al.* 2015). From the numerical perspective, and because the stellar halo constitutes just a few percent of the stellar budget of a galaxy, simulations with very high resolution are required. This is why most of the studies of substructure in stellar halos in a cosmological framework have been done by "tagging" dark matter particles in a very high resolution dark matter only simulation (White & Springel, 2000; De Lucia & Helmi 2008; Cooper *et al.* 2010). The work of Cooper *et al.* stands out because of the exceedingly high resolution provided by the Aquarius simulations (Springel *et al.* 2008). Even selecting a merely 1% of the dark matter particles has allowed the authors to study not only the global properties of halos, but also the amount of substructure both in space and in velocity (Cooper *et al.* 2011; Helmi *et al.* 2011; Gomez *et al.* 2013).

One of the results that has come out of these cosmological simulations is that at large distances, and as shown in Fig. 1, stellar halos are very lumpy, and that stars are anisotropically distributed on the sky. Their distribution in fact reflects the infall pattern of the objects in which they originate, which relates to the filamentary nature of the large scale structure in which galaxies are embedded. This anisotropic distribution should become apparent in the next generation of full sky surveys of the Milky Way.

We can be more quantitative and measure the amount of spatial substructure present in these stellar halos, for example in the form of an RMS contrast $< \rho^2 >^{1/2} / < \rho >$, and compare it to that of the Milky Way. Typically these halos depict a higher RMS contrast compared to a similar estimate obtained by using a sample of main sequence turn off stars in the SDSS over similar distances and area on the sky as for the simulations (Helmi *et al.* 2011). There are two points that need to be considered before firm conclusions about this difference can be drawn. Firstly, samples of main sequence turn off stars from surveys such as SDSS can suffer some amount of contamination from stars from e.g. the thick disk (and QSO's but this is easier to take into account because the distribution on the sky is much better known and relatively uniform). This contamination could provide as it were a screen on top of the underlying halo (whose relative normalisation and distribution on the sky are somewhat uncertain and model dependent), and thereby an RMS contrast $< \rho^2 >^{1/2} / < \rho >$ that will necessarily be reduced since $< \rho >$ becomes larger.

On the other hand, there is also a physical explanation, and that is that the modelled halos do not contain a smooth in-situ component. By construction, all halo stars are accreted in the Aquarius simulations. As argued by Helmi *et al.* (2011), a smooth component that accounts for 10% of the total stellar halo mass (or 30% in the inner ~ 10 kpc), will provide enough of a screen to actually reduce the RMS contrast and bring it in agreement with the observations.

The high RMS contrast is evidently due to the very same asymmetry discussed earlier: It is the imprint of an anisotropic pattern that provides regions on the sky that are basically devoid of halo stars. This statistic is more sensitive than for example a measurement of the RMS obtained by comparison to a smooth halo model fitted to the data: $< (\rho - \rho_m)^2 >^{1/2} / < \rho_m >$. Such a statistic will naturally lead to a smaller RMS value. This might partly explain why Bell *et al.* 2008 using also main sequence turn off stars in the SDSS found more consistency with the simulations of Bullock & Johnston (2005).

Which of the above listed solutions will bring better agreement between cosmological models and the observations remains to be determined. Fortunately this will soon be possible with the ongoing sky surveys (PANSTARRS, DES, ATLAS) and also with the Gaia mission. If the halo has an important spatially smooth in-situ component then there should be little to no substructure in the kinematics in the form of moving groups or streams in the inner halo of the Galaxy. As Gomez *et al.* (2013) have shown, the Aquarius halos depict significant amounts of kinematic substructure even though they were built fully hierarchically and in a somewhat "chaotic" fashion. That substructure remains is not surprising as it is a consequence of volume conservation in phase-space, but it actually also implies that streams stars are not on exponentially diverging trajectories that would preclude detection of moving groups above the Poisson noise level.

3. Streams as probes of mass distribution

Streams stars are excellent probes of the gravitational potential in which they move because they map close to a single orbit (Sanders & Binney, 2013). This characteristic, especially of thin streams, allows us in principle to determine the force field and its properties. Most of the work so far has focused on modelling the streams from the Sagittarius dwarf galaxy (e.g. Law & Majewski, 2010), although more recently also other streams have been employed (e.g. Koposov *et al.* 2010; Kupper *et al.* 2015). At the moment the field is hungry for data, but this will soon become available thanks to the Gaia mission. In parallel it is important not to neglect the modelling aspect, as much of the work so far has been based on using a single stream, and typically with simplifying assumptions about the gravitational potential.

In an attempt to address these issues, Sanderson *et al.* (2015) have put forward a new method using multiple streams (integrated in a simple gravitational potential, as well as streams from the Aquarius simulations) to establish their ability to recover the true parameters of the underlying potential. Their method relies on the assumption that in the best potential, the streams will be most strongly clumped in the space of actions. Their work shows that even in the case of the hierarchically grown dark matter halos from the Aquarius simulations they are able to recover the characteristic parameters of the spherically averaged mass distribution very well, if full phase space coordinates are available for their stars (Sanderson, Hartke & Helmi, in prep).

In the near future it will be important to incorporate more information (such as the angles, as argued by e.g. Sanders, 2014), but also establish if the shape or time-evolution are amenable to constraining, when realistic Gaia data samples (including errors) are available. Furthermore, exploration of a potential that has multiple components will also be necessary. The action-based approach seems very powerful, but will be expensive and difficult if the actions need to be computed for approximate potentials that may not provide uniformly good fits to the true potential at all locations in phase-space, or equivalently work equally well for all streams (although see Binney, 2012; Sanders & Binney, 2015).

4. Substructure and globular clusters

The halo of our neighbour M31 is a great example of accretion in action. It provides us with a panoramic view of ongoing stellar halo assembly. Thanks to panoramic surveys such as PAndAS (McConnachie *et al.*, 2009), many new globular clusters and dwarf galaxies in this system have been discovered (e.g. Mackey *et al.* 2010; Richardson *et al.* 2011). These have revealed their own surprises, such as what appears to be a giant disk structure of satellites, all rotating in the same sense (Ibata *et al.* 2013).

The spatial distribution and kinematics of the outer halo globular clusters of M31 have turned out to be extremely interesting. Often these halo clusters appear to be associated with streams and overdensities, possibly also dynamically because of their cold kinematics (Veljanoski *et al.* 2013). It is therefore very likely that these clusters have been accreted along with the progenitors of the streams seen in the halo of M31.

Also interesting is that as a population, the outer globular clusters depict rotation and that the axis of rotation is fairly aligned with the minor axis of M31's optical disk (Veljanoski *et al.* 2014). This is the true also for the inner globular clusters, although the rotation amplitude is smaller. Interestingly, the sense of rotation is the same as that of the giant disk of satellites, although the rotation axes are somewhat misaligned.

This large degree of coherence is astounding at first sight. It has been argued that it may be due to major gas-rich mergers in which many of the dwarfs could have formed (Hammer *et al.*, 2013). However, we have noted earlier in this contribution that the distribution of stars in stellar halos in cosmological simulations is anisotropic and reflects the filamentary pattern of the cosmic web, along which satellites and building blocks have fallen in. In this sense it might perhaps not be so surprising to find coherence in the globular cluster population.

To explore to some extent the likelihood of finding rotation in a cosmological setting, Veljanoski & Helmi (2016) have taken a simple approach. Using the Aquarius stellar halos, they have selected a subset of tagged particles and relabelled these as "globular clusters", an example of which is shown in Fig. 2. Using such Mock globular clusters systems, they asked: "how often is there a rotation signal in projection as large as observed in M31 globular cluster system, and how does this depend on the viewing angle"?

Fig. 3 shows some examples of how the rotation signal depends on viewing angle for the Mock systems in the Aquarius C halo. There are many angles for which the signal

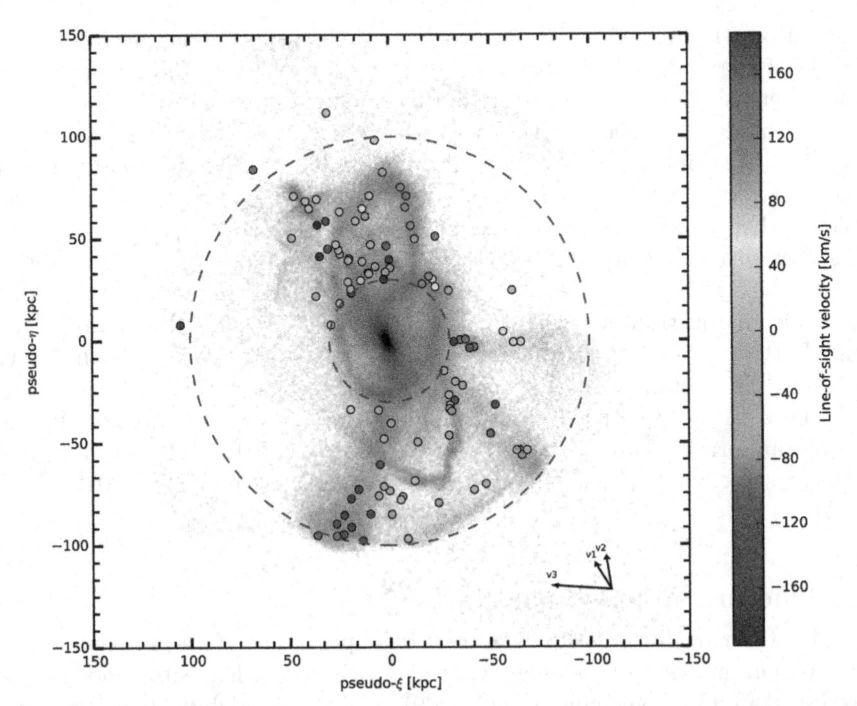

Figure 2. Sky distribution of stars (in grey) and "globular clusters" in the Aquarius-C halo, seen in projection from a distance of 1 Mpc. The colours indicate the line-of-sight velocities of the clusters, and for this particular viewing perspective the amplitude of the rotation signal is similar to that found in M31 (Veljanoski & Helmi, in prep.).

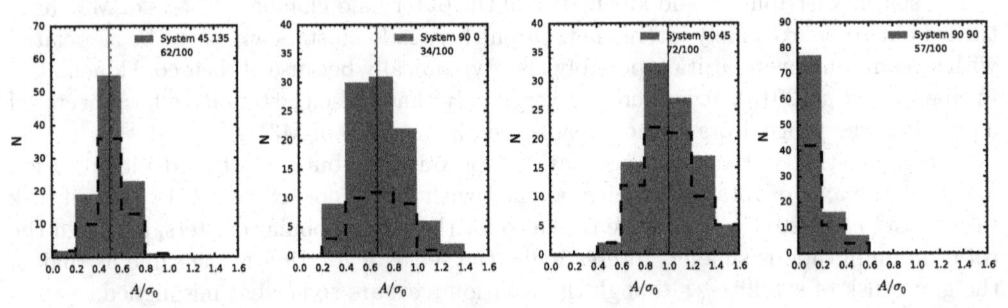

Figure 3. Distribution of the amplitude of the rotation signal A/σ for Mock samples of globular clusters from the Aquarius C halo, for different viewing perspectives. The value found for M31 is $A/\sigma = 0.67$ (Veljanoski & Helmi, 2016).

has a lower amplitude than in M31 (left and right most panels), there are some for which the amplitude is larger (third panel). Averaged over all perspectives, the probability of having a kinematic ratio as big as in M31 or larger is 0.2 for the Aquarius C system.

The angle for which a high amplitude signal is observed corresponds to one in which the true rotation axis of the system is aligned within 30 degrees with the axis determined from the projected positions and velocities of the Mock globular clusters. The true rotation axis also aligns well with the axis of rotation of the whole stellar halo, again within 40 degrees. The question now is how is this correlated with the orientation or rotation axis of the disk. In the Aquarius simulations there is no disk component. However, what we do know about these simulations is that there is preferential infall along filaments, which implies that both the satellites as well as their debris, but also the gas that comes into the system are all reflecting similar directions of infall and rotation. Therefore, one expects that this gas, when it settles, ought to have the same rotation sense as that of the stellar halo (albeit a much higher amplitude because it has collapsed to the center), and therefore also the same as that of the globular clusters.

Essentially this seems to imply that we see high rotation in the globular cluster system of M31 because we are looking at M31's disk not too far from edge-on.

5. Outlook

In the coming decade we will be able to learn much more about the assembly history of stellar halos, both from the observational perspective as well as from the theoretical point of view. Observationally, we expect a revolution to take place with the Gaia satellite data, now that phase-space information for millions of halo stars will become available. This will allow us in the case of the Milky Way, to establish the relative importance of in-situ formation vs accretion, but also to recover whenever possible, the building blocks of the halo as well as determine their star formation and chemical evolution history using upcoming instruments such as WEAVE and 4MOST.

Going beyond our Galaxy will also be very important. Although for the Milky Way we will be able to study the progenitor systems in great detail, by studying other halos we get a broader overview of the process, which is likely to have been rather stochastic and time-dependent. To establish the ubiquity of accretion events, as well as to make quantitative comparisons to cosmological predictions, it is necessary to map the halos of hundreds to thousands of galaxies. This is challenging but can be extremely rewarding, and may be soon within reach.

From the theoretical perspective, higher resolution hydrodynamical simulations of systems like the Milky Way are urgently needed. The comparisons presented here, and most of those in the literature, rely on the ability of using dark matter only simulations to tag particles. But baryonic effects, such as the presence of a disk, might change both the shape of the stellar halos as well as the amount of debris present and the timescales of survivability (see e.g. Bailin *et al.* 2014). Cosmological simulations in which the stellar halo has at least 5×10^5 particles seem necessary, but will hopefully become feasible soon. Since different galaxies should have undergone different merger histories, quantitative comparisons with suites of simulations should allow us to probe some of the expected stochasticity of the process.

Acknowledgements

I am very grateful to my collaborators, especially the Virgo Consortium for the Aquarius simulations, Robyn Sanderson and Jovan Veljanoski. This work has been financially supported by ERC Starting Grant GALACTICA-240271, NOVA and an NWO-Vici grant.

References

Bailin, J., Bell, E. F., Valluri, M., *et al.* 2014, *ApJ*, 783, 95
Bell, E. F., Zucker, D. B., Belokurov, V., *et al.* 2008, *ApJ*, 680, 295
Bullock, J. S., & Johnston, K. V. 2005, *ApJ*, 635, 931
Binney, J. 2012, *MNRAS*, 426, 1324
Cooper, A. P., Cole, S., Frenk, C. S., *et al.* 2010, *MNRAS*, 406, 744
Cooper, A. P., Cole, S., Frenk, C. S., & Helmi, A. 2011, *MNRAS*, 417, 2206
Deason, A. J., Belokurov, V., & Weisz, D. R. 2015, *MNRAS*, 448, L77

De Lucia, G., & Helmi, A. 2008, *MNRAS*, 391, 14
Duc, P.-A., Cuillandre, J.-C., Karabal, E., *et al.* 2015, *MNRAS*, 446, 120
Gómez, F. A., Helmi, A., Cooper, A. P., *et al.* 2013, *MNRAS*, 436, 3602
Hammer, F., Yang, Y., Fouquet, S., *et al.* 2013, *MNRAS*, 431, 3543
Helmi, A. 2008, *A&ARev*, 15, 145
Helmi, A., Cooper, A. P., White, S. D. M., *et al.* 2011, *ApJL*, 733, L7
Jofré, P., & Weiss, A. 2011, *A&A*, 533, A59
Ibata, R. A., Lewis, G. F., Conn, A. R., *et al.* 2013, *Nature*, 493, 62
Küpper, A. H. W., Balbinot, E., Bonaca, A., *et al.* 2015, *ApJ*, 803, 80
Koposov, S. E., Rix, H.-W., & Hogg, D. W. 2010, *ApJ*, 712, 260
Law, D. R., & Majewski, S. R. 2010, *ApJ*, 714, 229
Leaman, R., VandenBerg, D. A., & Mendel, J. T. 2013, *MNRAS*, 436, 122
Mackey, A. D., Huxor, A. P., Ferguson, A. M. N., *et al.* 2010, *ApJL*, 717, L11
Martínez-Delgado, D., Gabany, R. J., Crawford, K., *et al.* 2010, *AJ*, 140, 962
McConnachie, A. W., Irwin, M. J., Ibata, R. A., *et al.* 2009, *Nature*, 461, 66
Janowiecki, S., Mihos, J. C., Harding, P., *et al.* 2010, *ApJ*, 715, 972
Richardson, J. C., Irwin, M. J., McConnachie, A. W., *et al.* 2011, *ApJ*, 732, 76
Sanders, J. L., & Binney, J. 2013, *MNRAS*, 433, 1826
Sanders, J. L. 2014, *MNRAS*, 443, 423
Sanders, J. L., & Binney, J. 2015, *MNRAS*, 447, 2479
Sanderson, R. E., Helmi, A., & Hogg, D. W. 2015, *ApJ*, 801, 98
Springel, V., Wang, J., Vogelsberger, M., *et al.* 2008, *MNRAS*, 391, 1685
Veljanoski, J., Ferguson, A. M. N., Mackey, A. D., *et al.* 2013, *ApJL*, 768, L33
Veljanoski, J., Mackey, A. D., Ferguson, A. M. N., *et al.* 2014, *MNRAS*, 442, 2929
Veljanoski, J., & Helmi, A., 2016, arXiv:1602.04018 (submitted to A&A)
White, S. D. M., & Springel, V. 2000, The First Stars, 327

The General Assembly of Galaxy Halos: Structure, Origin and Evolution
Proceedings IAU Symposium No. 317, 2015
A. Bragaglia, M. Arnaboldi, M. Rejkuba & D. Romano, eds.

doi:10.1017/S1743921315009163

The early gaseous and stellar mass assembly of Milky Way-type galaxy halos

Gerhard Hensler[1,2] and Mykola Petrov[1]

[1]Dept. of Astrophysics, Univ. of Vienna, Tuerkenschanzstr. 17, 1180 Vienna, Austria
email: gerhard.hensler@univie.ac.at
[2]Nat. Astron. Obs. of Japan, 2-21-1 Osawa, Mitaka-shi, Tokyo 181-8588, Japan

Abstract. How the Milky Way has accumulated its mass over the Hubble time, whether significant amounts of gas and stars were accreted from satellite galaxies, or whether the Milky Way has experienced an initial gas assembly and then evolved more-or-less in isolation is one of the burning questions in modern astronomy, because it has consequences for our understanding of galaxy formation in the cosmological context. Here we present the evolutionary model of a Milky Way-type satellite system zoomed into a cosmological large-scale simulation. Embedded into Dark Matter halos and allowing for baryonic processes these chemo-dynamical simulations aim at studying the gas and stellar loss from the satellites to feed the Milky Way halo and the stellar chemical abundances in the halo and the satellite galaxies.

Keywords. Galaxy: formation, Galaxy: halo, Galaxy: stellar content, galaxies: halos, galaxies: formation, galaxies: abundances

---

1. Introduction

In the past half century, by means of more sensitive observations of the Milky Way (MWG) stellar halo its formation was interpreted by two different processes: The first prefers the monolithic collapse (Eggen *et al.* 1962), and the second is the accretion model by Searl & Zinn (1978). At the same time, White & Rees (1978) proposed their Cold Dark Matter (CDM) hierarchical clustering paradigm in which galaxies are results from cooling and fragmentation of residual gas within the transient potential wells provided by the DM. In this framework galaxy formation proceeds in a "bottom up" manner starting with the formation of small clumps of gas inside DM subhalos, which then merge hierarchically into larger systems (Blumenthal *et al.* 1984, Springel *et al.* 2005).

CDM simulations of cosmological structure and galaxy formation predict the existence of a large number of such DM subhalos surrounding massive DM gravitational potentials. These subhalos should serve as the DM progenitors of dwarf galaxies (DGs) which indeed permeate the Local Group (LG), most of them concentrated as satellites around the MWG and M31. The closer to a mature galaxy they live, the more gas-free they are like elliptical DGs and are, therefore, called dwarf spheroidals (dSphs). Because of their low surface brightness, though even close to the MWG, for a long time only a few of them could be observed in the range of M_V = -14^m to -10^m separated clearly from Globular Clusters. Their number increased over the last years thanks to systematic surveys like SDSS shifting the lower brightness limit by the recently discovered ultra-faint DGs (UFDs) to almost -2^m (see e.g. Belokurov *et al.* 2010, Belokurov *et al.* 2014) so that they extend the DG sequence to its faintest end.

Theoretically already expected and verified by numerical simulations (Johnston *et al.* 2008), satellites in the neighbourhood and, thereby, in the tidal field of mature galaxies lose continuously gas and stars, the later observable as tidal streams (Lynden-Bell[2] 1995,

Jerjen *et al.* 2013). Due to their loss of orbital energy and angular momentum their fate is the partly disruption and their death as individuals is the accretion by their mature galaxy. This scenario of tidal disruption is at present most strikingly demonstrated by the Sagittarius DG (Ibata *et al.* 1994) with its tidal tails wrapped around the MWG (Majewski *et al.* 2003).

The CDM merging hypothesis requires the infall and accumulation of the MW mass not only by DM subhalos and gas but also by stars. If this "mining" of the halo with dSph stars (Salvadori *et al.* 2008) has happened in the very early epoch with the first stars only, no differences at the low Z end will tell us about as long as the MW halo stars are formed from the same gaseous substrate. At larger metallicities the α/Fe ratio of dSphs, however, declines already (due to supernova type Ia (SNIa) enrichment) while the halo stars are systematically at the constant value of SNII enrichment at the same [Fe/H] (Shetrone *et al.* 2003, Tolstoy *et al.* 2003, Venn *et al.* 2004, Koch *et al.* 2008; see also reviews by Koch (2009) and Tolstoy *et al.* (2009)). This fact allows to pin-down that progenitors of present-day dSphs are not the expected building blocks of the galactic halo and to explain the lack of the observed number of stellar streams. From kinematics of halo stars, however, a dichotomy is found by Carollo *et al.* (2007) and Bell *et al.* (2008), one regular population in the inner region of about 10 kpc radius and an outermost heterogeneous and decoupled halo population most plausibly accreted from disrupted satellites. The metallicity distribution function (MDF) of the UFDs suggests that these tiny systems contain a larger fraction of extremely metal-poor stars than the MW halo does (Kirby *et al.* 2008) and witness the chemical imprint of the interstellar medium (ISM) when the Universe was less than 1 Gyr old.

Detections of hyper metal-poor stars (Beers & Christlieb 2005) in the galactic halo and their peculiar element abundances (see e.g. Frebel *et al.* (2005)) opened a new field of galactic archeology, namely, modelling the element production by the first stars in the halo as well as in the UFDs towards understanding the zero metallicity nucleosynthesis and studying the formation of the halo. Chemical evidences for this scenario, especially in the metal-poor stellar content of the galactic halo, is mentioned also by Frebel *et al.* (2010), Frebel & Bromm (2012), and others. On the other hand, it is currently unclear how the metal-poor MDF tail of the classical DGs, in which extremely metal-poor stars are absent, compares with that of the halo and the UFDs.

2. Modeling the Milky Way satellite system

In contrast to the DM evolution of subhalos treated by pure N-body simulations, the evolution of the baryonic component is much more complex because of the physical processes at work, such as star formation (SF), gas cooling, dissipation, energy and mass feedback. Baryonic gas loses kinetic energy dissipatively and thermal energy by radiation leading to cooling and gravitational collapse, while stellar radiation and winds as well as SNe lead to energy and chemical feedback. Almost all modelling up to now is dedicated to investigate the effects of various processes on the dSph evolution separately or the dSph evolution as an isolated system.

Although the gasdynamical simulations of dSphs advanced from 1D chemo-dynamical models by Hensler *et al.* (2004) to 3D (see e.g. Smooth-Particle Hydrodynamics (SPH) models by Revaz *et al.* (2009), Pasetto *et al.* (2011)), they mainly lack not only of a self-consistent treatment of various internal processes, but focussed on particular aspects only. Their results do not deviate too much from observational data, however, the system of satellites is exposed to a whole bunch of external processes also, like e.g. ram-pressure (Mayer *et al.* 2007) and tidal (Read *et al.* 2006) stripping, gas accretion, and further

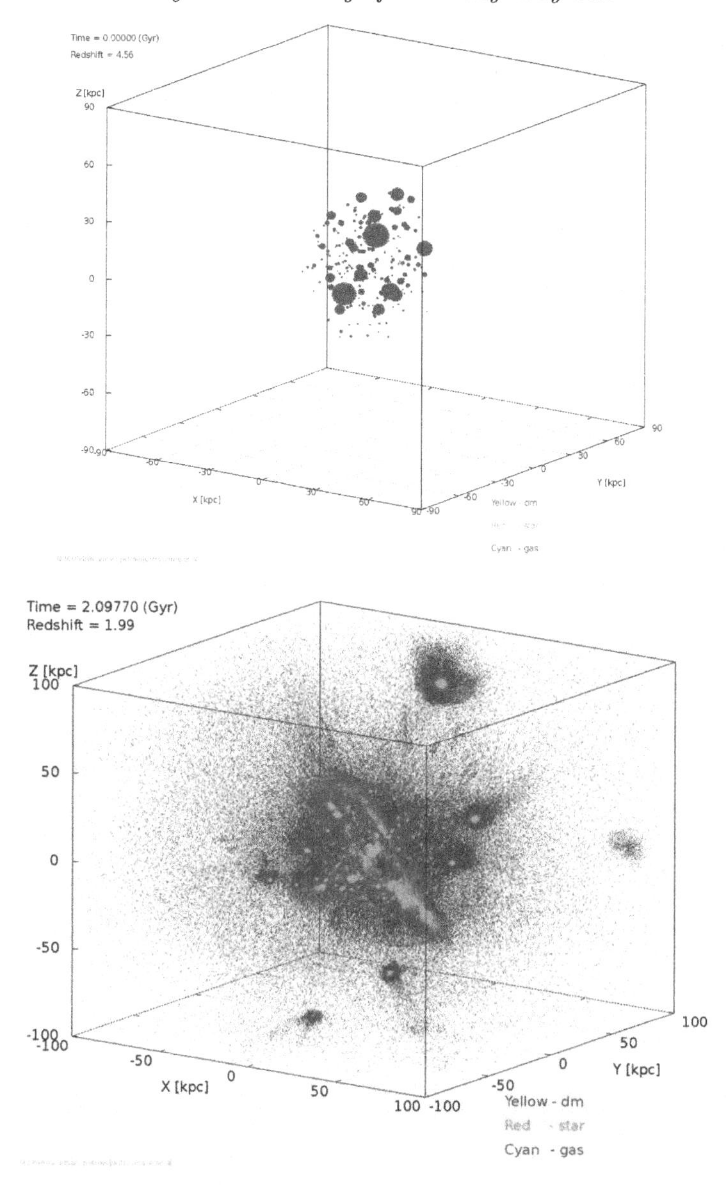

Figure 1. Cubes of 200 kpc length around the Milky Way (at their center). *upper panel:* Initial conditions of the Milky Way's satellite system: Distribution of Dark Matter (DM) subhalos within a sphere of 40 kpc radius around the Milky Way at redshift z=4.56. *lower panel:* Snapshot of the satellites' dynamical evolution 2.1 Gyr after the numerical onset, i.e. at redshift z=2. The DM subhalos are filled with 17% baryonic gas mass, form stars, and lose mass of all constituents due to various effects (see text).

more. If these are not taken into account but simulations are limited to isolation DGs the models cannot allow a reliable trace-back of the evolution of any dSph galaxy.

In addition, simple non-dynamical considerations as performed to understand the chemical evolution (Lanfranchi *et al.* 2006, Lanfranchi & Matteucci 2010, Prantzos 2008, Kirby *et al.* 2011a, Kirby *et al.* 2011b) provide only a limited understanding of the real evolution of dSphs.

Font *et al.* (2006) investigated the nature of the progenitors of the stellar halo for a set of MWG-type galaxies and studied the chemical enrichment patterns in the context of the CDM model with a combination of semi-analytic prescriptions. They concluded that the difference in chemical abundance patterns in local halo stars versus surviving satellites arises naturally from the predictions of the hierarchical structure formation in a CDM universe.

Here we present a model of the early evolution and mass assembly history of the MWG's halo by the system of satellite galaxies treated in the gravitational field of the MWG. For this purpose we select a MWG-like DM host halo from the cosmological ΛCDM simulation Via Lactea II (Diemand *et al.* 2008). The constraints are, that it does not undergo a major merger event over the Hubble time and that sufficient subhalos exist which allow the accretion by the host galaxy. For the simulations an advanced version of the single-gas chemo-dynamical SPH/N-body code is applied, treating the production and chemical evolution of 11 elements.

Since the acceptable computational time limits the number of gas particles to two million and the DM particles to the same order and because we aim at reaching a mass resolution of $10^3 M_\odot$ per SPH particle, only 250 subhalos in the total-mass range of $10^6 < M_{sat}/M_\odot < 6 \times 10^8$ could be followed from redshift $z = 4.56$ with its baryonic content. Unfortunately, this fact limits the radius of consideration to 40 kpc around the MWG's center of mass. In order to study the construction of the MWG halo by accretion of subhalos with both gas and stars, as a first step, the chemo-dynamical evolution of the dSph system is followed for the first 2 Gyr, i.e. until redshift $z = 2$ (see Fig.1).

Starting with a 10^4 K warm gas of 17% of the subhalo masses in virial equilibrium and under the assumption that re-ionization is improbable to have affected the LG dSphs (Grebel & Gallagher 2004), cooling allows the gas particles to achieve SF conditions in all satellites, but its efficiency directly depends on the mass of a satellite and its dynamical history (e.g. merging with other satellites or disruption by the MWG gravitational potential).

3. The Milky Way halo formation

Fig. 1 shows the evolution of the DM-gas-stars composit of the satellite galaxies. The SF starts in all satellites almost simultaneously, then ceases for the lowest mass objects, while it continues in more massive ones with fluctuations due to gas loss but also interactions with other objects. dSphs develop their stellar components and element abundances dependent on the distance from the MWG. Gas is pushed out from low-mass dSphs by their internal stellar energy release and lost from massive dSphs more by the tidal force. Both effects feed the MWG gas halo by pre-processed hot gas (fig. 2). Inherently as an additional effect, dSphs also get rid of their gas by their motion within the bath of their lost hot gas. Stars are also disrupted from the satellites and accumulate in the MWG halo at the early stages from all objects with low metallicity. Lateron, only stars from the massive satellites contribute due the cessation of SF in the less-massive systems. In total, $1.88 \times 10^8 M_\odot$ of gas and $9.53 \times 10^7 M_\odot$ of stellar mass are torn off from the satellite galaxies and got bound to the Galactic halo. From the same demolishing effect additionally $2.63 \times 10^9 M_\odot$ of DM mass fed the MWG.

For the first 0.1 Gyrs of the simulation there is a considerable variance of stellar oxygen abundance in the whole system ($-5. \leqslant [O/H] \leqslant -0.5$) reflecting the very inhomogeneous production and distribution of enriched gas. After 0.1 Gyrs merging of the satellites' ISM promotes the mixing of heavy elements. Finally, almost complete recycling of the gas erases the abundance inhomogeneities so that oxygen in stars converges to

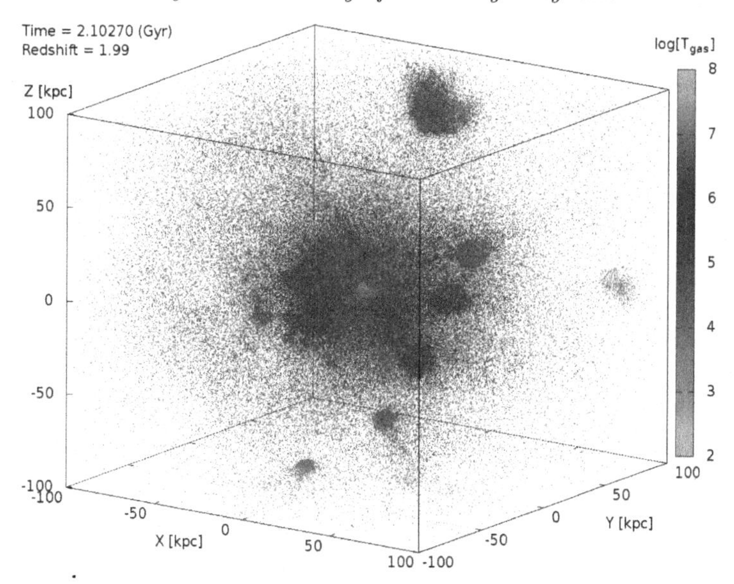

Figure 2. Snapshot of the gas distribution 2.1 Gyr after the numerical onset, i.e. at redshift z=2.76. The gas temperature is coloured according to the right colour panel.

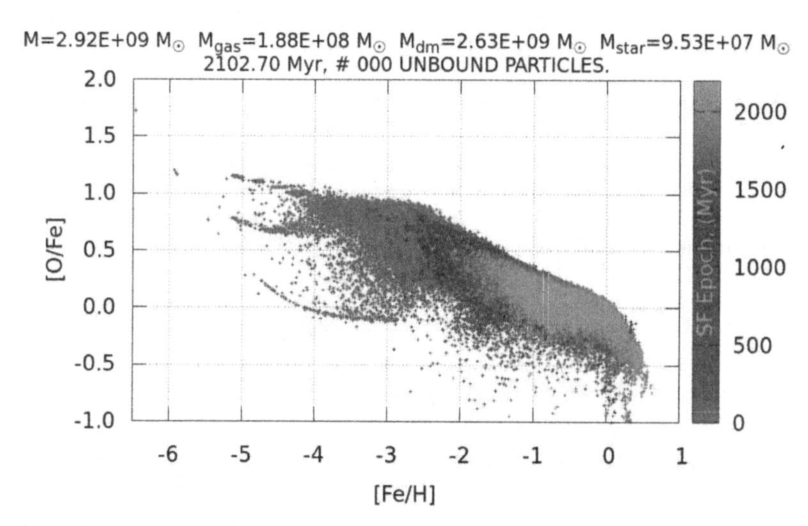

Figure 3. [O/Fe] vs. [Fe/H] distribution of star and gas that became unbound from the satellite galaxies within the first 2.1 Gyr after the model onset. The values of the element ratios are coloured according to the time when the stars and gas are dissolved from the satellites. The emerging stripes are artefacts by used values from tables during the first enrichment episode.

$-1. \leqslant [Fe/H] \leqslant 0.$ with a small dispersion (fig. 3). These high abundances show that the too efficient metal-enrichment in a single-gas phase treatment has to be relaxed by a more realistic chemo-dynamical multi-phase prescription of the ISM (Liu *et al.* 2015).

Acknowledgements

The authors are grateful to Simone Recchi for his contributions and continuous discussions on the chemical enrichment. This work was partly supported by the Austrian Science Foundation FWF under project no. P21097.

References

Beers, T. & Christlieb, N., 2005, *Ann. Rev. A&A*, 43, 531
Bell, E.F., Zucker, D.B., Belokurov, V., *et al.*, 2008, *Astroph.J.*, 680, 295
Belokurov, V., Walker, M.G., Evans, N.W., *et al.*, 2010, *Astroph.J.*, 712, L103
Belokurov, V., Irwin, M.J., Koposov, S.E., *et al.*, 2014, *MNRAS*, 441, 2124
Blumenthal, G.R., Primak, J.R., & Rees, M.J., 1984, *Nature*, 311, 517
Carollo, D., Beers, T.C., Lee, Y.S., *et al.*, 2007, *Nature*, 450, 1020
Diemand, J., Kuhlen, M., Madau, P., *et al.*, 2008, *Nature*, 454, 735
Eggen, O.J., Lynden-Bell, D., & Sandage, A.R., 1961, *Astroph.J.*, 136, 748
Font, A.S., Johnston, K.V., Bullock, J.S., & Robertson, B.E., 2006, *Astroph.J.*, 638, 585
Frebel, A., Aoki, W., Christlieb, N., *et al.*, 2005, *Nature*, 434, 871
Frebel, A., Kirby, E.N., & Simon, J.D., 2010, *Nature*, 464, 72
Frebel, A., & Bromm, V., 2012, *Astroph.J.*, 759, 115
Frebel, A., & Norris, J.E., 2015, *Ann. Rev. A&A*, 53, 631
Grebel, E.K., & Gallagher, J.S., 2004, *Astroph.J.*, 610, L89
Hensler, G., Theis, C., & Gallagher, J.S., 2004, *Astron.Astroph.*, 426, 25
Ibata, R.A., Gilmore, G., & Irwin, M.J., 1994, *Nature*,370, 194
Jerjen, H., Da Costa, G.S., Willman, B., *et al.*, 2013, *Astroph.J.*, 769, 14
Johnston, K.V., Bullock J.S., Sharma S., *et al.*, 2008, *Astroph.J.*, 689, 936 .
Kirby, E.N., Simon, J.D., Geha, M., *et al.*, 2008, *Astroph.J.*, 685, L43
Kirby, E.N., Lanfranchi, G.A., Simon, J.D., *et al.*, 2011a, *Astroph.J.*, 727, 78
Kirby, E.N., Cohen, J.G., Smith, G.H, *et al.*, 2011b, *Astroph.J.*, 727, 79
Koch, A., Grebel, E.K., Gilmore, G.F., *et al.*, 2008, *Astron.J.*, 135, 1580
Koch, A., 2009, *Rev. Modern Astron.*, 21, 9
Lanfranchi, G.A., Matteucci, F., & Cescutti, G., 2006, *Astron.Astroph.*, 453, 67
Lanfranchi, G.A., & Matteucci, F., 2010, *Astron.Astroph.*, 512, A85
Liu, L., Petrov, M., Hensler, G., *et al.*, 2015, *MNRAS*, submitted
Lynden-Bell, D., & Lynden-Bell, R.M., 1995, *MNRAS*, 275, 429
Majewski, S.R., Skrutskie, M.F., Gomez-Flechoso, M.A., *et al.* 2013, *Astroph.J.*, 599, 1082
Mayer, L., Kazantzidis, S., Mastropietro, C., & Wadsley, J., 2007, *Nature*, 445, 738
Pasetto, S., Grebel, E.K., Berczik, P., & Spurzem, R., 2011, *Astron.Astroph.*, 525, A99
Prantzos, N., 2008, *Astron.Astroph.*, 489, 525
Read, J.I., Wilkinson, M.I., Evans, N.W., *et al.*, 2006, *MNRAS*, 366, 429
Revaz, Y., Jablonka P., Sawala T., *et al.*, 2009, *Astron.Astroph.*, 501, 189
Salvadori, S., Ferrara, A., & Schneider, R., 2008, *MNRAS*, 386, 348
Searl, L., & Zinn, R., 1978, *Astroph.J.*, 225, 357
Shetrone, M., Venn, K.A., Tolstoy, E., *et al.*, 2003, *Astron.J.*, 125, 688
Springel, V., White, S.D.M., Jenkins, A., *et al.*, 2005, *Nature*, 435, 629
Tolstoy, E., Venn, K.A., Shetrone, M., *et al.*, 2003, *Astron.J.*, 125, 707
Tolstoy, E., Hill, V., & Tosi, M., 2009, *Ann. Rev. A&A*, 47, 371
Venn, K.A., Irwin, M, Shetrone, M.D., *et al.*, 2004. *Astron.J.*, 128, 1177
White, S.D.M., & Rees, M.J., 1978, *MNRAS*, 183, 341

The General Assembly of Galaxy Halos: Structure, Origin and Evolution
Proceedings IAU Symposium No. 317, 2015
A. Bragaglia, M. Arnaboldi, M. Rejkuba & D. Romano, eds.

doi:10.1017/S1743921315007140

Contributions to the Galactic halo from in-situ, kicked-out, and accreted stars

Allyson A. Sheffield[1], Kathryn V. Johnston[2], Katia Cunha[3], Verne V. Smith[4] and Steven R. Majewski[5]

[1]LaGuardia Community College, City University of New York, Department of Natural Sciences, 31-10 Thomson Ave., Long Island City, NY, 11101, USA
email: asheffield@lagcc.cuny.edu
[2]Columbia University, Dept. of Astronomy, Mail Code 5246, New York, NY, 10027, USA
[3]Observatório Nacional, Rua General José Cristino, 77, 20921-400 São Cristóvão, Rio de Janeiro, RJ, Brazil
[4]National Optical Astronomy Observatories, Tucson, AZ, 98765, USA
[5]University of Virginia, Dept. of Astronomy, P.O. Box 400325, Charlottesville, VA, 22904, USA

Abstract. We report chemical abundances for a sample of 66 M giants with high S/N high-resolution spectroscopy in the inner halo of the Milky Way. The program giant stars have radial velocities that vary significantly from those expected for stars moving on uniform circular orbits in the Galactic disk. Thus, based on kinematics, we expect a sample dominated by halo stars. Abundances are derived for α-elements and neutron capture elements. By analyzing the multi-dimensional abundance space, the formation site of the halo giants – in-situ or accreted – can be assessed. Of particular interest are a class of stars that form in-situ, deep in the Milky Way's gravitational potential well, but are "kicked out" of the disk into the halo due to a perturbation event. We find: (1) our sample is dominated by accreted stars and (2) tentative evidence of a small kicked-out population in our Milky Way halo sample.

Keywords. Galaxy: halo, Galaxy: structure, stars: abundances

1. Introduction

Where did stars in the Milky Way's halo form? The ΛCDM model predicts that the Milky Way's halo was built in a "bottom-up" fashion, and this view is now generally accepted due to overwhelming evidence of the relics of past mergers. It is still uncertain, however, what fraction of the halo is made up of such accreted debris. Close to the time of accretion, a group of stars formed in a particular satellite of the Milky Way will show coherence spatially, kinematically, and chemically. In the inner halo where dynamical timescales are short, spatial coherence will become blurred quickly, although kinematical and chemical coherence remain. Kinematics alone may still lead to ambiguity, as a merger event can cause stars formed in the Milky Way to redistribute into rings in the halo and these rings can be difficult to distinguish from accreted satellite stars. Thus, to get a more complete profile of a star's formation history, both kinematical and chemical information are needed.

Stars form in gas clouds, and the chemical patterns of stars reflect successive generations of stellar evolution within the cloud. Due to their formation in Type II supernovae, α-elements probe the relative timescale of formation for populations of stars and are sensitive to the star formation efficiency of the progenitor. Thus, a star that formed in a dSph progenitor will occupy a different region in the [α/Fe]-[Fe/H] plane than a star formed in a Milky Way-sized progenitor. The addition of r- and s-process elements gives a more complete evolutionary picture of the star's formation site. The yields of r- and s-process elements, which are synthesized in Type II supernovae and thermally pulsating AGB stars, respectively, are coupled to the Fe seed nuclei present in the formation site; thus, the yields of neutron capture elements vary with metallicity and provide further constraints on a star's formation scenario.

As detailed in Sheffield *et al.* (2012), three possible formation scenarios for halo stars can be identified: **in-situ halo** stars, which formed in the Milky Way's main dark matter (DM) halo; **kicked-out disk** stars, which formed much deeper within the potential well of the Milky Way's dominant DM halo and were subsequently ejected into the halo due to a perturbation event; **accreted** stars, which formed in a separate DM subhalo. Kicked-out disk stars will have chemical abundance patterns similar to the Milky Way's disk but halo-like kinematics. We note that an additional class is also possible, wherein stars that originated in a fairly massive satellite were accreted and dragged into the plane of the disk, thus forming a population of "accreted disk stars" with disk-like kinematics but chemistry reflective of the satellite's DM subhalo (Read *et al.* 2008).

In this study, we use M giants selected from 2MASS to assess the formation scenario of stars in the inner halo. M giants are a useful probe of the formation history of the nearby halo: they are old, metal-rich stars and thus are likely to be found in the thick disk and the metal-rich tail of the halo. They are also sensitive to more recent accretion (Sharma *et al.* 2011) so surveys isolating M giants will tend to pick up stars associated with luminous, metal-rich satellites. This was seen in the work of Majewski *et al.* 2003, in which they mapped the Galaxy in M giants and the Sgr core and northern and southern streams were strikingly revealed. We present an expanded study of Sheffield *et al.* (2012), with nearly double the sample size of M giants and the addition of neutron capture elements.

2. Methodology

2.1. *Chemical Tagging*

In this contribution, we present results for 69 M giants with high S/N, high-resolution spectroscopy, 3 of which are M giants selected as thick disk calibration stars. The stars were selected from a medium-resolution survey of 1799 M giants (Sheffield *et al.* 2012), and the selection criteria for medium-resolution spectroscopy and high-resolution followup is detailed in that work. Sheffield *et al.* (2012) presented the results for stars in the [Ti/Fe]-[Fe/H] plane; we have added the Ba II λ6496 line and the Eu II λ6645 line in this work (spectral synthesis was used to account for hyperfine/isotopic splitting). For metal-rich stars, Ba undergoes slow neutron capture; Eu is essentially a pure r-process neutron capture element.

To begin "tagging" the stars chemically, we first draw boundary lines by eye in the [Ti/Fe]-[Fe/H] plane, separating the general regions for kicked-out disk, accreted, and in-situ halo stars. These lines are shown in the bottom panel of Figure 1. The orange, purple, and green filled circles in Figure 1 are stars taken from the literature for disk, halo, and satellite stars, respectively. In all three chemical planes shown in Figure 1, there is a fairly distinct region for satellite stars, although there is clearly some scattering of accreted stars into the kicked-out disk and in-situ regions.

If a star is tagged as a kicked-out disk or in-situ halo star in the [Ti/Fe]-[Fe/H] plane, a more definitive classification can be made by seeing where it falls in the [Ba/Fe]-[Fe/H] and [Eu/Fe]-[Fe/H] planes. At these relatively high metallicities, both the [Ba/Fe]-[Fe/H] and [Eu/Fe]-[Fe/H] planes show a fairly distinct region for satellite stars, mainly from Fornax (Letarte *et al.* 2010) and the LMC (Pompéia *et al.* 2008). Using this technique, we first assess the stars classified as kicked-out disk stars in the [Ti/Fe]-[Fe/H] plane.

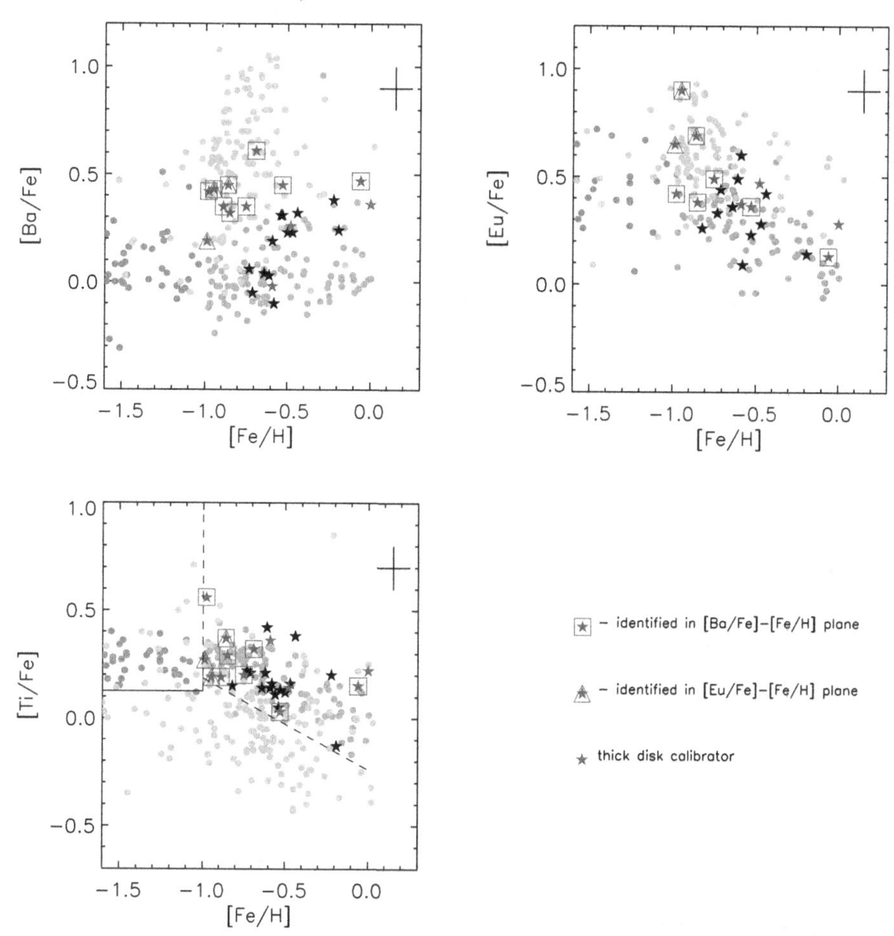

Figure 1. The α-element and neutron capture chemical planes, with disk, halo, and accreted stars from the literature shown as the orange, blue, and green filled circles, respectively. The program stars falling into the kicked-out disk region in the titanium plane are shown as black or red filled stars. The red filled stars fall into the accreted region in either barium space (these are enclosed by a square) or europium space (these are enclosed by a triangle). The violet filled stars are thick disk calibrators.

These results are shown in Figure 1, where stars falling into the kicked-out disk region are shown as either black or red filled stars. The red stars fall into the accreted regions in either the [Ba/Fe]-[Fe/H] or [Eu/Fe]-[Fe/H] planes (the caption provides further details on the symbols). Of the 26 stars that fall into the kicked-out disk region based on their [Ti/Fe]-[Fe/H] location, we reclassified 6 stars using this tagging technique as accreted rather than kicked-out disk stars.

The same technique was used to more accurately classify stars in the in-situ halo region, which can contain stars formed via any of the three scenarios (any kicked-out disk stars in this region would be due to a primordial merger). The results are summarized in Figure 2, and we reclassified 8 of the 13 stars in this region as accreted, based on their location in the [Ba/Fe]-[Fe/H] or [Eu/Fe]-[Fe/H] planes.

2.2. *Assessing formation scenarios*

After adding the [Ba/Fe]-[Fe/H] and [Eu/Fe]-[Fe/H] planes, we reclassified 14 stars as accreted. Of the 66 stars analyzed, we find 40 (61%) accreted, 21 (32%) kicked-out disk,

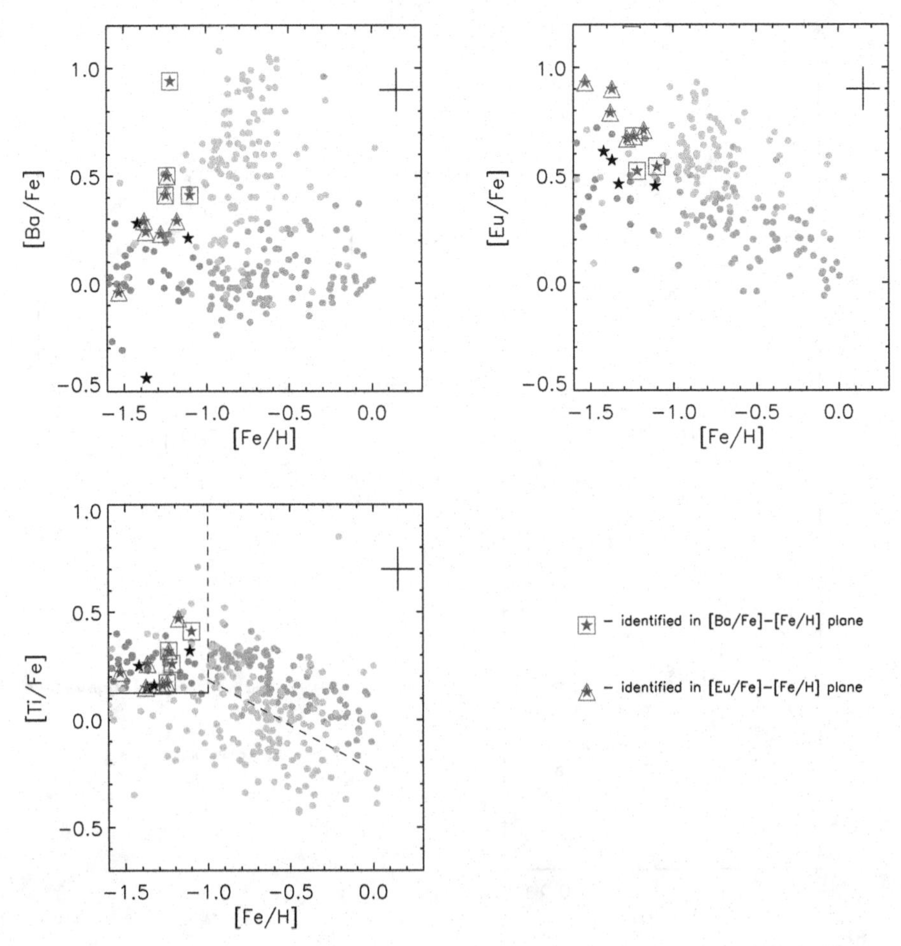

Figure 2. Same as Fig. 1, but now highlighting stars falling into the in-situ halo region in the [Ti/Fe]-[Fe/H] plane.

and 5 (7%) in-situ halo. The results show a significant fraction of accreted stars in the inner halo (distances were found for the stars using the Dartmouth isochrones and are in the range of ~1-10 kpc). This is similar to the results of Nissen & Schuster's (2010, 2011) studies of dwarfs in the very nearby halo, where they also detect a significant presence of accreted stars.

To assess how many stars in the kicked-out disk region are actually kicked-out disk stars, as opposed to thick disk stars with high velocities, we defined a velocity V' to take into account the asymmetric drift: $V' = v_{\rm hel} + v_{\rm asymm}\cos(b)\sin(l)$. The value of the dispersion that minimizes $|V'|$ was found to be σ=52.5 km/s (Sheffield *et al.* 2012). In Figure 3, we show the distributions of $|V'|$ for the entire medium-resolution sample and the 66 high-resolution stars for this study. The black dotted line is a Gaussian with a dispersion equal to 1.5σ, meant to represent the thick disk population. The kicked-out population is shown as the red distribution and, although we do find that roughly 10% are quite distinctly outside of the tail of the thick disk distribution, we still cannot claim conclusive evidence of a kicked-out disk population in the Milky Way without further statistical tests.

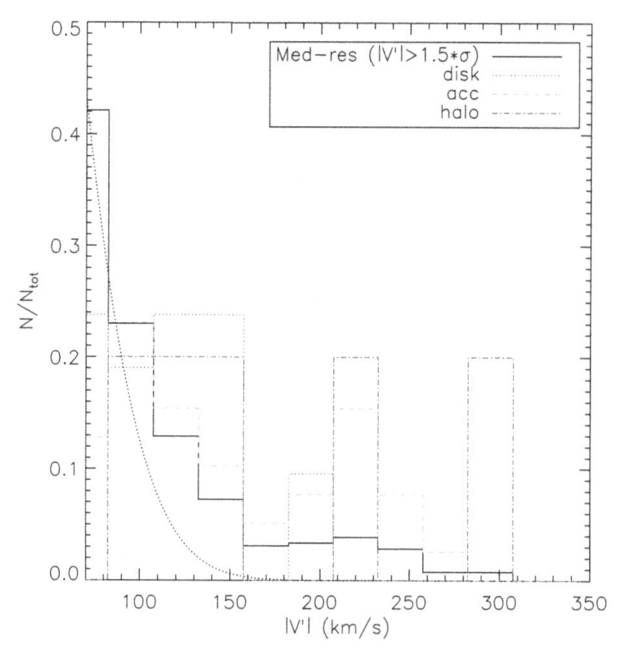

Figure 3. Distribution of the modulus of the asymmetric drift corrected radial velocities, V', for: all medium-resolution stars with $|V'| > 1.5$-σ (solid black line), stars classified as kicked-out disk (red dotted line), stars classified as in-situ halo (blue dot-dashed line), and stars classified as accreted (green dashed line).

3. Conclusions and Future Work

Using chemical tagging, we have classified the formation scenarios for a sample of 66 M giants in the inner halo. We find that the majority of the stars (61%) are classified as accreted. These results are consistent with the ΛCDM paradigm whereby galaxies are built up hierarchically through mergers. Merger events dynamically impact the structure of a galaxy: for example, a merger can create "rings" of stars in the Galaxy (see, e.g., Xu *et al.* 2015 and Price-Whelan *et al.* 2015). Stars in these rings are thus a kicked-out disk population. Kinematically, a kicked-out population and an accreted population can be challenging to distinguish; the chemical abundance patterns of stars are thus invaluable in making this distinction. A small kicked-out population (~5%) has been detected in Andromeda (Dorman *et al.* 2013), a finding which is consistent with predictions from cosmological simulations (e.g., Tissera *et al.* 2013 find kicked-out disk populations of 1-30% in their simulations of Milky Way-sized galaxies from the Aquarius Project). We find suggestive evidence for a small kicked-out disk population in the Milky Way.

The full high-resolution sample of 109 M giants with the addition of more elements will be presented in an upcoming work.

References

Dorman, C. E., Widrow, L. M., Guhathakurta, P., *et al.* 2013, *ApJ*, 779, 103
Letarte, B., Hill, V., Tolstoy, E., *et al.* 2010, *A&A*, 523, A17
Majewski, S. R., Skrutskie, M. F., Weinberg, M. D., & Ostheimer, J. C. 2003, *ApJ*, 599, 1082
Nissen, P. E. & Schuster, W. J. 2010, *A&A*, 511, L10
Nissen, P. E. & Schuster, W. J. 2011, *A&A*, 530, A1
Pompéia, L., Hill, V., Spite, M., *et al.* 2008, *A&A*, 480, 379

Price-Whelan, A. M., Johnston, K. V., Sheffield, A. A., Laporte, C. F. P., & Sesar, B. 2015, *MNRAS*, 452, 676
Read, J. I., Lake, G., Agertz, O., & Debattista, V. P. 2008, *MNRAS*, 389, 1041
Sharma, S., Johnston, K. V., Majewski, S. R., Bullock, J., & Muñoz, R. R. 2011, *ApJ*, 728, 106
Sheffield, A. A., Majewski, S. R., Johnston, K. V., *et al.* 2012, *ApJ*, 761, 161
Tissera, P. B., Scannapieco, C., Beers, T. C., & Carollo, D. 2013, *MNRAS*, 432, 3391
Xu, Y., Newberg, H. J., Carlin, J. L., *et al.* 2015, *ApJ*, 801, 105

The General Assembly of Galaxy Halos: Structure, Origin and Evolution
Proceedings IAU Symposium No. 317, 2015
A. Bragaglia, M. Arnaboldi, M. Rejkuba & D. Romano, eds.

doi:10.1017/S1743921315008613

Which processes shape stellar population gradients of massive galaxies at large radii?

Michaela Hirschmann

UPMC-CNRS, UMR7095, Institut d' Astrophysique de Paris, F-75014 Paris, France
email: hirschma@iap.fr

Abstract. We investigate the differential impact of physical mechanisms, mergers (stellar accretion) and internal energetic phenomena, on the evolution of stellar population gradients in massive, present-day galaxies employing a set of high-resolution, cosmological zoom simulations. We demonstrate that negative metallicity and color gradients at large radii (>2Reff) originate from the accretion of metal-poor stellar systems. At larger radii, galaxies become typically more dominated by stars accreted from satellite galaxies in major and minor mergers. However, only strong galactic winds can sufficiently reduce the metallicity content of the accreted stars to realistically steepen the outer metallicity and colour gradients in agreement with present-day observations. In contrast, the gradients of the models without winds are inconsistent with observations (too flat). In the wind model, colour and metallicity gradients are significantly steeper for systems which have accreted stars in minor mergers, while galaxies with major mergers have relatively flat gradients, confirming previous results. This analysis greatly highlights the importance of both energetic processes and merger events for stellar population properties of massive galaxies at large radii. Our results are expected to significantly contribute to the interpretation of current and up-coming IFU surveys (like MaNGA and Califa), which in turn can help to better constrain still uncertain models for energetic processes in simulations.

Keywords. methods: numerical, galaxies: evolution, galaxies: formation, galaxies: elliptical and lenticular, galaxies: stellar content

---

1. Introduction

It is a natural prediction of modern hierarchical cosmological models that the assembly of massive galaxies involves major and minor mergers although most stars in most galaxies have been made in-situ from accreted or recycled gas. Nonetheless, these mergers are expected to play a significant role for the structural and morphological evolution of the massive early-type galaxy population. One important structural galaxy property, which is thought to be strongly influenced by mergers, are the (in general negative) metallicity gradients observed early-on in massive, present-day early-type (e.g. McClure & Racine 1969), but also in late-type galaxies (e.g. Wyse & Silk 1989), typically within $1R_{\rm eff}$. Thanks to improved and more elaborated observational techniques, present-day metallicity gradients can nowadays be measured out to much larger radii, partly out to $8R_{\rm eff}$, (e.g. La Barbera *et al.* 2012) and even beyond reaching e.g. $\sim 25R_{\rm eff}$ in a few nearby galaxies like NGC 5128 or M31 (e.g. Rejkuba 2014; Gilbert *et al.* 2014).

Previous studies (e.g. Kobayashi 2004) only investigate the emergence of inner gradients (up to 3 $R_{\rm eff}$) at comparably poor spatial resolution. Here, we focus on the stellar accretion origin of metallicity & color gradients in high-resolution re-simulated massive galaxies *at large radii* ($2R_{\rm eff} < r < 6R_{\rm eff}$) in a full cosmological context. We particularly intend to explore the combined effect of energetic phenomena such as strong galactic, stellar-driven winds (and AGN feedback), and of the individual merger and accretion his-

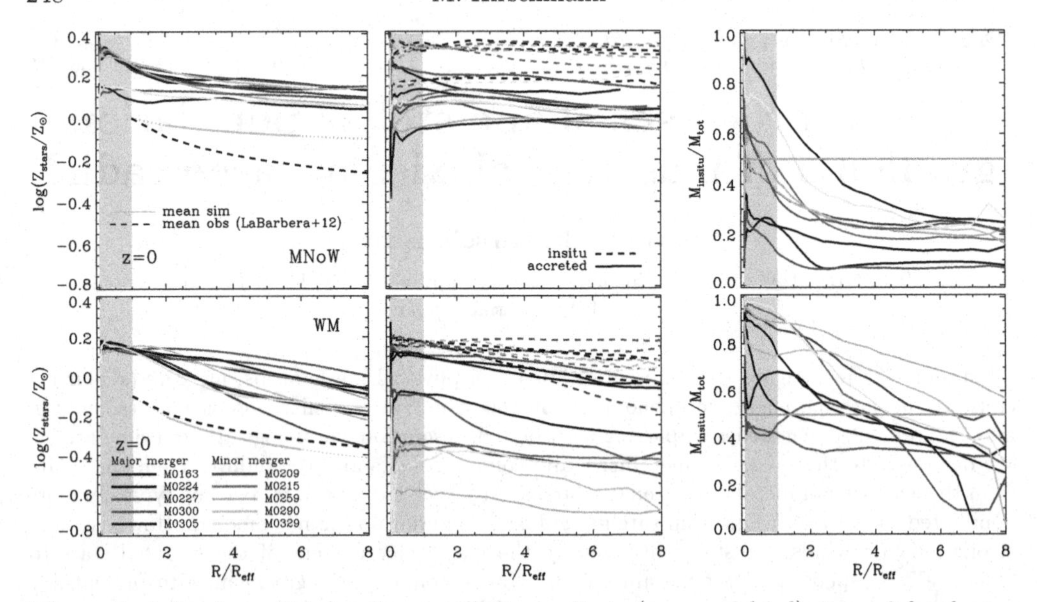

Figure 1. *Left column:* Total stellar metallicity gradients (mass weighted) at $z = 0$ for the ten main galaxies (different colors) simulated with the MNoW and WM model. The green solid lines indicate the average gradient at $2 < R/R_{\rm eff} < 6$. *Middle panels:* Metallicity gradients at $z = 0$ separated into stars formed in-situ (dashed lines) and accreted stars (solid lines). *Right panels:* Fraction of in-situ to total stellar mass as a function of radius. Figure taken from Hirschmann *et al.* (2015).

tories on the in-situ formed and accreted stellar fractions and on steepening or flattening of stellar population gradients at these large radii (Hirschmann *et al.* 2015).

2. High-resolution simulations of individual galaxy halos

We consider the 10 most massive halos (covering a mass range of $6 \times 10^{12} < M_{\rm halo} < 2 \times 10^{13} M_{\odot}$) of our high-resolution, cosmological zoom simulation sets presented in (Hirschmann *et al.* 2013) and (Hirschmann *et al.* 2015). The simulations include a treatment for metal enrichment (SNII, SNIa and AGB stars) and a phenomenological feedback scheme for galactic winds (Oppenheimer & Dave 2006, 2008). The dark matter/gas particles have a mass resolution of $m_{\rm dm} = 2.5 \cdot 10^7 M_{\odot} h^{-1}$ and $m_{\rm gas} = m_{\rm star} = 4.2 \cdot 10^6 M_{\odot} h^{-1}$, respectively with a co-moving gravitational softening length for the gas and star particles of 400 h^{-1}pc and for the dark matter particles of 890 h^{-1}pc (Oser *et al.* 2010).

These cosmological zoom simulations were shown to be successful in suppressing early star formation at $z > 1$, in predicting reasonable star formation histories for galaxies in present day halos of $\sim 10^{12} M_{\odot}$, in producing galaxies with high cold gas fractions (30 - 60 per cent) at high redshift, and in significantly reducing the baryon conversion efficiencies for halos ($M_{\rm halo} < 10^{12} M_{\odot}$) at all redshifts in overall good agreement with observational constraints. Due to the delayed onset of star formation in the wind models, the metal enrichment of gas and stars is delayed and is also found to agree well with observational constraints.

3. Metallicity gradients of present-day massive galaxies

In Fig. 1, we show the total (mass-weighted) stellar metallicity gradients out to 8 $R_{\rm eff}$ for the main galaxies at $z = 0$ in the MNoW and the WM model (as indicated in the legend). For the MNoW galaxies, the (most) central metallicity has values of $Z/Z_{\odot} \sim 0.4$

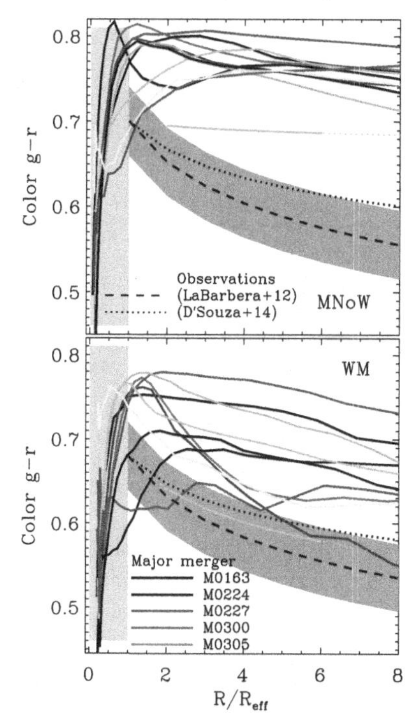

Figure 2. G-r color gradients (mass weighted) at $z = 0$ for the ten galaxies (different colors) in the MNoW (top panel) and WM model (bottom panel). The average simulated gradient at $2 < R/R_{\rm eff} < 6$ (green solid lines) is compared to observations from La Barbera *et al.* (2012) and D'Souza *et al.* (2014). Figure taken from Hirschmann *et al.* (2015).

and drops to $Z/Z_\odot \sim 0.1$ at large radii, the slopes reach a minimum value of −0.25 dex/dex. This gradient is mainly driven by the accreted stars (the in-situ distributions are almost all nearly flat), which have on average roughly solar metallicity (top middle panel of Fig. 1) and dominate most systems outside $1R_{\rm eff}$ (top right panel of Fig. 1).

The WM galaxies (bottom panels of Fig. 1) have lower central metallicities ($Z/Z_\odot$~0.2) with much steeper outer gradients down to −0.76 dex/dex with a mean of −0.35 dex/dex. The reason for the steeper gradients in the WM compared to the MNoW model is twofold: on the one hand, the steeper gradients originate from the accretion of metal-poorer stellar populations. On the other hand, also the in-situ components show metallicity gradients contributing to the overall gradients. The latter is most likely due to infall of (particularly re-infall of previously ejected) metal-poor gas onto the galaxy which can be then turned into metal-poor stars as a consequence of an inside-out growth, the same process causing the metallicity gradients in disk galaxies. Late re-accretion of previously ejected gas occurs typically in the WM model due to the strong galactic winds, but not in the MNoW model, where the in-situ gradients are, therefore, relatively flat (see top middle panel of Fig. 1). However, despite of the partly negative in-situ gradients, we find that on average, the in-situ gradients are reduced by ~ 0.2 dex due to accretion of metal-poor stellar populations.

4. Color gradients of massive, early-type galaxies

An important set of observables for galaxies are their colours, which are observationally more easy to measure (from images) than age and metallicity (requiring spectroscopic data). Colours are (degenerately) dependent on the intrinsic metallicity and ages of their

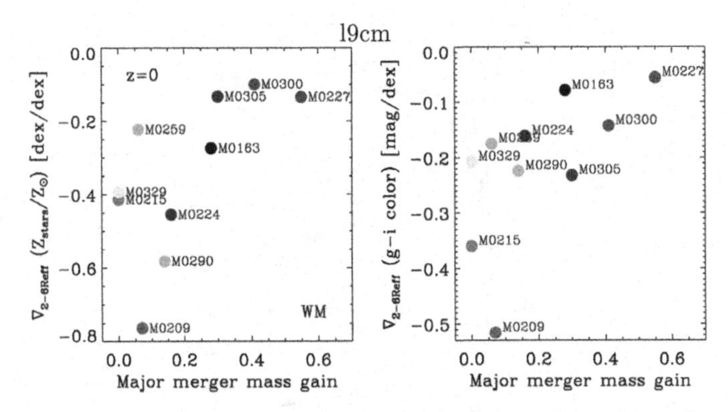

Figure 3. Stellar metallicity (left panel) and g-i color (right panel) gradients (between $2-6R_{\rm eff}$) versus the mass gain through major mergers for the ten re-simulated galaxies in the WM simulations. The higher the major merger mass gain, the smaller is the slope of the metallicity and color gradient. Figure taken from Hirschmann *et al.* (2015).

stellar populations. To facilitate a comparison of the simulated radial distributions of galaxy colours to present-day, observed (outer) *color* gradients we use the metal- and age-dependent models for the spectral evolution of stellar populations of Bruzual & Charlot (2003), assuming a Chabrier IMF to compute some photometric properties (g-i, u-g and g-r colours) of our simulated galaxies.

Fig. 2 shows the the g-r colour gradients (solid lines) for the MNoW and the WM galaxies (top and bottom panel, respectively) at $z = 0$. WM galaxies are generally slightly bluer than MNoW galaxies as a consequence of their younger stellar populations with lower metallicity. In addition, all galaxies of both models are significantly bluer at the centre, revealing steep positive colour gradients at $< 1\ R_{\rm eff}$ which is an unrealistic artefact of our models most likely due to missing AGN feedback resulting in too young central stellar populations.

At larger radii ($> 2\ R_{\rm eff}$), the two models behave differently. The MNoW galaxies have only very shallow colour gradients, either slightly decreasing or increasing with an average slope of -0.02 mag/dex for the g-r colors. The shallow colour gradients in the MNoW model stem from the relatively flat age and only slightly negative metallicity gradients. The former originates from a relatively old in-situ stellar population (compared to the accreted stars), and the latter from the accretion of relatively metal-rich stellar systems in the MNoW model. Instead with galactic winds, nearly all of the WM galaxies become continuously bluer with increasing radius and thus, reveal significantly steeper negative colour gradients than the MNoW galaxies (with mean slopes of -0.13 mag/dex for the g-r colours). This behaviour is entirely driven by the steeper metallicity gradients of the WM galaxies washing out any effect of the slightly increasing age gradients. Note that we find qualitatively similar results for the g-i and u-g color gradients (see Hirschmann *et al.* 2015).

5. The impact of the galaxy merger history

The black-blue lines in Fig. 1 and 2 indicate galaxies which experienced at least one major galaxy merger since $z = 2$, while the red-yellow lines illustrate those having undergone only minor galaxy mergers. Present-day galaxies having experienced a recent major merger have typically flatter gradients than those with a more quiet merger history.

Fig. 3 quantifies the connection between the galaxy merger history and the steepness of the metallicity and color gradients: we show the fitted metallicity (left panel) and color gradients (right panel) at $z = 0$ for WM galaxies versus the mass gain by major mergers. The mass gain by major mergers considers the entire stellar mass which was brought into the main galaxy by major mergers since $z = 2$ normalised to the present-day stellar mass.

Both the fitted metallicity and color gradients correlate with the past merger history: for a major merger mass gain above 20 %, the metallicity (color) gradients are flatter than -0.3 (-0.2) dex/dex. Instead, for lower x-values, the metallicity (color) gradients are mostly more negative than -0.4 (-0.2) dex/dex. This is due to the fact that the accreted stars show a huge variety of metallicities from $Z/Z_{\odot} \sim -0.6$ to $Z/Z_{\odot} \sim +0.1$ (see bottom middle panel of Fig. 1) depending on the exact merger history: in case of a recent major merger, the accreted metallicity is significantly larger than without a major merger (as more massive galaxies have higher metallicity). The higher metallicity of the accreted stars together with the different mixing behaviour (violent relaxation) in case of major mergers flattens the total metallicity gradients. The present-day color gradients are entirely driven by the behaviour of the metallicity gradients, leading to the same behaviour.

This result bears an important implication for observations, as it can help to reconstruct the past assembly history for observed present-day metallicity and color gradients: The relation between the steepness of the gradients and the individual merger history implies that observed massive galaxies having steep outer gradients most likely have not experienced any major merger event after $z = 1$, but instead several minor mergers.

6. Comparison to observations

Starting with the colour gradients, in the photometric study by La Barbera *et al.* (2012), they investigated colour gradients of early-type galaxies ($M_{\rm stellar} > 3 \times 10^{10} M_{\odot}$) even out to $8 \times R_{\rm eff}$ using the SDSS-based Spider survey. They measure average slopes for g-r colours of -0.16 ± 0.04 mag/dex (see black dashed line and grey shaded areas in Fig. 2). Compared to observed, g-r colour gradients, MNoW galaxies have on average too flat gradients (-0.02 mag/dex), while those of the WM galaxies are reasonably steep (-0.13 mag/dex) – in good agreement with the observations.

The recent work of D'Souza *et al.* (2014) show g-r colour profiles out to 100 kpc for stacked ellipticals (to be more precise, high concentration galaxies) with masses between $10^{10} - 10^{11.4} M_{\odot}$ using roughly 45.500 galaxies from the SDSS survey. They find values between -0.11 and 0.14 mag/dex with little mass trend, in perfect agreement with the g-r colour gradients predicted by our simulations (see black dotted lines in Fig. 2, where we have used the slope of -0.11 mag/dex measured for stellar masses of $10^{11.2} M_{\odot}$).

Turning to metallicity gradients, a recent study of Pastorello *et al.* (2014), using the SLUGGS survey, investigates metallicity gradients at least out to $2.5R_{\rm eff}$. For comparable stellar masses, their galaxies reveal slopes between -1.15 and $+0.18$ dex/dex. In addition, in the study by La Barbera *et al.* (2012), they derive metallicity gradients of their sample of early-type galaxies. For massive galaxies with $10^{11} < M_{\rm stellar} < 7 \times 10^{11} M_{\odot}$ they find outer metallicity gradients ($1-8 \times R_{\rm eff}$) in the range of -0.29 to -0.74 dex/dex depending on the stellar population model (illustrated by the black dashed lines in Fig 1).

Overall, for both colors and metallicities, our WM galaxies are able to cover such a broad range of slopes (e.g. from -0.8 to -0.1 dex/dex for metallicity) much better than the MNoW galaxies whose slopes are on average too flat (e.g. between -0.25 and $+0.03$ dex/dex for metallicity). The average metallicity and color gradient of the WM galaxies is in excellent agreement with the ones of La Barbera *et al.* (2012). *As expected a priori, this implies that a strong stellar feedback is a key mechanism to be consistent with observed steep metallicity gradients in massive galaxies in the local Universe.*

7. Conclusions

Analysing zoom simulations of 10 massive galaxies suggests that the outer negative metallicity and color gradients (at radii $r > 2\ R_{\rm eff}$) of present-day massive galaxies are mainly determined by the accretion of stars with lower metallicity (see Hirschmann *et al.* 2015). Towards low redshift, stars accreted in low metallicity galaxies become more and more dominant at large radii and the metallicity gradients of in-situ formed stars in the wind model are enhanced by ~ 0.2 dex/dex by accretion of metal-poor systems. The model with galactic winds predicts steeper total metallicity gradients (on average -0.35 dex/dex) as the accreted stellar systems are significantly more metal-poor and despite of the fact that much less stellar mass in total is accreted compared to the no-wind model. Note that the negative outer color gradients in our simulated present-day massive galaxies are entirely driven by the negative outer metallicity gradients, and therefore, originate from the same processes, the interplay of strong stellar feedback and the individual merger history.

Even if not explicitly discussed in this article, first, preliminary simulations including AGN-driven winds indicate that this additional process does qualitatively not alter the stellar accretion origin of metallicity gradients at large radii, but apparently mainly affects the inner metallicity gradients.

Overall, we can conclude that a combination of stellar accretion (in minor mergers) of low mass satellites (as predicted from idealized experiments by Villumsen 1983) and strong stellar feedback results in steep outer metallicity gradients successfully matching the broad range of observed metallicity profiles of local galaxies at large radii.

Acknowledgements

MH acknowledges financial support from the European Research Council via an Advanced Grant under grant agreement no. 321323NEOGAL.

References

Bruzual M. & Charlot 2003, *MNRAS*, 344, 1000
D'Souza R. *et al.* 2014, *MNRAS*, 443, 1433
Gilbert K. *et al.* 2014, *ApJ*, 796, 20
Hirschmann M. *et al.* 2013, *MNRAS*, 436, 2929
Hirschmann M. *et al.* 2015, 449, 528
Kobayashi C., 2004, *MNRAS*, 347, 740
La Barbera F. *et al.* 2012, *MNRAS*, 426, 2300
McClure R. & Racine R. 1969, *AJ*, 74, 1000
Oppenheimer B. & Dave R. 2006, *MNRAS*, 373, 1265
Oppenheimer B. & Dave R. 2008, *MNRAS*, 387, 577
Oser L. *et al.* 2010, *ApJ*, 725, 2312
Pastorello N. *et al.* 2014, *MNRAS*, 442, 1003
Rejkuba M. *et al.* 2014, *ApJL*, 791, 6
Villumsen J. 1983, *MNRAS*, 204, 219
Wyse R. & Silk J. 1989, *ApJ*, 339, 700

The General Assembly of Galaxy Halos: Structure, Origin and Evolution
Proceedings IAU Symposium No. 317, 2015
A. Bragaglia, M. Arnaboldi, M. Rejkuba & D. Romano, eds.

doi:10.1017/S1743921315009795

Dual Stellar Halos in Early-type Galaxies and Formation of Massive Galaxies

Myung Gyoon Lee and In Sung Jang
Departmemt of Physics and Astronomy, Seoul National University, Seoul, Korea
email:mglee@astro.snu.c.kr

Abstract. M105 in the Leo I Group is a textbook example of a standard elliptical galaxy. It is only one of the few elliptical galaxies for which we can study their stellar halos using the resolved stars. It is an ideal target to study the structure and composition of stellar halos in elliptical galaxies. We present photometry and metallicity of the resolved stars in the inner and outer regions of M105. These provide strong evidence that there are two distinct stellar halos in this galaxy, a metal-poor (blue) halo and a metal-rich (red) halo. Then we compare them with those in other early-type galaxies and use the dual halo mode formation scenario to describe how massive galaxies formed.

Keywords. galaxies: halos, galaxies: stellar content, galaxies: abundances, galaxies: elliptical and lenticular, cD galaxies: formation, etc.

---

1. Introduction

It has been long considered that radial surface brightness profiles of massive elliptical galaxies can be described in general by one component (for example, by the de Vaucouleurs $R^{1/4}$ law), indicating that each of them is composed of a single halo. However, deeper imaging in recent studies finds often much fainter substructures in the outskirts of these galaxies, indicating that there may be more than one components (for example, Watkins *et al.*(2014)).

On the other hand, the color distribution of the globular clusters in massive ealty-type galaxies (ETGs) is found to be bimodal, indicating the presence of two distinguishable populations: blue and red globular cluster systems. These two populations show significant differences in their spatial distribution in the sense that the distribution of the blue population is more extended and circular compared with the red population, and the distribution of diffuse stellar light is closer to that of the red population.

From these differences between the blue and red globular cluster systems Park & Lee(2013) suggested that massive ETGs may host dual halos and that the formation of massive ETG can be described by the dual halo mode formation scenario. The reality of dual halos in massive galaxies can be tested by studying the metallicity and spatial distribution of the resolved stars in nearby ETGs.

2. Dual Stellar Halos in ETGs

In 1997 Elson (1997) obtained photometry of resolved stars in a halo field of NGC 3115 (S0), finding that the $(V - I)$ color distribution of the resolved red giant is clearly bimodal. From this she suggested that NGC 3115 has two distinct halo populations: a metal-poor halo ([Fe/H] ≈ -1.3) and a metal-rich halo ([Fe/H] ≈ -0.0). Recently Peacock *et al.*(2015) confirmed it later in the study of the outer fields in NGC 3115 using deeper photometry.

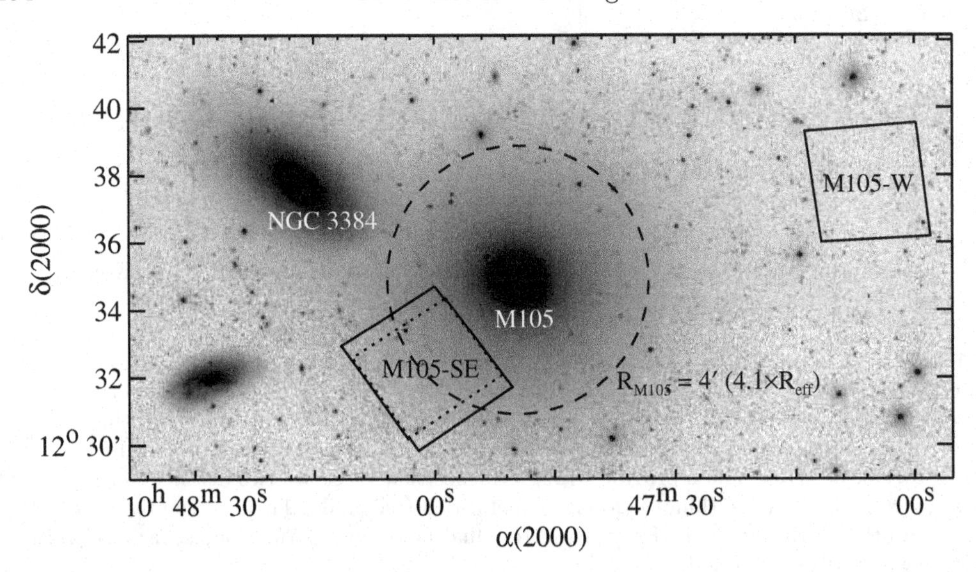

Figure 1. Locations of the two HST fields (SE and W) for M105 on the gray scale map of the Sloan Digital Sky Survey *i*-band image.

Since then photometry of resolved stars in several ETGs provided hints for the presence of dual stellar halos (see Peacock *et al.*(2015) and references therein): some examples are NGC 5128 (S0 pec), NGC 3377 (E5), and M105 (NGC 3379, E1) (Harris & Harris(2002), Harris *et al.*(2007a), Harris *et al.*(2007b), Rejkuba *et al.*(2011), Crnojević *et al.*(2013), and Bird *et al.*(2015)).

M105 in the Leo I Group is a textbook example of a standard elliptical galaxy. Old red giant stars in the halo of M105 are easily resolved in the HST images so that it is an ideal target to study the structure and composition of stellar halos in elliptical galaxies. It is only one of the few elliptical galaxies for which we can study their stellar halos using the resolved stars. In this study we present the current progress of our work on the resolved stellar populations in M105. We derive photometry and metallicity of the resolved stars in the inner and outer regions of M105. Then we compare them with those in other ETGs and suggest a dual halo mode formation scenario for massive ETGs (see Lee & Jang(2015) for details).

2.1. *Resolved Stars in the Standard Elliptical galaxy M105*

We derived deep photometry of the resolved stars from the F606W and F814W images of two fields available in the Hubble Space Telescope (HST) archive, as marked in Fig. 1. One field is located at $R \sim 4R_{\rm eff}$ and the other is at $R \sim 12R_{\rm eff}$. The resolved stars in the outer field were studied previously by Harris *et al.*(2007b). We used DAOPHOT (Stetson(1994)) for detection and point-spread-function-fitting photometry of the stars.

The color-magnitude diagrams (CMDs) of the resolved stars in the inner and outer regions in Fig. 2 show a broad red giant branch (RGB). Comparison with the theoretical isochrones for 12 Gyr age in the Dartmouth model (Dotter *et al.*(2008)) shows that the RGB stars have a large range of metallicity. The distance to M105 is estimated to be $d = 10.23 \pm 0.09$ Mpc using the tip of the RGB method (Lee *et al.*(1993)). The $(V - I)$ color histograms of the bright RGB stars ($M_{\rm bol} \leqslant -3.0$ mag) in the lower panels of Fig. 2 show the presence of two distinguishable components, as pointed by Elson(1997) in the case of NGC 3115. However, they show a significant difference between the inner and outer regions, in the sense that the outer region shows a much stronger narrow blue peak

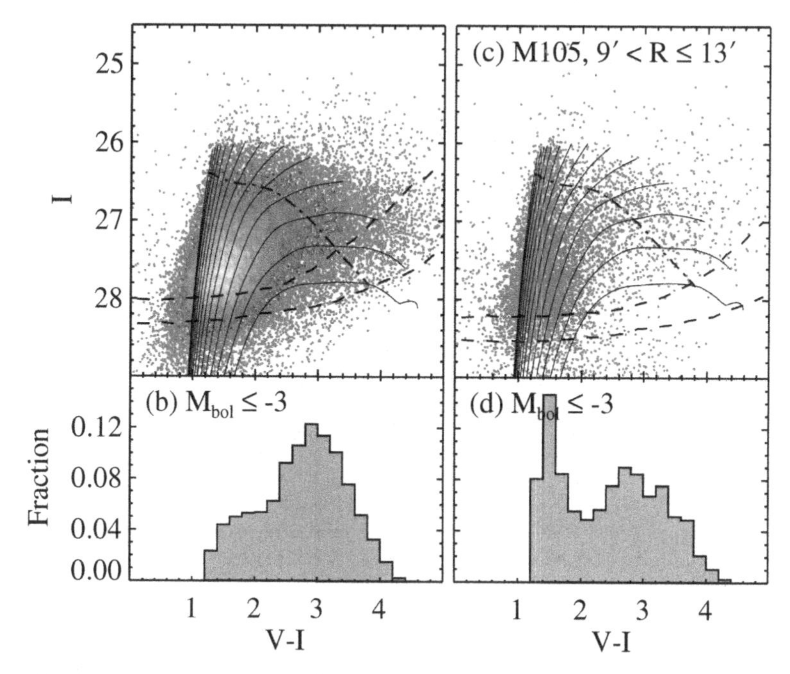

Figure 2. Color-magnitude diagrams (upper panels) of the resolved stars and $(V - I)$ color histograms (lower panels) of the bright RGB stars ($M_{\rm bol} < -3.0$) in the inner and outer regions of M105. Curved solid lines denote theoretical isochrones for 12 Gyr age in the Dartmouth models Dotter *et al.*(2008). They cover a range of metallicity ([Fe/H] = −2.4 to +0.2 ([M/H] = −2.2 to +0.4). The 50% and 20% completeness limits are shown by the dashed lines. Dot-dashed lines represent $M_{\rm bol} < -3.0$.

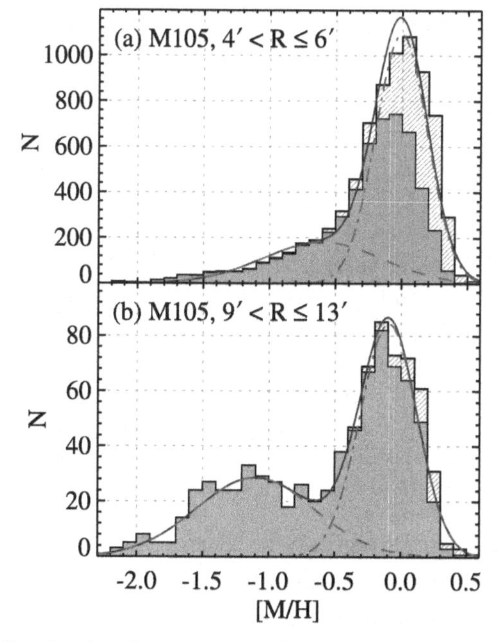

Figure 3. Metallicity distribution functions for the bright RGB stars ($M_{\rm bol} \leqslant -3.0$) in the inner and outer regions of M105. The filled and hatched histograms represent the MDFs before and after completeness correction. Note that the MDFs can be fitted well by double Gaussian components (curved lines).

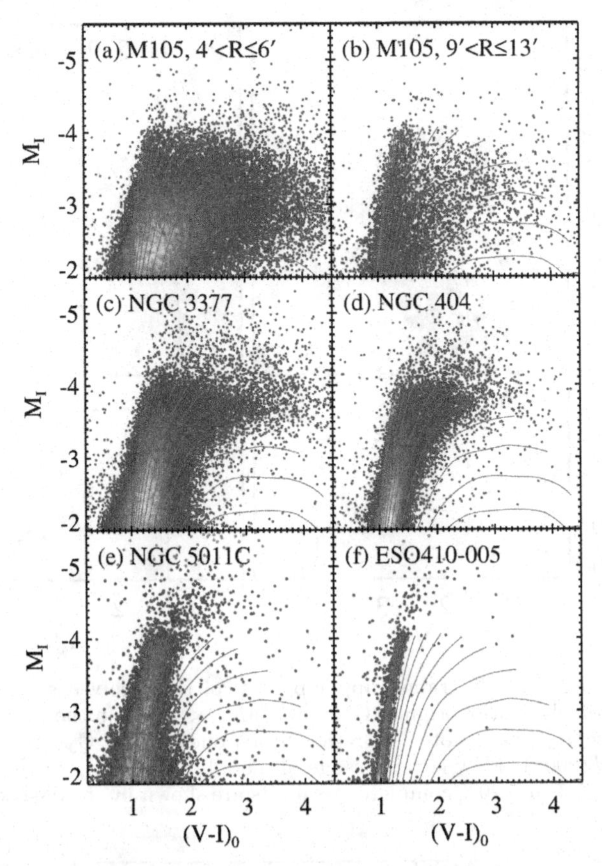

Figure 4. A comparison of the CMDs of the resolved stars in M105, NGC 3377 (E5, $M_V = -19.89$ mag), NGC 404 (S0, $M_V = -17.35$ mag), NGC 5011C (S0, $M_V = -14.74$ mag), and ESO410-005 (dE3, $M_V = -12.45$ mag). Same 12 Gyr isochrones as in Fig. 2 are also plotted (curved lines). Note the diversity of the RGBs.

than the inner region. We estimate the metallicity of the stars from the comparison with the theoretical isochrones and the positions of the stars in the CMDs. Fig. 3 shows the metallicity distribution functions (MDFs) of the bright RGB stars ($M_{\rm bol} < -3.0$ mag) in M105. The MDFs in Fig. 3 shows two distinguishable components as in the case of the color histograms, which can be fit reasonably well by double Gaussian functions. The metallicity of the metal-rich peak change little between the inner and outer regions, but the metal-poor component is more significant in the outer region than in the inner region. This shows that there are dual halos in M105: a dominant metal-rich (red) halo and a weaker metal-poor (blue) halo. Radial number density profiles of the resolved stars show that the metal-poor halo is significantly more extended than the metal-rich halo. It is noted that the MDF of each halo can be described by accretion box models.

2.2. *Resolved Stars in Bright to Faint ETGs*

We compare the CMDs of M105 with those in other ETGs with bright to faint luminosity in Fig. 4: NGC 3377 (E5, $M_V = -19.89$ mag), NGC 404 (S0, $M_V = -17.35$ mag), NGC 5011C (S0, $M_V = -14.74$ mag), and ESO410-005 (dE3, $M_V = -12.45$ mag). We obtained photometry of these stars from the HST images in the archive. Fig. 4 shows that the RGBs vary significantly depending on the luminosity of the galaxies. The RGBs get broader and mean RGB colors get redder as their host galaxies become brighter.

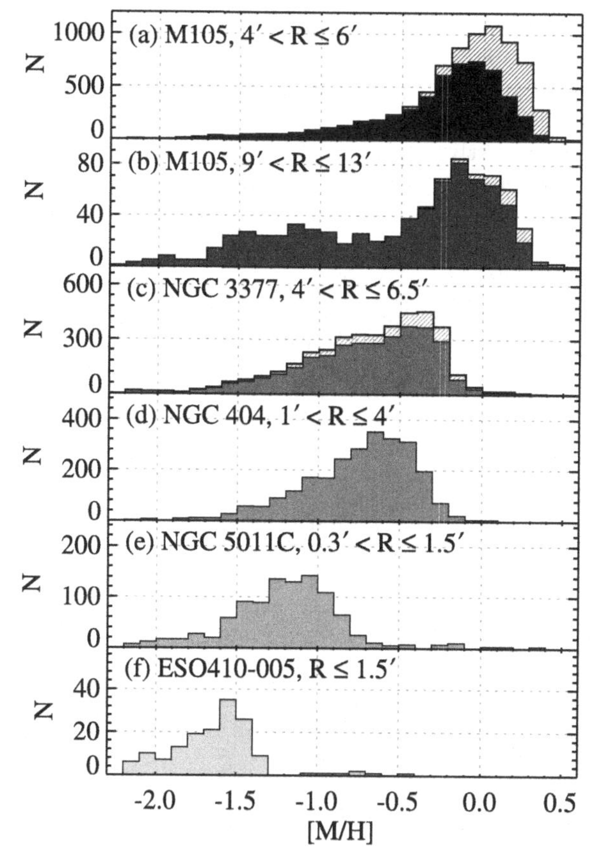

Figure 5. A comparison of the MDFs of the bright RGB stars in the same ETGs as in Fig. 4. Note that the metallicity of the metal-poor peak of the outer region in M105 is similar to that of NGC 5011C.

The MDFs of the resolved stars in these galaxies show two notable features. First, the peak metallicity decreases as the luminosity of their host galaxies decreases. Second, the MDF of the metal-poor component in the outer region of M105 is similar to that of NGC 5011C which has a magnitude of $M_V = -14.74$ mag. This implies that the origin of the metal-poor halo in M105 is probably a dwarf galaxy with $M_V \approx -15$ mag, and while the origin of the metal-rich halo in M105 is a massive galaxy which may be an outcome of major merging or monolithic collapse.

3. A Scenario for Formation of Massive Galaxies: Dual Halo Mode Formation

Considering the MDFs of the resolved stars depending on the galactocentric distance and the distinguishable properties of the blue and red globular cluster systems, we suggest a scenario to explain how massive ETGs might have formed, updating the previous version based on the study of the globular clusters in Park & Lee (2013). In this scenario, massive galaxies form through the metal-rich halo mode followed by the metal-poor halo mode.

First, a massive progenitor formed through in situ rapid collapse of a massive proto-galactic cloud or wet merging of less massive galaxies with gas. It will be dominated by metal-rich stars with a minor fraction of metal-poor stars so that it will correspond to a metal-rich halo. At the similar time numerous dwarf galaxies with low metallicity form in and around the massive galaxies. They will be the major source of low metallicity stars later. Second, the metal-rich massive galaxy grows further via dry merging/or accretion of less massive galaxies (mostly dwarf galaxies) so that its outer part will be dominated by metal-poor stars (originated from low mass stellar systems including dwarf galaxies and globular clusters). The resulting system corresponds to the metal-poor halo. Thus the metal-poor halo will consist mostly of the remnant stars from dwarf galaxies, and some of low metallicity stars from the original massive progenitor. This scenario can be tested further with simulations in the future (for example, Cooper *et al.*(2015)).

This work was supported by the National Research Foundation of Korea (NRF) grant funded by the Korea Government (MSIP) (No. 2012R1A4A1028713).

References

Harris, W. E. & Harris, G. L. H. 2002, *ApJ*, 123, 3108
Bird, S. A., Flynn, C., Harris, W. E., & Valtonen, M. 2015, *ApJ*, 575, 72
Cooper, A. P., Parry, O. H., Lowing, B. Cole, S., & Frenk, C. 2015, *MNRAS*, 454, 3185
Crnojević, D., Ferguson, A. M. N., Irwin, M. J., *et al.* 2013, *MNRAS*, 432, 832
Dotter, A., Chaboyer, B., Jevremović, D., *et al.* 2008, *ApJS*, 178, 89
Elson, R. A. W. 1997, *MNRAS*, 286, 771
Harris, W. E., Harris, G. L. H., Layden, A. C., & Stetson, P. B. 2007, *AJ*, 134, 43
Harris, W. E., Harris, G. L. H., Layden, A. C., & Wehner, E. M. H. 2007, *ApJ*, 666, 903
Lee, M. G., Freedman, W. L., & Madore, B. F. 1993, *ApJ*, 417, 553
Lee, M. G. & Jang, I. S. 2015, *ApJ*, submitted
Park, H. S. & Lee, M. G. 2013, *ApJL*, 773, L27
Peacock, M. B., Strader, J., Romanowsky, A. J., & Brodie, J. P. 2015, *ApJ*, 800, 13
Rejkuba, M., Harris, W. E., Greggio, L., & Harris, G. L. H. 2011, *A&Ap*, 526, AA123
Rejkuba, M., Harris, W. E., Greggio, L., *et al.* 2014, *ApJ* (Letter), 791, L2
Stetson, P. B. 1994, *PASP*, 106, 250
Watkins, A. E., Mihos, J. C., Harding, P., & Feldmeier, J. J. 2014, *ApJ*, 791, 38

The General Assembly of Galaxy Halos: Structure, Origin and Evolution
Proceedings IAU Symposium No. 317, 2015
A. Bragaglia, M. Arnaboldi, M. Rejkuba & D. Romano, eds.

doi:10.1017/S1743921315010947

IAU Symposium 317 Summary

Raffaele G. Gratton

INAF-Osservatorio Astronomico di Padova
email: raffaele.gratton@oapd.inaf.it

Abstract. The assembly of the halo yields fundamental information on the formation and evolution of galaxies: this was quite exhaustively discussed at this very important symposium. I present a brief personal summary of the meeting, outlining those points that I found more exciting and suggestive. I also remarked a few areas that were possibly not enough expanded. I found this research field extremely interesting and I think there are great expectations for new developments in the next few years, thanks to the new large spectroscopic surveys and the ESA GAIA satellite.

Keywords. Galaxy (the): halo

---

1. Introduction

First of all: it is an honour and also a very difficult task to summarize such a lively and wide meeting. Giving justice to all speakers is impossible, and I am certainly not able to give a single coherent speech that combines all have been said by many of you, often much more competent than me. In addition, such a coherent speech have already been given by our chairman, O. Gerhard, in his very nice general talk. In this summary to the meeting I will then present my personal view, trying to summarize what I have learnt in these days and those points that are not yet entirely clear to me.

The first thing I learnt is that there is a quite uniform approach within the field. For instance, all the simulations and interpretations presented at the meeting are based on the paradigm of hierarchical growth of galaxies in a Λ-CDM universe. While this might be perfectly reasonable, I think it should perhaps not be considered as obvious as it apparently seems to be. In general, I would encourage people to discuss this approach more in depth, and I would welcome and consider possibly more enriching a less uniform approach, though I understand it may be dangerous especially for young investigators to lose time following routes that may reveal sterile.

The content of my talk is the following: (i) Stellar halo structure; (ii) First stars; (iii) Globular clusters; (iv) Dwarf spheroidals; and (v) What is next?

2. Stellar halo structure

Many talks considered this issue, that was central to the meeting. Different tracers of the halo were considered, including red giant branch stars (A. McConnachie, M. Rejkuba, M. Steinmetz, D. Crnojevic, M. Hilker, K. Gilbert, A. Kupper, S. Koposov, E. Peng, M. Collins, J. Greene, A. Ferguson, A. Monachesi, R. de Jong, M.G. Lee, C. Grillmair, A. Sheffield), globular clusters (GCs: S. Larsen, M. Hilker, J. Brodie), planetary nebulae (PNe: M. Arnaboldi, W. Reid, J. Brodie, N. Napolitano), and RR Lyrae variables (G. Fiorentino. J. Cohen, P. Pietrukowicz). There were very interesting reviews and contributions about models (K. Johnston, J. Read, B. Cook, A. Helmi, M. Hirschmann, G. Hensler), while the talk by Jorge Penarrubia was more on methodology.

In general, a quite coherent pattern emerges from all this work. K. Johnston spoke about the origin of the halo stars and considered three main components: accreted, in situ, and kicked out from disk. This general view was shared by all speakers. The relative fractions of halo stars that are due to these different mechanisms depend on distance from the center. The in situ component prevails in the inner halo while the accreted one in the outer halo. Stars that originated in the disk and kicked out into the halo makes up some 1-30% of the total within $\sim$ 20 kpc (Zolotov *et al.* 2009, 2010, Tissera *et al.* 2012, 2013 etc). My comment here is that this view considers two main regions in the halo: inner and outer halo. Given the strong galactocentric density gradient existing in the halo, the inner halo is the vast majority of the halo. On the other hand, the outer halo has less dynamical evolution and better conserved traces of its origin. Likely for this reason, most of this meeting focused on the outer halo. A second issue concerns the nature of the accretion episodes, and in particular the distinction between dry and gas-rich mergers. Most theoretical work concerns dry mergers, likely because they are simpler to model. However, gas-rich mergers may imply much larger accreted masses, and might be more relevant to understand the formation of the bulk of the halo.

I have learnt a number of simple rules: (i) Since there is a Mass (actually Luminosity) Metallicity relation for dwarf galaxies (Kirby *et al.* 2013), then low metallicity stars are mainly produced by minor mergers while high metallicity ones are produced by major mergers (B. Cook talk). (ii) Thin/cold streams are related to GCs (C. Grillmair talk). (iii) The offset between leading/trailing tidal tails is determined by the position of the Lagrangian points (A. Kupper talk). (iv) Make often back ups of my notes. Surely I should have learnt also other things but likely I was not good enough to do so.

At this meeting, we have seen fantastic new results concerning the halo of the Milky Way and of M31: this shows that the research field is extremely active. Many new dwarf Spheroidals (dSph) are being discovered; the total of such objects known in the Milky Way is over 40 now. This increment might suggest revisiting of the missing satellite issue: is it really still a problem? The 40 faint satellites of the Milky Way discovered so far are consistent with the number of objects predicted from the SDSS brighter limit (Tollerud *et al.* 2008, ApJ 688, 277). Then there is still a missing satellite objects for the Milky Way. On the other hand, I notice that most new dSphs are very small while the problem of the missing satellites appears at rather large masses (M> 10^8 $M_\odot$). Then, a more attractive possible explanation of the missing satellite issue is on my opinion related to the issue of the mass distribution in dSph's, that is the question of cuspy vs core. In fact, as outlined in several talks, cores are more easily destroyed than cusps by tides during the dynamical interaction with the Milky Way. It is then possible that many satellites of the Milky Way (and of other galaxies too) are destroyed during the violent phases of the dynamical interaction.

Several studies support the view that the accretion history of the Milky Way was different from that of M31: the latter has more prominent substructures, and much more recent accretions. This shows that there is a considerable variance in accretion histories. This also results from simulations of the formation of stellar halos. This is a strong motivation to study (lots of) galaxy halos because our own is not enough and representative. The study of external halos is however difficult because it requires deep imaging to very low surface brightness and/or resolved populations.

If we now look more in detail to the Milky Way, talks in the meeting outlined the following facts. The census of surviving satellites is still incomplete. From existent data, the predicted number of satellites is between 50 to 500 within a galactocentric distance $R_{\rm G}$ of $R_{\rm G}$ < 280 kpc, the exact number depending on the assumed density profile. The density profile of halo stars from various tracers show a steepening at $R_{\rm G} \sim 25$ kpc. The velocity dispersion of halo stars may be obtained from data obtained by the SDSS: the velocity dispersion is independent of distance for $R_{\rm G}$ < 10 kpc showing that the inner halo is well mixed. There are few information for large distances, though in some cases we have data reaching as far as 150 kpc. This data suggests that the profile flattens at distances $>$ 100 kpc. There are four main substructures: the Sagittarius dSph and the related stream; the Monoceros ring; the Virgo cloud; and the Hercules-Aquila cloud. Summing up all these main substructures make $\sim 2\,10^8\ M_\odot$ in stars and probably about 20 GCs. Sagittarius dSph alone provides about 20% of halo debris. From the halo there is no evidence for late accretion of large satellites in the Milky Way (Justin Read). I notice that this was also derived from the characteristics of the thin disk, that would have been significantly heated by such an event.

At variance with the Milky Way, M31 is a good example of disk heating with a substantial fraction of the halo being attributed to the kicked out component. Johnston noticed that there is plenty of substructures around the disk. McConnachie showed wonderful images from the PAndAS survey that display clear structures over smooth halo (Ibata *et al.* 2014): he reported that only a small fraction of the M31 halo appears smooth; the smooth fraction is about 6% at high metallicity and 60% at low metallicity. In any case, it is clear that the halo of M31 is very extended and more metal-rich than that of the Milky Way. Ferguson noticed that most of the substructures in M31 are metal-rich ([Fe/H]> -1.1) and only a tiny fraction has dSph metallicity. Possibly the most striking structure seen in the M31 halo is the giant stream extending toward M33 that has a total luminosity M_V brighter than -14, may be $M_V = -16$. This can be modelled as a satellite of $10^9\ M_\odot$ falling on a radial orbit $\sim$ 1 Gyr ago (A. Ferguson)

There are possibly puzzling results concerning the M31 halo. Mackey *et al.* (2010) used data from the PAndAS survey to show that the GCs in M31 align with streams in halo. Outer halo GCs show net rotation, apparently aligned with the disk rotation: is this only a coincidence? In her talk Helmi discussed this point and suggested that it is possibly explained by a peculiar point of view. I think this issue requires further investigation.

Several talks discussed the structure of the halos of galaxies in the Virgo cluster. It is found that large ellipticals at a redshift of $z \sim 2$ are 3-4 times more compact than at $z = 0$; hence most of the halo has been accreted later. An important issue concerns the relation between the halos of large galaxies and the intracluster light (ICL: see talks by C. Mihos and M. Arnaboldi). The overall idea is that gravitational interaction in groups strips stars and feed intracluster light/halo. According to this scenario, the ICL grows with time, making up to 10-20% of cluster light at $z = 0$. The Virgo cluster is littered with signatures of all the events that give rise to ICL. An important question is if there is a dominant formation channel for ICL. Studies of the stellar population in the Virgo ICL shows that it is mostly composed of old stars with a considerable range in metallicity. For various reasons (consideration of cosmic variance and attempt to reconstruct the evolution of ICL in a statistical way) there is need to get information on farther systems. Unfortunately, at such large distances we lose the possibility to study discrete populations. In general, a major is issue is how to separate halo from ICL, and even if it is reasonable to do this.

3. First stars

First stars are very interesting for various reasons. Among others, I may cite their possible relevance to re-ionization and the possibility they offer of tracing the contribution by individual polluters, such as single supernovae (SNe).

In his talk, K. Schlaufman made a distinction between stars belonging to population III.1 (really first stars) and III.2 (star that formed from material with no metals but within the UV radiation from III.1 stars). In any case, he underlined that the very first stars are expected in very biased environments, such as possibly the core of giant ellipticals. In the local environment (Milky Way and its neighbours), the most biased ambient is the bulge. This is the place where the very first stars should be looked. I notice that in order to properly design the searches for the very "first stars", we should understand how many "first stars" do we really expect. Answering this question requires a dedicated study.

There were several very interesting talks on extremely metal-poor stars, stemming from the nice review by A. Frebel. Various contributions concerned observations (W. Aoki, D. Yong, H. Li, T. Hansen, K. Venn, V. Hill), but there were also suggestive talks on models (C. Kobayashi, D. Romano, E. Starkenburg). Several authors remarked that we are finally reaching the limit of what can be observed. Keller *et al.* (2014) documented a star with [Fe/H]< -7 that is actually an upper limit because no iron line was detected in the spectrum. This prompted the question about what is the limit that can be measured? As shown in the conference, the technical limits are [Ca/H]=-9.4, [Fe/H]=-7.2, [Mg/H]=-6, and we are now close to these limits.

Most attention during the symposium was focused on the carbon enhanced metal-poor stars with no enhancement of the s-process nuclei (the so-called CEMP-No star). Some of them are the most Fe-poor objects known. For this reason these objects are considered the best proxy for the "first stars". However, as a person quite far from the field I have a somewhat different point of view: in fact, I remark that metal-poor C-rich stars are not that metal-poor in terms of overall metal content Z. Rather, their composition indicates that the production of C and Fe are decoupled at low metallicity. However, I notice that this happens also at high metallicity. Finally I notice that also the production of Ba (and possibly Sr too) is decoupled from that of Fe at low metallicity.

4. Globular clusters

Globular clusters (GCs) are fundamental tracers of the halo. At the meeting there were nice observational reviews updating on the most recent investigations concerning both Milky Way (E. Carretta, G. Piotto, G. Da Costa, D. Geisler, A. Milone) and external galaxies (S. Larsen, M. Hilker, J. Brodie). These talks showed that there is a strong synergy between (UV) photometry and spectroscopy in understanding and quantifying the overall pattern. There were also interesting theoretical contributions by C. Charbonnel and O. Gnedin. Summarizing, it is now clear that (all?) GCs host multiple stellar populations with significant differences in He, CNO, Na, Mg, Al, and sometimes Si and K abundances. Most likely, stars in GCs show discrete or at least clumpy distributions in the abundances of these elements. Simplifying, there are two main populations of stars in GCs: those with a composition indistinguishable from that of field halo stars and those that are enriched in Na and Al, and depleted in O and Mg. The first group is thought to have the original composition from which the GC formed, and are then called First Generation (FG) stars, while the second group is found to have their composition altered by H-burning at high temperature in as yet not clearly identified group of FG stars, and is then thought to have formed later as a second generation (SG). SG stars make up about 2/3 of the stars of GCs, but are rare in the general field.

For what concern their relation with the halo, it was remarked that GCs now make up some 1.2% of the halo, but they were surely more massive in the past. A conservative estimate (based on the fraction of SG stars found in the field, about 2.5%, and on the assumption that they were all lost from GCs) is that $\sim 5\%$ of the halo stars comes from GCs. However, it was commented that this might represent a substantial underestimate of the contribution of GCs to the formation of the halo that maybe even as large as 50%

if FG had indeed initially 10 more mass then SG as given by most scenarios for multiple populations (only a fraction of the FG stars likely produced the material from which the SG stars formed).

Many points in this overall scenario are still not well understood. First of all, how GCs formed? Many cosmologists think of pre-galactic formation (at least for blue sequence GCs) linking them to reionization. However, no GC star so far was found with $Z < 0.003\ Z_{\odot}$. In his talk, Gnedin remarked that the formation of GCs seems to be related to gas-rich mergers. The case of the Antennae galaxies, where many very massive young stars clusters are observed, may support this view. Another interesting issue is the presence of two sequences of GCs (blue and red GCs) in a large fraction of the external galaxies. How to produce these blue and red sequences? Most authors explain the presence of these sequences as a consequence of a given merging history: blue, metal-poor GCs are generally related to the halo or dry mergers, while red, metal-rich clusters are thought to be related to the bulge or thick disk, and in general to gas-rich mergers. If this scenario is correct, why the presence of two sequences is so common? We might indeed expect a wide range of formation histories leading to a correspondent range in the properties of the GC systems. J. Brodie discussed this point in her talk and proposed a general two-phase formation scheme for galaxies. This is very intriguing; in my opinion we should also understand why the specific frequency of GCs in galaxies and in different populations in a given galaxy have such large variations as observed.

In addition, it seems that there is a relation between at least some GCs and the nuclear clusters (see the talk by G. Da Costa). On turn, this relates to how nuclear clusters form (in situ or migrated) and on what is the relation between the nuclear star clusters and (super?)-massive BHs in the central regions of galaxies. Of course, it is possible that GCs are only related to nuclear clusters in dwarf galaxies, however this point certainly merits further investigations.

Though this symposium was devoted to the halo, I think it worth opening here a parenthesis on the bulge. Gnedin suggested that possibly a single large late gas-rich merger have produced the metal-rich GCs in the Milky Way. Since the metal-rich GCs in the MW are mainly in the bulge, is this indicating a merger origin for the bulge? This looks quite different from what Gerhard said in his talk: "The bulge/bar formed from perturbation of the (thick?) disk with little or no evidence for a previous bulge". Is there a tension between these two points of view? Still about the bulge: not mentioned during the meeting is the evidence that the bulge is old from the colour-magnitude diagram (Zoccali *et al.* 2003).

Summarizing, most have still to be understood about the early evolution of GCs. The observations (Carretta, Piotto, Milone) indicate that once they formed they had a complex evolution. C. Charbonnel observed that powerful explosions are required to empty of gas young massive clusters (observed) and GCs (presumed from lack of contribution by SNe to next generations). I note here that GCs are so massive that the high end of the stellar initial mass function is very well sampled. This leaves ample room for the formation of stars of very high mass, the most probable precursors of such powerful explosions.

5. Dwarf spheroidals

Dwarf spheroidals are another major constituents of the (outer) galactic halo. Many new dSph's have been discovered recently both in the Milky Way and in nearby galaxies. The talk by J. Simon presented an updated census and discussed the main parameters of these clusters. G. Battaglia discussed the very hot issue of the structure of the dSphs: are they cusped (as predicted for simple dark matter structures) or rather have isothermal cores (as possible if baryon feedback play a very important role)? She discussed how the impact of anisotropy should be properly understood before any conclusion can be drawn. Her conclusion is that at the current status cores seem favoured. This is likely very important for an understanding of the missing satellite issue because cores are more easily destroyed than cusps (talks by Gerhard and Simon).

V. Hill showed that metal-rich populations tend to be more centrally concentrated in dSph's. This suggested to me the question if some of the dSph have formed itself from accretion of smaller fragments. Personally, I think that good candidates for a similar scenario are Sagittarius and Fornax dSph's, not only because they are larger, but also because they host several GCs that might be the signature of a dramatic past (see previous Section). Indeed the Sagittarius dSph seem to have been in origin a medium-sized galaxy (original $M_V \sim -15$: A. Ferguson). Also the history of Carina dSph seem to include accretion (Hill).

The chemical composition of dSph's was also discussed. Metal-rich stars in dSph's have a composition clearly different from the bulk of the (inner) halo; they are also clearly too young (V. Hill). There is a wide diversity of dSph's but yet scaling relations are valid. The colour-magnitude diagrams allow to derive the star formation histories, showing sometimes evidence for different star formation episodes. On the whole, however, there is a tight $[\mathrm{Fe/H}]_{\mathrm{bulk}} - L/L_{\odot}$ relation (Kirby *et al.* 2013). The internal history of metal-enrichment shows that the decrease in [alpha/Fe] ratio does not occur at the same [Fe/H] for all dSph's: the knee follows the luminosity (V. Hill). While there is almost no dispersion in the metal-to-metal abundance ratios at high metallicity ($[\mathrm{Fe/H}] > -2$) for the massive dSph, there is for small ones.

K. Venn discussed the increased evidence for the presence of extremely metal-poor stars in dSph's. She showed that the dispersion in heavy elements, e.g. Ba, is huge al low metallicity. Given the small mass of the dSph's considered in the study she presented, this can be attributed to the fact that we are sampling the enrichment of a single SN in each case. In a few cases very low Mg abundances are found; they are interpreted as due to enrichment only from SNIa (Fe-enriched from SNIa pockets). In addition, J. Simon presented the discovery of CEMP-No stars in dSph. D. Romano discussed modelling of these results. The chemical composition of dSph's stars requires the presence of large outflows, but these outflows maybe Fe-rich (Recchi *et al.* 2001). The contribution by single SNe may explain outliers.

6. What's next

First I notice that there was much more at this meeting. P. Schechter provocatively asked why are galaxies on a 2-D fundamental plane while DM halos are on a 1-D? J. Knapan presented cases of truncation in face-on galaxies. J. Penarrubia presented a new approach to the study of dynamics that he called "gravitational mechanics", based on dynamical invariants, introducing Gaussians diffusion coefficients. B. Elmegreen discussed the accretion vs star formation in dwarf galaxies. L. Greggio presented the case for studies of stellar populations with future instruments (ELT, JWST).

6.1. *The halo and the disk*

There are then a few points not discussed in the Symposium that personally found of interest. For instance, it is well known that the Milky Way disk metallicity did not change significantly in the last ~ 10 Gyr; this implies that the gas fraction stayed roughly constant at about 10%. But gas is transformed into stars at about 1 $\mathrm{M}_{\odot}/yr$ rate. On the other hand, there is strong evidence that the Milky Way has not experienced major mergers in the last $\sim 8-10$ Gyr. This requires continuous infall of gas at roughly this rate: about 10^{10} $M_\odot$ of gas should have been accreted by the disk during this epoch. Most of this gas must have been in the form of small clouds. From where all this gas came?

6.2. *The Magellanic Clouds*

I also found rather surprising that there was almost no mention during the Symposium of the fact that by far the two most conspicuous objects in the Milky Way halo are the Magellanic Clouds: in total, they contain $\sim 10^{10}$ $M_\odot$ of baryonic matter, that is more than 10 times the rest of the halo. Though they are probably losing gas (possibly contributing to the infall in our galaxy) they are still gas-rich objects. This possibly indicates that they were not so close to the MW up to a quite recent past. However, I expect the destiny of the Magellanic Clouds is to be accreted by the MW on a Gyr or less scale. This will be a significant merger event, and will possibly change completely the MW halo.

6.3. *List of additional questions*

Finally, what about the (thick) disk/halo interface and the origin of the stellar disk? Was there a smooth transition between early accretion and disk formation (what I call a dissipative component)? Was not there a late bombardment phase? Are dry mergers a good approximation for the early formation of the halo (M. Hirschman)? Why most satellites are on well defined planes (J. Simon, A. Helmi)? What instruments for the future (L. Greggio)?

Summarizing, this meeting was of exceedingly interest, but the potential of development of or knowledge about the halo of our Galaxy is far from been exhausted. In the next few years we expect dramatic progresses from the new spectroscopic survey such as APOGEE, LAMOST, GES, and moreover from the Gaia mission. Also modelling may progress thanks to a better understanding of the feedback mechanisms in galaxy formation and evolution. We expect then a revolution in the field and we should be ready to accept new ideas and scenarios. So, good work to every one!

References

Ibata, R.A., Lewis, G.F., McConnachie, A.W., Martin, N.F., Irwin, M.J., *et al.* 2014 ApJ 780, 128

Keller, S. C., Bessell, M. S., Frebel, A., Casey, A. R., Asplund, M., *et al.* 2014, *Nature*, 506, 463

Kirby, E. N., Cohen, J. G., Guhathakurta, P., Cheng, L., Bullock, J. S., & Gallazzi, A. 2013, ApJ, 779, 102

Mackey, A. D., Huxor, A. P., Ferguson, A. M. N., Irwin, M. J., Tanvir, N. R., *et al.* 2010, *MNRAS*, 401, 533

Recchi, S., Matteucci, F., & D'Ercole, A, 2001, *MNRAS*, 322, 800

Tissera, P. B., White, S. D. M., & Scannapieco, C. 2012, *MNRAS*, 420, 255

Tissera, P. B., Scannapieco, C., Beers, T. C., & Carollo, D. 2013, *MNRAS*, 432, 3391

Tollerud, E. J., Bullock, J. S., Strigart, L. E., & Willman, B. 2008, ApJ 688, 277

Zoccali, M., Renzini, A., Ortolani, S., Greggio, L., Saviane, I., Cassisi, S., *et al.* 2003, A&A, 399, 931

Zolotov, A., Willman, B., Brooks, A. M., Governato, F., Brook, C. B., *et al.* 2009, ApJ, 702, 1058

Zolotov, A., Willman, B., Brooks, A. M., Governato, F., Hogg, D. W., *et al.* 2010, ApJ, 721, 738

The General Assembly of Galaxy Halos: Structure, Origin and Evolution
Proceedings IAU Symposium No. 317, 2015
A. Bragaglia, M. Arnaboldi, M. Rejkuba & D. Romano, eds.

doi:10.1017/S174392131600003X

The Milky Way, the Galactic Halo, and the Halos of Galaxies

Ortwin Gerhard

Max-Planck-Institut für Extraterrestrische Physik, Postsach 1312, 85741 Garching, Germany

Abstract. The Milky Way, "our" Galaxy, is currently the subject of intense study with many ground-based surveys, in anticipation of upcoming results from the Gaia mission. From this work we have been learning about the full three-dimensional structure of the Galactic box/peanut bulge, the distribution of stars in the bar and disk, and the many streams and substructures in the Galactic halo. The data indicate that a large fraction of the Galactic halo has been accreted from outside. Similarly, in many external galaxy halos there is now evidence for tidal streams and accretion of satellites. To study these features requires exquisite, deep photometry and spectroscopy. These observations illustrate how galaxy halos are still growing, and sometimes can be used to "time" the accretion events. In comparison with cosmological simulations, the structure of galaxy halos gives us a vivid illustration of the hierarchical nature of our Universe.

Keywords. Galaxies: general, Milky Way, halos, formation, kinematics and dynamics, structure.

---

1. Introduction: the Milky Way as a galaxy

Our Galaxy and its satellites together with the similarly massive Andromeda galaxy and its satellite system constitute the Local Group, a loose galaxy group in the outer reaches of the Virgo cluster. The Milky Way is a barred spiral galaxy with stellar mass $M_* \simeq 6 \times 10^{10}\, M_\odot$, near the massive end of the distribution (Lange *et al.* 2015). It has relatively low gas content and star formation rate, which places it in the green valley of the distribution of similar galaxies found in the SDSS survey (Licquia & Newman 2015). The central bulge is dominated by a box/peanut structure showing that most of the bulge formed from the disk through dynamical evolution processes. Together with the embedded disk the stellar mass in the bulge region is $\sim 1.6 \times 10^{10}\, M_\odot$, nearly 30% of the total (Portail *et al.* 2015). The box/peanut bulge is the three-dimensional part of the Galactic bar, which extends outwards in the disk to $R \sim 5$ kpc and has a mass of order additional $10^{10} \mathrm{M}_\odot$ (Wegg *et al.* 2015). The Galactic stellar disk includes almost all of the remaining stellar mass. Its dynamical structure is well-studied in the solar neighbourhood, located at $\simeq 8.2$ kpc from the center, but its radial scale-length, while relatively short, 2-3 kpc, is not accurately known. Beyond the solar radius, the outer disk has a prominent warp. The disk is surrounded by a halo of old, low-metallicity stars with only $\sim 1\%$ of the stellar mass, and some 50 satellite galaxies including the prominent Magellanic clouds and Sagittarius dwarf. See Bland-Hawthorn & Gerhard (2016) for a more extensive discussion and additional references.

2. The Galactic halo - observations and predictions

The Milky Way's stellar halo was first identified as a population of old, high-velocity, metal poor stars near the Sun, similar to the stars in globular clusters. The halo stars showed large random motions, little if any rotation, and a spheroidal to spherical spatial distribution. Following the influential paper of Eggen *et al.* (1962), a classical view of the halo developed as a smooth envelope of ancient stars from the time when the Galaxy first collapsed. Today, thanks to large stellar surveys such as SDSS, the stellar halo is seen as a complex structure with multiple components and unrelaxed substructures, which still accretes matter in the form of smaller galaxies. These are then tidally disrupted in the gravitational field, as is well-illustrated in the famous field of streams (Belokurov *et al.* 2006).

This complex structure is consistent with hierarchical galaxy formation models which predict that the Milky Way should have accreted of order ~ 100 satellite galaxies. Irregular density distributions are predicted in the outer halo due to shells and tidal streams from the satellite galaxies disrupting in the tidal field. At those radii, the time required for stars to complete multiple orbits and spread throughout the halo volume is a significant fraction of the age of the Galaxy. In simulations, the majority of the halo is often built at early times, ~ 10 Gyr ago, and most of the stellar halo stars come from the disruption of one or a few massive satellites accreted early-on. The outer halo is built more recently than the inner halo and the halo properties evolve as it builds up by the accretion (Bullock & Johnston 2005; Cooper *et al.* 2010). Some of the inner halo may have formed *in situ*, i.e., within the main body of the Galaxy (Abadi *et al.* 2006). Recent simulations suggest that a fraction of stars formed in the early Galactic disk could have been ejected into the inner halo, and further in situ halo stars could have formed from gas stripped from infalling satellites (McCarthy *et al.* 2012; Pillepich *et al.* 2015). Observational evidence for different properties of the inner and outer halo is presented by Carollo *et al.* (2007); Beers *et al.* (2012).

Various samples of tracer stars with photometric distances show that the overall density profile of the stellar halo steepens at $r \sim 25$ kpc. The data favour a double power law model, with an inner slope $\alpha_{\rm in} \simeq -2.5$ and an outer slope $\alpha_{\rm out} \simeq -(3.7 - 5.0)$, depending somewhat on the tracers used (Faccioli *et al.* 2014; Pila-Díez *et al.* 2015). The halo is slightly flattened, with axial ratio increasing from $q_{\rm in} \simeq 0.65$ to $q_{\rm out} \simeq 0.8$. Studies of a large number of stars with radial velocities and proper motions find that the stellar velocity ellipsoid of nearby halo stars is aligned with spherical coordinates, and quite strongly radially anistropic (Smith *et al.* 2009; Bond *et al.* 2010). The radial velocity dispersion profile falls quickly from a local $\sigma_r = 140$ km s^{-1} to $\sigma_r = 100$ km s^{-1} at $r \sim 20$ kpc, then remains approximately flat to $R \sim 50$ kpc, and finally decreases to $\sigma_r \sim 35$ km s^{-1} at ~ 100 kpc (Battaglia *et al.* 2005; Kafle *et al.* 2014), suggesting perhaps a tidal truncation at the largest radii.

From the density profile and kinematics of these tracer stars the mass of the Milky Way's dark matter halo can be determined. Other methods involve the kinematics of globular clusters and dwarf galaxies, estimation of the local escape velocity, and modelling of halo streams. Overall, the resulting mass of the Galaxy is still quite uncertain; in Bland-Hawthorn & Gerhard (2016) we estimate $M_{\rm vir} = 1.3 \pm 0.3 \times 10^{12}$ M$_\odot$ from the various studies.

In the stellar density maps, many streams, tails, and overdensities are seen. The four largest structures are the Sgr stream, the Galactic anticentre structure, the Virgo cloud, and the Hercules-Aquila cloud. Together these structures have a mass of 2-3 $\times\, 10^8\, M_\odot$ (Belokurov *et al.* 2013). The Sgr stream alone provides $\sim 20\%$ of all the halo debris, including probably ~ 20 globular clusters. Bell *et al.* (2008) found with SDSS main sequence turn-off stars that the rms deviation of the star counts around a smooth model is $\sim 40\%$ of the model itself, suggesting that a large fraction of the halo stars are in substructures. However, using blue horizontal branch stars as tracers, Deason *et al.* (2011) find a lower ratio, $\sim$5-20%. The total stellar mass in the halo is estimated $M_s = 4\text{-}7 \times 10^8\, M_\odot$ (Bland-Hawthorn & Gerhard 2016), about one percent of the Galaxy's stellar mass.

In simulations of stellar halos such as the Aquarius suite (Cooper *et al.* 2010), a significant galaxy-to-galaxy variance in the halo properties is observed, which is due to the different accretion histories of the simulated galaxies. Thus a full picture of the halo properties requires studying external galaxies.

3. Stellar halos in nearby galaxies

Deep images of stellar halos in spiral galaxies (Martínez-Delgado *et al.* 2010) show a variety of features indicating accretion of satellite galaxy stars, such as streams, spikes, wedges, giant plumes, partially disrupted satellites, umbrella-shaped tidal debris and other tidal clouds. With deep imaging it is possible to detect emission down to $\mu_V \sim 27$-29 mag arcsec^{-2} (see van Dokkum *et al.* 2014), however, various uncertainties due to flat fielding, sky background, bright star halos, or Galactic cirrus need to be controlled. Spectroscopy is generally possible only to $\mu_V \sim 25.5$ mag arcsec^{-2}, so information on the accreted stellar population is limited to colours, and kinematics is often only available in the inner halos.

For the nearest galaxies where resolved stars can be observed ($\sim$5 Mpc from the ground, $\sim$12 Mpc from space), one can probe to very deep surface brightness levels ($\mu_V \sim 33$ mag arcsec^{-2}), constrain the stellar population from colour-magnitude diagrams, and obtain kinematics for bright stars. A beautiful example is the Pandas survey of Andromeda (M31), showing that the M31 halo is dominated by debris substructure (e.g., Ferguson *et al.* 2002). Combining a large number of fields out to 200 kpc radius, Ibata *et al.* (2014) constructed stellar density and metallicity maps. These show an outward gradient in the overall metallicity of the M31 halo, a strong dependence of substructure on metallicity, and an ancient, low-metallicity ([Fe/H]< -1.7), approximately smooth, and near-spherical halo component.

Results from the GHOSTS survey of nearby spiral galaxies with the Hubble Space Telescope were recently published by Monachesi *et al.* (2016). In these more distant galaxies, information is generally restricted to a number of fields along the major and minor axes of the galaxy, reaching distances of typically $\sim$50 kpc from the galaxy center. The color profiles determined from these data show considerable diversity; field-to-field variations indicate stellar population variations such as would be expected if the halos had been built from several diverse accreted objects. The color and inferred metallicity profiles can be flat or falling, without clear correlation with galaxy stellar mass or rotation velocity, consistent with the galaxy-to-galaxy scatter predicted by simulations. By comparison with this sample of 6 galaxies, the Milky Way halo has a low metallicity and a weak gradient, while the M31 halo has high metallicity and a prominent gradient.

The seminal deep imaging study of Mihos *et al.* (2005) of the extended halos of early-type galaxies (ETGs) in the Virgo cluster core has been extended to larger galaxy samples by Tal *et al.* (2009) and Duc *et al.* (2015). The deep surface brightness maps show a variety of streams, shells, and other tidal features at large radii, indicating on-going accretion also in these more massive galaxies. Simulations such as those of Cooper *et al.* (2013) using particle-tagging methods predict that the fraction of accreted stars in galaxies is a strong function of mass, ranging from a few percent for Milky Way galaxies to 80-90% in the most massive ETGs. Indeed, the stellar halo in the Milky Way is a minor component by mass, but in giant galaxies like M87 in the Virgo cluster, almost all stars may have been accreted.

This is related to the observed size evolution of massive galaxies between $z=2$ and the present epoch (Trujillo *et al.* 2007; van Dokkum *et al.* 2010). The latter authors construct stacked surface density profiles for samples of galaxies at constant number density to show that the profiles grow more extended with decreasing redshift, from Sersic index $n \simeq 2$ at $z=2$ to $n \simeq 6$ at $z=0$, corresponding to an increase in the effective radius by a factor

4-5. The size evolution is believed to be driven by minor merger driven accretion (Naab *et al.* 2009; Oser *et al.* 2010).

In the outer halos kinematic measurements are possible only for bright tracers, planetary nebulae (PNs) and globular clusters (GCs). PNs are direct descendants of halo stars and generally follow light; they reproduce stellar kinematics in regions of overlap (Coccato *et al.* 2009). For very old, metal-rich stellar populations however the specific PN-frequency can be 2-3 times lower (Longobardi *et al.* 2015a). GCs split into red and blue populations, where the blue GCs trace halo and dwarf galaxy light, and the red GCs bulge light. The kinematics of red GCs is often similar to stellar kinematics (Pota *et al.* 2013) but not always (Coccato *et al.* 2013). In M31 the outer halo GCs lie preferentially on streams but are underrepresented on the diffuse halo light (Veljanoski *et al.* 2014). Thus PNs and GCs appear to trace somewhat different modes of accretion.

A tantalizing case of accretion has recently been uncovered in the Virgo-central ETG M87. In the projected phase-space of some 300 PNs Longobardi *et al.* (2015b) found a chevron structure extending to a radius of $\simeq 90$ kpc where both sides of the chevron meet at the systemic velocity. Reexamination of deep images from Mihos *et al.* (2005) revealed a giant crown-like substructure extending over 20 kpc $\times$ 60 kpc just inside this radius, which consists of bluer stars than the main M87 halo, and locally increases the halo surface brightness by $\sim 60\%$. From the PN kinematics the satellite appears to have come in on a near-radial orbit extending across much of the galaxy, about ~ 1 Gyr ago (based on simulations by Weil *et al.* 1997), such that the crown is located in the region where the debris orbits reach apocenter. This is direct evidence that the M87 outer halo is presently growing by accretion, and suggests that beyond $r \sim 70$ kpc it is no longer in dynamical equilibrium. A somewhat similar chevron was found in the GCs by Romanowsky *et al.* (2012); this appears to reach to larger radii and no photometric correspondent has been found, so that the relation to the crown remains unclear. At yet larger radii several streamers are seen in the deep image, apparently falling into M87; these are too faint for PNs but GCs have been found on one. Beyond about $r \sim 50$ kpc the PN velocity distribution shows an additional kinematic component with velocity dispersion ~ 1000 km s^{-1}. This intracluster component has a much shallower density profile than the M87 halo proper, and dominates beyond $r \sim 150$ kpc (Longobardi *et al.* 2015a).

PNs and GCs have been independently used to determine the angular momentum and mass distributions in ETG halos. λ_R-profiles from the PN.S survey (Arnaboldi *et al.* in prep.) show that ETGs which are slow rotators at small radii also rotate slowly in their halos, whereas fast rotators can have rising, constant, and sometimes falling λ_R-halo profiles (Coccato *et al.* 2009). GCs from the SLUGGS project (Brodie *et al.* 2014) show very similar λ_R-profiles (Pota *et al.* 2013), and both are in good agreement with predicted λ_R-profiles from cosmological resimulations of ETG halos (Wu *et al.* 2014). Dynamical modelling of the halo kinematic data has shown that dark matter dominates beyond $\sim 5R_e$ (de Lorenzi *et al.* 2008; Morganti *et al.* 2013; Cappellari *et al.* 2015), again in good agreement with the simulated ETGs.

Stellar population studies from integrated light show metallicity gradients to $\sim 2.5R_e$, consistent with the accretion of old, metal-poor stars into the halos of ETGs (Greene *et al.* 2012). In the halo of the central galaxy NGC 3311 in the Hydra I cluster, dissolution of two satellite galaxies is on-going (Arnaboldi *et al.* 2012), and deep spectra show that the accreted dwarf galaxy HCG 026 indeed adds very old and metal-poor stars to the BCG halo (Coccato *et al.* 2011). Such studies will be greatly advanced by deep integral-field spectroscopy with the planned E-ELT and TMT giant telescopes.

4. Conclusions

The study of galaxy halos is challenging because of the faint surface brightnesses in the halos, but offers much interesting insight into the late accretion and formation histories of galaxies. In the Milky Way halo streams, tails, and substructures show the importance of satellite accretion as predicted by hierarchical galaxy formation models. Smooth spatial and velocity structures in the inner halo indicate efficient mixing, while the steep outer density profile and declining velocity dispersion lead to a relatively low virial mass. In the near future, we expect substantial progress based on the astrometric data currently assembled by the Gaia mission. From the distance and proper motion measurements for halo stars it will be possible to study stellar streams in the halo in much greater detail than before and obtain unprecedented constraints on the Galaxy's accretion history and halo mass distribution.

The Milky Way is a massive spiral galaxy in a relatively isolated environment, going through the late stages of its evolution. It has low gas content and star formation rate, and a prominent box/peanut bulge and bar in the disk reaching 5 kpc radius. Simulations predict significant cosmic variance in galaxy accretion histories and stellar halo structure even at constant dark matter mass. Studies of external galaxy halos have confirmed some of this variance, e.g., in the difference between the Milky Way and M31 halos, and have revealed signatures of accretion at faint surface brightness around both spiral and elliptical galaxies. In massive galaxies, accreted halos are predicted to dominate over in situ stars, consistent with the size evolution observed in luminous ETGs. The few accreted satellites studied in ETG halos have been found to be old and metal-poor. Much work remains to be done to obtain a more complete picture of the accretion histories in galaxy halos and of the properties of the accreted satellites and their stars.

References

Abadi, M. G., Navarro, J. F., & Steinmetz, M. 2006, *MNRAS*, 365, 747
Arnaboldi, M., *et al.* 2012, *Astron. Astrophys.*, 545, A37
Battaglia, G., *et al.* 2005, *MNRAS*, 364, 433
Beers, T. C., *et al.* 2012, *Ap. J.*, 746, 34
Bell, E. F., *et al.* 2008, *Ap. J.*, 680, 295
Belokurov, V., *et al.* 2006, *Ap. J.*, 642, L137
—. 2013, *MNRAS*, 437, 116
Bland-Hawthorn, J. & Gerhard, O. 2016, *ARAA*, eprint arXiv:1602.07702
Bond, N. A., *et al.* 2010, *Ap. J.*, 716, 1
Brodie, J. P., *et al.* 2014, *Ap. J.*, 796, 52
Bullock, J. S. & Johnston, K. V. 2005, *Ap. J.*, 635, 931
Cappellari, M., *et al.* 2015, *Ap. J.*, 804, L21
Carollo, D., *et al.* 2007, *Nature*, 450, 1020
Coccato, L., Arnaboldi, M., & Gerhard, O. 2013, *MNRAS*, 436, 1322
Coccato, L., Gerhard, O., Arnaboldi, M., & Ventimiglia, G. 2011, *Astron. Astrophys.*, 533, A138
Coccato, L., *et al.* 2009, *MNRAS*, 394, 1249
Cooper, A. P., *et al.* 2010, *MNRAS*, 406, 744
—. 2013, *MNRAS*, 434, 3348
de Lorenzi, F., *et al.* 2008, *MNRAS*, 385, 1729
Deason, A. J., Belokurov, V., & Evans, N. W. 2011, *MNRAS*, 416, 2903
Duc, P.-A., *et al.* 2015, *MNRAS*, 446, 120
Eggen, O. J., Lynden-Bell, D., & Sandage, A. R. 1962, *Ap. J.*, 136, 748
Faccioli, L., *et al.* 2014, *Ap. J.*, 788, 105
Ferguson, A. M. N., *et al.* 2002, *Astron. J.*, 124, 1452

Greene, J. E., *et al.* 2012, *Ap. J.*, 750, 32
Ibata, R. A., *et al.* 2014, *Ap. J.*, 780, 128
Kafle, P. R., Sharma, S., Lewis, G. F., & Bland-Hawthorn, J. 2014, *Ap. J.*, 794, 59
Lange, R., *et al.* 2015, *MNRAS*, 447, 2603
Licquia, T. C. & Newman, J. A. 2015, *Ap. J.*, 806, 96
Longobardi, A., Arnaboldi, M., Gerhard, O., & Hanuschik, R. 2015a, *Astron. Astrophys.*, 579, A135
Longobardi, A., Arnaboldi, M., Gerhard, O., & Mihos, J. C. 2015b, *Astron. Astrophys.*, 579, L3
Martínez-Delgado, D., *et al.* 2010, *Astron. J.*, 140, 962
McCarthy, I. G., *et al.* 2012, *MNRAS*, 420, 2245
Mihos, J. C., Harding, P., Feldmeier, J., & Morrison, H. 2005, *Ap. J.*, 631, L41
Monachesi, A., *et al.* 2016, *MNRAS*, 457, 1419
Morganti, L., *et al.* 2013, *MNRAS*, 431, 3570
Naab, T., Johansson, P. H., & Ostriker, J. P. 2009, *Ap. J.*, 699, L178
Oser, L., *et al.* 2010, *Ap. J.*, 725, 2312
Pila-Díez, B., *et al.* 2015, *Astron. Astrophys.*, 579, A38
Pillepich, A., Madau, P., & Mayer, L. 2015, *Ap. J.*, 799, 184
Portail, M., Wegg, C., Gerhard, O., & Martinez-Valpuesta, I. 2015, *MNRAS*, 448, 713
Pota, V., *et al.* 2013, *MNRAS*, 428, 389
Romanowsky, A. J., *et al.* 2012, *Ap. J.*, 748, 29
Smith, M. C., *et al.* 2009, *MNRAS*, 399, 1223
Tal, T., van Dokkum, P. G., Nelan, J., & Bezanson, R. 2009, *Astron. J.*, 138, 1417
Trujillo, I., *et al.* 2007, *MNRAS*, 382, 109
van Dokkum, P. G., Abraham, R., & Merritt, A. 2014, *Ap. J.*, 782, L24
van Dokkum, P. G., *et al.* 2010, *Ap. J.*, 709, 1018
Veljanoski, J., *et al.* 2014, *MNRAS*, 442, 2929
Wegg, C., Gerhard, O., & Portail, M. 2015, *MNRAS*, 450, 4050
Weil, M. L., Bland-Hawthorn, J., & Malin, D. F. 1997, *Ap. J.*, 490, 664
Wu, X., *et al.* 2014, *MNRAS*, 438, 2701

The General Assembly of Galaxy Halos: Structure, Origin and Evolution
Proceedings IAU Symposium No. 317, 2015
A. Bragaglia, M. Arnaboldi, M. Rejkuba & D. Romano, eds.

doi:10.1017/S1743921315008352

r-Process Elements as Tracers of Enrichment Processes in the Early Halo

Johannes Andersen[1,2], Birgitta Nordström[1,2] and Terese T. Hansen[3]

[1]Dark Cosmology Centre, The Niels Bohr Institute, University of Copenhagen, Juliane Maries Vej 30, DK-2100 Copenhagen, Denmark
email: ja@nbi.ku.dk, birgitta@nbi.ku.dk

[2]Stellar Astrophysics Centre, Department of Physics and Astronomy, Aarhus University, DK-8000 Aarhus C, Denmark

[3]Landessternwarte, ZAH, Heidelberg University, Königstuhl 12, Heidelberg, D-69117, Germany
email: thansen@lsw.uni-heidelberg.de

Abstract. Significant minorities of extremely metal-poor (EMP) halo stars exhibit dramatic excesses of neutron capture elements. The standard scenario for their origin is mass transfer and dilution in binary systems, but requires them to *be* binaries. If not, these excesses must have been implanted in them from birth by processes that are not included in current models of SN II chemical enrichment. The binary population of such EMP subgroups is a test of this scenario.

Keywords. Galaxy: halo – Stars: chemically peculiar – binaries: spectroscopic – ISM: structure

---

1. Why might peculiar halo stars be interesting?

The bulk of extremely metal-poor (EMP) halo stars ([Fe/H] < -2.5) exhibit extremely well-defined abundance ratios between all elements up to Fe – see Frebel & Norris (2015). This is commonly taken to indicate that the early halo was enriched by a single class of sources and was promptly and well mixed, simplifying models of galaxy evolution. Yet, subgroups of EMP stars exist with dramatic excesses of neutron capture elements produced by the r and/or s processes and – even more importantly! – of carbon.

A solution, which was supported by literature data and has since been taken for granted, is that the excess elements were synthesized in an initially more massive binary companion, which has transferred the processed material to the surviving star by Roche-lobe overflow and/or a stellar wind. But *are* these stars *all* binaries? If not, the early halo was more complex – and more interesting! – than current models assume.

2. What we did

Our observing programme was simple in design: Systematic (~monthly), precise radial velocity monitoring with the stable, bench mounted, fibre-fed échelle spectrograph FIES at the 2.5m Nordic Optical Telescope (NOT). Spectra of high resolution ($R \sim 45,000$) and low S/N ratio enable us to determine radial velocities with errors of ~100 m s^{-1} over eight years and confidently discriminate between single stars and binaries.

3. What we found, and why it matters

Our results were simple, but surprising: The frequency (~17±5%) and orbital properties of binaries enriched in r-process elements or in carbon but, unlike CEMP-s stars, not in s-process elements (CEMP-no stars) are *completely normal* (Fig. 1). In contrast, CEMP-s stars are ~80% binaries; ~20% remain single – see Hansen *et al.* (2015abc).

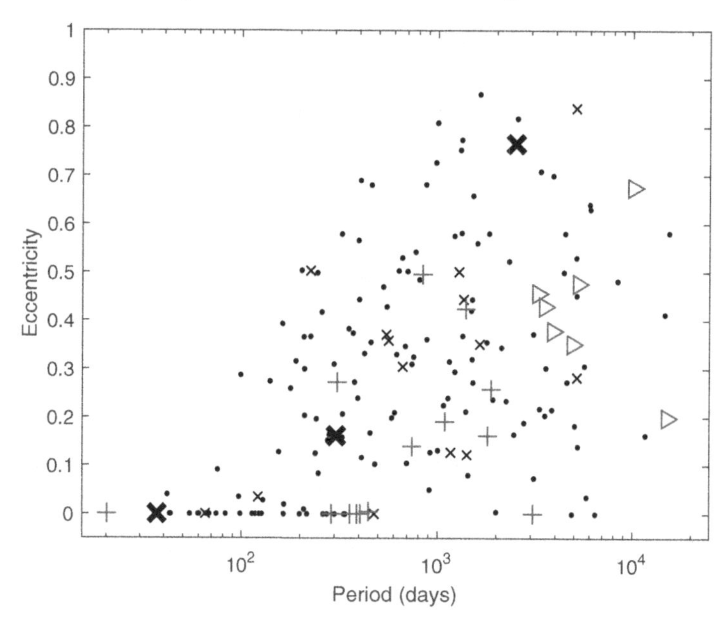

Figure 1. Period–eccentricity diagram for the binary systems observed in this project. Bold black crosses: Our three r-process enhanced binaries; red and blue crosses: CEMP-s binaries with and without(CEMP-no) s-process element enhancements, respectively (systems with $P > 3,000$ days are shown as right-pointing triangles since their true periods may be even longer). Black dots and small crosses: Comparison samples of giant binaries in Population I clusters: Mermilliod *et al.* (2007), Mathieu *et al.* (1990) and in Population II: Carney *et al.* (2003).

So the binary mass-transfer scenario can account for one class of peculiar halo stars, but by far not for all: the excess elements in the single stars must have been produced elsewhere and transported across interstellar distances in the ISM to pollute some, but not all of the natal clouds of long-lived EMP stars. For details, see Hansen *et al.* (2015abc). Yet another complexity in the life of galaxy modellers!

Acknowledgements:
This paper is based on observations made with the Nordic Optical Telescope, operated by the Nordic Optical Telescope Scientific Association at the Observatorio del Roque de los Muchachos, La Palma, Spain, of the Instituto de Astrofísica de Canarias. We thank many NOT staff members and students for readily obtaining most of the observations for us in service mode. We also thank our co-authors on the full papers on this work.
JA and BN gratefully acknowledge financial support from the Carlsberg Foundation and the Danish Natural Science Research Council, and TTH was supported by Sonderforschungsbereich SFB 881 "The Milky Way System" (subproject A4) of the German Research Foundation (DFG).

References

Carney, B. W., Latham, D. W., Stefanik, R. P., *et al.*, 2003, *AJ*, 125, 293
Frebel, A. & Norris, J. E., 2015, *ARAA*, 53, 631
Hansen, T. T., Andersen, J., Nordström, B., *et al.* 2015abc, *A&A*, in press
Mathieu, R. D., Latham, D. W., & Griffin, R. F. 1990, *AJ*, 100, 1899
Mermilliod, J.-C., Andersen, J., Latham, D. W., & Mayor, M., 2007, *A&A*, 473, 829

The General Assembly of Galaxy Halos: Structure, Origin and Evolution
Proceedings IAU Symposium No. 317, 2015
A. Bragaglia, M. Arnaboldi, M. Rejkuba & D. Romano, eds.

doi:10.1017/S1743921315008571

Origin of strong magnetic fields in Milky Way-like galaxies

Alexander M. Beck

University Observatory Munich
Scheinerstr. 1, D-81679 Munich, Germany
email: abeck@usm.lmu.de

Abstract. Magnetic fields are observed on all scales in the Universe (see e.g. Kronberg 1994), but little is known about the origin and evolution of those fields with cosmic time. Seed fields of arbitrary source must be amplified to present-day values and distributed among cosmic structures. Therefore, the emergence of cosmic magnetic fields and corresponding dynamo processes (see e.g. Zel'dovich *et al.* 1983; Kulsrud *et al.* 1997) can only be jointly understood with the very basic processes of structure and galaxy formation (see e.g. Mo *et al.* 2010).

Keywords. methods: numerical, galaxies: formation, galaxies: magnetic fields, early Universe.

Models and simulations of the magnetic field growth

The magnetic field is described by the induction equation of magnetohydrodynamics

$$\frac{\partial \mathbf{B}}{\partial t} = \nabla \times [\mathbf{v} \times \mathbf{B} - \eta \nabla \times \mathbf{B}] + \left.\frac{\partial \mathbf{B}}{\partial t}\right|_{\mathrm{seed}}, \tag{0.1}$$

where the description of astrophysical processes enters via the velocity field $\mathbf{v}$ and the plasma microphysics via the resistivity η. Furthermore, a magnetic seed field is required.

The dashed lines in Figure 1 show the analytical model for the primordial field growth (see Beck *et al.* 2012, for details) during galaxy formation with a solution of

$$B(a) = \frac{1}{a^2}\left[(4\pi\rho v_{\mathrm{turb}}^2)^{-1} + B_0^{-2} e^{-2\gamma t(a)}\right]^{-\frac{1}{2}}, \tag{0.2}$$

where a is the cosmological scale-factor, γ the growth rate, B_0 the initial field amplitude and $1/2\rho v_{\mathrm{turb}}^2$ the turbulent pressure as saturation level.

Another possibility is to create magnetic seed fields within the very first star forming objects by supernovae. This approach proposed by Rees (1987) and simulated by Beck *et al.* (2013a) gives very promising results. The dotted line in Figure 1 shows the contribution resulting only from supernovae without subsequent amplification within a Milky Way-like galactic halo. Given typical properties of the interstellar medium it approximates to

$$\dot{B}_{\mathrm{seed}} \approx B_{\mathrm{SN}} \left(\frac{r_{\mathrm{SN}}}{r_{\mathrm{SB}}}\right)^2 \left(\frac{r_{\mathrm{SB}}}{r_{\mathrm{Inj}}}\right)^3 \frac{\sqrt{\dot{N}_{\mathrm{SN}}\Delta t}}{\Delta t} \approx 10^{-9}\,\frac{\mathrm{G}}{\mathrm{Gyr}}. \tag{0.3}$$

The solid line in Figure 1 shows the results of the numerical simulations of Milky Way-like halo formation including magnetic fields of supernova origin. The simulations are based on the GADGET-3 code (Springel 2005) and a simple prescription for star formation (Springel & Hernquist 2003). We conclude that the magnetic fields of galaxies can be described as a result of the very basic processes of star and galaxy formation. The seed fields can be supplied by supernovae and the formation of structures yields

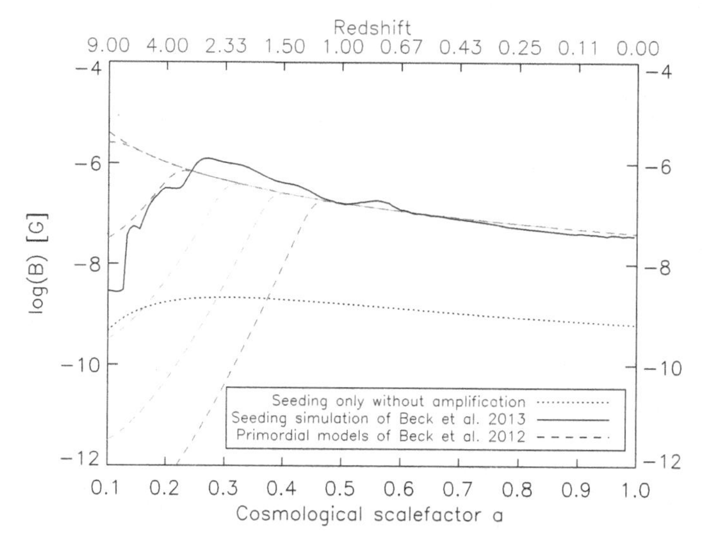

Figure 1. Mean amplitude of the magnetic field versus cosmic time composed from the analytical models of Beck *et al.* (2012) and numerical simulations of Beck *et al.* (2013a).

sufficient dynamo action to explain strong field amplitudes. However, with magnetic fields only present within collapsing and virialising structures, the question now arises how these fields can be transported into the large-scale structure and possible voids (Beck *et al.* 2013b).

Acknowledgments

This work is supported by the DFG Cluster of Excellence 'Universe', DFG Research Unit '1254' and by the 'Magneticum' project (http://www.magneticum.org).

References

Beck, A. M., *et al.*, 2012, *MNRAS*, 422, 2152
Beck, A. M., *et al.*, 2013a, *MNRAS*, 435, 3575
Beck, A. M., *et al.*, 2013b, *MNRAS*, 429, L60
Kronberg, P. P., 1994, Reports on Progress in Physics, 57, 325
Kulsrud, R. M., *et al.*, 1997, ApJ, 480, 481
Mo, H., *et al.*, 2010, Galaxy Formation and Evolution, Cambridge University Press
Rees, M. J., 1987, *QJRAS*, 28, 197
Springel, V. & Hernquist L., 2003, *MNRAS*, 339, 289
Springel V., 2005, *MNRAS*, 364, 1105
Zel'dovich, Y. B., *et al.*, 1983, Magnetic Fields in Astrophysics, Gordon and Breach, New York

The General Assembly of Galaxy Halos: Structure, Origin and Evolution
Proceedings IAU Symposium No. 317, 2015
A. Bragaglia, M. Arnaboldi, M. Rejkuba & D. Romano, eds.

doi:10.1017/S1743921315008716

Resolved Stellar Halos of M87 and NGC 5128: Metallicities from the Red-Giant Branch

Sarah A. Bird

LAMOST Fellow, Shanghai Astronomical Observatory,
80 Nandan Road, Shanghai, 200030, China, email: sarahbird@shao.ac.cn

Abstract. We have searched halo fields of two giant elliptical galaxies: M87, using *HST* images at 10 kpc from the galactic center, and NGC 5128 (Cen A), using VIMOS VLT images at 65 kpc from the center and archival *HST* data from 8 to 38 kpc from the center. We have resolved thousands of red-giant-branch (RGB) stars in these stellar halo fields using V and I filters, and, in addition, measured the metallicity using stellar isochrones. The metallicity distribution function (MDF) of the inner stellar halo of M87 is similar to that of NGC 5128's stellar halo.

Keywords. galaxies: halo, galaxies: elliptical, galaxies: stellar content, galaxies: individual: M87, galaxies: individual: NGC 5128

1. Introduction

The stellar halos of galaxies trace the earliest star-forming histories of galaxies and resolved stellar photometry of the red-giant population allow the measurement of the MDF. Currently, only galaxies within 20 Mpc can be studied by direct stellar photometry. Within this distance are only a few giant ellipticals, making them especially noteworthy to study. We compare in Table 1 the basic properties of two of such galaxies, M87 and NGC 5128, and the two halo fields which we have studied using space and ground-based images, respectively (Bird *et al.* 2010, 2015).

We have used the observed V and I magnitudes to derive the bolometric magnitude $M_{\rm bol}$ and to make cuts in magnitude and color, $-3.22 < M_{\rm bol} < -3.00$ and $1.5 < V-I < 2.3$, for the purpose of isolating a pure sample of RGB stars from the challenging ground-based observations of NGC 5128. We have applied the same selection window to the space-based observations of M87 and NGC 5128 and show the MDFs in Fig. 1.

Table 1. Comparison of M87 and NGC 5128 and our stellar halo observations.

Description	M87	NGC 5128
$I_{\rm TRGB}$ [mag]	27	24
Distance [Mpc]	16.7	3.8
Hubble Type	cD-gE	cD-gE/S0pec
Environment	Virgo Cluster	Centaurus Group
RA [$^{\rm h\ m\ s}$]	12 30 49.4	13 27 59
Dec [$^{\circ\ \prime\ \prime\prime}$]	+12 23 28	−42 14 50
$R_{\rm gc}$ [kpc]	12	65
$R_{\rm eff}$ [kpc]	6.3	5.8
Telescope	*HST*	VLT
Instrument	ACS WFC	VIMOS
Area [arcmin^2]	2	224
RGB	33890	1581

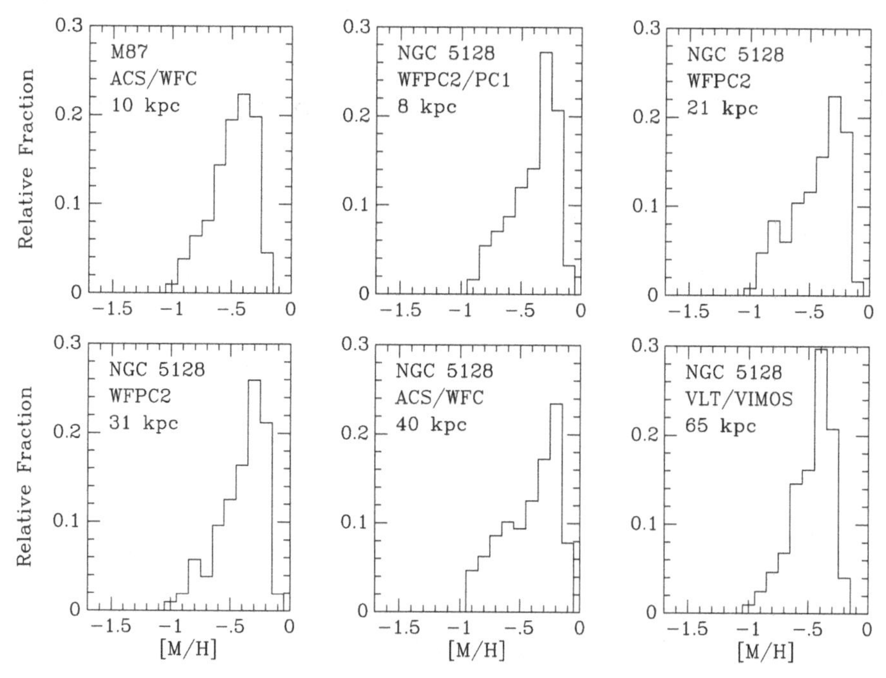

Figure 1. Normalized MDFs in [M/H] for the stellar halo fields of two elliptical galaxies: one 10 kpc field from M87 (Bird *et al.* 2010), and five fields from NGC 5128 including four *HST* fields at 8 kpc (Harris & Harris 2002), 21 kpc (Harris *et al.* 1999), 31 kpc (Harris & Harris 2000), and 40 kpc (Rejkuba *et al.* 2005), and one 65 kpc VLT/VIMOS field (Bird *et al.* 2015).

2. Results and Discussion

M87's inner halo MDF is similar to NGC 5128's halo MDFs, sharing peaks near [M/H] ~ -0.5. Super-solar metallicity stars are dimmer and substantially missed in the color magnitude diagrams and MDFs.

We have measured the metallicities from the RGB stars located in the halos of two elliptical galaxies, M87 and NGC 5128. The similarity of their MDFs reinforce the view that the oldest stellar populations in giant elliptical galaxies are fundamentally similar, regardless of whatever more recent accretion events may have happened to them. Further details of the studies presented here can be found in Bird *et al.* (2010) and Bird *et al.* (2015).

References

Bird, S., Harris, W. E., Blakeslee, J. P., & Flynn, C. 2010, *A&A*, 524, A71
Bird, S. A., Flynn, C., Harris, W. E., & Valtonen, M. 2015, *A&A*, 575, A72
Harris, G. L. H., & Harris, W. E. 2000, *AJ*, 120, 2423
Harris, G. L. H., Harris, W. E., & Poole, G. B. 1999, *AJ*, 117, 855
Harris, W. E., & Harris, G. L. H. 2002, *AJ*, 123, 3108
Rejkuba, M., Greggio, L., Harris, W. E., Harris, G. L. H., & Peng, E. W. 2005, *ApJ*, 631, 262

The General Assembly of Galaxy Halos: Structure, Origin and Evolution
Proceedings IAU Symposium No. 317, 2015
A. Bragaglia, M. Arnaboldi, M. Rejkuba & D. Romano, eds.

doi:10.1017/S1743921315006924

Subaru Hyper Suprime Cam Survey of the Andromeda Halo

Masashi Chiba[1], **Mikito Tanaka**[1] **and Yutaka Komiyama**[2]

[1]Astronomical Institute, Tohoku University, Sendai, Japan
email: chiba@astr.tohoku.ac.jp

[2]NAOJ, Mitaka, Tokyo, Japan

Abstract. We present a progress report on our deep and wide-field imaging survey of the Andromeda halo with Hyper Suprime Cam (HSC) mounted on Subaru. HSC is the upgraded prime focus camera after Suprime-Cam, having a field of view of 1.77 square degree (1.5 degree in diameter), namely about 10 times larger than that of Suprime-Cam. This camera will thus offer us great opportunities to explore unique and legacy surveys for the Andromeda halo, as well as for other Galactic Archaeology science cases.

Keywords. Galactic Archaeology, Subaru/HSC, M31

---

1. Deep and wide-field survey of the M31 halo

We are now carrying out intensive observing programs for a deep and wide-field imaging survey of M31's stellar halo using an HSC imager. This extremely wide-field camera in combination with Subaru will reach a depth of about 1.5 mag fainter than the Horizontal Branch (HB) magnitude ($\sim$ 27.4 and $\sim$ 26.4 mag in g and i bands, respectively) to obtain numerous HB stars, which will allow us to identify and characterize low surface brightness features as faint as μ of 33.5 mag arcsec^{-2}. This surface brightness is significantly lower than was possible with previous surveys that were based entirely on the selection of bright red giant branch (RGB) stars. It is thus possible to find and map out faint stellar streams even in the outer part of the halo at R beyond 90 kpc, where the very low density of bright RGB stars makes finding streams very difficult. Similarly, this Subaru/HSC survey will allow us to find lower luminosity dwarf satellites than hitherto possible through member HB stars, thereby providing important insight into the missing satellite problem.

This extremely wide-field camera will cover about 80 kpc x 80 kpc area (6 deg x 6 deg) in the north-west part of the M31 halo (Fig. 1), which is wide enough to contain a large portion of the outer halo. As back-up observations (for the nights with non-ideal seeing conditions), short-exposure imaging surveys for the rest of the halo areas in M31 are made to obtain the data of bright RGB stars in combination of the use of the narrow-band filter, NB515. The stars selected in this mode will be targets for our future spectroscopic follow-up using Prime Focus Spectrograph to be mounted on Subaru (see Chiba *et al.* 2015 in this volume).

2. Current status

In period S14B, we succeeded to obtain g and i deep imaging only for two HSC fields and yet slightly incomplete exposures for one field with $\sim$ 0.5 mag deficient compared with other two fields (filled circles in Fig. 1), due to the bad weather. The left panel of Fig. 2 shows the CMD of the stars in these three fields. It is clear that Subaru/HSC is quite powerful for imaging M31 stars as faint as g = 27.4 and i = 26.4, and successful

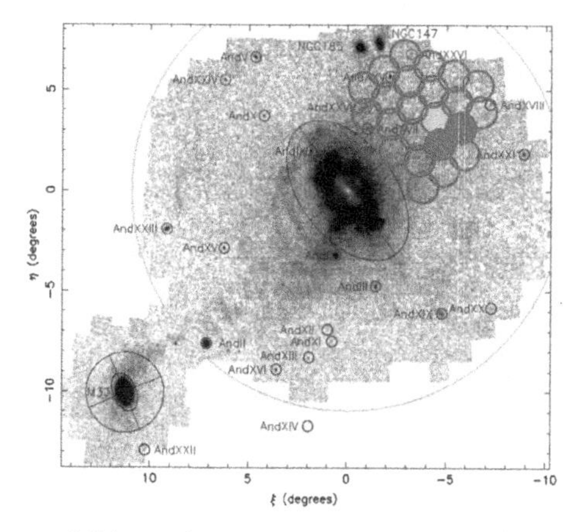

Figure 1. Proposed survey fields with HSC in M31's halo overlaid onto the RGB map of PAndAS. Filled circles denote the fields for which g- and i-band deep imagings were completed and partially completed, respectively, in S14B.

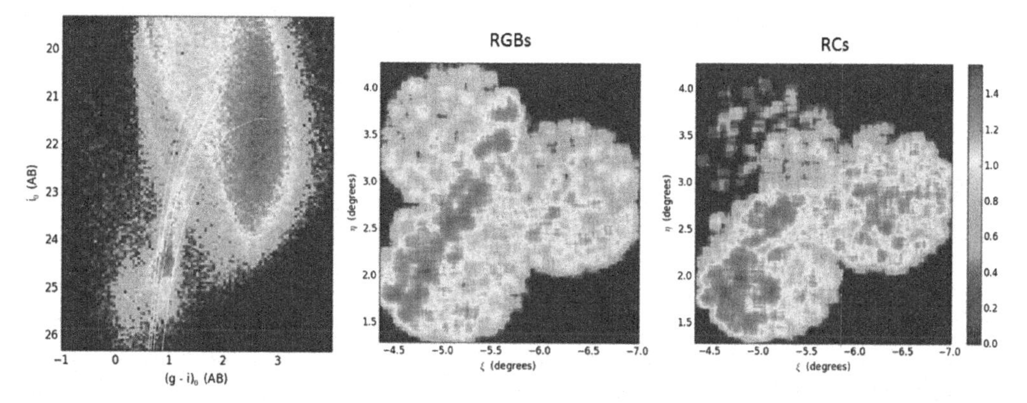

Figure 2. Left: i vs. $(g - i)$ color-magnitude diagram obtained from three HSC fields (filled circles in Fig. 1) in S14B. Middle: RGB map in these three fields selected with $22.7 < i < 24.0$. Right: RC map selected with $24.1 < i < 24.7$. Note that in the RC map, the northern stream in the upper-left HSC pointing disappears due to yet inefficient exposures and S/N.

for hunting Red Clump (RC) stars with $i \sim 24.5$ in HB; clearly the previous survey data with $i < 24$ using 4m-class telescopes are too shallow and largely affected by the contamination of the MW foreground stars (as seen in the upper right region, peaked at $i < 24$ and $2 < g - i < 3$), thus showing only the brightest part of the M31 halo with some uncertainties associated with this contamination issue. The middle and right panels in Fig. 2 show the spatial distributions of RGB and RC stars, respectively. It is remarkable that while both of these maps show stream-like features (from lower left to upper middle) corresponding to the part of the northern stream, the RC map shows additional faint substructures, which do not appear in the RGB map. Further HSC observations of M31's halo are scheduled to assess and limit these overdensities. We have also identified some systematic difference in RC magnitudes between the stars in and outside the northern stream, which reflects the different metallicities and ages. Deep imaging indeed enables us to discover new important features in M31's halo.

The General Assembly of Galaxy Halos: Structure, Origin and Evolution
Proceedings IAU Symposium No. 317, 2015
A. Bragaglia, M. Arnaboldi, M. Rejkuba & D. Romano, eds.

doi:10.1017/S1743921315006912

Galactic Archaeology with the Subaru Prime Focus Spectrograph

Masashi Chiba[1], Judith Cohen[2] and Rosemary F. G. Wyse[3]

[1]Astronomical Institute, Tohoku University, Sendai, Japan
email: chiba@astr.tohoku.ac.jp

[2]Caltech, Pasadena, CA, USA
email: jlc@astro.caltech.edu

[3]Johns Hopkins University, Baltimore, MD, USA
email: wyse@jhu.edu

Abstract. We present an overview of our Galactic Archaeology (GA) survey program with the Prime Focus Spectrograph (PFS) for Subaru. Following successful design reviews, the instrument is now under construction with first light anticipated in 2018. Main characteristics of PFS and the science goals in our PFS/GA program are described.

Keywords. Galactic Archaeology, Subaru/PFS

1. What is PFS?

PFS is a massively-multiplexed, fiber-fed optical and near-infrared 3-arm spectrograph, to be mounted on Subaru (Fig.1). Main characteristics of PFS are summarized as follows.

- Funding started in 2010 (PI: H. Murayama).
- Developed based on international collaboration: Kavli IPMU (U. of Tokyo), NAOJ, ASIAA, LNA, Caltech/JPL, LAM, Princeton, JHU, IAG/Uni de Sao Paulo and MPA.
- Field of view: 1.4 degree in diameter.
- Number of fibers: 2400.
- Spectrograph: 3-arm design to cover 380-1260 nm simultaneously, i.e., blue (3800-6700 Å), red (6500-10000 Å) and IR arms (9700-13000 Å).
- Resolution: LR mode ($R = 2,000 - 3,000$) and MR mode ($R = 5,000$)
- First light: tentatively scheduled in 2018.

2. GA survey

GA is one of three main science components of the Subaru Strategic Program for PFS (see Takada *et al.* 2014 for details). We expect to have 300 to 360 nights for the entire PFS program, of which about 100 nights are expected to go to the GA program. With PFS, we will measure radial velocities and chemical abundances of stars in the Milky Way and M31 to infer the past assembly histories of these galaxies and the structure of their dark matter halos. Science requirements are summarized in Table 1. Data will be secured for numerous stars in the Galactic thick-disk, halo and tidal streams as faint as $V = 22$ mag with a low-resolution mode with $R = 2,000$ to 3,000, including stars with $V < 20$ mag to complement the goals of the Gaia mission. A medium-resolution mode with $R = 5,000$ to be implemented in the red arm will allow the measurement of multiple alpha-element abundances (Left panel of Fig.2) and more precise velocities for Galactic stars, elucidating the detailed chemo-dynamical structure and evolution of each of the main stellar components of the Milky Way Galaxy and of its dwarf spheroidal galaxies (Right panel of Fig. 2). The M31 halo campaign will target red giant branch stars with

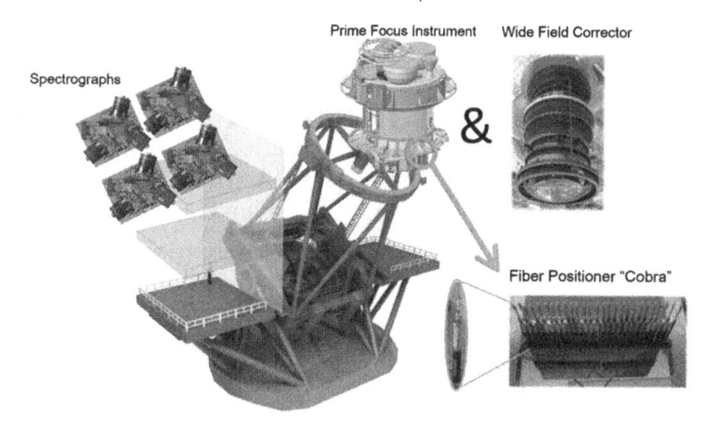

Figure 1. An overview of the baseline design of Subaru/PFS (taken from `http://pfs.ipmu.jp/instrumentation.html`).

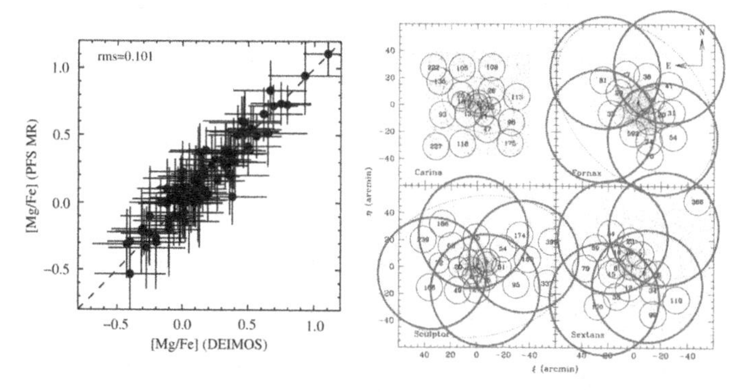

Figure 2. Left: Comparison of [Mg/Fe] derived from PFS-MR spectra to those of the DEIMOS data. Right: Proposed PFS pointings (solid circles) for the wide-field velocity measurements of some Galactic dSphs (Fornax, Sculptor and Sextans) reaching their tidal radii (dotted lines) (adapted from Walker *et al.* 2009).

Table 1. Summary of Galactic Archaeology Requirements for PFS

Mode	Requirements & Comments
LR	For the Milky Way stars ($V < 22$) and M31 halo ($21.5 < V < 22.5$)
(R =2000-3000)	Velocity precision of 5–10 km s^{-1}
	[Fe/H] to ~ 0.2 dex
	λ =3800 Å to 1 μm incl. Ca II HK, Ca I, Mgb/MgH, CaT
MR	For the Milky Way stars ($V < 19$ in bright and $V < 21$ in grey/dark time)
(R =5000)	Velocity precision of 3 km s^{-1}
	[Fe/H] to 0.15 dex, [X/Fe] (X=Mg, Si, Ca, Ti) to ~ 0.2 dex
	λ =7100 to 8850 Å incl. CaT and α-element lines

$21.5 < V < 22.5$ mag, obtaining radial velocities and spectroscopic metallicities over an unprecedented large area of its stellar halo. In synergy with these planned PFS survey, the coordinated imaging surveys with Hyper Suprime Cam are currently going on.

References

Takada, M. *et al.* 2014, *PASJ*, 66, 1
Walker, M. G. *et al.* 2009, *AJ*, 137, 3100

The General Assembly of Galaxy Halos: Structure, Origin and Evolution
Proceedings IAU Symposium No. 317, 2015
A. Bragaglia, M. Arnaboldi, M. Rejkuba & D. Romano, eds.

doi:10.1017/S1743921315006808

Clues on the first stars from CEMP-no stars

Arthur Choplin[1], **Georges Meynet**[1], **André Maeder**[1], **Raphael Hirschi**[2], **Sylvia Ekström**[1] **and Cristina Chiappini**[3]

[1]Geneva Observatory, Geneva University,
CH–1290 Sauverny, Switzerland
email: arthur.choplin@unige.ch
[2]Astrophysics group, Keele University,
Lennard-Jones Lab., Keele, ST5 5BG, UK
[3]Leibniz-Institut fuer Astrophysik,
An der Sternwarte 16, 14482 Potsdam, Germany

Abstract. The material used to form the CEMP-no stars presents signatures of processing by the CNO cycle and by He-burning from a previous stellar generation called spinstars. We compare the composition of the ejecta (wind + supernova) of a spinstar model to observed abundances of CEMP-no stars. We show that observed abundances as well as the isotope ratio $^{12}C/^{13}C$ may be reproduced by the spinstar ejecta if we assume different mass cuts when adding the supernova material to the wind ejecta.

Keywords. stars: evolution, rotation, massive, abundances, nucleosynthesis, chemically peculiar

---

1. Introduction

CEMP-no stars (Carbon Enhanced Metal Poor stars with no signature of s or r processes) are chemically peculiar objects that dominate the stellar populations at [Fe/H] < -3 (Aoki *et al.* 2010, Norris *et al.* 2013). The "spinstar scenario" (Meynet *et al.* 2010), suggests that CEMP-no stars formed in a region previously enriched by a fast rotating, low metallicity, massive star, experiencing mixing, mass loss and eventually a supernova at the end of its life.

We discuss a 32 $M_\odot$ spinstar model computed with the Geneva code. Absolute amounts of C, N, O, F, Ne, Na, Mg, Al as well as isotope ratios $^{12}C/^{13}C$, $^{24}Mg/^{25}Mg$ and $^{24}Mg/^{26}Mg$ in the ejecta are compared to observed CEMP-no abundances.

2. Results

Fig.1 shows the [X/H] ratios (left panel) and 3 isotope ratios (right panel). The grey line is the initial composition of the model which is a modified α-enhanced mixture (α-mod) : initial abundances of α-elements are enhanced and [C/N], [O/N] and $^{12}C/^{13}C$ ratios are set to 2, 1.6 and 30, according to suggestions of Maeder *et al.* (2014) for [C/N] and [O/N] and to prediction of galactic chemical evolution models at low metallicity of Chiappini *et al.* (2008) for $^{12}C/^{13}C$. The black lines are patterns in the ejecta when considering either the wind only, or the wind plus supernova obtained for various M_{cut}, M_{cut} being the mass coordinate inside the star delimiting the expelled part from the part which is kept into the remnant.

The effects of the CNO cycle and the Ne-Na Mg-Al chains are visible in every pattern (except in the $M_{cut} = 10M_\odot$ one). When the CNO cycle operates, ^{12}C and ^{16}O are transformed into ^{14}N and ^{13}C during the evolution, explaining the higher [N/H] ratios compared to [C/H] and [O/H]. $^{12}C/^{13}C$ ratios are close to the CNO equilibrium value

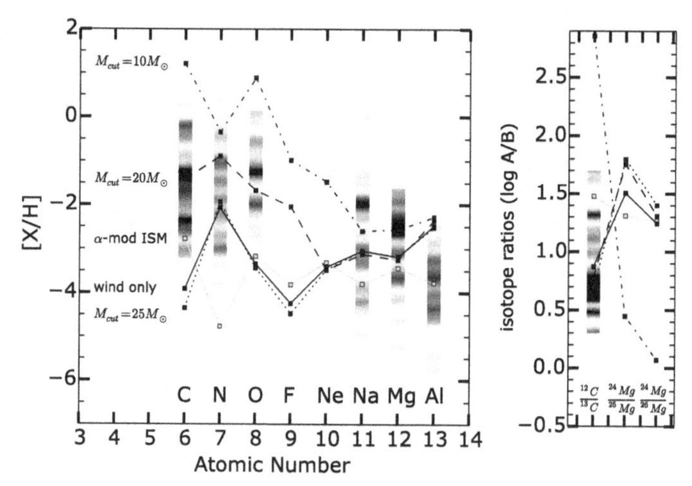

Figure 1. Predicted and observed [X/H] ratios (left) and isotope ratios $^{12}C/^{13}C$, $^{24}Mg/^{25}Mg$ and $^{24}Mg/^{26}Mg$ (right). Density map of observed CEMP-no are represented by rectangles colored from white (no CEMP-no at this value) to black. Black lines are predicted patterns in the ejecta considered : wind only (full line), wind + supernova with M_{cut} = 25 (dotted), 20 (dashed) and 10 $M_{\odot}$ (dot-dashed). The grey line corresponds to the initial composition of the spinstar.

($\log(^{12}C/^{13}C) \sim 0.7$), showing also that the major part of those ejecta was processed by the CNO cycle. Also some Ne and Mg were transformed into ^{23}Na and ^{27}Al owing to the Ne-Na Mg-Al chains so that [Na/H] > [Ne/H] and [Al/H] > [Mg/H] in the final ejecta.

About 2 $M_{\odot}$ of the He-burning shell was ejected in the fourth case ($M_{cut} = 10M_{\odot}$). The associated pattern bears indeed the signature of He-burning : [C/H], [O/H] and $^{12}C/^{13}C$ are several dex higher than previous patterns. It is worthwhile to remark that $^{12}C/^{13}C$ is a relevant isotope ratio to constrain M_{cut} : if too deep layers are expelled, part of the He-burning region is expelled and $^{12}C/^{13}C$ increases a lot, lying clearly outside of the observed values. Interesting also is the [Ne/H] ratio which have raised by $\sim$ 2 dex. This is due to the reaction $^{16}O(\alpha,\gamma)^{20}Ne$ and to the destruction of ^{14}N in the He-core through $^{14}N(\alpha,\gamma)^{18}F(,e^{+}\nu_{e})^{18}O(\alpha,\gamma)^{22}Ne$. Isotopes ratios $^{24}Mg/^{25}Mg$ and $^{24}Mg/^{26}Mg$ are lowered by $\sim$ 1 dex in this case because of the synthesis of ^{25}Mg and ^{26}Mg through $^{22}Ne(\alpha,n)^{25}Mg$ and $^{22}Ne(\alpha,\gamma)^{26}Mg$ in the He-core.

The models can explain large parts of the observed scatter in [X/H] and $^{12}C/^{13}C$ ratios, except for [Al/H], which is always overestimated by at least 1 dex. Since [Al/H] ~ -3.8 in the ISM, dilution of the ejecta with the initial ISM would allow [Al/H] values of -3.8 at best, but not lower, where more than half of the observed CEMP-no are lying. Aluminum surproduction is the biggest discrepancy between models and observations and should be investigated in a future work.

References

Aoki, W. 2010, *Carbon-Enhanced Metal-Poor (CEMP) stars*, Proc. IAU Symposium No. 265, p. 111

Norris, J. E., Yong, D., Bessell, M. S., *et al.* 2013, *ApJ*, 762, 28

Meynet, G., Hirschi, R., Ekstrom, S., *et al.* 2010, *A&A*, 521, A30

Maeder, A., & Meynet, G. and Chiappini, C. 2014, *A&A*, 576, A56

Chiappini, C., Ekström, S., Meynet, G., *et al.* 2008, *A&A* (Letters) 479, L9-L12

The General Assembly of Galaxy Halos: Structure, Origin and Evolution
Proceedings IAU Symposium No. 317, 2015
A. Bragaglia, M. Arnaboldi, M. Rejkuba & D. Romano, eds.

doi:10.1017/S1743921315008649

Formation and evolution of sub-galactic structures in a cosmological context

Kyungwon Chun[1], Jihye Shin[2] and Sungsoo S. Kim[1,3]

[1]School of Space Research, Kyung Hee University,
Yongin, Gyeonggi, 446-701, Korea
email: kwchun@khu.ac.kr
[2]Kavli Institute for Astronomy and Astrophysics, Peking University,
5 Yiheyuan Road, Haidian District, Beijing 100871, China
email: jhshin.jhshin@gmail.com
[3]Department of Astronomy and Space Science, Kyung Hee University,
Yongin, Gyeonggi, 446-701, Korea
email: sungsoo.kim@khu.ac.kr

Abstract. In this study, we aim to trace formation of the primordial globular cluster, ultra faint dwarf galaxy, and ultra compact dwarf in a cosmological context of a high-resolution hydrodynamic zoom-in simulation. We show that the baryon-dominated systems have experienced more interactions with the mini halos before infalling to the main halo.

Keywords. Galaxy: formation, methods: numerical, globular clusters: general

---

1. Simulation

Our goal is to trace formation of sub-galactic structures in ΛCDM cosmology. For this, we modify GADGET-3 (Springel 2005) code to include realistic baryonic physics. We calculate radiative heating/cooling rates using CLODY90 package (Ferland *et al.* 1998). Global reionization is considered in a whole simulation volume at redshift $z_{re} = 8.9$ (Haardt & Madau 1996). We assume that dense gas clouds of $n_H > 0.014\mathrm{cm}^3$ are shielded from the universal UV radiation (Sawala *et al.* 2010). Stars form when gas particles satisfy star formation criteria of Saitoh *et al.* (2008).

Initial conditions are generated by a MUSIC software (Hahn & Abel 2011). Selected cosmological parameters are $\Omega_m = 0.3, \Omega_\Lambda = 0.7, \Omega_b = 0.048$, and $h = 0.68$. We perform a zoom-in simulation of a cubic box of side length of 1 Mpc/h with 170 million dark matter (DM) and gas particles from redshift $z = 49$ to $z = 1.765$. Particle mass for the DM and gas is $M_{DM} \approx 4 \times 10^3$ M$_\odot$, and $M_{gas} \approx 8 \times 10^2$ M$_\odot$, respectively.

In order to identify halos, we used an Amiga Halo Finder software (AHF; Knollmann & Knebe 2009). The main halo is a dwarf galaxy sized halo with a virial mass of $M_{vir} = 5.75 \times 10^9$ M$_\odot$/h at $z = 1.765$. We find 75 mini halos around the main halo, which can be formation sites for the globular clusters, ultra compact dwarfs, and the ultra faint dwarfs. They satisfy the following conditions: (1) the maximum baryon mass during the evolution is between 10^5 M$_\odot$ and 10^8M$_\odot$ (2) they are located in virial radius of main halo at $z = 1.765$.

2. Evolution of mini halos

We classify the mini halos with their closest distances to the main halo center as follows: $\mathrm{R}_{peri}/\mathrm{R}_{vir}$ >0.15 mini halos belong to Group 1, while that of $\mathrm{R}_{peri}/\mathrm{R}_{vir}$ <0.15 for Group 2, where R_{peri} and R_{vir} is a pericenter distance and a viral radius of the main halo, respectively.

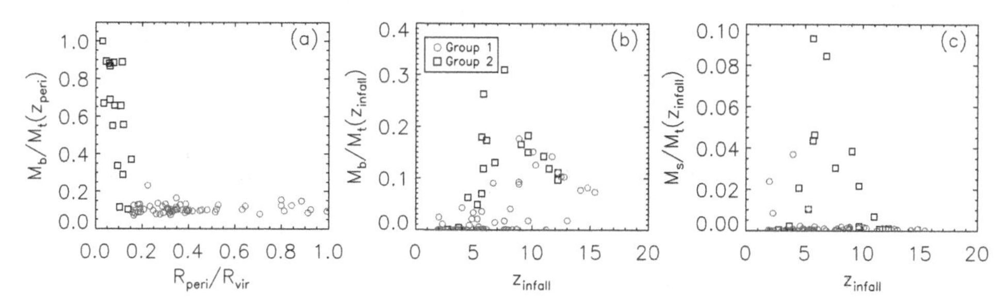

Figure 1. Diagram of baryon mass fraction and scaled distance to the main halo center (a). Baryon (or star) mass fractions of the mini halos when they infall to the main halo (b) and (c). The red diamonds and the black squares indicate the Group 1 and Group 2, respectively.

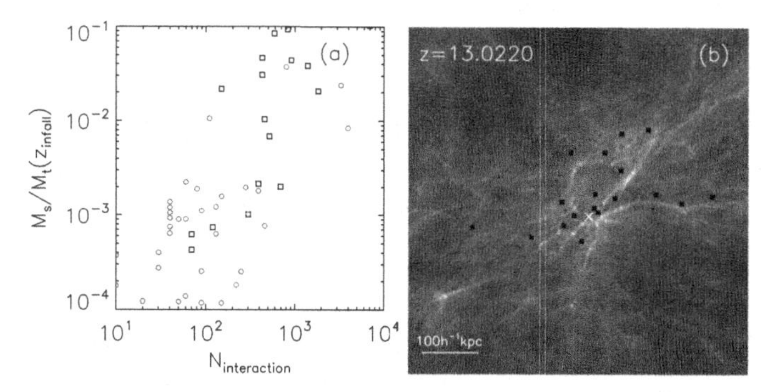

Figure 2. Diagram of star mass fractions of the mini halos when they infall to the main halo and number of interactions with the other mini halos (a). Distribution of mini halos on filamentary structures around the main halo (b).

We find that the Group 2 tends to lose more DM halos by stronger tidal force of the main halos (see Fig. 1a). They can evolve to the more baryon-dominated systems compared to the Group 1. The high baryon fraction of the Group 2 is also appeared even when the mini halos infall to the main halo (see Fig. 1b and 1c). Therefore, we infer that the different evolutionary tendency between the Groups is originated by the different environments that the mini halos have experienced before the infall.

To quantify the environmental effects on the mini halos before the infall, we count how many times the mini halos experience interactions with the other mini halos ($N_{interactions}$). Figure 2 shows that more interactions trigger the more stars formation, and that the Group 2 experiences the more interactions compared to the Group 1. Figure 2b shows that the mini halos of the Group 2 are preferentially located along the filamentary structures, where interactions between mini halos are more frequent.

References

Ferland, G. J., Korista, K. T., Verner, D. A., Ferguson, J. W., Kingdon, J. B., & Verner, E. M. 1998, *PASP*, 110, 761

Haardt, F. & Madau, P. 1996, *ApJ*, 461, 20

Hahn, O. & Abel, T. 2011, *MNRAS*, 415, 2101

Knollmann, S. R. & Knebe, A. 2009, *ApJs*, 182, 608

Saitoh, T. R., Daisaka, H., Kokubo, E., Makino, J., Okamoto, T., Tomisaka, K., Wada, K., & Yoshida, N. 2008, *PASJ*, 60, 667

Sawala, T., Scannapieco, C., Maio, U., & White, S. 2010, *MNRAS*, 402, 1599

Springel, V. 2005, *MNRAS*, 364, 1105

The General Assembly of Galaxy Halos: Structure, Origin and Evolution
Proceedings IAU Symposium No. 317, 2015
A. Bragaglia, M. Arnaboldi, M. Rejkuba & D. Romano, eds.

doi:10.1017/S1743921315006936

The extended stellar substructures of four metal-poor globular clusters in the Galactic bulge

Sang-Hyun Chun[1] and Young-Jong Sohn[2]

[1]Yonsei University Observatory, 120-749 Seoul, Korea
email: shchun@galaxy.yonsei.ac.kr

[2]Department of Astronomy, Yonsei University, 120-749, Seoul, Korea
email: sohnyj@yonsei.ac.kr

Abstract. We investigated the stellar density substructures around four metal-poor globular clusters (NGC 6266, NGC 6626, NGC 6642, and NGC 6723) in the Galactic bulge. Wide-field near-infrared (JHK_s) imaging data were obtained from WFCAM of UKIRT telescope. Field stars contamination around the globular clusters was reduced by using a statistical weighted filtering algorithm. Tidal stripping stellar substructures in the form of tidal tail (NGC 6266 and NGC 6626) or small density lobes/chunk (NGC 6642 and NGC 6723) were found around the four globular clusters in the two-dimensional density contour maps. We also find the overdensity features, which deviate from the theoretical models, in the outer region of radial density profiles. The observed results imply that the four globular clusters have experienced a strong tidal force or the bulge/disk shock effect of the Galaxy.

Keywords. Globular clusters, Tidal tail

1. Introduction

According to the hierarchical scenario of the Galaxy formation, galaxies are developed by merging and accretion of small fragments like dwarf galaxies and globular clusters (Moore *et al.* 1999, Bullock *et al.* 2001). In this process, accreted satellite systems are disrupted by tidal force and shock, and then leave stellar substructures such as tidal tails and tidal streams (Bullock & Johnston 2005). Thus, the study of the stellar substructures can provide the observational evidence for merging and accretion of satellites. We investigated the tidal substructures around four metal-poor globular clusters in the bulge region.

2. Extended stellar substructures

We obtained wide-field ($45' \times 45'$) near-infrared photometric data for four globular clusters (NGC 6266, NGC 6626, NGC 6642, and NGC 6723) using the WFCAM array of UKIRT. A statistical weighted filtering algorithm (Chun *et al.* 2012, Grillmair *et al.* 1995, Odenkirchen *et al.* 2003) was applied to the CMDs of the clusters in order to minimize the field star contamination.

Figure 1 shows the two-dimensional contour maps and radial density profile of the four globular clusters. We find that NGC 6266 and NGC 6626 show extended stellar substructures beyond tidal radius, while NGC 6642 and NGC 6723 have small density chunks or lobes near tidal radius. The extended stellar substructures seem to be aligned with the proper motion direction of cluster and the direction of the Galactic center.

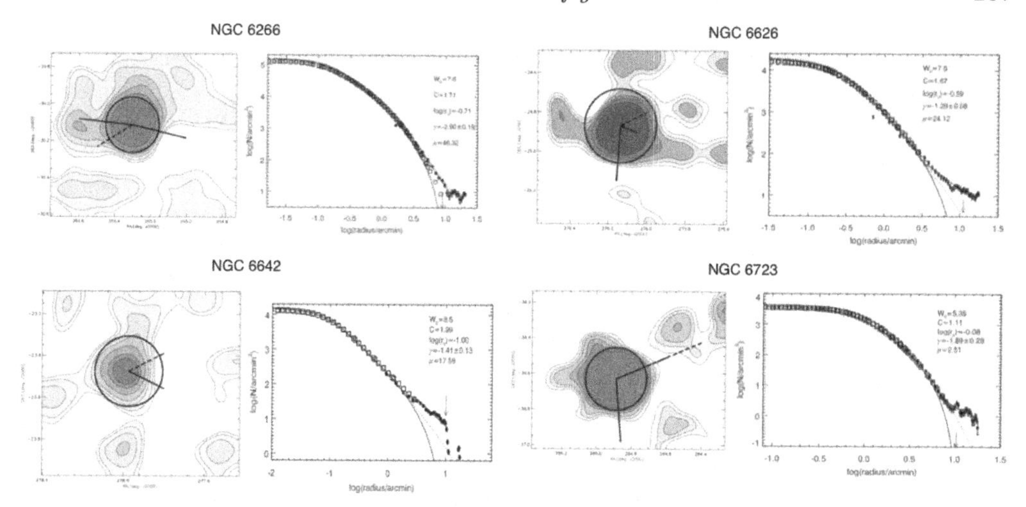

Figure 1. Iso-density maps and radial profiles for the four globular clusters. *In contour maps:* Arrows indicate the proper motion of cluster. The solid line and dashed lines indicate the direction of the Galactic center and that perpendicular to the Galactic plane, respectively. *In radial profiles:* the solid curve and dotted curved show the King model and Wilson models. Open rectangules represent the surface brightness profile of Trager *et al.* (1995). Filled circles are our measurements, and arrows indicate the tidal radius of the clusters.

The distorted stellar substructures are also well represented in the radial profile as an overdensity feature which departs from theoretical King (King 1966) and Wilson (Wilson 1975) models at the outer region of the clusters.

3. Discussion

We found that tidally extended stellar features are present in the vicinity of four metal-poor globular clusters. The stellar substructures seem to be associated with the interaction of the Galactic tidal force and proper motion of clusters. Our results indicate that strong tidal force or bulge/disk shock have affected the shape of the four globular clusters in the bulge region, and provide further constraints on the formation of the globular clusters in the bulge region.

References

Bullock, J. S., Kratsov, A. V., & Weinberg, D. H. 2001, *ApJ*, 548, 33
Bullock, J. S. & Johnston, K. V. 2005, *ApJ*, 635, 931
Chun, S.-H., Kim, J.-W., Kim, M.-J., *et al.* 2012, *AJ*, 144, 26
Grillmair, C. J., Freeman, K. C., Irwin, M., & Quinn, P. J. 1995, *AJ*, 109, 2553
King, I. R. 1966, *AJ*, 71, 64
Moore, B., Ghigna, S., Governato, F., *et al.* 1999, *ApJ*, 524, L19
Odenkirchen, M., Grebel, E. K., Dehnen, W., *et al.* 2003, *AJ*, 126, 2385
Trager, S. C., King, Ivan R., & Djorgovski, S. 1995, *AJ*, 109, 218
Wilson, C. P. 1975, *AJ*, 80, 175

The General Assembly of Galaxy Halos: Structure, Origin and Evolution
Proceedings IAU Symposium No. 317, 2015
A. Bragaglia, M. Arnaboldi, M. Rejkuba & D. Romano, eds.

doi:10.1017/S1743921315009151

Measuring the Stellar Halo Velocity Anisotropy With 3D Kinematics

Emily C. Cunningham[1], Alis J. Deason[1,2], Puragra Guhathakurta[1], Constance M. Rockosi[1], Roeland P. van der Marel[3], S. Tony Sohn[4]

[1]Department of Astronomy & Astrophysics, UC Santa Cruz
1156 High Street, Santa Cruz, CA 95060, USA

[2] Physics Department, Stanford University
382 Via Pueblo Rd, Stanford, CA 94305, USA

[3] Space Telescope Science Institute
3700 San Martin Drive, Baltimore, MD 21218, USA

[4] Department of Physics and Astronomy, The Johns Hopkins University
Baltimore, MD 21218, USA

Abstract. We present the first measurement of the anisotropy parameter β using 3D kinematic information outside of the solar neighborhood. Our sample consists of 13 Milky Way halo stars with measured proper motions and radial velocities in the line of sight of M31. Proper motions were measured using deep, multi-epoch HST imaging, and radial velocities were measured from Keck II/DEIMOS spectra. We measure $\beta = -0.3^{+0.4}_{-0.9}$, which is consistent with isotropy, and inconsistent with measurements in the solar neighborhood. We suggest that this may be the kinematic signature of a relatively early, massive accretion event, or perhaps several such events.

Global kinematic properties, such as the relative pressure between tangential and radial velocity components, otherwise known as the velocity anisotropy, can provide important insight into the formation of the stellar halo. Using the 6D phase space information from our survey, we have made the first direct measure of the velocity anisotropy outside of the solar neighborhood (Cunningham *et al.* 2016).

In Deason *et al.* (2013b), we published the first proper motions of distant Milky Way halo stars; in Cunningham *et al.* (2016), we present the line-of-sight (LOS) velocities for those same stars. Our 13 objects are the first stars outside of the solar neighborhood to have 3D kinematic information. The left panel of Figure 1 shows a cumulative LOS velocity histogram of the 13 stars in our sample, and the right panel shows the proper motions of the halo stars color coded by LOS velocity (velocities are shown in the Galactocentric frame). We estimated the parameters of the velocity ellipsoid ($\langle v_l \rangle, \sigma_{\rm LOS}, \sigma_l, \sigma_b$) using Markov Chain Monte Carlo (Figure 2), and made the first measurement of the anisotropy parameter β using 3D kinematic information outside of the solar neighborhood. We find $\beta = -0.3^{+0.4}_{-0.9}$, consistent with isotropy and inconsistent with solar neighborhood measurements, which find a highly radial anisotropy ($\beta \sim 0.5 - 0.7$). Simulations of accreted stellar haloes indicate that β should increase outward (e.g. Abadi *et al.* 2006), and yet we observe a “dip” in the radial anisotropy profile. This dip is coincident with the break radius in the Milky Way's stellar density profile. We argue that the presence of large, dynamically “fluffy” global substructre, such as a shell (or multiple shells), is one explanation for both the steep fall-off in stellar density beyond the break radius and the decrease in anisotropy at that radius. Deason *et al.* (2013a) argued that a break in the Milky Way stellar density profile could be created by the build-up of stars at apocenter from either one relatively massive accretion event or several, synchronous accretion events. In this

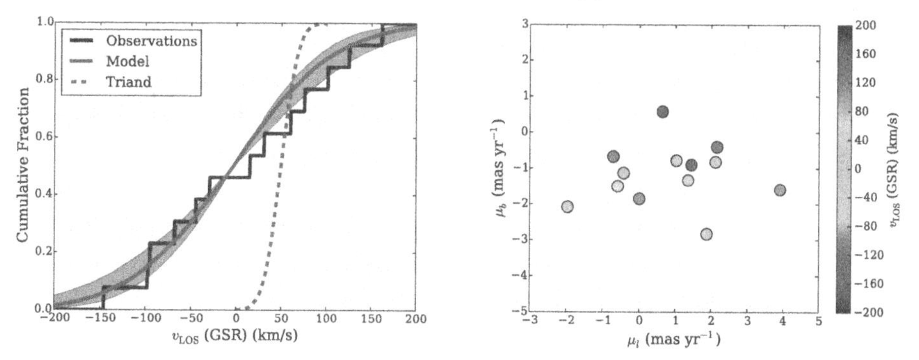

Figure 1. Left: Cumulative histogram of LOS velocities of the 13 halo stars in our sample (black). The overplotted blue line shows the cumulative distribution function (CDF) for the most likely value for $\sigma_{\rm LOS}$, with the shaded blue region indicating the 68% confidence region. An approximate CDF for TriAnd is shown in red ($v_0 \sim 50$ km s^{-1}, $\sigma \sim 15$ km s^{-1}). Right: Proper motions in galactic coordinates of our 13 halo stars, color coded by LOS velocity relative to the Galactic Standard of Rest (GSR).

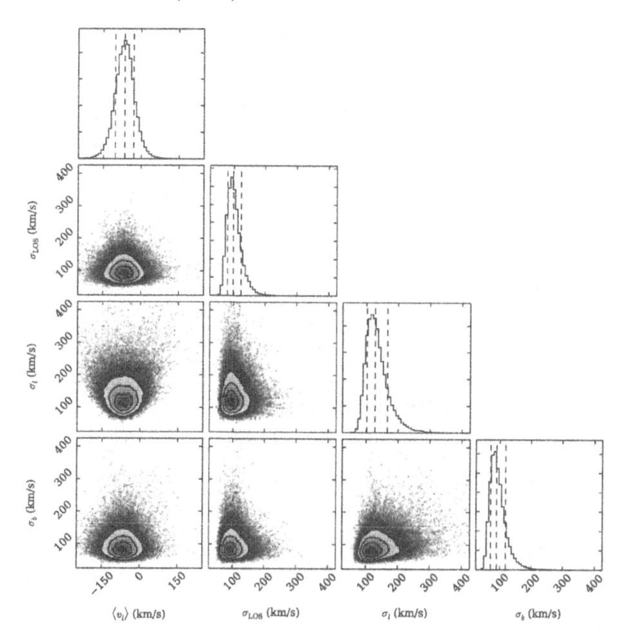

Figure 2. Projections of the posterior probability distribution for the four free parameters of the velocity ellipsoid in our model: $\langle v_l \rangle, \sigma_{\rm LOS}, \sigma_l$ and σ_b. Contours are shown at 0.5, 1, 1.5 and 2 σ, respectively. The top panel in each column shows the 1D marginalized PDF for each parameter, with 0.16, 0.5 and 0.84 quantiles indicated by dashed vertical lines.

scenario, we would expect the stars to have an increase in tangential motion relative to radial motion at the turnaround radius, and thus a more isotropic β, just as we observe.

References

Abadi, M.G., Navarro, J.F., & Steinmetz, M.2006, *MNRAS*, 365, 747

Cunningham, E. C., Deason, A. J., Guhathakurta, P., *et al.* 2016, arXiv:1602.03180

Deason, A. J., Belokurov, V., Evans, N. W., & Johnston, K. V. 2013a, *ApJ*, 763, 113

Deason, A. J., Van der Marel, R. P., Guhathakurta, P., Sohn, S. T., & Brown, T. M. 2013b, *ApJ*, 766, 24

The General Assembly of Galaxy Halos: Structure, Origin and Evolution
Proceedings IAU Symposium No. 317, 2015
A. Bragaglia, M. Arnaboldi, M. Rejkuba & D. Romano, eds.

doi:10.1017/S1743921315007279

Proper-Motion Based Kinematics Study of Galactic RR Lyraes

Andrei K. Dambis[1], **Leonid N. Berdnikov**[1,2], **Alexei S. Rastorguev**[1] **and Marina V. Zabolotskikh**[1]

[1]Sternberg Astronomical Institute, M.V.Lomonosov Moscow State University, Universitetskii pr. 13, Moscow, 119992 Russia
email: `dambis@yandex.ru`

[2]Astronomy and Astrophysics Research division, Entoto Observatory and Research Center, P.O. Box 8412, Addis Ababa, Ethiopia

Abstract. We use the UCAC4 and SDSS proper motions of about 7500 RR Lyrae type variables located within $\sim$ 10 kpc from the Sun to study the dependence of their velocity ellipsoid on Galactocentric distance in the R_G = 3-17 kpc interval. The radial velocity dispersion, σ_{VR}, decreases from $\sim$ 190 km/s at R_G=3.5–5.5 kpc down to $\sim$ 100 km/s at R_G=13–15 kpc, and the σ_{VT}/σ_{VR} ratio remains virtually constant ($\sigma_{VT}/\sigma_{VR} \sim$ 0.54–0.64) in the Galactocentric distance interval from R_G=4.5 kpc to R_G=10.5 kpc increasing to $\sim$ 0.9 both toward the Galactic center and beyond R_G=11 kpc.

Keywords. stars: kinematics, stars: RR Lyrae, Galaxy: kinematics and dynamics

---

Our kinematic tracer sample consists of 7464 RR Lyrae type variables located within $\sim$ 10 kpc from the Sun drawn from the lists of Szczygiel *et al.* (2009) (based on ASAS survey data), Torrealba *et al.* (2015), and Drake *et al.* (2013) (based on the Catalina Sky Survey data) with the proper motions adopted from the UCAC4 (Zacharias *et al.* 2013) and SDSS DR12 (Alam *et al.* 2015) catalogs. We compute the photometric distances to these RR Lyraes using mid-IR WISE W1-band intensity-mean magnitudes determined from ALLWISE data (Wright *et al.* 2010) and our W1-band period-metallicity-luminosity relation

$$< M_{W1} >= -0.814 + 0.106[\mathrm{Fe/H}] - 2.381\log P$$

(Dambis *et al.* 2014).

We use the maximum-likelihood method to determine the average kinematical parameters of the entire sample – the bulk-motion components (U_0, V_0, W_0) = (-16 ± 5, -212 ± 4, +3 ± 5) km/s and the velocity-ellipsoid parameters σ_{VR} = 178 ± 4 km/s and σ_{VT}/σ_{VR} =0.66± 0.02 – from proper-motion data exclusively. We also determine the velocity-ellipsoid parameters within 1 kpc-thick Galactocentric distance bins (Figs. 1–2). The radial velocity dispersion decreases steadily from $\sim$ 190 km/s near the Galactic center to $\sim$ 100 km/s beyond $R_G \sim$ 13 kpc, whereas the σ_{VT}/σ_{VR} velocity dispersion ratio remains almost constant between R_G=4.5 kpc and R_G=10.5 kpc and increases both toward the Galactic center and beyond $R_G \sim$ 13 kpc. The results become highly uncertain beyond $R_G \sim$ 12 kpc.

Acknowledgments

This research was supported by the Russian Scientic Foundation (project 14-22-00041).

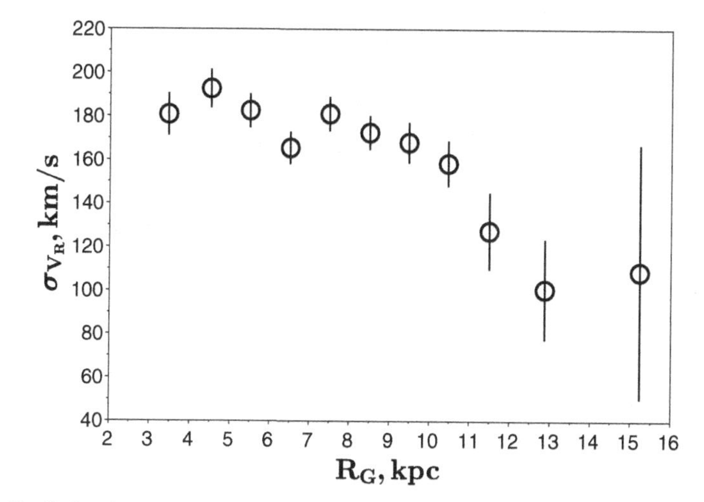

Figure 1. Radial velocity dispersion for halo RR Lyraes as a function of Galactocentric distance.

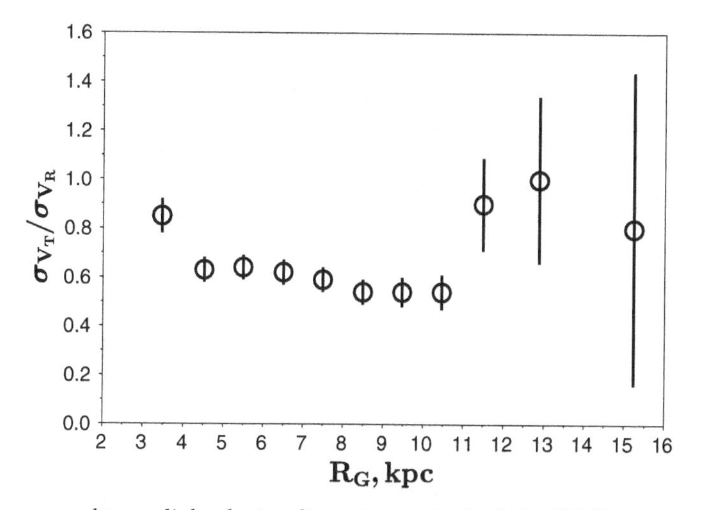

Figure 2. Transversal-to-radial velocity dispersion ratio for halo RR Lyraes as a function of Galactocentric distance.

References

Alam, S. *et al.* 2015, *arXiv1501.00963*
Dambis, A. K., Rastorguev, A. S., & Zabolotskikh, M. V. 2014, *MNRAS*, 439, 3765
Drake, A. J. *et al.* 2013, *ApJ*, 763, 32
Szczygiel, D. M., Pojmanski, G., & Pilecki, B. 2009, *AcA*, 59, 137
Torrealba, G. *et al.* 2015, *MNRAS*, 446, 2251
Wright, E. L. *et al.* 2010, *AJ*, 140, 1868
Zacharias, N. *et al.* 2013, *AJ*, 145, 44

The General Assembly of Galaxy Halos: Structure, Origin and Evolution
Proceedings IAU Symposium No. 317, 2015
A. Bragaglia, M. Arnaboldi, M. Rejkuba & D. Romano, eds.

doi:10.1017/S1743921315010467

From the Outskirts of Galaxies to Intra Cluster Light

Klaus Dolag, Rhea-Silvia Remus and Adelheid F. Teklu

Universitäts-Sternwarte München, Scheinerstrasse 1, München, Germany
email: dolag@usm.uni-muenchen.de

Abstract. We use the Magneticum Pathfinder (**www.magneticum.org**) hydro-dynamical cosmological simulation set to investigate the buildup of the stellar component within cosmological structures. These simulations result in the self-consistent formation of ICM, AGNs, and both spheroidal and disk galaxy populations, which properly reproduce the observed properties.

Keywords. hydrodynamics, numerical, galaxy: formation, clusters, dynamics

---

1. The Magneticum Pathfinder Simultions

The simulations treat metal-dependent radiative cooling, heating from a uniform time-dependent ultraviolet background, star formation and the chemo-energetic evolution of the stellar population as traced by SNIa, SNII and AGB stars with the associated feedback processes, as well as formation and evolution of super-massive black holes and the associated quasar and radio-mode feedback processes. For a detailed description see Dolag *et al.* (in prep), Hirschmann *et al.* (2014) and Teklu *et al.* (2015).

2. Intra Cluster Light

In galaxy clusters, the velocities of the stars in the cD galaxy and the diffuse component (DSC) have dynamically well-distinct kinematic distributions, which can be well characterized by two Maxwellian distributions. While the velocity dispersion of the stars in the cD galaxy represents the central mass of the stars, the velocity dispersion of the DSC is much larger and typically reaches almost the values of the overall dark matter halo, see Dolag *et al.* (2010) for the predictions from hydrodynamical simulations and Bender *et al.* (2015) or Longobardi *et al.* (2015) for observational confirmation. The density distributions, however, can in most cases be described by a single, radial profile and only in rare cases need to be described by the sum of two components with different radial shapes.

3. Universality of Outer Stellar Halo Profiles

Our simulations show that not only the dark matter radial density profile but also the outer stellar radial density profiles can be well described by a universal profile. This profile, $\rho(r) = \rho_{-2}\ \exp\left\{-\frac{2}{\alpha}\left[\left(\frac{r}{r_{-2}}\right)^{\alpha} - 1\right]\right\}$ (where α controls the curvature and ρ_{-2} is the density and r_{-2} the radius at which $\rho(r) \propto r^{-2}$), is also known as Einasto profile (Retana-Montenegro *et al.* 2012). We find that the fitting parameters of those Einasto profiles are not independent but closely correlated for fixed total mass, from galaxies to clusters, as can be seen as Fig. 2. This universal profile exists over a large range of total masses and is found to be independent of galaxy type. This strongly indicates that the formation of the outer stellar halo is dominated by accretion, which occurs in all halos.

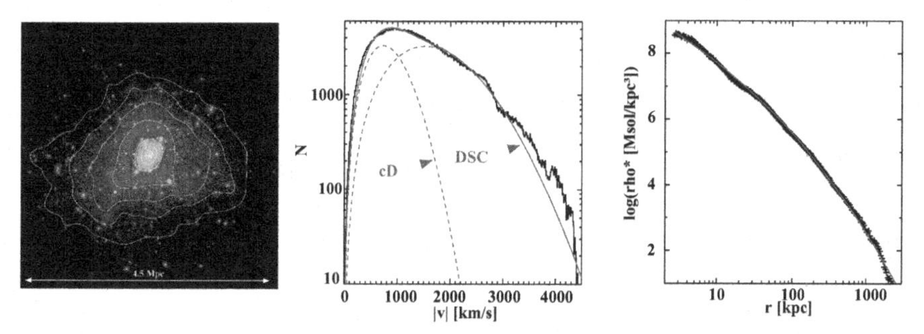

Figure 1. Stellar density map of the most massive cluster (color coded). The white contours show the diffuse stellar component plus the central galaxy (cD) after subtracting the stars from the other cluster member galaxies. *Middle:* Distribution of the stellar velocities for the DSC and the cD galaxy (black histogram) with a double Maxwellian fit (red line). *Right:* Radial stellar density profile of DSC and cD (black symbols) with a single Sérsic profile fit (red line).

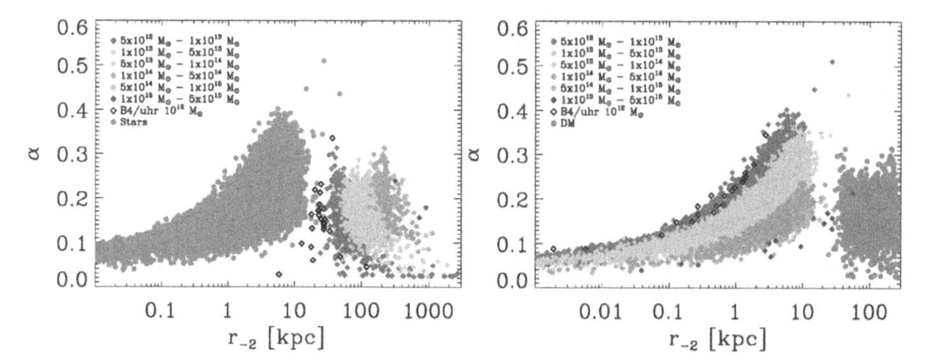

Figure 2. Best-fit parameters for Einasto profile fits to the outer halos of galaxies, ranging from halo masses of $10^{12} M_{\odot}$ up to massive clusters with several times $10^{15} M_{\odot}$ (color coded). *Left panel:* Fits to the dark matter components. *Right panel:* Fits to the stellar outer halos.

4. Conclusions

The amount of curvature of the outer stellar halo density profile is therefore a diagnostic for the merging history of galaxies. Since the Einasto profile is strongly curved at the outskirts, it can locally always be represented by a power law. For Milky Way mass halos, for example, we find slopes in the range of -3 to -6 at radii between 40–100kpc, as done in observations (e.g. Deason *et al.* (2014)). This can well explain the observed differences between the stellar outer halo density profiles observed for Milky Way and Andromeda.

References

Bender, R., Komedy, J., Cornell, M. F., & Fischer, D. B. 2015, *ApJ*, 807,56

Deason, A.J., Belokurov, V., Koposov, S.E., Rockosi, C.M. 2014 *ApJ*, 787, 30

Dolag, K., Murante, G., & Borgani, S. 2010, *MNRAS*, 405, 1544

Hirschmann, M., Dolag, K., Saro, A., *et al.* 2014, *MNRAS*, 442, 2304

Longobardi, A., Arnaboldi, M., Gerhard, O., & Rinhard, H. 2015, *A&A*, 579, 135

Retana-Montenegro, R., van Hese, E., Gentile, G., *et al.* 2012 *A&A*, 540, 70

Teklu, A.F., Remus, R.-S., Dolag, K., *et al.* 2015 *ApJ*, in press

The General Assembly of Galaxy Halos: Structure, Origin and Evolution
Proceedings IAU Symposium No. 317, 2015
A. Bragaglia, M. Arnaboldi, M. Rejkuba & D. Romano, eds.

doi:10.1017/S1743921315007097

Photometric Metallicity of the Sagittarius Stream in the south Galactic cap

Cuihua Du [1], Jiayin Gu [1], Yunpeng Jia[2], Xiyan Peng[2], Zhenyu Wu[2], Jun Ma[2], Xu Zhou[2], Yanchun Liang[2]

[1]School of physics sciences, University of the Chinese Academy of Sciences, Beijing 100049, China
email: ducuihua@ucas.ac.cn

[2]Key Laboratory of Optical Astronomy, National Astronomical Observatories, Chinese Academy of Sciences, Beijing 100012, China

Abstract. Based on SDSS and South Galactic Cap U-band Sky Survey (SCUSS) photometry, we try to study the photometric metallicity of the Sagittarius (Sgr) stream in the south Galactic cap. We find that the Sgr stream has a wider metallicity distribution, and that its median metallicity is richer than that of the field halo stars. The neighboring field halo stars in our studied fields can be modeled by a two-Gaussian model, with peaks at [Fe/H]$= -1.9$ and [Fe/H]$= -1.5$. The metallicity distribution function (MDF) of the mixed population (Sgr stream and halo stars) has peaks at [Fe/H]$= -1.9$, [Fe/H]$= -1.5$ and [Fe/H]$= -0.5$, respectively.

Keywords. Galaxy: metallicity - Galaxy: stellar content - Galaxy: halo

1. Introduction

It is now generally believed that the Galaxy was formed through repeated aggregation with dwarf galaxies and this merging process left behind many streams, satellites and substructures in the Galactic halo. The chemistry and kinematics of the substructures in the halo preserve a detailed record of its formation or the Galaxy's formation. Some large surveys provided astronomers a great opportunity to study the substructures of the Galaxy photometrically and spectroscopically (Ivezić *et al.* 2000, Newberg *et al.* 2002, 2007, Belokurov 2006, 2007, Grillmair 2006, 2009). Over the past decades, many works have appeared on the metallicity of substructures such as the Sgr stream that are based either on high-resolution spectra of a small amount of stars or on low-resolution spectra of some giant stars. But the limited number of spectra is far from enough to study the metallicity of a vast region of the Galaxy. In this regard, we attempt to estimate the photometric metallicity distribution function (MDF) of the Sgr stream in south Galactic cap. This is an application of the SCUSS u-band photometric metallicity calibration (Gu *et al.* 2015).

2. Photometric metallicity estimation of the Sgr stream in the south Galactic cap

In this study, we use mainly the SCUSS u-band and SDSS Dr8 data. The SCUSS is an international cooperative project, which is undertaken by National Astronomical Observatories, Chinese Academy of Sciences and Steward Observatory, University of Arizona, USA. It is an u band (3538 Å) imaging survey program with the 90 inch (2.3 m) Bok telescope located on Kitt Peak. This survey has imaged ~ 4000 deg^2 in the south Galactic cap ($30° < l < 210°$, $-80° < b < -20°$). The limiting magnitude of SCUSS u band

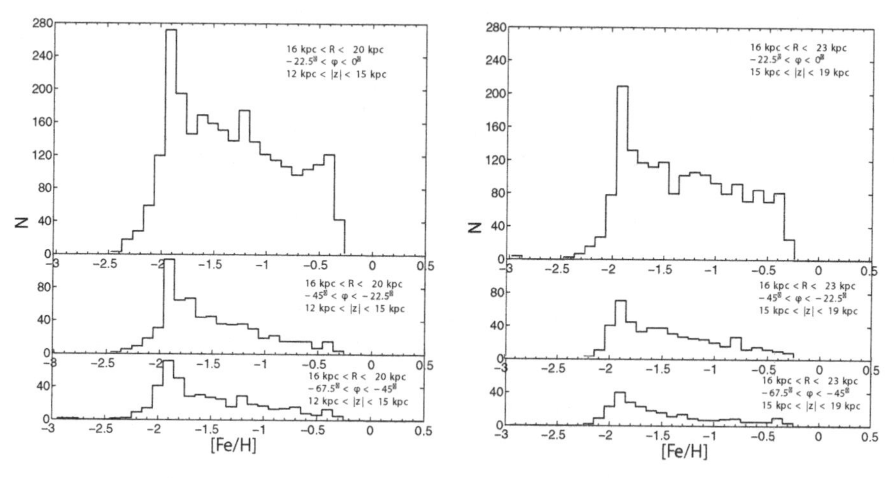

Figure 1. The metallicity distribution of the F/G ($0.2 < g - r < 0.4$) main sequence stars with $g < 21$ in different space volume. The top two histograms correspond to the Sgr stream region. The middle and the bottom histograms correspond to the vicinities of the Sgr stream region.

may be 1.5 mag deeper than that of SDSS u (Jia *et al.* 2014). Due to its relative deep magnitude and high accuracy from SCUSS data, we expect a wide application range of the photometric estimate. Gu *et al.* (2015) give the SCUSS photometric metallicity calibration. Since the photometric metallicity calibration can be used up to a faint magnitude of g=21, we can apply it to the Sgr stream stars with $20.5 < g < 21$ to analyze its MDF.

We find that a three-Gaussian model is appropriate for the MDF of the mixed population (Sgr stream and halo stars), and a two-Gaussian model for the MDF of neighboring stars (mainly halo stars). The three Gaussians of the mixed population have peaks at [Fe/H]=−1.9, [Fe/H]=−1.5 and [Fe/H]=−0.5 respectively. The two-Gaussian model representing the MDF of halo stars in the studied field has peaks at [Fe/H]=−1.5 and [Fe/H]=−1.9. This shows that the Sgr stream has a wider metallicity distribution, and that its median metallicity is richer than that of the field halo stars.

References

Belokurov, V., *et al.*, 2006, *ApJ*, 642, L137
Belokurov, V., *et al.*, 2007, *ApJ*, 657, L89
Grillmair, C. J. 2006, *ApJ*, 651, L29
Grillmair, C. J. 2009, *ApJ*, 693, 1118
Gu, J. Y., *et al.*,2015, *MNRAS*, 452, 3092
Ivezić, Ž., *et al.*, 2000, *AJ*, 120, 963
Jia, Y. P., *et al.*, 2014, *MNRAS*, 441, 503
Newberg, H., *et al.*, 2002, *ApJ*, 569, 245
Newberg, H., Yanny, B., Cole, N., Beers, T., Re Fiorentin, P., Schneider, D., & Wilhelm, R. 2007, *ApJ*, 668, 221

The General Assembly of Galaxy Halos: Structure, Origin and Evolution
Proceedings IAU Symposium No. 317, 2015
A. Bragaglia, M. Arnaboldi, M. Rejkuba & D. Romano, eds.

doi:10.1017/S1743921315008625

Tracing the Galactic Halo: Obtaining Bayesian mass estimates of the Galaxy in the presence of incomplete data

Gwendolyn Eadie[1], William Harris[1], Lawrence Widrow[2] and Aaron Springford[3]

[1]Dept. of Physics & Astronomy, McMaster University,
1280 Main St. W, Hamilton, ON, L8S 4M1, Canada
email: eadiegm@mcmaster.ca
email:harris@physics.mcmaster.ca
[2]Dept. of Physics, Engineering Physics, & Astronomy, Queen's University
Stirling Hall, 64 Bader Lane, Kingston, ON, K7L 3N6
email: widrow@astro.queensu.ca
[3]Dept. of Mathematics & Statistics, Queen's University
Jeffrey Hall, University Drive, Kingston, ON, K7L 3N6, Canada
email: aaron.springford@queensu.ca

Abstract. The mass and cumulative mass profile of the Galaxy are its most fundamental properties. Estimating these properties, however, is not a trivial problem. We rely on the kinematic information from Galactic satellites such as globular clusters and dwarf galaxies, and this data is incomplete and subject to measurement uncertainty. In particular, the complete 3D velocity vectors of objects are sometimes unavailable, and there may be selection biases due to both the distribution of objects around the Galaxy and our measurement position. On the other hand, the uncertainties of these data are fairly well understood. Thus, we would like to incorporate these uncertainties and the incomplete data into our estimate of the Milky Way's mass. The Bayesian paradigm offers a way to deal with both the missing kinematic data and measurement errors using a hierarchical model. An application of this method to the Milky Way halo mass profile, using the kinematic data for globular clusters and dwarf satellites, is shown.

Keywords. dark matter, Galaxy: general, halo, kinematics and dynamics; methods: statistical.

---

1. Introduction

The mass and mass profile of the Galaxy can be estimated using the kinematic data of tracer objects such as globular clusters (GCs) and dwarf galaxies (DGs). Armed with the tracers' position and velocity data, one can assume models for the gravitational potential and tracer density profile, and obtain parameter estimates for the mass profile. One problem with using the kinematic data of tracers, however, is that we do not always have their 3-dimensional velocity vectors; often, the proper motions of tracers have not been measured, rendering their velocity data incomplete.

Eadie, Harris, & Widrow (2015) (hereafter EHW) developed a way to use both complete and incomplete data simultaneously when estimating the mass of the Milky Way. They used a Bayesian method that employs a model's Galactocentric distribution function (DF)— similar to the method first suggested by Little & Tremaine (1987)— and that treats the unknown tangential velocities as parameters. Using a Hernquist (1990) model, they found a total mass estimate that was in agreement with many other studies (see Wang *et al.* 2015). Although measurement uncertainties were not included in the analysis, they performed a sensitivity analysis which revealed that uncertainties may contribute up to half of the uncertainty in the parameter estimates. Due to the latter finding, we now incorporate the measurements uncertainties in the analysis via a hierarchical model.

2. Method

We use the same general method as outlined in EHW to determine the mass of the Milky Way under the assumption of the isotropic Hernquist (1990) model DF, which is in the Galactocentric frame (GF). The measurement uncertainties of the kinematic data are in the Heliocentric reference frame, and are assumed to be independent and approximately Gaussian distributed. Thus, we incorporate the uncertainties into the Bayesian analysis via a likelihood in the Heliocentric frame, where the assumption of probability independence can be applied.

We use $\boldsymbol{y} = (r, v_{los}, \mu_\delta, \mu_\alpha \cos\delta)$ to denote the Heliocentric observations, and the vector $\boldsymbol{\Delta} = (\boldsymbol{\Delta v_{los}}, \boldsymbol{\Delta\mu_\delta}, \boldsymbol{\Delta(\mu_\alpha \cos\delta)})$ to denote the *known* measurement uncertainties of the velocities. We assume that $\boldsymbol{y}$ are drawn from Gaussian distributions centered on the true values $\boldsymbol{\vartheta} = (\boldsymbol{r}, \boldsymbol{v_{los}}, \boldsymbol{\mu_\delta}, \boldsymbol{\mu_\alpha} \cos\boldsymbol{\delta})$, with standard deviation equal to the measurement uncertainties $\boldsymbol{\Delta}$. The likelihood, with $\boldsymbol{\vartheta}$ as parameters and $\boldsymbol{\Delta}$ as fixed, is then,

$$\mathcal{L}(\boldsymbol{y}|\boldsymbol{\vartheta},\boldsymbol{\Delta}) = p(r|\boldsymbol{r},\boldsymbol{\Delta r})p(v_{los}|\boldsymbol{v_{los}},\boldsymbol{\Delta v_{los}})p(\mu_\delta|\boldsymbol{\mu_\delta},\boldsymbol{\Delta\mu_\delta})p(\mu_\alpha\cos\delta|\boldsymbol{\mu_\alpha}\cos\boldsymbol{\delta},\boldsymbol{\Delta\mu_\alpha}\cos\boldsymbol{\delta}). \tag{2.1}$$

Equation 2.1 is then used in a hierarchical Bayesian paradigm,

$$p(\boldsymbol{\theta}|\boldsymbol{y},\boldsymbol{\Delta}) \propto \prod_i^N \mathcal{L}(\boldsymbol{y_i},\boldsymbol{\Delta_i}|\boldsymbol{\vartheta_i})p(h(\boldsymbol{\vartheta_i})|\boldsymbol{\theta})p(\boldsymbol{\theta}) \tag{2.2}$$

where $p(h(\boldsymbol{\vartheta_i})|\boldsymbol{\theta})$ represents the model DF given parameters of interest, $\boldsymbol{\theta}$. The $\boldsymbol{\vartheta}$ parameters are transformed to the GF through a function $h(\boldsymbol{\vartheta})$, following Johnson & Soderblom (1987) (using updated values for the Solar motion). A more detailed explanation of this method will be available in Eadie, Harris, & Springford (2015) and in future works.

3. Results

The preliminary results shown in the lower part of the poster (see supplementary material), are noticeably different than the results presented in EHW. Their estimate for the total mass, assuming the isotropic Hernquist model, was $1.55 \times 10^{12} M_\odot$ with a 95% credible interval of $(1.73, 1.42) \times 10^{12} M_\odot$. Here, when the same model is assumed but the hierarchical method is adopted, the estimate is $0.78 \times 10^{12} M_\odot$, with a 95% credible interval $(0.69, 0.90) \times 10^{12} M_\odot$. The discrepancy in these two results may be explained by the incorporation of measurement uncertainties. EHW already showed that high-velocity objects such as Pal 3 have significant influence on the mass estimate— when these objects' velocity uncertainties are taken into account, the tracers may carry less weight.

References

Eadie, G., Harris, W., & Widrow, L. M. 2015, *ApJ*, 806, 54
Eadie, G., Harris, W., & Springford, A. 2015, *Joint Statistical Meetings Proceedings*, in press.
Hernquist, L. 1990 *ApJ*, 356, 359
Johnson, D. & Soderblom, D. 1987 *AJ* 93, 4
Little, B. & Tremaine, S. 1987, *ApJ*, 320, 493
Wang, W., Han, J., Cooper, A., Shaun, C., Frenk, C., & Lowing, B. 2015 *MNRAS*, 453, 377

The General Assembly of Galaxy Halos: Structure, Origin and Evolution
Proceedings IAU Symposium No. 317, 2015
A. Bragaglia, M. Arnaboldi, M. Rejkuba & D. Romano, eds.

doi:10.1017/S1743921315006894

Halo formation and evolution: unification of structure and physical properties

Allan D. Ernest and Matthew P. Collins

Charles Sturt University,
Locked Bag 588, Wagga Wagga, Australia, 2678
email: aernest@csu.edu.au

Abstract. The assembly of matter in the universe proliferates a wide variety of halo structures, often with enigmatic consequences. Giant spiral galaxies, for example, contain both dark matter and hot gas, while dwarf spheroidal galaxies, with weaker gravity, contain much larger fractions of dark matter, but little gas. Globular clusters, superficially resembling these dwarf spheroidals, have little or no dark matter. Halo temperatures are also puzzling: hot cluster halos contain cooler galaxy halos; dwarf galaxies have no hot gas at all despite their similar internal processes. Another mystery is the origin of the gas that galaxies require to maintain their measured star formation rates (SFRs). We outline how gravitational quantum theory solves these problems, and enables baryons to function as weakly-interacting-massive-particles (WIMPs) in Lambda Cold Dark Matter (LCDM) theory. Significantly, these dark-baryon ensembles may also be consistent with primordial nucleosynthesis (BBN) and cosmic microwave background (CMB) anisotropies.

Keywords. halo formation and evolution, dark matter, gravitational quantum theory

---

1. Introduction

The application of quantum mechanics to gravity and its experimental verification is now well established (Jenke *et al.* 2011). Using this approach, Ernest (2009) has shown that quantum theory predicts the existence of vast numbers of extremely long lived, well-bound, stable and electromagnetically invisible Rydberg-type gravitational eigenstates in large-scale deep gravity wells like those of galactic halos. Baryons with wavefunction ensembles containing large fractions of these dark states will exhibit WIMP-like behavior. Furthermore, the larger a gravitational well is, the greater its "percentage" of well-bound dark states. Quantum statistical analysis also shows that these dark baryons are more highly favored on approach to equilibrium (Ernest & Collins 2014): larger, deeper wells with more relaxed halos contain the largest fractions of dark baryons.

2. Aligning halo structure and properties in the quantum approach

It can be shown (Ernest, 2006) that baryons could have rapidly formed dark-baryonic-ensemble halos in the wells of super-massive primordial black holes (SMPBHs) formed by direct collapse (Carr, 1975) at the last phase transition (e^+/e^-), very early in universal history. These dark-eigenspectral-ensemble baryons made up the dark fraction (~4/5) of matter in the universe, the remaining, more-strongly-interacting ensembles forming the traditional baryonic component. Importantly, because they behave like WIMPS in most other ways, dark-state baryons can potentially be consistent with BBN and CMB anisotropies. (Dark-ensemble baryons have different BBN cross sections to lab-based, localized-particles and will not strongly participate in baryon-photon oscillation.) Examples of unifying structure with properties using the quantum approach are given below.

(1) *Dark matter-hot gas ratio correlated with halo type and evolutionary path*: Dwarf elliptical galaxies and globular clusters seem like very different objects but in the quantum scenario they are the un-merged descendants of the original primordial dark-eigenspectral-ensemble halos, whose structural differences relate to their evolutionary paths. Since quasi-equilibrium favors dark states, an un-merged, relaxing halo will have its non-coalesced dark-baryon fraction grow larger with time. Dwarf spheroidal galaxies are examples of these relaxed systems. Their eigenstate equivalent kinetic energies ($\propto$ total mass to radius ratio M_0/R_0) are $<< 1$ eV, so that their eigenspectrally-visible baryonic component does not emit X-rays, and their large dark matter component is actually due their large reservoir of dark-ensemble baryonic gas. Globular clusters are initially similar structures but have their visible-eigenspectral baryon fraction stripped by passage through the denser galactic medium. Particle-particle interactions (expected from theory to be stronger than electromagnetic ones) act to equilibrate the dark-visible imbalance, eventually resulting in stripping the dark component as well, leaving globular clusters with observationally little gas or dark matter. Massive spiral and elliptical galaxies, formed from merging, have halos which also gradually darken, as any remaining un-coalesced baryons shift to statistically-favored dark eigenspectral ensembles. Hence, left undisturbed, active galaxies eventually become quiescent ellipticals. The dark eigenspectral ensembles can function as hidden repositories of baryons however, and quiescent ellipticals with baryons in these reservoirs can be reborn as active galaxies on merging or disruption. Merging shifts particle eigenspectra away from the dark eigenspectral repository, and provides a fresh supply of coalescent-compliant baryons that increase SFRs.

(2) *Correlation of halo gas temperature with effective quantum state kinetic energy*: The quantum approach predicts a clear correlation of halo gas temperature T with eigenspectral equivalent kinetic energy. For clusters of galaxies, $M_0/R_0 \sim 2 \times 10^{22}$ kg/m, implying $T_{theor.} \sim 5 \times 10^3$ eV, while $T_{obs.} \sim 3 - 10 \times 10^3$ eV. There is equally good correlation for massive spirals: 2×10^2 eV (*theor.*), $1 - 2.5 \times 10^2$ eV (*obs.*); and ellipticals: 8×10^2 eV (*theor.*), $1 - 10 \times 10^2$ eV (*obs.*). Dwarf ellipticals show no sign of hot gas ($T_{theor.} \sim$ 0.03 eV), yet the same hot-gas production mechanisms occur in these dwarf galaxies as in larger galaxies, and its observational absence strengthens the voracity of the quantum approach. Lastly, the Sun's hot corona surrounding its relatively cool photosphere is also explained: in the quantum approach the corona's high storage temperature is predicted from the average of its longer-lived eigenstate kinetic energies ($T_{theor.} \sim 350$ eV).

3. Summary

If one accepts the universal truth of quantum theory, then it is clear that it both predicts and requires un-coalesced baryons to be dark on large scales. In this context it is difficult to see how a quantum approach does *not* provide the solution to the dark matter problem. A universe having WIMPs as baryons in a dark disguise adds a new dimension to understanding galaxy evolution and other astrophysical problems by providing opportunity for dark and visible matter inter-conversion throughout cosmic history.

References

Carr, B. 1975, *ApJ*, 201, 1

Ernest, A. D. 2006, in J. Val Blain (ed.), *Dark Matter: New Research*, (New York: Nova) p. 91

Ernest, A. D. 2009, *J. Phys. A: Math. Gen.*, 42, 115207, 115208

Ernest, A. D. & Collins, M. P. 2014, *AIP Congress, 7-11 Dec., 2014 Canberra, Australia*

Jenke, T., Geltenbort,P., Lemmel, H.,& Abele, H. 2011, *Nature Phys*, 7, 468

The General Assembly of Galaxy Halos: Structure, Origin and Evolution
Proceedings IAU Symposium No. 317, 2015
A. Bragaglia, M. Arnaboldi, M. Rejkuba & D. Romano, eds.

doi:10.1017/S1743921315007073

Lithium evolution from Pre-Main Sequence to the Spite plateau: an environmental solution to the cosmological lithium problem

Xiaoting Fu[1], Alessandro Bressan[1], Paolo Molaro[2], Paola Marigo[3]

[1]SISSA - International School for Advanced Studies,
via Bonomea 265, 34136 Trieste, Italy
email: xtfu@sissa.it, sbressan@sissa.it

[2]INAF - Osservatorio Astronomico di Trieste,
via G. B. Tiepolo 11, 34143 Trieste, Italy
email: molaro@oats.inaf.it

[3]Dipartimento di Fisica e Astronomia, Università di Padova,
Vicolo dell'Osservatorio 2, I-35122 Padova, Italy
email: paola.marigo@unipd.it

Abstract. Lithium abundance derived in metal-poor main sequence stars is about three times lower than the primordial value of the standard Big Bang nucleosynthesis prediction. This disagreement is referred to as the lithium problem. We reconsider the stellar Li evolution from the pre-main sequence to the end of main sequence phase by introducing the effects of overshooting and residual mass accretion. We show that ^{7}Li could be significantly depleted by convective overshooting in the pre-main sequence phase and then partially restored in the stellar atmosphere by residual accretion which follows the Li depletion phase and could be regulated by EUV photo-evaporation. By considering the conventional nuclear burning and diffusion along the main sequence we can reproduce the Spite plateau for stars with initial mass $m_0 = 0.62 - 0.80\ M_\odot$, and the Li declining branch for lower mass dwarfs, e.g, $m_0 = 0.57 - 0.60\ M_\odot$, for a wide range of metallicities (Z=0.00001 to Z=0.0005), starting from an initial Li abundance $A(\mathrm{Li}) = 2.72$.

Keywords. stars: abundances, stars: pre–main-sequence, stars: Population II

---

1. Model

Envelope overshooting. Overshooting (OV) may occur at the borders of any convectively unstable region. As the star moves toward the zero age main sequence (ZAMS), we vary the overshoot efficiency proportionally to the mass size of the outer unstable convective region (f_{cz}), until the value estimated for the Sun $\Lambda_e = 0.3H_p$ (Christensen-Dalsgaard *et al.* 2011) is reached: $\Lambda_e = 0.3 + (1.5 - 0.3) * f_{cz}$.

Residual accretion. After the main accretion during the proto-stellar phase, the star leaves the stellar birth-line, while a residual accretion keeps going on. We assume that the residual accretion begins at τ_{acc} when deuterium burning ends. The accretion rate $\dot{M}_{acc} = 2 \times 10^{-8}(t/\tau_{acc})^{-0.9}$ $[M_\odot/\mathrm{yr}]$ is applied to recover most of the observed residual accretion rates. Fig. 1 illustrates the effect of late accretion on the Li evolution for a $m_0 = 0.75M_\odot$ star. The accretion material falling onto the star contains initial ^{7}Li. Even if accretion is very small, it restores the surface ^{7}Li towards the initial value.

EUV-photoevaporation. Late accretion will last until the remaining gas reservoir is consumed or some feedback mechanism from the star itself is able to clean the nearby disk. Extremely UV (EUV) photons have energy high enough to heat the disk surface gas and flow it away. The disk mass loss is given by Dullemond *et al.* (2007): $\dot{M}_{EUV} \sim$

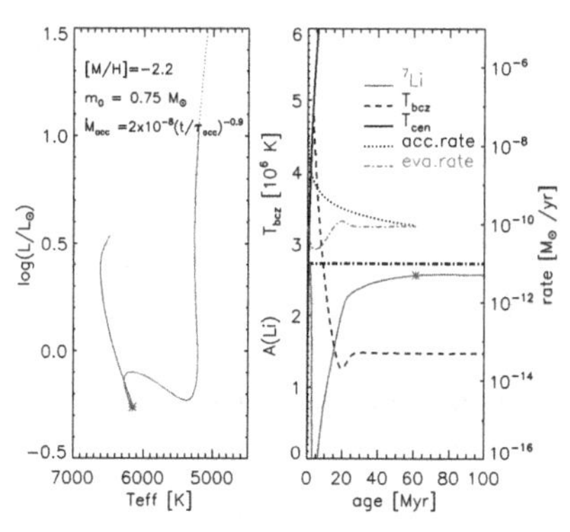

Figure 1. PMS evolution for a $m_0 = 0.75 M_\odot$, [M/H]=-2.2 star. Left panel: H-R diagram of the evolutionary track. The solid line starts when deuterium burning ends. The asterisk marks the end of the residual accretion. Right panel: Li evolution starting from A(Li)=2.72 (horizontal line). The temperatures at the center (T_{cen}, dark red dot dot dashed line) and at the base of the convection zone (T_{bcz}, dark red dashed line) are also shown. The accretion rate is the dark blue dotted line while the EUV photo-evaporation is the dark green dot dashed line.

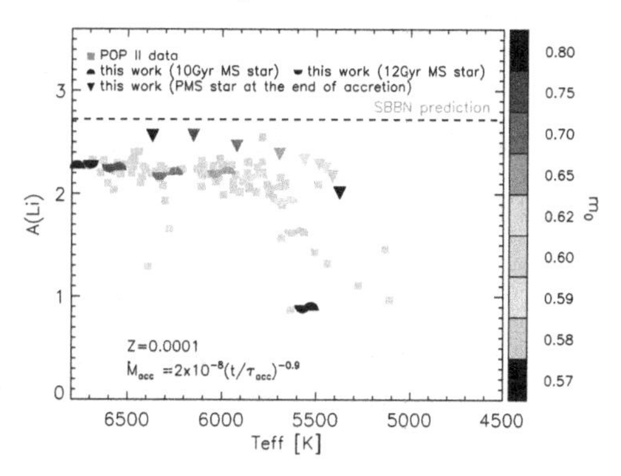

Figure 2. Our results in comparison with the lithium abundance measurements in Pop II stars. The grey filled squares are Pop II data from Molaro *et al.*(2012). Our predictions are shown for stars at the end of the late accretion phase (filled triangles), and on the main sequence at 10 Gyr (filled upper circle) and 12 Gyr (filled lower circle). Symbols are color-coded according to the initial stellar mass (legend on the right). The black dashed line marks the primordial Li abundance according to the SBBN.

$4 \times 10^{-10}(\Phi_{EUV}/(10^{41}s^{-1}))^{0.5}(M_*/M_\odot)^{0.5}$ $[M_\odot/\mathrm{yr}]$, where Φ_{EUV} is the EUV photon luminosity produced by the central star as a black body. When the evaporation rate is larger than the accretion rate, the Li restore together with the accretion effectively stops.

Main sequence diffusion and Li burning. During the MS phase Li could be burned at the convective zone base. For masses larger than $m_0 = 0.60$ M$_\odot$, Li burning is insignificant. Another effect, microscopic diffusion, leads to a depletion of the surface elements.

2. Results

By considering the PMS and MS Li evolution, at the age of the Pop II stars (10-12 Gyr), $m_0 = 0.62$ M$_\odot$ - 0.80 M$_\odot$ stars nicely populate the Spite plateau (Fig.2). The Li declining branch is also reproduced by stars with $m_0 = 0.57$ M$_\odot$ - 0.60 M$_\odot$. Both PMS Li depletion and MS diffusion/burning contribute to the total A(Li) decrease, with the former process playing the main role. With the same parameters, we can reproduce the plateau and the Li declining branch, over a wide range of metallicities (from Z=0.00001 to Z=0.0005, which is from [M/H]=-3.2 to [M/H]=-1.5). This environmental Li evolution model (Fu *et al.* 2015) also offers the possibility to interpret the decrease of Li abundance in extremely metal-poor stars, the Li disparities in spectroscopic binaries and the low Li abundance in planet hosting stars.

References

Christensen-Dalsgaard, J., Monteiro, M. J. P. F. G., Rempel, M., *et al.* 2011, *MNRAS*, 414, 1158

Dullemond, C. P., Hollenbach, D., Kamp, I., & D'Alessio, P. 2007, in: B. Reipurth, D. Jewitt, & K. Keil (eds.), *Protostars and Planets V*, (Tucson, AZ: Univ. Arizona Press), p. 555

Fu, X., Bressan, A., Molaro, P., & Marigo, P. 2015, *MNRAS*, 452, 3256

Molaro, P., Bressan, A., Barbieri, M., *et al.* 2012, *MemSAIt*, 22, 233

The General Assembly of Galaxy Halos: Structure, Origin and Evolution
Proceedings IAU Symposium No. 317, 2015
A. Bragaglia, M. Arnaboldi, M. Rejkuba & D. Romano, eds.

doi:10.1017/S174392131500681X

Hot subdwarf stars in the Galactic halo Tracers of prominent events in late stellar evolution

Stephan Geier[1,2], Thomas Kupfer[3], Veronika Schaffenroth[1,4], Ulrich Heber[4] and the MUCHFUSS collaboration

[1]European Southern Observatory, Karl-Schwarzschild-Str. 2, 85748 Garching, Germany

[2]Dr. Karl Remeis-Observatory & ECAP, Astronomical Institute, Friedrich-Alexander University Erlangen-Nürnberg, Sternwartstr. 7, D 96049 Bamberg, Germany

[3]Department of Astrophysics/IMAPP, Radboud University Nijmegen, P.O. Box 9010, 6500 GL Nijmegen, The Netherlands

[4]Institute for Astro- and Particle Physics, University of Innsbruck, Technikerstr. 25/8, 6020 Innsbruck, Austria

Abstract. Hot subdwarf stars (sdO/Bs) are the stripped cores of red giants located at the bluest extension of the horizontal branch. They constitute the dominant population of UV-bright stars in old stellar environments and are most likely formed by binary interactions. We perform the first systematic, spectroscopic analysis of a sample of those stars in the Galactic halo based on data from SDSS. In the course of this project we discovered 177 close binary candidates. A significant fraction of the sdB binaries turned out to have close substellar companions, which shows that brown dwarfs and planets can significantly influence late stellar evolution. Close hot subdwarf binaries with massive white dwarf companions on the other hand are good candidates for the progenitors of type Ia supernovae. We discovered a hypervelocity star, which not only turned out to be the fastest unbound star known in our Galaxy, but also the surviving companion of such a supernova explosion.

Keywords. binaries: spectroscopic, binaries: eclipsing, stars: subdwarfs, stars: brown dwarfs

---

Hot subdwarf stars (sdO/Bs) are evolved core helium-burning stars with very thin hydrogen envelopes. About half of the sdB stars are in close binaries and are formed by common envelope ejection. The companions are in most cases either late main sequence stars of spectral type M or compact objects like white dwarfs (WDs). Subdwarf binaries with massive CO-WD companions are candidates for supernova type Ia (SN Ia) progenitors. The project Massive Unseen Companions to Hot Faint Underluminous Stars from SDSS (MUCHFUSS) aims at finding the sdB binaries with the most massive compact companions like massive WDs, neutron stars or black holes as well as the least massive companions like substellar objects (e.g. Geier *et al.* 2015b).

We selected and classified about $\sim$ 1400 hot subdwarf stars from the Sloan Digital Sky Survey (SDSS DR7) by colour selection and visual inspection of their spectra. Stars with high velocities have been reobserved and individual SDSS spectra have been analysed. In total 177 radial velocity variable subdwarfs have been discovered and 1914 individual radial velocities measured. We constrain the fraction of close massive companions of H-rich hot subdwarfs to be smaller than $\sim$ 1.3% (Geier *et al.* 2015b). Light curves with a duration of typically 2-3 hours have been obtained of 66 subdwarf binaries from our target list. We found three eclipsing systems, two of them with the first confirmed brown dwarf companions, and one hybrid sdB pulsator with reflection effect.

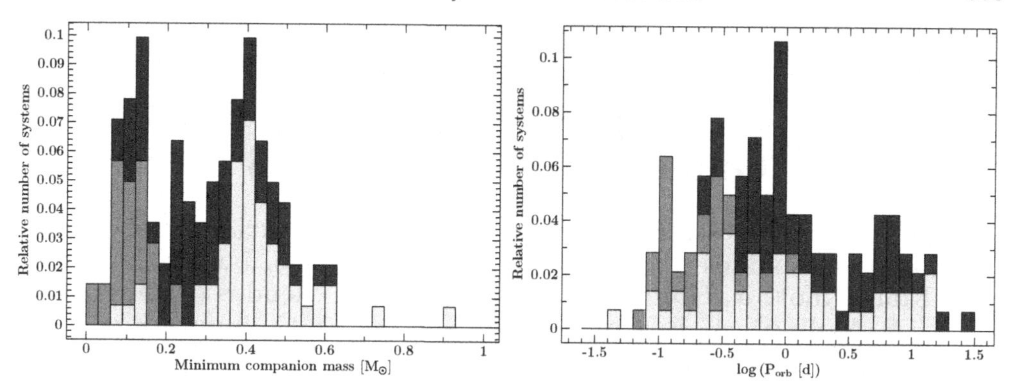

Figure 1. Histogram of minimum companion masses and orbital periods (light grey: WD companions, grey: dM companion, dark grey: unknown companion type).

We determined orbital parameters of 30 close sdB binaries and studied the known population. The distribution of minimum companion masses is bimodal. One peak around $0.1\,M_\odot$ corresponds to the low-mass main sequence (dM) and substellar companions. The other peak around $0.4\,M_\odot$ corresponds to the WD companions. The derived masses for the WD companions are significantly lower than the average mass for single carbon-oxygen WDs (See Fig. 1, Kupfer *et al.* 2015).

In the course of the MUCHFUSS project, we discovered the first close sdB systems with brown dwarf companions and, subsequently, another even closer system (Geier *et al.* 2011; Schaffenroth *et al.* 2014, 2015). Such binaries are important to probe common envelope evolution and study the yet unknown influence of substellar objects like brown dwarfs or giant planets on stellar evolution. We derive a number fraction of $> 4\%$ close sdB binaries with brown dwarf companions and a similar fraction of close binaries with M dwarf companions. Substellar companions are therefore as frequent as low-mass stellar companions at short orbital periods.

We detected high RV-variability of the bright sdB CD-30°11223. Photometric follow-up revealed both shallow transits and eclipses. The binary system, which is composed of a carbon/oxygen WD ($\sim 0.76\,M_\odot$) and an sdB ($\sim 0.51\,M_\odot$), has the shortest orbital period ($P \simeq 0.049$ d) ever measured for a hot subdwarf binary (Geier *et al.* 2013). In the future the system will transfer mass from the helium star to the WD. After a critical amount of helium is deposited on the surface, the helium is ignited. Modelling this process shows that the detonation in the accreted helium layer is sufficiently strong to trigger the explosion of the core. Thermonuclear supernovae have been proposed to originate from this so-called double-detonation of a WD. The surviving donor hot subdwarf star will then be ejected with its orbital velocity. The properties of such a remnant match the hypervelocity star US 708, a He-sdO moving with $\sim 1200\,\mathrm{km\,s^{-1}}$. This object is the fastest unbound star known in our Galaxy (Geier *et al.* 2015a).

References

Geier, S., Schaffenroth, V., Drechsel, H., *et al.* 2011, *ApJ*, 731, L22

Geier, S., Marsh, T. R., Wang, B., *et al.* 2013, *A&A*, 554, 54

Geier, S., Fürst, F., Ziegerer, E., *et al.* 2015a, *Science*, 347, 1126

Geier, S., Kupfer, T., Heber, U., *et al.* 2015b, *A&A*, 577, 26

Kupfer, T., Geier, S., Heber, U., *et al.* 2015, *A&A*, 576, 44

Schaffenroth, V., Geier, S., Heber, U., *et al.* 2014, *A&A*, 564, 98

Schaffenroth, V., Barlow, B. N., Drechsel, H., & Dunlap, B. H. 2015, *A&A*, 576, 123

The General Assembly of Galaxy Halos: Structure, Origin and Evolution
Proceedings IAU Symposium No. 317, 2015
A. Bragaglia, M. Arnaboldi, M. Rejkuba & D. Romano, eds.

doi:10.1017/S1743921315008856

Searching for planetary nebulae at the Galactic halo via J-PAS

Denise R. Gonçalves[1], T. Aparício-Villegas[2,3], S. Akras[1], A. Cortesi[4], M. Borges-Fernandes[2], S. Daflon[2], C. B. Pereira[2], S. Lorenz-Martins[1], W. Marcolino[1], A. Kanaan[5], K. Viironen[6], C. Mendes de Oliveira[4], A. Molino[4], A. Ederoclite[6] and the J-PAS Collaboration

[1]Observatório do Valongo - Universidade Federal do Rio de Janeiro, Brazil; email: denise@astro.ufrj.br [2]Observatório Nacional - Ministério de Ciência e Tecnologia, Brazil; [3]Instituto de Astrofísica de Andalucía, Spain [4]IAG - Universidade de São Paulo, Brazil; [5]Departamento de Física - Universidade Federal de Santa Catarina, Brazil; [6]Centro de Estudios de Física del Cosmos de Aragón, Spain.

Abstract. The Javalambre-Physics of the Accelerating Universe Astrophysical Survey (J-PAS) is a narrow-band imaging, very wide field cosmological survey. It will last 5 years and will observe 8500 sq. deg. of the sky. There will be 54 contiguous narrow-band filters of 145Å FWHM, from 3,500 to 10,000Å. Two broad-band filters will be added at the extremes, UV and IR, plus the 3 –g, r, and i– SDSS filters. Thus, J-PAS can be an important tool to search for new planetary nebulae (PNe) at the halo, increasing their numbers, because only 14 of them have been convincingly identified in the literature. Halo PNe are able to reveal precious information for the study of stellar evolution and the early chemical conditions of the Galaxy. The characteristic low continuum and intense emission lines of PNe make them good objects to be searched by J-PAS. Though covering a significantly smaller sky area, data from the ALHAMBRA survey were used to test our J-PAS strategy to search for PNe. Our first results are shown in this contribution.

1. Halo PNe properties versus J-PAS magnitudes

Halo PNe like BoBn 1, DdDm 1 and PS 1 are located somewhere between 11 and 24 kpc from the Sun, and have B magnitudes of 16, 14 and 13.4, respectively (Howard *et al.* 1997). Such values are easily encompassed by the J-PAS limits (Benítez *et al.* 2014).

We have used the J-PAS narrow-band photometry to study the possibility of detecting PNe (Benítez *et al.* 2014). From the observed spectra of known halo PNe, we got the J-PAS synthetic photometry. In Fig. 1 we show the observed spectrum of a halo PN (top panel), as well its J-PAS "SED" (in magnitudes; bottom-left panel), in which it is straightforward to see the two conspicuous emissions (associated with the intense [O III] and H_α emission lines). Therefore, halo PNe can easily be detected by the J-PAS survey.

2. ALHAMBRA: PNe in the Galactic halo or in nearby galaxies?

Using the average characteristics of halo PNe, we constructed a grid of CLOUDY models to simulate the spectra of these objects. With these simulated spectra, other 4 observed spectra (MWC 574, Pereira & Miranda 2007; DdDm 1, Kwitter *et al.* 1998; NGC 2242, Kwitter *et al.* 2003; and H 4-1, SDSS, respectively), and the Viironen *et al.* (2009) colour-colour diagram, we defined the location where halo PNe are better separated from other emission line objects, in the $r' - H_\alpha$ versus $r' - i'$ plane.

Though covering a significantly smaller sky area, data from the ALHAMBRA survey (Moles *et al.* 2008) were used to test our strategy. The ALHAMBRA photometric system (Aparıcio-Villegas *et al.* 2010) consists of 20 optical medium-band filters, plus the

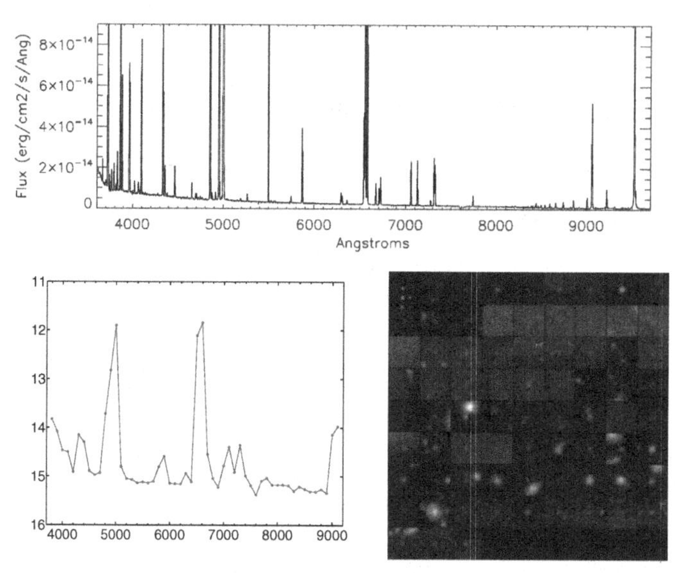

Figure 1. *Top*: spectrum of DdDm 1 (Kwitter *et al.* 1998). *Bottom – left*: J-PAS "SED" based on this spectrum. *Bottom – right*: composite (Subaru-B, Subaru-R and F814-HST) images of the selected PN candidates.

three standard *JHKs* near-infrared (NIR) bands. Because of the availability of the NIR magnitudes, the Viironen *et al.* (2009) near-IR ($J - H$ versus $H - K_s$, left panel of their Fig. 1) diagram was also used to better define the PN candidates. Using the optical diagram we identified 75 possible PN candidates in the ALHAMBRA fields, number which was significantly reduced by using the NIR plane. Candidates found this way were then filtered by having low ODDS (that measures how well a source fits to a template of any kind of galaxy; see Molino *et al.* 2014). This fact should help eliminating high-redshift galaxies from the candidate list (in ALHAMBRA the galaxy distribution peaks at z=0.3, with only a few reaching z=1). High-z (3.1 to 3.3) AGNs and QSOs, with Lyα 1216Å and CIV 1549Å mimicking the [O III] and H$_\alpha$ emission-lines, could appear among our candidates (Ciardullo *et al.* 2002) – we are testing other colour-colour diagrams, for J-PAS, to avoid these high-z impostors. We ended up with 16 bona-fide ALHAMBRA PN candidates. Their nature is being confirmed via follow-up spectroscopy (proposals were submitted for 2016A). The candidates we identified are faint (from 19 to 24 in V mag), therefore, instead of halo PNe, they are most probably extragalactic PNe. If proposals are successful, results will be available before the start of the J-PAS survey.

References

Aparicio-Villegas, T., Alfaro, E. J., Cabrera-Caño, J., Moles, M., *et al.* 2010, *AJ*, 139, 1242

Benítez, N., Dupkeb, R., Moles, M., Sodré, L., Cenarro, A. J., *et al.* 2014, *astro-ph/*, arXiv, 1403.5237

Ciardullo, R, Feldmeier, J. J., Krelove, K., *et al.* 2002, *ApJ*, 566, 784

Howard, J. W., Henry, R. B. C., & McCartney, S. 1997, *MNRAS*, 284, 654

Kwitter K. B. & Henry R. B. C. 1998, *ApJ*, 493, 247

Kwitter, K. B., Henry, R. B. C., & Milingo, J. B. 2003, *PASP*, 115, 80

Moles M., Benítez N., Aguerri J. A. L., Alfaro E. J., Broadhurst T., *et al.* 2008, *AJ*, 136, 1325

Molino A., Benítez N., Moles M., *et al.* 2014, *MNRAS*,441, 2891

Pereira, C. B. & Miranda, L. F. 2007, *A&A*, 467, 124

Viironen, K., Mampaso, A., Corradi, R. L. M., Rodríguez, M., *et al.* 2009, *A&A*, 502, 465

The General Assembly of Galaxy Halos: Structure, Origin and Evolution
Proceedings IAU Symposium No. 317, 2015
A. Bragaglia, M. Arnaboldi, M. Rejkuba & D. Romano, eds.

doi:10.1017/S1743921315006717

A universality of dark-halo surface density for the Milky Way and Andromeda dwarf satellites as a probe of the coldness of dark matter

Kohei Hayashi[1] **and Masashi Chiba**[2]

[1]Kavli Institute for the Physics and Mathematics of the Universe, University of Tokyo, 5-1-5 Kashiwa-no-ha, Kashiwa, 277-8568, Japan
email: kohei.hayashi@ipmu.jp

[2]Astronomical Institute, Tohoku University, 6-3 Aoba-ku, Sendai, 980-8578, Japan
email: chiba@astr.tohoku.ac.jp

Abstract. We propose a new astrophysical test on the nature of dark matter based on the properties of dark halos associated with dwarf spheroidal galaxies. The method adopts a mean surface density of a dark halo defined within a radius of maximum circular velocity, which is derivable for a wide variety of galaxies with any dark-matter density profiles. We find that even though dark halo density profiles are derived based on the different assumptions for each galaxy sample, this surface density is generally constant across a wide mass range of galaxy. We find that at higher halo-mass scales, this constancy for real galaxies can be naturally reproduced by both cold and warm dark matter (CDM and WDM) models. However, at low-mass scales, for which we have estimated from the Milky Way and Andromeda dwarf satellites, the mean surface density derived from WDM models largely deviates from the observed constancy, whereas CDM models are in reasonable agreement with observations.

1. Mean Surface Density of Dark Halo

In order to obtain severer limits on dark matter models, we define the mean surface density of a dark matter halo within the radius of maximum circular velocity,

$$\Sigma_{V_{\rm max}} = \frac{M(r_{\rm max})}{\pi r_{\rm max}^2}, \tag{1.1}$$

where

$$M(r_{\rm max}) = \int_0^{r_{\rm max}} 4\pi\rho_{\rm dm}(r')r'^2 dr', \tag{1.2}$$

$\rho_{\rm dm}(r)$ indicates *any* dark matter density profiles, and $r_{\rm max}$ is a radius at maximum circular velocity, $V_{\rm max}$, of assumed dark halo profiles, $\rho_{\rm dm}$.

Using this definition (1.1), we estimate $\Sigma_{V_{\rm max}}$ for late- and early-type spirals with pseudo-isothermal dark halos (de Blok *et al.* 2008; Spano *et al.* 2008), dwarf irregulars with Burkert dark halos (Gentile *et al.* 2005, 2007), galaxy–galaxy weak lensing sample of spiral and elliptical galaxies with Burkert profiles (Donato *et al.* 2009), and MW's and M31's dwarf spheroidal galaxies with double-power law density profiles (Hayashi & Chiba 2015). The symbols with error bars in Figure 1 show the estimated mean surface density, $\Sigma_{V_{\rm max}}$, of the above data sample as a function of $V_{\rm max}$. It is found that even though dark halo density profiles are derived based on the different assumptions for each

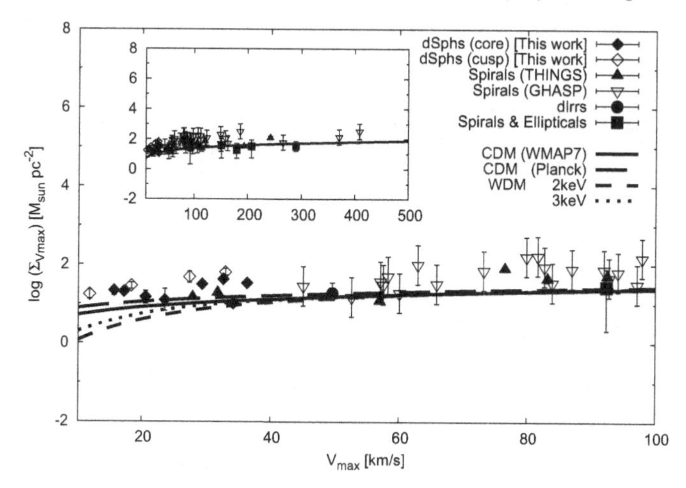

Figure 1. Mean surface density of a dark halo, $\Sigma_{V_{\max}}$ as a function of $V_{\max}$ for different type of galaxies (each point with error bars) and for theoretical prediction from ΛCDM (solid lines) and ΛWDM-based (dashed lines) N-body simulations.

galaxy sample and estimated by several independent methods, this surface density is sufficiently constant across a wide range of $V_{\max}$ of about 10 to 400 km s^{-1}.

2. Constraints on dark matter models

This mean surface density can be directly computed from the theoretical predictions based on cosmological N-body simulations within the Λ-dominated CDM and WDM models. First, we adopt the NFW dark matter density profile, which can reproduce CDM (Navarro *et al.* 1997) and WDM (Lovell *et al.* 2014) dark halos. We then evaluate the scale density and the scale length in the cosmological context and the mass-concentration relation at $z = 0$. Finally, using dark matter density profiles obtained by the above procedure, we calculate $\Sigma_{V_{\max}}$ for CDM (WMAP7 and Planck) and WDM (particle masses with 2 and 3 keV) models.

The solid and dashed lines in Figure 1 indicate the predicted $\Sigma_{V_{\max}}$ versus $V_{\max}$ for CDM and WDM models, respectively. It is found that at higher halo-mass scales, this constancy for real galaxies can be naturally reproduced by both dark matter models, even though we do not perform any fitting to the data. However, at dwarf-galaxy mass scales the mean surface density derived from WDM models deviates systematically from the observed universality, whereas CDM models are in reasonable agreement with observations. Thus, this test supports CDM models.

References

de Blok, W. J. G., Walter, F., Brinks, E., *et al.* 2008, *AJ*, 136, 2648
Donato, F., Gentile, G., Salucci, P., *et al.* 2009, *MNRAS*, 397, 1169
Gentile, G., Burkert, A., Salucci, P., Klein, U., & Walter, F. 2005, *ApJL*, 634, L145
Gentile, G., Salucci, P., Klein, U., & Granato, G. L. 2007, *MNRAS*, 375, 199
Hayashi, K. & Chiba, M. 2015, *ApJ*, 810, 22
Lovell, M. R., Frenk, C. S., Eke, V. R., *et al.* 2014, *MNRAS*, 439, 300
Navarro, J. F., Frenk, C. S., & White, S. D. M. 1997, *ApJ*, 490, 493
Spano, M., Marcelin, M., Amram, P., *et al.* 2008, *MNRAS*, 383, 297

The General Assembly of Galaxy Halos: Structure, Origin and Evolution
Proceedings IAU Symposium No. 317, 2015
A. Bragaglia, M. Arnaboldi, M. Rejkuba & D. Romano, eds.

doi:10.1017/S1743921315007139

Chemo-dynamical evolution model: Enrichment of r-process elements in the Local Group dwarf galaxies

Yutaka Hirai[1,2,3], Yuhri Ishimaru[4,5], Takayuki R. Saitoh[6], Michiko S. Fujii[2], Jun Hidaka[7,2] and Toshitaka Kajino[2,1]

[1]Department of Astronomy, Graduate School of Science, The University of Tokyo, 7-3-1 Hongo, Bunkyo-ku, Tokyo 113-0033, Japan
email: yutaka.hirai@nao.ac.jp

[2]Division of Theoretical Astronomy, National Astronomical Observatory of Japan, 2-21-1 Osawa, Mitaka, Tokyo 181-8588, Japan

[3]Research Fellow of Japan Society for the Promotion of Science

[4]Department of Natural Sciences, International Christian University, 3-10-2 Osawa, Mitaka, Tokyo 181-8585, Japan

[5]Institut d'Astrophysique de Paris, 98bis Boulevard Arago, 75014, Paris, France

[6]Earth-Life Science Institute, Tokyo Institute of Technology, 2-12-1 Ookayama, Meguro-ku, Tokyo 152-8551, Japan

[7]School of Science and Engineering, Meisei University, 2-1-1 Hodokubo, Hino, Tokyo 191-0042, Japan

Abstract. Neutron star mergers are one of the candidate astrophysical site(s) of r-process. Several chemical evolution studies however pointed out that the observed abundance of r-process is difficult to reproduce by neutron star mergers. In this study, we aim to clarify the enrichment of r-process elements in the Local Group dwarf galaxies. We carry out numerical simulations of galactic chemo-dynamical evolution using an N-body/smoothed particle hydrodynamics code, ASURA. We construct a chemo-dynamical evolution model for dwarf galaxies assuming that neutron star mergers are the major source of r-process elements. Our models reproduce the observed dispersion in [Eu/Fe] as a function of [Fe/H] with neutron star mergers with a merger time of 100 Myr. We find that star formation efficiency and metal mixing processes during the first $\lesssim$ 300 Myr of galaxy evolution are important to reproduce the observations. This study supports that neutron star mergers are a major site of r-process.

Keywords. galaxies: abundances — galaxies: dwarf — galaxies: evolution — Local Group — methods: numerical

---

Neutron star mergers (NSMs) are one of the promising astrophysical sites of r-process. Argast *et al.* (2004) suggested that it is difficult to reproduce the observed star-to-star scatters in extremely metal-poor stars due to the long merger time ($\sim$ 100 Myr) and low rate ($\sim 10^{-4}$ yr^{-1} for a Milky Way size galaxy) of NSMs. On the other hand, Ishimaru *et al.* (2015) pointed out that this problem can be solved if the Milky Way halo was formed from sub-haloes with low star formation efficiency. Enrichment of r-process elements (e.g., Eu) in dwarf galaxies which would be the building blocks of the Milky Way is not understood yet. We aim to clarify the enrichment of r-process elements in dwarf galaxies with a high-resolution chemo-dynamical evolution model assuming NSMs are the major site of r-process.

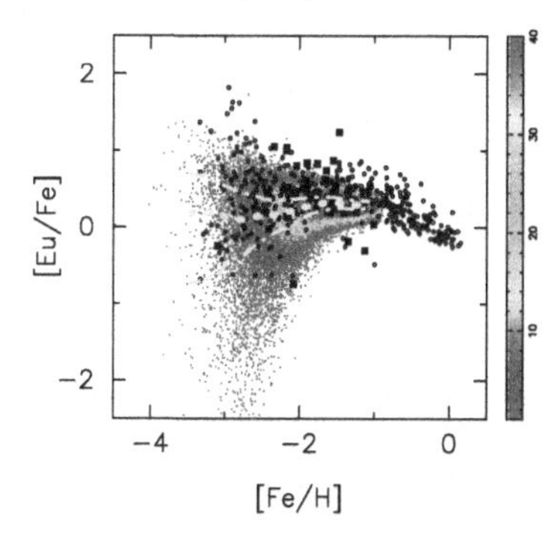

Figure 1. [Eu/Fe] as a function of [Fe/H]. Different shades of grey refer to the number of stars produced in our model (legend on the right). The dotted curve is the median of the model predictions. The dash-dotted curves are the first and third quartiles, respectively. Circles are observed values for Galactic halo stars (SAGA database, Suda *et al.* 2008). Squares are the observed value of stars in Carina, Draco, Leo I, Sculptor, and Ursa Minor dSphs (SAGA database, Suda *et al.* 2014).

We perform a series of simulations using an N-body/smoothed particle hydrodynamics code, ASURA (Saitoh *et al.* 2008; Saitoh *et al.* 2009). This code includes metallicity dependent cooling, star formation, and supernova feedback. We adopt metal mixing assuming the average metallicity of 32 surrounding gas particles to a newly formed star particle. For dwarf galaxy models, both dark matter and gas particles are initially distributed along with the pseudo isothermal profile following Revaz & Jablonka (2012). Details of implementation to the code and models are discussed in Hirai *et al.* (2015).

Figure 1 shows [Eu/Fe] as a function of [Fe/H] in our model assuming NSMs with a merger time of 100 Myr. As shown in this figure, star-to-star scatters of [Eu/Fe] in extremely metal-poor stars produced by NSMs with a merger time of 100 Myr are consistent with the observations. This is because the average metallicity of stars is constant during the first $\sim$ 300 Myr from the beginning of the star formation in the galaxy. Due to the low star formation efficiency of the galaxy, the spatial distribution of metallicity is highly inhomogeneous in the first $\lesssim$ 300 Myr. Since most gas particles are enriched by a single supernova in this epoch, the metallicity of the stars is mainly determined by the distance from each supernova to the gas particles, which formed the stars. NSMs with merger time of 100 Myr can therefore account for the observations of extremely metal-poor stars. In contrast, metallicity is well correlated with the galactic age after $\sim$ 300 Myr, irrespective of the distance from each supernova to the gas particle. Since supernova products have already been well mixed in a galaxy, the stellar metallicity is determined by the number of supernovae.

References

Argast, D., Samland, M., Thielemann, F.-K., & Qian, Y.-Z. 2004, *A&A*, 416, 997

Hirai, Y., Ishimaru, Y., Saitoh, T. R., Fujii, M. S., Hidaka, J., & Kajino, T. 2015, *ApJ*, 814, 41

Ishimaru, Y., Wanajo, S., & Prantzos, N. 2015, *ApJ*, (Letter), 804, L35

Revaz, Y. & Jablonka, P. 2012, *A&A*, 538, A82

Saitoh, T. R., Daisaka, H., Kokubo, E., Makino, J., Okamoto, T., Tomisaka, K., Wada, K., & Yoshida, N. 2008, *PASJ*, 60, 667

Saitoh, T. R., Daisaka, H., Kokubo, E., Makino, J., Okamoto, T., Tomisaka, K., Wada, K., & Yoshida, N. 2009, *PASJ*, 61, 481

Suda, T., Katsuta, Y., Yamada, S., *et al.* 2008, *PASJ*, 60, 1159

Suda, T., Hidaka, J., & Ishigaki, M.,*et al.* 2014, *MmSAI*, 85, 600

The General Assembly of Galaxy Halos: Structure, Origin and Evolution
Proceedings IAU Symposium No. 317, 2015
A. Bragaglia, M. Arnaboldi, M. Rejkuba & D. Romano, eds.

doi:10.1017/S1743921315009552

Chemical evolution of r-process elements in the Draco dwarf spheroidal galaxy

M. N. Ishigaki[1], T. Tsujimoto[2], T. Shigeyama[3] and W. Aoki[2]

[1]Kavli Institute for the Physics and Mathematics of the Universe, University of Tokyo
5-1-5 Kashiwanoha, Kashiwa, 277-8583, Japan
email: miho.ishigaki@ipmu.jp

[2]National Astronomical Observatory of Japan,
2-21-1 Osawa, Mitaka, Tokyo 181-8588, Japan
email: taku.tsujimoto@nao.ac.jp

[3]Research center for the early universe, University of Tokyo,
Bunkyo-ku, Tokyo, 113-0033, Japan
email: shigeyama@resceu.s.u-tokyo.ac.jp

Abstract. A dominant astrophysical site for r-process, which is responsible for producing heavy neutron-capture elements, is unknown. Dwarf spheroidal galaxies around the Milky Way halo provide ideal laboratories to investigate the origin and evolution of r-process elements. We carried out high-resolution spectroscopic observations of three giant stars in the Draco dwarf spheroidal galaxy to estimate their europium abundances. We found that the upper-limits of [Eu/H] are very low in the range [Fe/H]< -2, while this ratio is nearly constant at higher metallicities. This trend is not well reproduced with models which assume that Eu is produced together with Fe by SNe, and may suggest the contribution from other objects such as neutron-star mergers.

Keywords. stars:abundances, galaxies:dwarf, galaxies:individual (Draco)

1. Motivation

Heavy neutron-capture elements are mainly originated from r-process. These elements are ubiquitous among very metal-poor stars in the Milky Way halo, which suggests that the r-process has been operating since the early Universe (e.g. Sneden *et al.* 2008, Roederer *et al.* 2014). Although the physical conditions required for the r-process are well understood, the corresponding astrophysical sites are still unknown. Theoretical studies suggest that core-collapse supernovae (SNe) and/or mergers neutron-stars (or neutron star-black hole systems) could be possible sites for the r-process (e.g. Wanajo & Ishimaru 2006). These events are expected to have different enrichment timescales and thus can be tested by observed chemical evolution of r-process elements in old stellar populations.

We carried out high-resolution spectroscopic observations to measure europium (Eu) abundances in three giant stars in the Draco dwarf spheroidal galaxy (dSph), which is known to host an old, metal-poor and relatively simple stellar population and thus is a particularly ideal laboratory to study individual r-process events in the early universe. Europium is almost purely produced in the r-process, with little contamination from s-process (c.f. Ba), and thus provides a clean test for the r-process enrichment in this galaxy.

2. Results and discussion

The observations of the three Draco stars were carried out with the High-Dispersion Spectrograph (HDS; Noguchi *et al.* 2002) mounted on the Subaru telescope during Aug

Table 1. Summary of the atmospheric parameters and abundances.

	$T_{\rm eff}$	$\log g$	ξ	[Fe/H]	[Eu/H]
Irwin 19826	4028	0.6	1.7	-1.45 ± 0.12	-1.33 ± 0.21
Irwin 20751	4278	0.6	1.7	-2.12 ± 0.13	< -2.0
Irwin 21275	4406	0.7	2.3	-2.51 ± 0.09	< -2.1

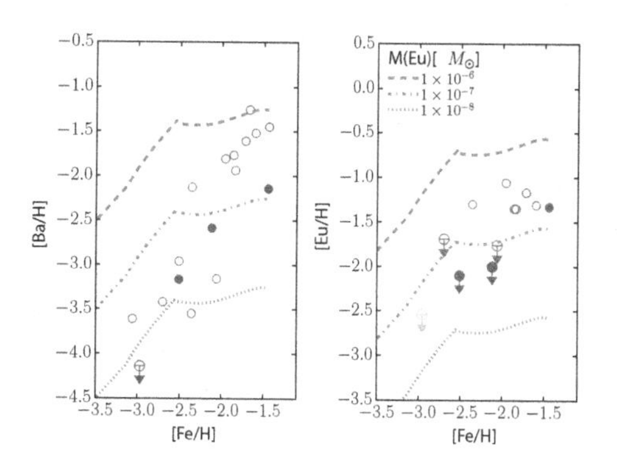

Figure 1 Observed [Ba/H] and [Eu/H] abundances in this study (filled circles) and those from literature. The gray lines are predictions of chemical evolution models assuming that Eu is produced in every SNe (10-40$M_\odot$) by an amount 10^{-6}, 10^{-7}, $10^{-8} M_\odot$. They are calculated based on the model of Kirby *et al.* (2011), whose model parameters have been obtained to reproduce observed α-elements and [Fe/H] distributions in Draco. Mass of Ba, assumed to be produced in SNe, is adopted according to the Eu yields with the r-process ratio [Ba/Eu]r-process= -0.7. Note that contribution from s-process is not taken into account.

16-18, 2014. An optical wavelength range was covered with a spectral resolution $R \sim 30000$.

Results of the Eu measurements are summarized in Table 1. An absorption line of Eu II (6645 Å) was weakly detected in the most metal-rich star in our sample, Irwin 19826, while only upper limits were obtained for the two more metal-poor stars. Figure 1 shows [Ba/H] and [Eu/H] abundances plotted against [Fe/H] for the sample stars and for stars from literature (Shetrone *et al.* 2001, Fulbright *et al.* 2004, Cohen & Huang 2009). The [Eu/H] abundances in the Draco stars are characterized by the very low values at [Fe/H]< -2, and an apparent plateau ([Eu/H]~ -1.3 dex) at higher metallicities. This trend is not well reproduced by canonical chemical evolution models under the simple assumption that Eu is exclusively produced in SNe together with Fe (gray lines). These models generally predict a steep increase of [Eu/H] with [Fe/H] at [Fe/H]< -2, which becomes shallower at higher metallicities, due to the onset of Type Ia SNe, and thus are unable to simultaneously explain the observed features over the whole [Fe/H] range.

This result may indicate that r-process elements were produced, at least in part, in neutron-star mergers (Tsujimoto & Shigeyama 2014; Tsujimoto *et al.* 2015). Further observations are needed to test whether similar [Eu/H]-[Fe/H] trends are observed in other faint dSphs around the Milky Way.

References

Cohen, J. & Huang, W. 2009, *ApJ*, 701, 1053
Fulbright, J. P., Rich, R. M., & Castro, S. 2004, *ApJ*, 612, 447
Kirby, E., Cohen, J. G., & Smith, G. H., *et al.* 2011, *ApJ*, 727, 79
Noguchi, K., Aoki, W., Kawanomoto, S., *et al.* 2002, *PASJ*, 54, 855
Roederer, I. U., Cowan, J. J., & Preston, G. W., *et al.* 2014, *MNRAS*, 445, 2970
Shetrone, M. D., Côté, P., & Sargent, W. L. W. 2001, *ApJ*, 548, 592
Sneden, C., Cowan, J. J., & Gallino, R. 2008, *ARAA*, 46, 241
Tsujimoto, T. & Shigeyama, T. 2014, *A&A*, 565, L5
Tsujimoto, T., Ishigaki, M. N., Shigeyama, T., & Aoki, W. 2015, *PASJ*, 67, L3
Wanajo, S. & Ishimaru, Y. 2006, *Nuclear Physics A*, 777, 676

The General Assembly of Galaxy Halos: Structure, Origin and Evolution
Proceedings IAU Symposium No. 317, 2015
A. Bragaglia, M. Arnaboldi, M. Rejkuba & D. Romano, eds.

doi:10.1017/S174392131500705X

Connection between cusp-core problem and too-big-to-fail problem in CDM model

Kazuki Kato[1⋆], Masao Mori[1,2] and Go Ogiya[3,4]

[1]Faculty of Pure and Applied Sciences, University of Tsukuba,
[2]Center for Computational Sciences, University of Tsukuba,
[3]Ludwig-Maximilians-Universität München,
[4] Max-Planck-Institut für extraterrestrische Physik,
⋆email: katokzk@ccs.tsukuba.ac.jp

Abstract. The standard paradigm of structure formation in the universe, the cold dark matter cosmology, contains several crucial unsolved problems such as "cusp-core problem" and "too-big-to-fail problem". To solve these problems, we study about the dynamical response of a virialized system with a central cusp to the energy feedback driven by periodic supernova feedback using collisionless *N*-body simulations with the Nested-Particle-Mesh code. The resonance between dark matter particles and the density wave excited by the oscillating potential plays a significant role in the cusp-core transition of dark matter halos. Furthermore, we show that the cusp-core transition with periodic supernova feedback can solve the too-big-to-fail problem.

Keywords. galaxies: dwarf, galaxies: evolution, galaxies: halos, galaxies: structure, Local Group, cosmology: theory, dark matter

1. Introduction

Cold dark matter (CDM) cosmology explains the statistical property and evolution process the large-scale structure of the universe. However, it has some issues on the small scale such as cusp-core (CC) problem and too-big-to-fail (TBTF) problem. The CC problem is that a central density profile of dark matter halo (DMH) predicted by simulations is inconsistent with observational one. Cosmological simulations (e.g., Navarro *et al.* 1996) always show that the density profile of DMH diverges at the center. On the other hand, observed DMH in dwarf galaxies has revealed almost constant density structures at their center (e.g., Burkert 1995). The TBTF problem (Boylan-Kolchin *et al.* 2011) is that cosmological simulations produce more massive satellite galaxies around a Milky Way (MW) -like galaxy than observed.

To solve these problems, we focus on CC transition driven by supernova (SN) feedback. Ogiya & Mori (2014) showed that periodic SN feedback can solve the CC problem, and the core size depends completely on the frequency of the oscillation. However, Garrison-Kimmel *et al.* (2013) reported that SN feedback failed to solve the CC problem. This contradicting situation motivates us to reexamine this problem using a detailed simulation.

2. Numerical simulations of cusp-core transition

We perform collisionless *N*-body simulations using the Nested-Particle-Mesh code. In this simulation, we use 16,777,216 particles and the smallest mesh size is 17 pc. DMH is represented by Hernquist sphere (Hernquist 1990) with a total mass of $2.28 \times 10^9\ M_\odot$ and a scale radius of 2.2 kpc. To model the periodic SN feedback, we adopt time-varying potentials with different oscillation periods into the DMH. This time-varying potential

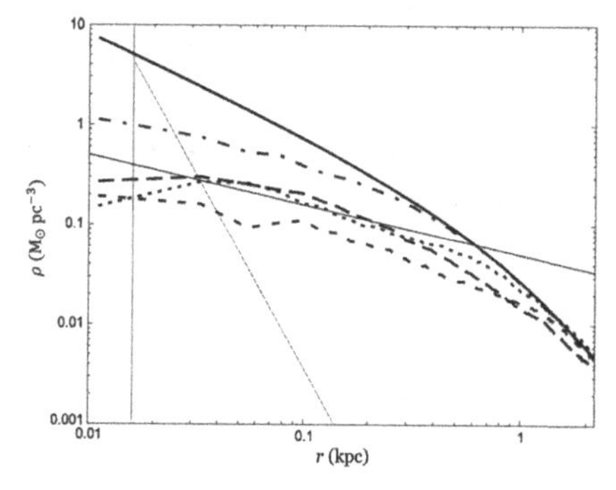

Figure 1. Density profiles of DMH with different oscillation periods: initial state (thick solid line); 500 Myr (long-dashed line); 100 Myr (short-dashed line); 50 Myr (dotted line); 13 Myr (dotted-dashed line). $\rho \propto r^{-0.5}$ (thin solid line); limit of two-body relaxation effect (left side of the thin dotted line); smallest mesh size (vertical thin solid line).

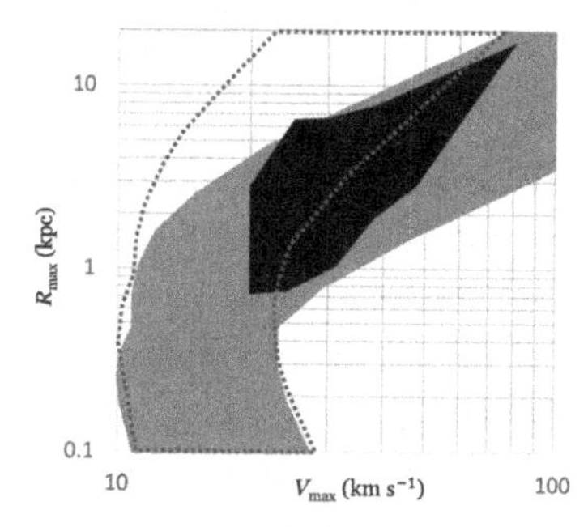

Figure 2. Maximum circular velocity, $V_{\max}$, and that position, $R_{\max}$, relation of DMHs. The region surrounded by a dotted-line corresponds the observed massive MW satellites (Wolf *et al.* 2010) assuming NFW profile. The prediction of CDM simulations (black region) done by Springel *et al.* (2008) and Diemand *et al.* (2008) indicates the over production of massive satellites.

is also represented by Hernquist sphere with a total baryon mass of $10^8\ M_\odot$ and a scale radius of 210 pc (cf. Garrison-Kimmel *et al.* 2013), and analyze the quasi-equilibrium state of DMH after 10 oscillation times.

The result of simulations shows that the periodic SN feedback can drive a transition from a cuspy dark matter profile to a core like profile, and the core size directly connects with the oscillation period. This result is consistent with Ogiya & Mori (2014). Furthermore, we found newly that the power-law index of the DMH central density is independent of the oscillation period. We also calculated ellipsoidal perturbation and found that DMH central density strongly depends on the shape of perturbation. This result will be reported in a forthcoming paper.

3. Application to the too-big-to-fail problem

We assumed that DMH experienced the CC transition at the starburst epoch (Ogiya *et al.* 2014). Then, the density profile of DMH has changed from NFW profile (cusp) to Burkert profile (core). In contrast to Boylan-Kolchin *et al.* (2011), our model provides consistency between the CDM prediction (black region) and the observations of MW satellites (gray region). We conclude that the CC transition driven by periodic SN feedback can simultaneously solve both the CC problem and the TBTF problem.

References

Boylan-Kolchin, M., Bullock, J. S., & Kaplinghat, M. 2011, *MNRAS*, 415, L40
Burkert, A. 1995, *Apj*, 447, L25
Diemand, J. *et al.* 2008, *Nature*, 454, 735
Garrison-Kimmel, S. *et al.* 2013, *MNRAS*, 433, 3539
Hernquist, L. 1990, *Apj*, 356, 359
Navarro, J. F., Frenk, C. S., & White, S. D. M. 1996, *ApJ*, 462, 563
Ogiya, G., Mori, M., Ishiyama, T., & Burkert, A. 2014, *MNRAS*, 440, L71
Ogiya, G. & Mori, M. 2014, *ApJ*, 793, 46
Springel, V. *et al.* 2008, *MNRAS*, 391, 1685
Wolf, J. *et al.* 2010, *MNRAS*, 406, 1200

The General Assembly of Galaxy Halos: Structure, Origin and Evolution
Proceedings IAU Symposium No. 317, 2015
A. Bragaglia, M. Arnaboldi, M. Rejkuba & D. Romano, eds.

doi:10.1017/S1743921315007103

Two New Ultra-Faint Star Clusters in the Milky Way Halo

Dongwon Kim

Research School of Astronomy and Astrophysics, The Australian National University, Mt Stromlo Observatory, via Cotter Rd, Weston, ACT 2611, Australia
email: dongwon.kim@anu.edu.au

Abstract. Kim 1 & 2 are two new star clusters discovered in the Stromlo Missing Satellite Survey. Kim 1, located at a heliocentric distance of 19.8 ± 0.9 kpc, features an extremely low total luminosity ($M_V = 0.3 \pm 0.5$ mag) and low star concentration. Together with the large ellipticity ($\epsilon = 0.42 \pm 0.10$) and irregular isophotes, these properties suggest that Kim 1 is an intermediate mass star cluster being stripped by the Galactic tidal field. Kim 2 is a rare ultra-faint outer halo globular cluster located at a heliocentric distance of 104.7 ± 4.1 kpc. The cluster exhibits evidence of significant mass loss such as extra-tidal stars and mass-segregation. Kim 2 is likely to follow an orbit confined to the peripheral region of the Galactic halo, and/or to have formed in a dwarf galaxy that was later accreted into the Galactic halo.

Keywords. Galaxy, halo, globular clusters, dwarf galaxies.

---

1. Discovery of Kim 1 & 2

Kim 1 & 2 are ultra-faint stellar systems recently discovered by Kim & Jerjen (2015a) and Kim *et al.* (2015a) in the Stromlo Missing Satellite (SMS) survey (PI: Helmut Jerjen) (see also Kim *et al.* 2015b and Kim & Jerjen 2015b). Kim 1 was first detected in the pre-existing Sloan Digital Sky Survey Data Release 10 (Ahn *et al.* 2014) and Kim 2 in an independent 500 sqr degree survey using the Dark Energy Camera (DECam) at the 4m Blanco telescope at CTIO as a part of the SMS survey. Their true identities were confirmed by deep follow-up imaging using DECam and the Gemini-South 8-m telescope respectively (Fig. 1). Both objects are located in the Vast Polar Structure (Pawlowski *et al.* 2015).

2. Properties of the Star Clusters and Implications

Kim 1 features an old ($12^{+1.5}_{-3.0}$ Gyr) and metal-poor ([Fe/H]$=-1.7^{+0.5}_{-0.2}$) stellar population at a heliocentric distance of 19.8 ± 0.9 kpc. Its small physical size ($r_h = 6.9 \pm 0.6$ pc) and the extremely low luminosity ($M_V = 0.3 \pm 0.5$) are comparable to those of the faintest known star clusters such as Segue 3, Koposov 1 & 2, and Muñoz 1 (Fadely *et al.* 2011; Koposov *et al.* 2007; Muñoz *et al.* 2012). However, Kim 1 exhibits an usually low star concentration, large ellipticity ($\epsilon = 0.42 \pm 0.10$) and irregular outer isophotes, and is lacking a well defined center. The observed properties suggest that Kim 1 is most likely an intermediate mass star cluster being stripped by the Galactic tidal field.

Kim 2 lies at a heliocentric distance of 104.7 ± 4.1 kpc. With a half-light radius of 12.8 ± 0.6 pc and ellipticity of $\epsilon = 0.12 \pm 0.10$, the cluster features the properties of typical outer halo globular clusters, except for the rather high metallicity ([Fe/H]$= -1.0^{+0.18}_{-0.21}$) and unusually low luminosity ($M_V = -1.5 \pm 0.5$). These parameters are similar to those for the outer halo globular cluster AM 4, which is thought to be associated with the Sagittarius dwarf spheroidal galaxy. In addition, the cluster exhibits evidence of significant mass loss

Figure 1. Left panel: $6 \times 6\,\text{arcmin}^2$ DECam cutout g-band image of Kim 1 (from Kim & Jerjen 2015a). Member stars are highlighted with circles. Right panel: $4 \times 4\,\text{arcmin}^2$ GMOS cutout g-band image of Kim 2 (from Kim *et al.* 2015a). The cluster is located at the centre of the image. North is up, east is to the left.

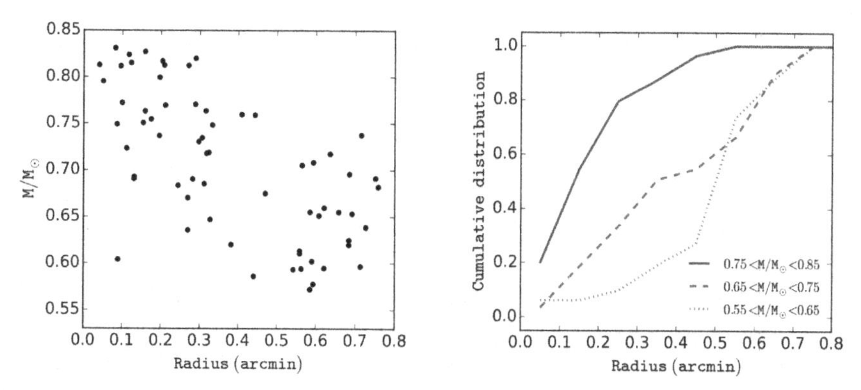

Figure 2. Mass segregation in Kim 2 (from Kim *et al.* 2015a). Left panel: stellar mass distribution of Kim 2 main sequence stars within $2r_h$ (~ 0.8 arcmin) as a function of the distance from the cluster center. Right panel: cumulative distribution function of Kim 2 main sequence stars for three different mass intervals.

such as extra-tidal stars and mass-segregation (Fig. 2). An outer halo globular cluster with such low star density is likely to follow an orbit confined to the peripheral region of the Milky Way halo, and/or to have formed in a dwarf galaxy that was later accreted into the halo (Sollima *et al.* 2011). Consequently, the cluster could have avoided major tidal disruption and survived until the present epoch.

References

Ahn, C. P., Alexandroff, R., Allende Prieto, C., *et al.* 2014, *ApJS*, 211, 17

Fadely, R., Willman, B., Geha, M., *et al.* 2011, *AJ*, 142, 88

Kim, D. & Jerjen, H. 2015a, *ApJ*, 799, 73

Kim, D., Jerjen H., Milone, A. P., Mackey, D., & Da Costa, G. S. 2015a, *ApJ*, 803, 63

Kim, D., Jerjen H., Mackey, D., Da Costa, G. S., & Milone, A. P. 2015b, *ApJL*, 804, L44

Kim, D. & Jerjen, H. 2015b, *ApJL*, 808, L39

Koposov, S., de Jong, J. T. A., Belokurov, V., *et al.* 2007, *ApJ*, 669, 337

Muñoz, R. R., Geha, M., Côté, P., *et al.* 2012, *ApJL*, 753, L15

Pawlowski, M.S., McGaugh, S.S., & Jerjen, H. 2015 *MNRAS*, 453, 1047

Sollima, A., Martínez-Delgado, D., Valls-Gabaud, D., & Peñarrubia, J. 2011, *ApJ*, 726, 47

The General Assembly of Galaxy Halos: Structure, Origin and Evolution
Proceedings IAU Symposium No. 317, 2015
A. Bragaglia, M. Arnaboldi, M. Rejkuba & D. Romano, eds.

doi:10.1017/S1743921315007127

Disk dwarf galaxy as the progenitor of the Andromeda giant stream

Takanobu Kirihara[1], Yohei Miki[2], Masao Mori[1,2] and Toshihiro Kawaguchi[3]

[1]Faculty of Pure and Applied Physics, University of Tsukuba, Tennodai 1-1-1, Tsukuba, Ibaraki, Japan
email: kirihara@ccs.tsukuba.ac.jp

[2]Center for Computational Sciences, University of Tsukuba, Tennodai 1-1-1, Tsukuba, Ibaraki, Japan

[3]Sapporo Medical University, S1W17, Chuoh-ku, Sapporo, Hokkaido, Japan

Abstract. We present a study of the morphology of a progenitor galaxy that has been disrupted and formed a giant southern stellar stream in the halo of Andromeda galaxy(M31). N-body simulations of a minor merger of M31 with a dwarf galaxy suggest that the progenitor's rotation plays an important role in the formation of an asymmetric surface brightness distribution of the stream.

Keywords. galaxies: individual (M31), galaxies: interactions, galaxies: kinematics and dynamics

---

1. Introduction

As represented by a giant southern stream (GSS), recent deep and wide observations with optical wavelength have revealed a wealth of substructures in the halo of M31 (Ibata *et al.* 2001; Martin *et al.* 2013). Observed properties of the GSS and north-eastern and western stellar shells have been reproduced in detail with N-body simulations of a minor merger that has occurred about 700 Myr ago (Fardal *et al.* 2007; Mori & Rich 2008; Miki *et al.* 2014; Kirihara *et al.* 2014). The total stellar mass of the progenitor has been estimated $M_* > 10^8$ $M_\odot$ (Font *et al.* 2006).

We focus on an asymmetric surface brightness distribution of the GSS that has been firstly suggested by McConnachie *et al.* 2003. We analyzed RGB star count map (Irwin *et al.* 2005) and obtained azimuthal surface brightness profile of the GSS (Fig. 1). The stellar density of the GSS decreases sharply at the eastern side of the GSS from the most luminous direction. On the other hand, the western side of the GSS extends widely. So far, any simulations assuming a merger of a spherical non-rotating progenitor failed to reproduce this characteristic structure of the GSS (e.g., Fig. 1).

2. Methods and Results

We examine the collision of the disk satellite galaxy and M31 using N-body simulation. The initial stable model of satellite galaxies with a disk, a bulge, and a dark matter halo is generated by GalactICS (Kuijken & Dubinski 1995). In this paper, we especially focus on the inclination (θ, ϕ) of the spin axis of the progenitor's disk. Following previous studies, a fixed-potential of M31 and an orbit of the progenitor are taken from Fardal *et al.* 2007. N-body simulations are carried out on the T2K-Tsukuba, HA-PACS, and COMA system in Center for Computational Sciences, University of Tsukuba.

We compare the observed asymmetric spatial structures of the GSS with that of the results of our simulations using χ^2 analysis. Fig. 2 shows a χ^2 map of the comparison

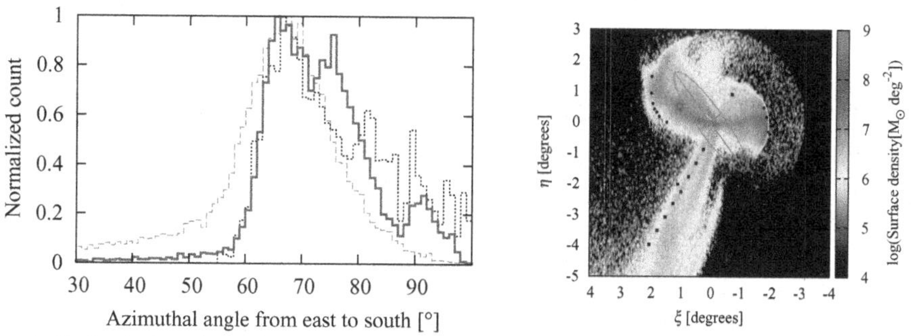

Figure 1. Left: Azimuthal star counts distribution of the GSS. Each line corresponds the observed data (black dotted), best-fitting model (blue solid), and spherical progenitor model (magenta dashed), respectively. Right: Snapshot of the surface density distribution of the disrupted progenitor. The inclined elliptical line describes the shape of the M31 disk. Square symbols and circles are observed fields of the GSS and edges of the shells (Font *et al.* 2006; Fardal *et al.* 2007) , respectively.

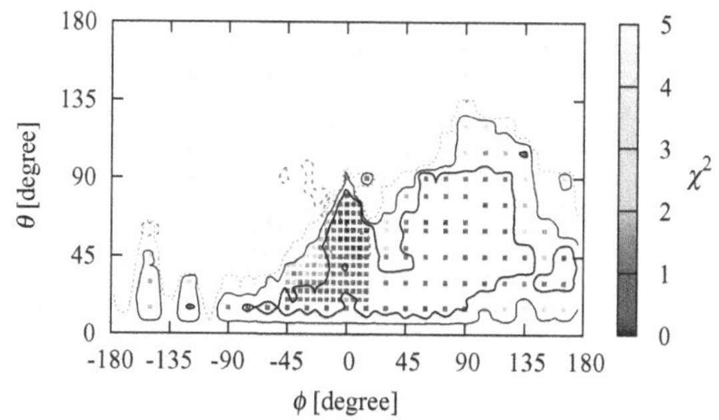

Figure 2. χ^2 map for the eastern width of the GSS on the plane (θ, ϕ). The squares indicate the simulated parameters. Black lines describe 1, 2, and 3σ confidence intervals, and the dotted red line is 1 σ level, in which the brightest angle of the GSS is reproduced.

for the narrow eastern width from the brightest direction of the GSS. The most region that the χ^2 value is smaller than 1.0 ($\nu = 2$; 1 σ confidence interval) corresponds to the disk with anti-clockwise rotation. Fig. 1 presents one of the best-fitting models; the left panel shows an azimuthal density profile of the GSS, and the right panel describes a snapshot of surface density distribution. It is clear that our simulation nicely reproduced the observed asymmetric structure of the GSS.

We conclude that the rotation of the progenitor was crucial for the formation of the asymmetric structure of the GSS, and most likely the progenitor of the GSS was a disk dwarf galaxy.

References

Fardal, M. A., Guhathakurta, P., Babul, A., & McConnachie, A. W. 2007, *MNRAS*, 380, 15

Font, A. S., Johnston, K. V., Guhathakurta, P., Majewski, S. R., & Rich, M. 2006, *AJ*, 131, 1436

Ibata, R., Irwin, M., Lewis, G., Ferguson, A., & Tanvir, N. 2001, *Nature*, 412, 49

Kirihara, T., Miki, Y., & Mori, M. 2014, *PASJ*, 66, L10

Kuijken, K. & Dubinski, J. 1995, *MNRAS*, 277, 1341

Martin, N. F. *et al.*, 2013, *ApJ*, 776, 80

McConnachie, A. W., Irwin, M., Ibata, R., Ferguson, A., Lewis, G., & Tanvir, N., 2003, *MNRAS*, 343, 1335

Miki, Y., Mori, M., Kawaguchi, T., & Saito, Y. 2014, *ApJ*, 783, 87

Mori, M. & Rich, M. 2008, *ApJ*, 674, 77

The General Assembly of Galaxy Halos: Structure, Origin and Evolution
Proceedings IAU Symposium No. 317, 2015
A. Bragaglia, M. Arnaboldi, M. Rejkuba & D. Romano, eds.

doi:10.1017/S1743921315008662

Chemical Evolution of R-process Elements in the Hierarchical Galaxy Formation

Yutaka Komiya and Toshikazu Shigeyama

Research center for the early universe, University of Tokyo
Hongo, 7-3-1, Bunkyo-ku, Tokyo, 113-0033, Japan
email: komiya@resceu.s.u-tokyo.ac.jp

Abstract. The main astronomical source of r-process elements has not yet been identified. One plausible site is neutron star mergers (NSMs). From the perspective of Galactic chemical evolution, however, it has been pointed out that the NSM scenario is incompatible with observations. Recently, Tsujimoto & Shigeyama (2014) pointed out that NSM ejecta can spread into much larger volume than ejecta from a supernova. We re-examine the chemical evolution of r-process elements under the NSM scenario considering this difference in propagation of the ejecta. We find that the NSM scenario can be compatible with the observed abundances of the Milky Way halo stars.

Keywords. stars: Population II, nuclear reactions, nucleosynthesis, abundances, Galaxy: evolution, Galaxy: halo

1. Introduction

The rapid neutron capture process (r-process) is thought to be a main process to synthesise elements heavier than the iron group in extremely metal-poor stars. But the dominant r-process element source in the universe is still matter of debate. Two possible sites have been proposed; core-collapse supernovae and neutron star mergers (NSMs).

From the viewpoint of chemical evolution, some difficulties of the NSM scenario have been pointed out (Argast *et al.* 2004). Since NSMs have a very low event rate and very large r-process yield per event, the NSM scenario predicts too large abundance scatter and many stars lacking r-process elements, in contradiction to observations (Roederer 2013, Komiya *et al.* 2014).

Recently, however, Tsujimoto & Shigeyama (2014) pointed out that the ejecta from a NSM can spread into a very large volume since they have very large velocities of $10-30\%$ of the speed of light. The stopping length, l_s, of ^{153}Eu at the speed of $0.2c$ is estimated to be $\sim 2.6(n/1\mathrm{cm}^{-3})^{-1}$kpc, where n is the number density of neutral hydrogen.

In this paper, we revisit the chemical evolution of r-process elements considering this difference of propagation between NSM ejecta and supernova ejecta.

2. The hierarchical chemical evolution model

We use the chemical evolution model considering the hierarchical galaxy formation based on the extended Press-Schechter theory (Komiya *et al.* 2014, Komiya *et al.* 2015). The Galactic halo is formed through merger of proto-galaxies and the chemical abundance of each proto-galaxy evolves almost independently. In this work, we modify the chemical evolution model taking into account that NSM ejecta can go out from a host proto-galaxy and be captured by other proto-galaxies.

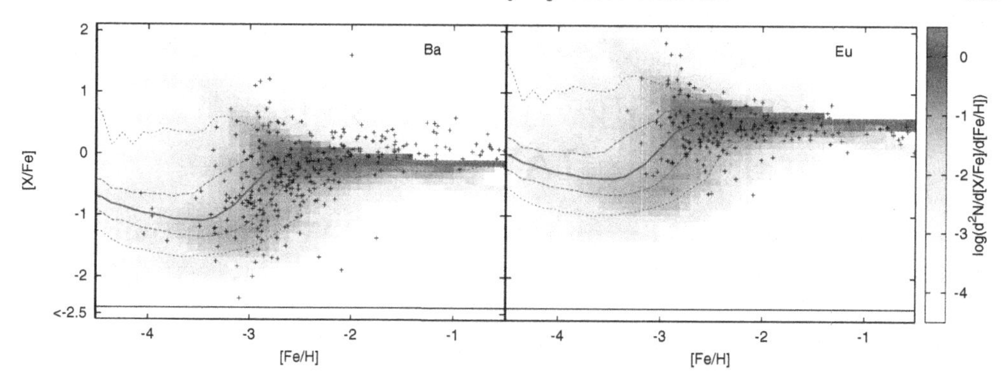

Figure 1. The predicted distribution of halo stars on the [Fe/H]-[Ba/Fe] (left panel) and [Fe/H]-[Eu/Fe] (right panel) planes is color coded. Black crosses are the observational sample from the SAGA database (Suda *et al.* 2008). Blue lines denote percentile curves for 5%, 25%, 50%, 75% and 95% of the predicted distribution.

The escape fraction, $f_{\rm esc}$, of NSM ejecta is assumed to be $f_{\rm esc} = \exp(-\alpha R_{g,0}/l_s)$, where $R_{g,0}$ is the radius of the host proto-galaxy, and α is a free parameter. If r-process elements move straight, $\alpha = 1$, but if it moves along a winding path due to a magnetic field, $\alpha > 1$. We assume that the capture rate by other proto-galaxies is $f_{\rm cap} = f_{\rm esc}(R_{g,n}^2/4\pi d^2)\{1 - \exp(-\alpha R_{g,n}/l_s)\}$, where d is the distance between galaxies and $R_{g,n}$ is the radius of a capturing galaxy. We set the NSM event rate to be 1/1000 times the SN event rate and the coalescence timescale is $10^6 - 10^8$yr (Belczynski *et al.* 2006).

3. Results and Discussion

When we do not consider the escape of NSM ejecta from the host proto-galaxies, the majority of the stars with [Fe/H] < -3 show very low r-process abundance with [Ba/Fe] < -2.5. On the other hand, if NSM ejecta directly go away from the proto-galaxy ($\alpha = 1$), most of the NSM ejecta go to the intergalactic space from small proto-galaxies.

We show the results of the model assuming $l_s/\alpha = 4$ kpc in Figure 1. In this case, a small fraction of NSM ejecta escapes from its host proto-galaxy and pollutes other proto-galaxies. This model well reproduces the observations of halo stars, as shown.

This result indicates that the NSM scenario is not rejected by chemical evolution, when we consider the propagation of NSM ejecta beyond the proto-galaxies.

References

Argast, D., Samland, M, Thielemann, F.-K., & Qian, Y.-Z. 2004, *A&A*, 416, 997
Belczynski, K., Perna, R., Bulik, T., *et al.* 2006, *ApJ*, 648, 1110
Komiya, Y., Yamada, S., Suda, T., & Fujimoto, M. Y. 2014, *ApJ*, 783, 132
Komiya, Y., Suda, T., & Fujimoto, M. Y. 2015, *ApJ Letters*, 808, L47
Tsujimoto, T., & Shigeyama, T. 2014, *A&A*, 565, L5
Roederer, I. U. 2013, *AJ*, 145, 26
Suda, T., Katsuta, Y., Yamada, S., *et al.* 2008, *PASJ*, 60, 1159

The General Assembly of Galaxy Halos: Structure, Origin and Evolution
Proceedings IAU Symposium No. 317, 2015
A. Bragaglia, M. Arnaboldi, M. Rejkuba & D. Romano, eds.

doi:10.1017/S1743921315008686

The Milky Way evolution under the RAVE perspective

Georges Kordopatis[1] on behalf of the RAVE consortium
[1]Leibniz-Institut für Astrophysik Potsdam, An der Sternwarte 16, 14482 Potsdam, Germany
email: gkordopatis@aip.de

Abstract. The RAdial Velocity Experiment (RAVE) collected from 2003 to 2013 medium resolution spectra for $5 \cdot 10^5$ low-mass stars of our Galaxy, improving our understanding of the Milky Way evolution and of its properties outside the Solar neighbourhood. This proceeding gives an overview of RAVE results obtained in the last two years.

---

1. Overview of the survey and of the public catalogue

The *Radial Velocity Experiment* project (Steinmetz *et al.* 2006) used the 6dF instrument mounted on the 1.2 m Schmidt telescope of the Anglo-Australian Observatory, to obtain spectra of 480,000 stars in the magnitude range $8 < I < 12$ mag (up to ~ 3 kpc from the Sun). The fourth data release (DR4 Kordopatis *et al.* 2013a) publishes for these stars the atmospheric parameters, chemical abundances, line-of-sight velocities and distances, as well as cross-correlations with photometric and proper motion catalogues. Typical uncertainties of DR4 are $\sim 150\,K$ for $T_{\rm eff}$, 0.2 dex for $\log g$, and $\sim 0.1 - 0.2$ dex for metallicity and chemical abundances. Distances are derived with errors better than $\sim 20\%$ and radial velocities with mean accuracy of $1.5\,{\rm km\,s^{-1}}$. These uncertainties lead to median errors on 3D Galactocentric velocities of $15\,{\rm km\,s^{-1}}$.

2. Description of the Milky Way morphology: structure and dynamics

Combining *RAVE* with the Geneva-Copenhagen Survey, Sharma *et al.* (2014) constrained the three-component age-velocity dispersion relations, the radial dependence of the velocity dispersions, the Solar peculiar motion ($U_\odot = 10.96, V_\odot = 7.53, W_\odot = 7.54\,{\rm km\,s^{-1}}$), the circular speed at the Sun ($\Theta_0 = 232.8\,{\rm km\,s^{-1}}$), and the fall of mean azimuthal motion with height above the mid-plane. The authors found that the radial scale length of the velocity dispersion profile of the thick disc is smaller than that of the thin disc, in agreement with recent high-resolution spectroscopic studies separating the discs based on their chemical composition (e.g. Kordopatis *et al.* 2015). On a similar topic, Binney *et al.* (2014) studied the velocities of the disc stars and highlighted the non-Gaussianity of the velocity distribution functions. This led to the publication of formulae from which the shape and orientation of the velocity ellipsoid can be determined at any location.

The north-south differences in the stellar kinematics have been studied in Williams *et al.* (2013). Among others, a clear vertical rarefaction-compression pattern up to 2 kpc above the plane has been found, originating either by accretions (e.g. Widrow *et al.* 2014), or by the effect of the spiral arms on the stellar orbits (Faure *et al.* 2014).

The Galactic escape speed has been evaluated by Piffl *et al.* (2014b) to be $V_{\rm esc} = 533^{+54}_{-41}\,{\rm km\,s^{-1}}$. The authors further found that the dark matter and baryon mass interior to three virial radii is $1.3^{+0.4}_{-0.3} \times 10^{12} M_\odot$, in good agreement with recently independently published mass estimates. In addition, by modelling the kinematics of giant stars up to ~ 1.5 kpc from the Sun, Piffl *et al.* (2014a) found that the dark mass contained within the iso-density surface of the dark halo that passes through the Sun, and the surface density within 0.9 kpc of the plane, are almost independent of the halo's axis ratio q and estimated that the baryonic mass is at most 4.3% of the total Galaxy mass.

Bienaymé *et al.* (2014) estimated the vertical force (K_z) at 1 kpc and 2 kpc from the plane, and found an unexpectedly large amount of dark matter at distances greater than 2 kpc. This could be evidence of either (i) a flattening of the dark halo of the order of 0.8, (ii) a spherical cored dark matter profile whose density does not drop sharply radially or (iii) the presence of a secondary dark component resulting from dark matter accretion.

The chemo-dynamical properties of the high-velocity stars in *RAVE* have been investigated in Hawkins *et al.* (2015). The authors report the discovery of a metal-rich halo star that has likely been dynamically ejected from the thick disc, which could aid in explaining the assembly of the most metal-rich component of the Galactic halo.

Finally, Antoja *et al.* (2014) modelled the observed radial dependence of V_ϕ of the Hercules moving group to determine the pattern speed for the Galactic bar. Then, extending the analysis of the bar's influence to larger distances from the Galactic mid-plane, Antoja *et al.* (2015) found asymmetries in the $V_R - V_\phi$ velocity distributions of the disc stars, which suggest that the nature of reported thick disc substructures might be associated with dynamical resonances rather than to accretion events.

3. Milky Way internal evolution and accretion history

Kos *et al.* (2013, 2014) studied the ISM at the extended Solar neighbourhood by measuring the spatial variations of the strength of the absorption line at 8620 Å, associated to a diffuse interstellar band (DIB). They found that the DIB density follows the interstellar dust spatial distribution, with however a scale height significantly larger. The youngest stellar populations, presumably formed from a similar ISM, have been addressed in Conrad *et al.* (2014), where members of 110 open clusters have been identified, updating the radial velocities and metallicities of present cluster catalogues.

Boeche *et al.* (2013, 2014) derived the chemical gradients in the Galaxy and confirmed that the iron and $\alpha-$elements radial gradients flatten as a function of the distance from the plane. This result suggests a different chemical enrichment history for the thick disc compared to the thin disc.

Kordopatis *et al.* (2013b) investigated the properties of the thick disc, down to metallicities of -2 dex. They identified the typical correlation between metallicity and azimuthal velocity of the canonical thick disc stars (e.g. Kordopatis *et al.* 2011, 2013c), which suggests that radial migration could not have been the main mechanism at the origin of the formation of this structure.

Minchev *et al.* (2014) analysed the velocity dispersion of giant disc stars as a function of chemistry and found that the metal-poor stars having [Mg/Fe] > 0.4 dex (the oldest in the sample) have lower velocity dispersions than same metallicity stars with lower [Mg/Fe] abundances. According to the authors this is a signature of a past merger event in the Galaxy, followed by a quiescent period dominated by radial migration.

First direct evidence of stellar radial migration happening at the co-rotation resonances with the spiral arms has been found in Kordopatis *et al.* (2015a). In that study, the orbital and spatial properties of stars that have [M/H] $\gtrsim +0.2$ are studied. It is found that more than half of these super metal-rich stars have migrated through mechanisms that maintain the circularity of the orbits. Given the spatial distribution of these stars, the authors also put constraints on the history of the spiral arms of the Milky Way, that should have been only few and of large structure at least for the last 6 Gyr.

Finally, two recent works addressed globular clusters in *RAVE*. Using solely radial velocities, Anguiano *et al.* (2015) identified members of NGC 3201, ω Cen and NGC 362, to validate the *RAVE*–DR4 stellar parameter and distances. Kunder *et al.* (2014) investigated the chemo-dynamical properties of the available stars around M 22, NGC 1851 and NGC 3201, and reported some stars belonging to these clusters being at projected distances of $\sim$ 10 degrees away from their respective cores. Furthermore, in the radial velocity histograms of the regions surrounding NGC 1851 and NGC 3201, a peak of stars at 230 km s^{-1} is seen, consistent with extended tidal debris from ω Cen.

4. Perspectives and relation with Gaia

Future RAVE data releases are planned to improve, among others, the calibration of the metal-rich end of the metallicity distribution. This will be achieved thanks to the constant addition of metallicity measurements coming from Gaia-benchmark stars (e.g. Jofré *et al.* 2014) and high-resolution spectra of super-solar metallicity stars (see Kordopatis *et al.* 2015a).

The Gaia satellite will obtain proper motions of exquisite quality and distances derived by parallax, overtaking RAVE's precisions by orders of magnitude. However, Gaia will not publish parameter for its stars before ~2017. Until then, RAVE is one of the most significant samples from which Galactic archaeology is possible.

Acknowledgements

Funding for RAVE has been provided by: the Australian Astronomical Observatory; the Leibniz-Institut für Astrophysik Potsdam (AIP); the Australian National University; the Australian Research Council; the French National Research Agency; the German Research Foundation (SPP 1177 and SFB 881); the European Research Council (ERC-StG 240271 Galactica); the Instituto Nazionale di Astrofisica at Padova; The Johns Hopkins University; the National Science Foundation of the USA (AST-0908326); the W. M. Keck foundation; the Macquarie University; the Netherlands Research School for Astronomy; the Natural Sciences and Engineering Research Council of Canada; the Slovenian Research Agency; the Swiss National Science Foundation; the Science & Technology Facilities Council of the UK; Opticon; Strasbourg Observatory; and the Universities of Groningen, Heidelberg and Sydney. The research leading to these results has received funding from the European Research Council under the European Union's Seventh Framework Programme (FP7/2007-2013)/ERC grant agreement no. 321067. The RAVE web site is at `http://www.rave-survey.org`.

References

Anguiano, B., Zucker, D. B., Scholz, R.-D., *et al.* 2015, *MNRAS*, 451, 1229
Antoja, T., Helmi, A., Dehnen, W., *et al.* 2014, *A&A*, 563, A60
Antoja, T., Monari, G., Helmi, A., *et al.* 2015, *ApJL*, 800, L32
Bienaymé, O., Famaey, B., Siebert, A., *et al.* 2014, *A&A*, 571, A92
Binney, J., Burnett, B., Kordopatis, G., *et al.* 2014, *MNRAS*, 439, 1231
Boeche, C., Siebert, A., Piffl, T., *et al.* 2014, *A&A*, 568, A71
Boeche, C., Siebert, A., Piffl, T., *et al.* 2013, *A&A*, 559, A59
Conrad, C., Scholz, R.-D., Kharchenko, N. V., *et al.* 2014, *A&A*, 562, A54
Faure, C., Siebert, A., & Famaey, B. 2014, *MNRAS*, 440, 2564
Hawkins, K., Kordopatis, G., Gilmore, G., *et al.* 2015, *MNRAS*, 447, 2046
Kordopatis, G., Binney, J., Gilmore, G., *et al.* 2015a, *MNRAS*, 447, 3526

Kordopatis, G., Gilmore, G., Steinmetz, M., *et al.* 2013a, *AJ*, 146, 134
Kordopatis, G., Gilmore, G., Wyse, R. F. G., *et al.* 2013b, *MNRAS*, 436, 3231
Kordopatis, G., Hill, V., Irwin, M., *et al.* 2013c, *A&A*, 555, A12
Kordopatis, G., Recio-Blanco, A., de Laverny, P., *et al.* 2011, *A&A*, 535, A107
Kordopatis, G., Wyse, R. F. G., Gilmore, G., *et al.* 2015, *A&A*, 582, 122
Kos, J., Zwitter, T., Grebel, E. K., *et al.* 2013, *ApJ*, 778, 86
Kos, J., Zwitter, T., Wyse, R., *et al.* 2014, Science, 345, 791
Kunder, A., Bono, G., Piffl, T., *et al.* 2014, *A&A*, 572, A30
Minchev, I., Chiappini, C., Martig, M., *et al.* 2014, *ApJL*, 781, L20
Piffl, T., Binney, J., McMillan, P. J., *et al.* 2014a, *MNRAS*, 445, 3133
Piffl, T., Scannapieco, C., Binney, J., *et al.* 2014b, *A&A*, 562, A91
Sharma, S., Bland-Hawthorn, J., Binney, J., *et al.* 2014, *ApJ*, 793, 51
Steinmetz, M., Zwitter, T., Siebert, A., *et al.* 2006, *AJ*, 132, 1645
Widrow, L. M., Barber, J., Chequers, M. H., & Cheng, E. 2014, *MNRAS*, 440, 1971
Williams, M. E. K., Steinmetz, M., Binney, J., *et al.* 2013, *MNRAS*, 436, 101

The General Assembly of Galaxy Halos: Structure, Origin and Evolution
Proceedings IAU Symposium No. 317, 2015
A. Bragaglia, M. Arnaboldi, M. Rejkuba & D. Romano, eds.

doi:10.1017/S1743921315006663

Imaging of NGC 5907's stellar stream

Seppo Laine[1], Carl J. Grillmair[1], David Martínez–Delgado[2], Aaron J. Romanowsky[3], Peter L. Capak[1], Richard G. Arendt[4], Matthew L. N. Ashby[5], James E. Davies[6], Steven R. Majewski[7] and R. Jay GaBany[8]

[1]Spitzer Science Center—Caltech, MS 314-6, Pasadena, CA 91125, USA, email: `seppo@ipac.caltech.edu` [2]Astronomisches Rechen-Institut, Zentrum für Astronomie der Universität Heidelberg, Mönchhofstr. 12-14, D-69120 Heidelberg, Germany, [3]Department of Physics and Astronomy, San José State University, One Washington Square, San Jose, CA 95192, [4]CRESST/UMBC/NASA GSFC, Code 665, Greenbelt, MD 20771, USA, [5]Harvard-Smithsonian Center for Astrophysics, 60 Garden St., Cambridge, MA 02138, USA, [6]Minor Planet Center, Harvard-Smithsonian Center for Astrophysics, 60 Garden St, MS-18, Cambridge, MA 02138. USA, [7]Department of Astronomy, University of Virginia, Charlottesville, VA 22904, USA, [8]Black Bird Observatory, 5660 Brionne Drive, San Jose, CA 95118, USA

Abstract. We have obtained deep g, r, and i-band Subaru and ultra-deep 3.6 μm IRAC images of parts of the multiply-wrapped stellar stream around the nearby edge-on galaxy NGC 5907. We have fitted the surface brightness measurements of the stream with FSPS stellar population synthesis models to derive the metallicity and age of the brightest parts of the stream. The resulting relatively high metallicity ([Fe/H] $= -0.3$) is consistent with a major merger scenario but a satellite accretion event cannot be ruled out.

Keywords. galaxies: evolution, galaxies: halos, galaxies: individual (NGC 5907), galaxies: interactions, galaxies: structure

---

1. Introduction

The detection of stellar streams around nearby galaxies (e.g., Martínez–Delgado *et al.* 2008, 2009, 2010) provides unequivocal evidence of galaxy accretion events. These features reveal the mechanism e.g. for the hierarchical growth of galaxy halos. However, little is known about the progenitors of stellar streams, including their mass and metallicity. While most of the evidence of minor interactions has come from observations of the Milky Way and M31, nearby galaxies outside the Local Group provide even more spectacular evidence of such galaxy disruption events. In this contribution we examine the multiply-wrapped stellar stream around the nearby ($D \approx 17$ Mpc) edge-on disk galaxy NGC 5907 with visible light and near-infrared observations.

2. Observations, Data Reduction and Results

NGC 5907 was observed with Subaru's Suprime-Cam imager in g-, r-, and i-bands, and with Spitzer's IRAC camera at 3.6 μm. We used an iterative scheme to subtract the background light in Subaru images. We ran the IRAC data through Fixen self-calibration (Fixen *et al.* 2000) and through one iteration of the GOODS pipeline (Grumm *et al.* 2005) artifact mitigation and then mosaicked the frames. We masked the foreground and background sources by hand before measuring the surface brightness along the brightest part of the stream. We fitted the surface brightnesses with FSPS SED models from Conroy *et al.* (2009) and Conroy & Gunn (2010). The best fit (metallicity

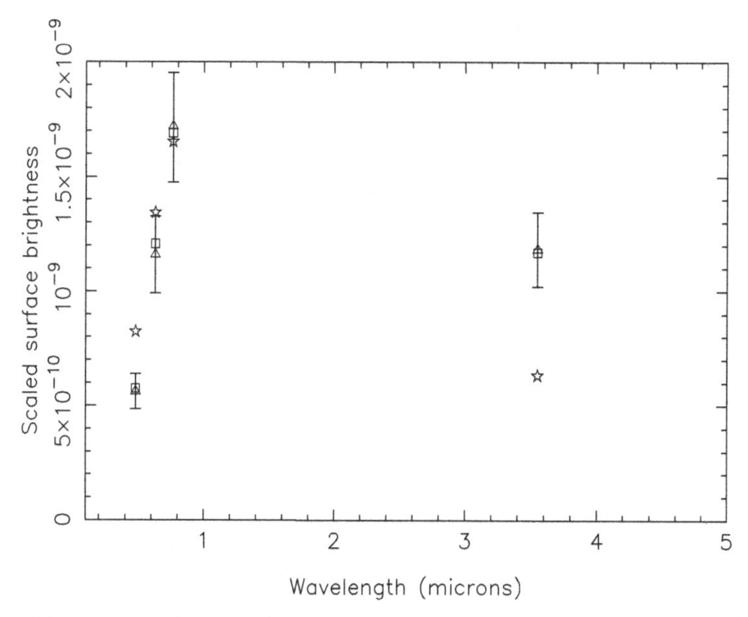

Figure 1. Observations (triangles), best-fit model (squares) and a low metallicity model (stars) in the brightest part of the stellar stream of NGC 5907.

[Fe/H] = −0.3, age = 14.96 Gyrs) is shown in Figure 1 where we also show a model with a low metallicity ([Fe/H] = −1.98, age = 9.44 Gyrs).

Martínez–Delgado *et al.* (2008) modeled the stellar stream with a satellite companion having a stellar mass of $\approx 3.5\times 10^8$ $M_\odot$. In contrast, Wang *et al.* (2012) modeled the stream with a major merger (mass ratio 1:3 – 1:12). While our relatively high metallicity value suggests a major merger origin of the stream, it is possible that the metallicity is consistent with a satellite accretion event such as the Sagittarius Stream (Chou *et al.* 2007) or the giant stream in M31 (Kalirai *et al.* 2006). This work is based in part on observations made with the Spitzer Space Telescope, operated by JPL/Caltech under a contract with NASA and with the Subaru Telescope, operated by NAOJ. Support for this work was provided by NASA through an award issued by JPL/Caltech.

References

Chou, M.-Y., *et al.* 2007, *ApJ*, 670, 346

Conroy, C., Gunn, J. E., & White, M. 2009, *ApJ*, 699, 486

Conroy, C. & Gunn, J. E. 2010, *ApJ*, 712, 833

Fixen, D. J., Moseley, S. H., & Arendt, R. G. 2000, *ApJS*, 128, 651

Grumm, D., Casertano, S., Dickinson, M., & Holfeltz, S. 2005, in: P. Shopbell, M. Britton & R. Ebert (eds.) *ASP-CS* 347, *Astronomical Data Analysis Software and Systems XIV*, (San Francisco: ASP), p. 454

Kalirai, J., *et al.* 2006, *ApJ*, 641, 268

Martínez–Delgado, D., Peñarrubia, J., Gabany, R. J., Trujillo, I., Majewski, S. R., & Pohlen, M. 2008, *ApJ*, 689, 184

Martínez–Delgado, D., Pohlen, M., Gabany, R. J., Majewski, S. R., Peñarrubia, J., & Palma, C. 2009, *ApJ*, 692, 955

Martínez–Delgado, D., *et al.* 2010, *AJ*, 140, 962

Wang, J., Hammer, F., Athanassoula, E., Puech, M., Yang, Y., & Flores, H. 2012, *A&A*, 538, A121

The General Assembly of Galaxy Halos: Structure, Origin and Evolution
Proceedings IAU Symposium No. 317, 2015
A. Bragaglia, M. Arnaboldi, M. Rejkuba & D. Romano, eds.
doi:10.1017/S1743921315008789

What can isolated elliptical galaxies tell us about Cold Dark Matter?

Richard R. Lane[1]†, Tom Richtler[1] and Ricardo Salinas[2]

[1]Departamento de Astronomía, Universidad de Concepción, Casilla 160 C, Concepción, Chile

[2]Department of Physics and Astronomy, Michigan State Univ., East Lansing, MI 48824, USA

Abstract. Due to their environment isolated elliptical galaxies (IEs) should not be undergoing extant evolutionary processes yet many IEs have interacting dwarf companions, and where no merger remnants are visible IEs are often dynamically young. Furthermore, some IEs do not require dark matter to explain their dynamics. However, according to Cold Dark Matter (CDM) simulations all elliptical galaxies should be dark matter dominated, even if isolated, and IEs are much rarer in nature than predicted by CDM. Moreover, merging at the $\sim 10^7$ M$_\odot$ level was recently discovered in the M31 system, showing that hierarchical merging may indeed be scale-free, as predicted by CDM. It seems a natural question to ask: what can IEs tell us about CDM? Here we analyse several IEs as probes of CDM. Our results spawn many new questions.

Keywords. galaxies: elliptical and lenticular, cD; galaxies: kinematics and dynamics; galaxies: structure; galaxies: evolution; galaxies: dwarf; galaxies: star clusters; dark matter

1. Introduction

Cold Dark Matter (CDM) theory has performed exceedingly well, and the existence of dark matter in spiral galaxies, or the possibility of a gravitational interaction which differs from that of Newtonian gravity, is not generally disputed. While it seems that central cluster galaxies and some of the most massive galaxies in the Fornax and Virgo clusters are embedded in massive dark haloes (Kelson *et al.* 2002, Schuberth *et al.* 2012), isolated elliptical galaxies (IEs) and galaxies in less dense environments are only beginning to be studied in detail.

IEs are important probes of CDM because, while galaxies in dense environments have been shown to behave dynamically as CDM predicts, some IEs have been shown not to require any DM to explain their dynamics (Richtler *et al.* 2015, Lane *et al.* 2015), in opposition to CDM, and CDM simulations predict many more IEs than we find in nature (Niemi *et al.* 2010). Furthermore, due to their environment the evolution of IEs should have stopped. However, many IEs have interacting dwarf companions (Lane *et al.* 2013, Richtler *et al.* 2015) or have dynamically young stellar populations (Lane *et al.* 2015), and merging at the $\sim 10^7$ M$_\odot$ level was recently discovered in the M31 system (Amorisco *et al.* 2014). These latter results are directly predicted by CDM simulations.

It is, therefore, vitally important that the dynamics of IEs are reconciled with CDM theories if we are to further our understanding of dark matter.

2. Results

One of the major results of this work is the extension of the study by Salinas *et al.* (2012) who found no requirement for DM in the inner $\sim 1R_e$ in NGC 7507. In Lane

† Current email address: rlane@astro.puc.cl

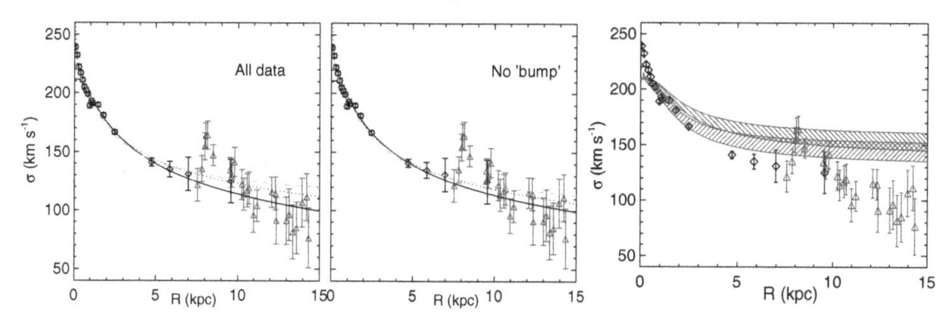

Figure 1. Velocity dispersion versus galactocentric radius for NGC 7507. Filled squares are from GMOS longslit spectra (Salinas *et al.* 2012), open squares are from GMOS slitmask spectra (Lane *et al.* 2015). *Left and middle panels:* models which consider all the velocity dispersion points. The solid black curve is the stars-only model. The red dotted curve is the best-fitting model with $\beta = 0.3$ and NFW halo with $c = 10$. The green dashed curve is the model under the anisotropy by Mamon & Łokas (2005) allowing the maximum amount of dark mater. *Middle panel:* same as left panel, but excluding the velocity dispersion "bump" between $70'' < R < 80''$ for the fits. *Right panel:* MOND models. Symbols as in the other panels. The blue hatched area covers isotropic models with the range of accepted a_0 values from Famaey *et al.* (2007), while the red area represents the same range for a_0 but under Mamon & Łokas (2005) anisotropy.

et al. (2015), using Jeans model fitting to velocities resulting from multi-object spectroscopic data, we extended the measured velocity dispersion to $\sim 3R_e$, and still find no requirement for dark matter to explain the dynamics of this system and, furthermore, our MOND models do not fit the data at any radii (see Figure 1).

Another interesting IE is NGC 7796 which has a dwarf companion with three separate young ($\sim 10^9$ Gyr) blue cores (Richtler *et al.* 2015). This appears to be evidence for hierarchical clustering at the $\sim 10^6\,M_\odot$ level, the lowest mass merging yet uncovered. Furthermore, the companion is tidally disrupting, the first evidence of hierarchical merging at two different mass scales in the same system – something that is clearly predicted by CDM. A massive dark halo is, however, excluded by currently available data, although *some* dark matter, as required for the baryonic Tully-Fisher relation, might be present.

3. Conclusions

The more we search the more IEs we find that do not necessarily require any DM to describe their dynamics. This is puzzling because CDM tells us that all massive galaxies should be DM dominated. What is special about IEs that make them good candidates for this behaviour? That is a question that does not, as yet, have an answer.

References

Amorisco, N. C., Evans, N. W., & van de Ven, G. 2014, *Nature*, 507, 335
Famaey, B., Gentile, G., Bruneton, J.-P., & Zhao, H. 2007, *Phys. Rev. D*, 75, 063002
Kelson, D. D., Zabludoff, A. I., Williams, K. A., *et al.* 2002, *ApJ*, 576, 720
Lane, R. R., Salinas, R., & Richtler, T. 2013, *A&A*, 549, A148
Lane, R. R., Salinas, R., & Richtler, T. 2015, *A&A*, 574, A93
Mamon, G. A. & Łokas, E. L. 2005, *MNRAS*, 363, 705
Niemi, S.-M., Heinämäki, P., Nurmi, P., & Saar, E. 2010, *MNRAS*, 405, 477
Richtler, T., Salinas, R., Lane, R. R., Hilker, M., & Schirmer, M. 2015, *A&A*, 574, A21
Salinas, R., Richtler, T., Bassino, L. P., Romanowsky, A. J., & Schuberth, Y. 2012, *A&A*, 538, A87
Schuberth, Y., Richtler, T., Hilker, M., *et al.* 2012, *A&A*, 544, A115

The General Assembly of Galaxy Halos: Structure, Origin and Evolution
Proceedings IAU Symposium No. 317, 2015
A. Bragaglia, M. Arnaboldi, M. Rejkuba & D. Romano, eds.

doi:10.1017/S1743921315009801

Reconstructing the Accretion History of the Galactic Halo Using Stellar Chemical Abundance Ratio Distributions

Duane M. Lee[1], Kathryn V. Johnston[2], Bodhisattva Sen[3] and Will Jessop[3]

[1]Research Center for Galaxies and Cosmology, Shanghai Astronomical Observatory, Shanghai, China 200030
email: duane@shao.ac.cn

[2]Dept. of Astronomy, Columbia University, New York, NY, 10027 USA
email: kvj@astro.columbia.edu

[3]Dept. of Statistics, Columbia University, New York, NY, 10027 USA
email: bodhi@astro.stat.edu

Abstract. In this study we tested the prospects of using 2D chemical abundance ratio distributions (CARDs) found in stars of the stellar halo to determine its formation history. First, we used simulated data from eleven "MW-like" halos to generate satellite template sets of 2D CARDs of accreted dwarf satellites which are comprised of accreted dwarfs from various mass regimes and epochs of accretion. Next, we randomly drew samples of $\sim 10^{3-4}$ mock observations of stellar chemical abundance ratios ([α/Fe], [Fe/H]) from those eleven halos to generate samples of the underlying densities for our CARDs to be compared to our templates in our analysis. Finally, we used the expectation-maximization algorithm to derive accretion histories in relation to the satellite template set (STS) used and the sample size. For certain STS used we typically can identify the relative mass contributions of all accreted satellites to within a factor of 2. We also find that this method is particularly sensitive to older accretion events involving low-luminous dwarfs e.g. ultra-faint dwarfs — precisely those events that are too ancient to be seen by phase-space studies of stars and too faint to be seen by high-z studies of the early Universe. Since our results only exploit two chemical dimensions and near-future surveys promise to provide $\sim 6-9$ dimensions, we conclude that these new high-resolution spectroscopic surveys of the stellar halo will allow us (given the development of new CARD–generating dwarf models) to recover the luminosity function of infalling dwarf galaxies — and the detailed accretion history of the halo — across cosmic time.

Keywords. Galaxy: abundances – Galaxy: evolution – Galaxy: formation – Galaxy: halo – Galaxy: stellar content – galaxies: dwarf – galaxies: stellar content – methods: statistical – stars: abundances

---

1. Brief Introduction

What is the nature, evolution, and origin of the Galactic halo? Observations of the Galactic halo support the theory that accreted dwarf galaxies built up (most) of the halo over time via hierarchical merging (Searle & Zinn 1978). If so, then what is the merger history of the MW halo? If hierarchical merging of satellites constitutes the prevailing mode of building the stellar halo as suggested by both observations (see above) and simulations (N-body, SPH, AMR, etc. — both with & without gas; see, e.g., Bullock & Johnston 2005 and Cooper *et al.* 2010), then devising a way to recover its accretion history may be paramount to a better understanding galactic evolution. It may also place further constraints on the nature of dark matter halos (see, e.g., Diemand & Moore 2011).

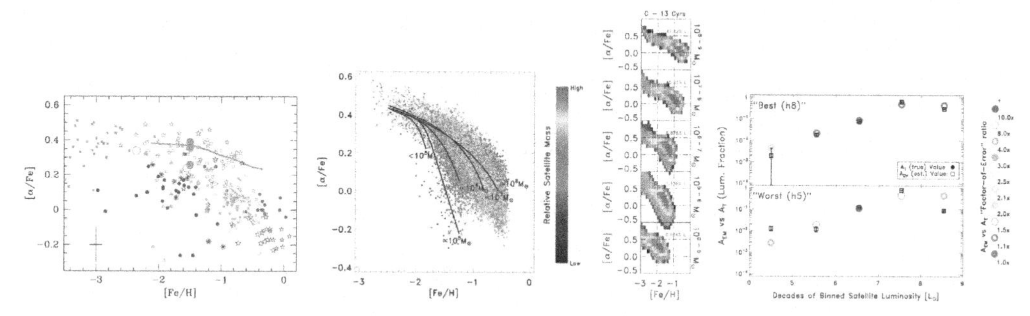

Figure 1. *Left*: Plot is a reproduction of Figure 12 from Geisler *et al.* (2007). *Middle-left*: Plot of [α/Fe] vs. [Fe/H] for $\sim 3 \times 10^4$ "star particles." Each particle is color-coded to represent the relative stellar mass/luminosity of its parent satellite. *Middle-right*: Plot of 1×5 STS in the $t_{acc} - M_{sat}$ plane. *Right*: Plot of 1×5 AHP.

In this work we apply a method that leverages "statistical chemical tagging" (e.g., Schlaufman *et al.* 2012) via the expectation-maximization (EM) algorithm on chemical abundance data from the Bullock & Johnston (2005) simulations (also see Robertson *et al.* 2005 and Font *et al.* 2006) which substantially expands on the idea that Unavane *et al.* (1996) introduced nearly two decades earlier. Unlike stars born in the same cluster, stars born in the same dwarf galaxy do not necessarily share the same chemical composition. However, pioneering studies in the past decade have shown that stars in different dwarfs do have distinct (while overlapping) chemical abundance ratio distributions (CARDs; see, e.g., Geisler *et al.* 2007 and references therein). Hence, the CARD of the Galactic halo should, in theory, reflect the relative number of halo progenitors that were accreted with different galactic masses and accretion epochs.

2. Some Models & Results

The four panels of Figure 1 represents the basic approach and results of the study. The CARDs from observations (*left*) are similar to CARDs from the simulations (*middle-left*). The CARDs of the accreted satellites in the simulations can be marginalized over accretion time to produce the satellite template set (STS) displayed (*middle-right*). Finally, the EM algorithm can be applied to the simulation data using the STS to estimate the accretion history profile (AHP) shown (*right*). Full details of the method, models, and results are provided in Lee *et al.* (2015).

References

Bullock, J. S. & Johnston, K. V. 2005, *ApJ* 635, 931

Cooper, A. P. , Cole, S. , Frenk, C. S. , White, S. D. M. , Helly, J. , Benson, A. J. , De Lucia, G. , Helmi, A. , Jenkins, A. , Navarro, J. F. , Springel, V. , & Wang, J. 2010, *MNRAS* 406, 744

Diemand, J., & Moore, B. 2011, *Adv. Sc. Lett.* 4, 297

Font, A. S.,Johnston, K. V.,Bullock, J. S., & Robertson, B. E. 2006, *ApJ*

Geisler, D., Wallerstein, G., Smith, V. V., & Casetti-Dinescu, D. I. 2007, *PASP*

Lee, D. M.,Johnston, K. V.,Sen, B., & Jessop, W. 2015, *ApJ* 802, 48

Robertson, B., Bullock, J. S., Font, A. S., Johnston, K. V., & Hernquist, L. 2005, *ApJ* 632, 872

Schlaufman, Kevin C., Rockosi, Constance M., Lee, Young Sun, Beers, Timothy C., & Prieto, Carlos Allende, Rashkov, Valery, Madau, Piero, Bizyaev, Dmitry 2012, *ApJ* 749, 77

&Searle, L. & Zinn, R. 1978, *ApJ* 225, 357

&Unavane, M. and Wyse, R. F. G., & Gilmore, G. 1996, *ApJ* 278, 727

The General Assembly of Galaxy Halos: Structure, Origin and Evolution
Proceedings IAU Symposium No. 317, 2015
A. Bragaglia, M. Arnaboldi, M. Rejkuba & D. Romano, eds.

doi:10.1017/S1743921315009667

Measure the local dark matter density with LAMOST spectroscopic survey

Chao Liu[1], Qiran Xia[2] and Shude Mao[3,2,4]

[1]Key Lab of Optical Astronomy, National Astronomical Observatories, CAS, Beijing China
email: liuchao@nao.cas.cn

[2] National Astronomical Observatories, CAS, 20A Datun Road, Chaoyang District, 100012, Beijing, China

[3]Center for Astrophysics , Department of Physics, Tsinghua University, 10086, Beijing, China

[4] Jodrell Bank Centre for Astrophysics, The University of Manchester, Alan Turing Building, Manchester M13 9PL, UK

Abstract. The local dark matter density plays the key role in the distribution of the dark matter halo near the Galactic disk. It will also answer whether a dark matter disk exists in the Milky Way. We measure the local dark matter density with LAMOST observed stars located at around the north Galactic pole. The selection effects of the observations are well considered and corrected. We find that the derived DM density, which is around $0.0159^{+0.0047}_{-0.0057}$ $M_{\odot}$ pc^{-3} providing a flat local rotation curve.

Keywords. Galaxy: kinematics and dynamics – Galaxy: disc – dark matter

1. Introduction and observation data

The local dark matter density is important not only for the astronomers in constraining the total mass of the dark matter halo of the Milky Way, but also for the physicists in the searching of the dark matter particles. Since Oort (1932) firstly measured this value, a great amount of works have been done to constrain the quantity of the dark matter in the solar neighborhood (Read 2014; Piffl *et al.* 2014; Bienaymé *et al.* 2014). However, the results are quite diverse and sometimes are not in agreement with each other. Comparisons of the various results of the measurement are difficult in the sense that either dynamical models or the observed samples are different.

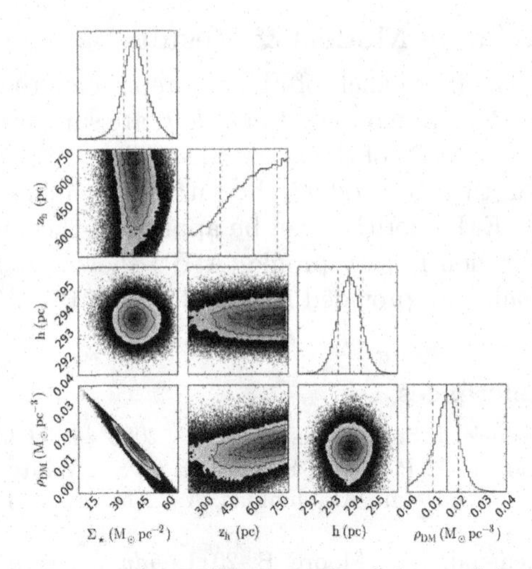

Figure 1. The MCMC result of the model parameters. The parameters from the left to right are $\Sigma_{\star}$, z_h, h, and ρ_{DM}, respectively. The parameters from top to bottom are z_h, h, and ρ_{DM}, respectively. The contours display the 0.5, 1, 1.5, and 2 σ levels. The solid lines in the histogram panels indicate the median values, and the dashed lines indicate the 1-σ region.

In this work, we carefully select more than 1400 G/K dwarf stars located at around the north Galactic pole (NGP) from the LAMOST DR2 catalog (Cui *et al.* 2012, Zhao *et al.* 2012, Deng *et al.* 2012). Because the Galactic latitude of these stars are larger than 85°, the line-of-sight velocities approximately equal to the vertical velocities. The uncertainty of the measured line-of-sight velocity is about 4-5 $\mathrm{km s^{-1}}$. We adopt the distance obtained by Carlin *et al.* (2015) for the sample with uncertainty of 20%.

The selection effect is considered for the selected stars. Because the LAMOST survey does not explicitly bias to specific color index for the high Galactic latitude area in its targeting strategy, we can safely assume that the luminosity function of the spectroscopic stars is same as the photometric sample within a small sky region and color–magnitude diagram. Then we can use the number of the local photometric stars to correct the selection effect of the spectroscopic stars.

2. Dynamical model and result

We assume that

(*a*) The star tracers are in dynamical equilibrium.

(*b*) The Milky Way disk is axisymmetric.

(*c*) The rotation curve is flat and the local dark matter density is constant with z in the solar neighborhood.

From Jeans equation, we obtain

$$\frac{d}{dz}[\nu(z)\sigma_z^2(z)] = -\nu\frac{d\Phi(z)|_{R_\odot}}{dz}, \tag{2.1}$$

where ν, σ_z, and Φ are the vertical density profile, velocity dispersion, and gravitational potential, respectively. According to the Poisson's equation, we obtain

$$4\pi G(\rho_{disc}(z) + \rho_{gas}(z) + \rho_{\rm DM}(z))|_{R_\odot} = \frac{d^2\Phi(z)|_{R_\odot}}{d^2 z}, \tag{2.2}$$

where ρ_{disc}, ρ_{gas}, are $\rho_{\rm DM}$ the mass volume densities for the stellar disk, gas disk, and the dark matter halo, respectively. Finally, we give an analytical form for K_z force as (Zhang *et al.* 2013)

$$K_z(z) \equiv -\frac{d\Phi(z)}{dz} = -2\pi G\left\{\Sigma_\star\left[1-\exp\left(-\frac{\rm z}{\rm z_h}\right)\right] + \Sigma_{gas} + 2\rho_{\rm DM}z\right\} \tag{2.3}$$

We apply the Markov chain Monte Carlo (MCMC) simulation to derive the parameters based on the above equations and show the result in Figure 1. The surface density of the gas disk is adopted as 13 $M_\odot$ $\mathrm{pc^{-2}}$ (Zhang *et al.* 2013). We finally find that the local dark matter density is $0.0159^{+0.0047}_{-0.0057}$ $M_\odot$ $\mathrm{pc^{-3}}$.

References

Bienaymé, O., Famaey, B., Siebert, A., *et al.* 2014, *A&A*, 571, 92
Carlin, J. L., Liu, C., Newberg, H., *et al.* 2015, *AJ*, 150, 4
Cui, X. Q., Zhao, Y. H., Chu, Y. Q. *et al.* 2012, *RAA*, 12, 1197
Deng, L. C., Newberg, H., Liu, C., *et al.* 2012, *RAA*, 12, 735
Oort, J. H., 1932, Bulletin of the Astronomical Institutes of the Netherlands, 6, 249
Piffle, T., Binney, J., McMillan, P. J., *et al.* 2014, *MNRAS*, 445, 3133
Read, J. I., 2014, arXiv:1404.1938
Zhang, L., Rix, H.-W., van de Ven, G., Bovy, J., Liu, C., & Zhao, G., 2013, *ApJ*, 772, 108
Zhao, G., Zhao, Y. H., Chu, Y. Q., Jing, Y. P., & Deng, L. C. 2012, *RAA*, 12, 723

The General Assembly of Galaxy Halos: Structure, Origin and Evolution
Proceedings IAU Symposium No. 317, 2015
A. Bragaglia, M. Arnaboldi, M. Rejkuba & D. Romano, eds.

doi:10.1017/S1743921315007085

Halo Mass Estimation for Galaxy Groups: The Role Of Magnitude Gaps

Yi Lu[1], **Xiaohu Yang**[1,2] **and Shiyin Shen**[1]

[1]Key Laboratory for Research in Galaxies and Cosmology,
Shanghai Astronomical Observatory, Nandan Road 80, Shanghai 200030, China
email: luyi@shao.ac.cn

[2]Center for Astronomy and Astrophysics,
Shanghai Jiao Tong University, Shanghai 200240, China
email: xyang@sjtu.edu.cn

Abstract. We find that for the galaxy groups, the luminosity gap between the brightest and the subsequent brightest member galaxies in a halo (group) can be used to significantly reduce the scatter in the halo mass estimation based on the luminosity of the brightest galaxy alone. These corrections can significantly reduce the scatter in the halo mass estimations by $\sim 50\%$ to $\sim 70\%$ in massive halos.

1. Introduction

Galaxy groups provide an important step in understanding processes in galaxy formation and cosmology. There are many methods to identify galaxy groups. Yang *et al.*(2005a) developed a halo-based group finder which has been successfully applied to 2dFGRS (Yang *et al.* 2005a), SDSS DR4 and DR7 (Yang *et al.* 2007). One of the key steps in the halo-based group finder is the estimation of halo masses of candidate galaxy groups. Usually, group total luminosity (e.g. Yang *et al.* 2005a; 2007) is considered to be a reliable halo mass indicator. Unfortunately, for shallow and high redshift surveys, only a few brightest member galaxies can be observed in each dark matter halo. In case the survey volume is difficult to calculate because of the bad survey geometry, the halo mass estimation based on the total luminosity may become unachievable.

To estimate the halo masses for poor galaxy systems, one may make use of the central-host halo relation. As shown in Yang *et al.* (2008), for massive halos $L_c \propto M_h^{\sim 0.25}$, the typical scatter is about 0.15. The central (or the brightest) galaxy alone cannot provide a reliable estimation. Thus we improve this relation by using luminosity gap, defined as $\log L_{\rm gap} = \log L_c - \log L_i$, where L_i is the luminosity of the i-th brightest member galaxies.

2. Overview

We make use of four sets of mock galaxy catalogs obtained from conditional luminosity function, the subhalo abundance, and semi-analytical model respectively. In Fig. 1, blue open circles show the median and the 68% confidence levels (error bars) of halo masses, $\log M_h$, as a function of central galaxy luminosity $\log L_c$. We fit the median $M_h(L_c)$ relations with the following functional form:

$$\log M_h = \exp(\log L_c - \log M_a) + \log M_b \,. \tag{2.1}$$

To tighten the errorbars by using luminosity gap, we formally write

$$\log M_h(L_c, L_{\rm gap}) = \log M_h(L_c) + \Delta \log M_h(L_c, L_{\rm gap}) \,, \tag{2.2}$$

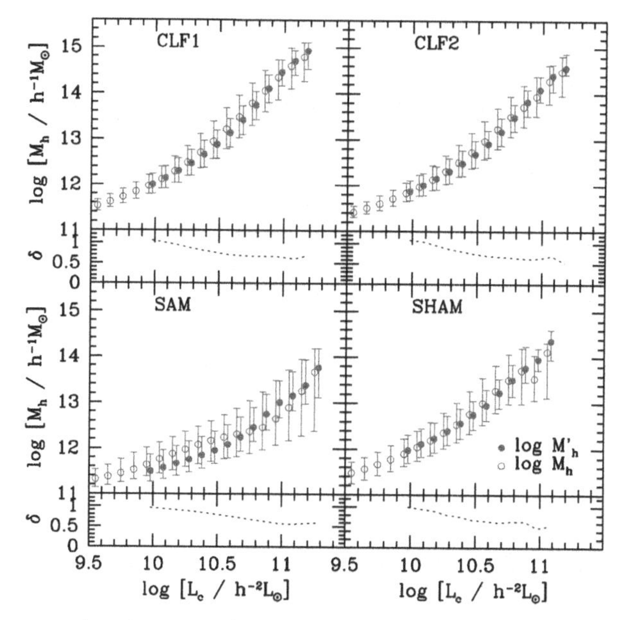

Figure 1. The comparison between the scatter in the original $\log M_h(L_c)$ relation and the new $\log M_h(L_c, L_{\rm gap})$ model.

where the first term on the right side is the empirical relation described by Eq. (2.1). And the 'luminosity gap' here, defined as the luminosity ratio between the central and the second brightest member galaxies in the same dark matter halo, $L_{\rm gap} = L_c/L_2$. We use the following functional form to model $\Delta \log M_h(L_c, L_{\rm gap})$,

$$\Delta \log M_h(L_c, L_{\rm gap}) = \eta_a \exp(\eta_b \log L_{\rm gap}) + \eta_c \,, \tag{2.3}$$

where parameters η_a, η_b and η_c may all depend on L_c:

$$\begin{aligned} \eta_a(L_c) &= \exp(\log L_c - \beta_1) \\ \eta_b(L_c) &= \alpha_2(\log L_c + \beta_2)\,, \\ \eta_c(L_c) &= -(\log L_c - \beta_3)^{\gamma_3} \end{aligned} \tag{2.4}$$

which in total has five free parameters. For comparison, we define a 'pre-corrected' halo mass

$$\log M_h' = \log M_h - \Delta \log M_h(L_c, L_{\rm gap})\,, \tag{2.5}$$

and check if the scatter in the $M_h'(L_c)$ relation is significantly reduced relative to that in the $M_h(L_c)$. If the correction by $\Delta \log M_h(L_c, L_{\rm gap})$ were perfect, the scatter in the $M_h'(L_c)$ would be reduced to 0.

In Fig. 1, we define the ratio between the corrected and original errorbars as δ in the sub-panels. It is clear that the scatter in $M_h'(L_c)$ is significantly reduced, especially for massive halos/groups where the scatter is reduced by about 50%.

References

Yang, X., Mo, H. J., van den Bosch, F. C., & Jing, Y. P. 2005a, *MNRAS*, 356, 1293

Yang, X., Mo, H. J., van den Bosch, F. C., *et al.* 2007,*ApJ*, 671, 153

Yang, X., Mo, H. J., & van den Bosch, F. C. 2008, *ApJ*, 676, 248

The General Assembly of Galaxy Halos: Structure, Origin and Evolution
Proceedings IAU Symposium No. 317, 2015
A. Bragaglia, M. Arnaboldi, M. Rejkuba & D. Romano, eds.

doi:10.1017/S1743921315006791

Impact of NLTE on research of early chemical enrichment of the dwarf galaxies

Lyudmila Mashonkina[1], Pascale Jablonka[2], Pierre North[2] and Tatyana Sitnova[1]

[1]Institute of Astronomy, Russian Academy of Sciences
Pyatnitskaya st. 48, RU-119017 Moscow, Russia
email: lima@inasan.ru
[2]Laboratoire d' Astrophysique, Ecole Polytechnique Fédérale de Lausanne (EPFL)
Observatoire de Sauverny, CH-1290 Versoix, Switzerland

Abstract. Based on high-resolution observed spectra, the non-local thermodynamic equilibrium (NLTE) line formation, and precise stellar atmosphere parameters, we present the first complete sample of dwarf spheroidal galaxies (dSphs) with accurate chemical abundances in the very metal-poor (VMP) regime. The obtained stellar elemental ratios are compared with chemical enrichment models, and we show that NLTE is a major step forward for studies of the dSph and the Milky Way (MW) chemical evolution.

Keywords. stars: abundances, galaxies: abundances, galaxies: dwarf, galaxies: evolution

---

Studies of chemical abundances of VMP stars provide important clues for better understanding the early chemical enrichment processes of the host galaxy and the onset of star formation. Our research concerns the dSphs orbiting the Milky Way, where individual stars accessible for high-resolution spectroscopy are all giants. The classical LTE assumption commonly applied to stellar abundance analyses is, in particular, questionable for such objects. We aim to revise the chemical abundances of a complete sample of VMP stars in the classical dSphs Sculptor, Sextans, and Fornax and in the ultra-faint dwarf (UFD) galaxy Boötes I based on the NLTE line formation and to test chemical enrichment models. The same methods were used to derive abundances of the MW halo comparison sample.

NLTE calculations were performed for a number of chemical species, using our original model atoms. Inelastic collisions with H I atoms were treated applying accurate rate coefficients for Na I, Mg I, Al I, and Si I and the scaled Drawin formula for the remaining species. In the relevant stellar parameter range, NLTE leads to strengthened lines and negative NLTE abundance corrections for Na I, but to weakened lines and positive corrections for all other chemical species. The amount of NLTE correction varies according to the different species and depends on atmospheric parameters.

The stellar sample includes 10 members of the Scl, Fnx, and Sex dSphs observed by Tafelmeyer *et al.* (2010) and Jablonka *et al.* (2015, JNM15) and 7 Boötes I stars from Norris *et al.* (2010) and Gilmore *et al.* (2013). For comparison with the MW halo, nine cool VMP giants were selected from Cohen *et al.* (2013, CCT13), and three stars from our previous studies. For Boötes I and CCT13 we rely on the published equivalent widths.

Stellar atmosphere parameters. This study is based on non-spectroscopic $T_{\rm eff}$ and log g. The effective temperatures were derived from photometry in the original papers. For the Scl, Fnx, and Sex stars the accurate determination of log g takes advantage of the known distance. We show that, when applying NLTE, the Fe I/Fe II and Ti I/Ti II ionisation equilibria are fulfilled using the adopted $T_{\rm eff}$ /log g for all stars, except for the two most

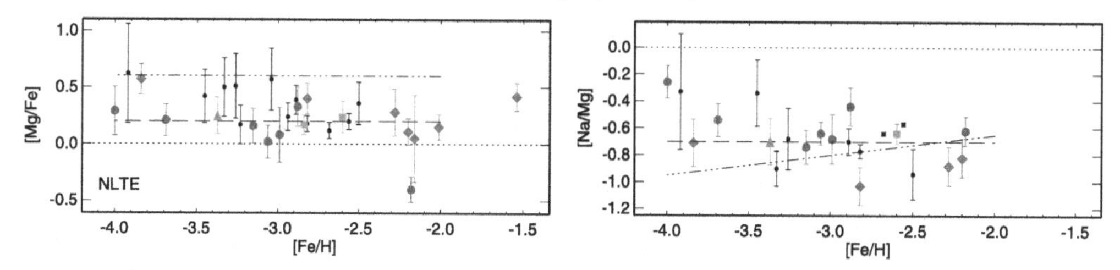

Figure 1. [Mg/Fe] and [Na/Mg] NLTE ratios of stars in the Sculptor (circles), Sextans (squares), Fornax (triangles), and Boötes I (rhombi) dSphs and the MW halo (small circles). The dashed and dash-three-dotted lines indicate the yields of a single supernova of 14.4 $M_{\odot}$ (Lai *et al.* 2008) and the MW chemical evolution calculations of Kobayashi *et al.* (2011), respectively.

MP stars, with [Fe/H] $\leqslant -3.7$. The problem may lie with only two Fe II lines detected and the uncertainty in $T_{\rm eff}$. To make the Fe I and Fe II NLTE abundances agree, their $T_{\rm eff}$ should be increased by 170 and 200 K respectively. The log g of the Boötes I and CCT13 stars were estimated from the YY isochrones at 12 Gyr. We revised log g upward by 0.28 dex for Boo-94 and by 0.1-0.2 dex for the CCT13 stars according to the Fe I/Fe II ionisation equilibrium in NLTE. With respect to the metallicity scale derived from Fe I lines under the LTE assumption, NLTE leads to a shift of +0.2-0.34 dex.

NLTE elemental ratios. Part of our results are displayed in Fig. 1. In line with the previous LTE studies, we find that the dSphs show enhanced Mg and Ca relative to Fe in the VMP regime, with Mg/Fe and Ca/Fe ratios that are similar for different dSphs and similar to those for the MW halo. We improve the average ratios for each galaxy: [Mg/Fe] = 0.18, 0.20, 0.25, 0.26, and [Ca/Fe] = 0.52, 0.36, 0.42, 0.36 for the Scl, Sex, Fnx, Boö I dSphs, respectively. Note that the Scl star ET0381 ([Fe/H] = −2.18) was excluded from the mean ratios (see JNM15 for detailed discussion). Unlike previous studies, Ti is found to follow Fe and reveal no enhancement, within the error bars. NLTE is crucial for accurate determination of elemental ratios involving Na and Al. In NLTE, galaxies with different masses show very similar [Na/Mg] ratios at [Fe/H] > -3.8, while in LTE, they show much less consistent ratios, and the scatter of [Na/Mg] is twice larger in each individual galaxy. Furthermore, [Na/Al] ratios obtained in NLTE are close to solar for galaxies with different masses, while in LTE, Na is strongly enhanced relative to Al, with [Na/Al] > 0.5.

Comparing our results with the nucleosynthesis models (Fig. 1), we conclude that the intermediate-mass (14.4 $M_{\odot}$) SNeII were the dominant source of early chemical enrichment in classical dSphs. For the Boötes I UFD, the statistics of the [Fe/H] $\simeq -3$ stars should be increased.

Acknowledgements. L.M. thanks a support from the International Astronomical Union, the Swiss National Science Foundation (SCOPES project IZ73Z0-152485), the Russian Foundation for Basic Research (14-02-91153-GFEN) of the participation at IAUS317.

References

Cohen, J. G., Christlieb, N., Thompson, I., *et al.* 2013, *ApJ*, 778, 56
Gilmore, G., Norris, J. E., Monaco, L., *et al.* 2013, *ApJ*, 763, 61
Jablonka, P., North, P., Mashonkina, L., *et al.* 2015, *A&A*, 583, A67
Kobayashi, C., Karakas, A. I., & Umeda, H. 2011, *MNRAS*, 414, 3231
Lai, D. K., Bolte, M., Johnson, J. A., *et al.* 2008, *ApJ*, 681, 1524
Norris, J. E., Yong, D., Gilmore, G., & Wyse, R. F. G. 2010, *ApJ*, 711, 350
Tafelmeyer, M., Jablonka, P., Hill, V., *et al.* 2010, *A&A*, 524, A58

The General Assembly of Galaxy Halos: Structure, Origin and Evolution
Proceedings IAU Symposium No. 317, 2015
A. Bragaglia, M. Arnaboldi, M. Rejkuba & D. Romano, eds.

doi:10.1017/S1743921315007000

Very Metal-poor Stars Observed by the RAVE Survey

Gal Matijevič[1] and the RAVE Collaboration
[1]Leibniz-Institut für Astrophysik Potsdam (AIP),
An der Sternwarte 16, 14482 Potsdam, Germany
email: gmatijevic@aip.de

Abstract. Radial Velocity Experiment (RAVE) observed ~500,000 southern sky stars between 2003 and 2013 in the infra-red calcium triplet (CaII) spectral region. In this study we extended the analysis of RAVE very metal-poor stars ([Fe/H] < -2) presented by Fulbright *et al.* (2010). We employed a novel method for identifying the metal-poor stars and developed a tool for modeling CaII lines where we also modeled the background noise to avoid systematical biases in the equivalent width (EW) measurements. Final metallicity values were derived with a flexible calibration approach using only 2MASS photometric data and EW measurements obtained from the RAVE spectra.

Keywords. Metal-poor stars, RAVE survey, Gaussian processes

1. Introduction

Metal-poor stars have been extensively used for stellar and galactic archeology studies over the last few decades (Beers & Christlieb 2005). They offer an insight into the nucleosynthesis in the early Galaxy and subsequent metal enrichment by the supernovae, and allow us to study the conditions in the young Universe in which they were born. Large-scale all-sky spectroscopic surveys like RAVE offer a great opportunity to discover new very-metal poor stars and measure their metallicities.

2. Metal-poor spectra identification

RAVE's 4th data release (Kordopatis *et al.* 2013, DR4) includes atmospheric parameters including metallicities for most of the observed stars. However, due to the lack of significant features in the spectra of very metal-poor stars, the parameters become less reliable with decreasing metallicity and the spectra need to be reanalyzed. The first step is to correctly identify all metal-poor candidates. We used a method called t-SNE (van der Maaten 2014) to compute a 2D projection of all spectra in the database (similarly as in Matijevič *et al.* 2012). The method groups morphologically similar spectra together so it is easy to identify and isolate the group of $\sim$ 3000 metal-poor FGK-type candidates for which we computed better metallicity estimates (Fig. 1).

3. Line profile modeling

Deriving abundances requires measuring the equivalent widths. Usually this is done by defining a line profile and fitting it to the spectral lines by minimizing the χ^2 statistic. The EW is measured by integrating the profile. Problems arise if the conditions under which the χ^2 is a good estimator are violated, i.e. when noise is not independent and

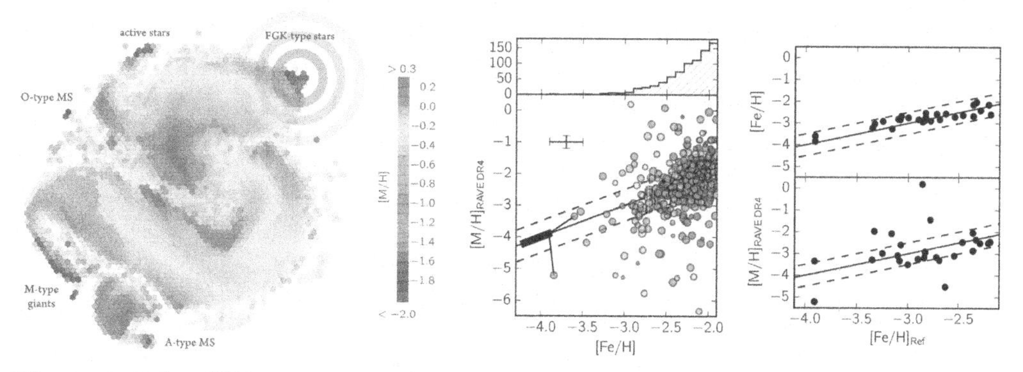

Figure 1. *Left:* t-SNE projection of RAVE spectra with $S/N > 10$. Circled region is occupied by the spectra that were reanalyzed. *Center:* Comparison between the new metallicities and those provided by DR4. *Right:* Comparison between new, DR4, and high resolution reference [Fe/H] values of a subset of stars.

identically distributed. This often happen because a) the profile of the targeted line also includes weaker lines, and b) reduction and normalization introduce variations or offsets in the continuum. Ignoring these problems can lead to biased EW measurements and underestimated uncertainties because the solution is too stiff. We corrected for these difficulties by introducing additional terms in the likelihood function to model the correlated noise using Gaussian processes (Rasmussen & Williams 2006). This way we computed *less* precise but *more* accurate EWs.

4. Metallicity calibration

Equivalent widths of CaII triplet lines correlate very well with metallicity, but also depend strongly on the luminosity which introduces degeneracy: spectra of less metal-poor dwarfs resemble spectra of more metal-poor giants. For clusters this can be resolved by adding a term to the calibration relation that describes the distance to the cluster's horizontal branch. Unfortunately, this method cannot be used for field stars at unknown distances. Instead, we relied on measured EWs and 2MASS JHK_S colors. Motivated by The Cannon (Ness *et al.* 2015), we defined the calibration relation by training the model on a subset for which we have high-resolution data form Ruchti *et al.* (2013) and Carrera *et al.* (2013). Afterwards, we evaluated the measured EWs and colors of the remaining stars in the metal-poor sample to obtain their metallicities. This was done with Gaussian processes to give the model enough flexibility. We were able to identify several hundred new very-metal poor stars. Comparison with the metallicities measured form the high resolution spectra confirms that new values are indeed much more accurate (Fig. 1).

References

Beers, T. C. & Christlieb, N. 2005, *ARA&A*, 43, 531
Carrera, R., Pancino, E., Gallart, C., *et al.* 2013, *MNRAS*, 434, 1681
Fulbright, J. P., Wyse, R. F. G., & Ruchti, G. R., *et al.* 2010, *ApJ*, 724, L104
Kordopatis, G., Gilmore, G., Steinmetz, M., *et al.* 2013, *AJ*, 146, 134
Matijevič, G., Zwitter, T., Bienaym, O., *et al.* 2012, *ApJS*, 200, 14
Ness, M., Hogg, D. W., & Rix, H. -W., *et al.* 2015, *ApJ*, 808, 16
Rasmussen, C. E. & Williams, C. 2006, *MIT Press*
Ruchti, G. R., Bergemann, M., Serenelli, A., *et al.* 2013, *MNRAS*, 429, 126
van der Maaten, L. J. P.. 2014, *Journal of Machine Learning Research*, 15, 3221 ?

The General Assembly of Galaxy Halos: Structure, Origin and Evolution
Proceedings IAU Symposium No. 317, 2015
A. Bragaglia, M. Arnaboldi, M. Rejkuba & D. Romano, eds.

doi:10.1017/S1743921315007012

Near-Field Cosmology with RR Lyrae Variable Stars: A First View of Substructure in the Southern Sky

C. Navarrete[1,2], S. Duffau[2,1], A. K. Vivas[3], M. Catelan[1,2], G. Hajdu[1,2], G. Torrealba[4], C. Cortés[5,2], V. Belokurov[4], S. Koposov[4] and A. J. Drake[6]

[1]Instituto de Astrofísica, Pontificia Universidad Católica de Chile, Av. Vicuña Mackenna 4860, 782-0436 Macul, Santiago, Chile
email: `[cnavarre, sduffau]@astro.puc.cl`

[2]Millennium Institute of Astrophysics, Santiago, Chile

[3]Cerro Tololo Inter-American Observatory, Casilla 603, La Serena, Chile

[4]Institute of Astronomy, Madingley Road, Cambridge CB3 0HA, UK

[5]Departamento de Física, Facultad de Ciencias Básicas, Universidad Metropolitana de Ciencias de la Educación, Av. José Pedro Alessandri 774, 776-0197 Ñuñoa, Santiago, Chile

[6]California Institute of Technology, 1200 East California Boulevard, Pasadena, CA 91225, USA

Abstract. We present the current status of the spectroscopic follow-up of a large number of RR Lyrae (RRL) halo overdensity candidates recently found by Torrealba *et al.* (2015) using southern-hemisphere data from the Catalina Real-time Transient Survey (CRTS). Characterizing the individual RRL stars in these overdensities is crucial to confirm them as real halo substructures. Low-resolution spectra have been obtained for RRL stars in 11 different overdensities, using the SOAR and Magellan telescopes. Radial velocities and metallicities have been derived so far for 123 and 99 RRL stars, respectively.

Keywords. Galaxy: halo, Galaxy: structure, stars: variables: RR Lyrae

---

A complete census of substructure in the Milky Way halo is essential to understand the history of formation of our Galaxy, and thereby provide constraints on models of hierarchical galaxy formation. Until recently, however, the southern celestial hemisphere has remained virtually uncharted territory, with halo substructure having mostly been identified in the north. The situation is now beginning to change; in particular, Torrealba *et al.* (2015, hereinafter T15) have recently provided a list of southern overdensity candidates based on RRL stars identified using CRTS data. Spectroscopic follow-up of the individual RRL stars in these overdensity candidates is now required to confirm their true nature, or to discard them as artifacts or random clumps among halo field stars.

The main goal of our project is to perform spectroscopic follow-up of the T15 candidates, thereby deriving radial velocities (RVs) and metallicities ([Fe/H]) of individual RRL contained in these overdensities. This, together with their distances and other information from the literature (when available), will help us to properly establish the physical properties of each overdensity. By comparing with data for other previously known substructures, including those from the northern hemisphere, we also hope to establish any possible links that may exist amongst them.

Low-resolution ($R \sim 2000$) spectra have been obtained using LDSS3@LCO and Goodman@SOAR for ≈ 200 stars covering 11 of the 27 overdensities from the T15 catalog, including Sgr 1, Crv 1, Hya 1, and Hya 2. Figure 1 displays the RRL from T15 and those

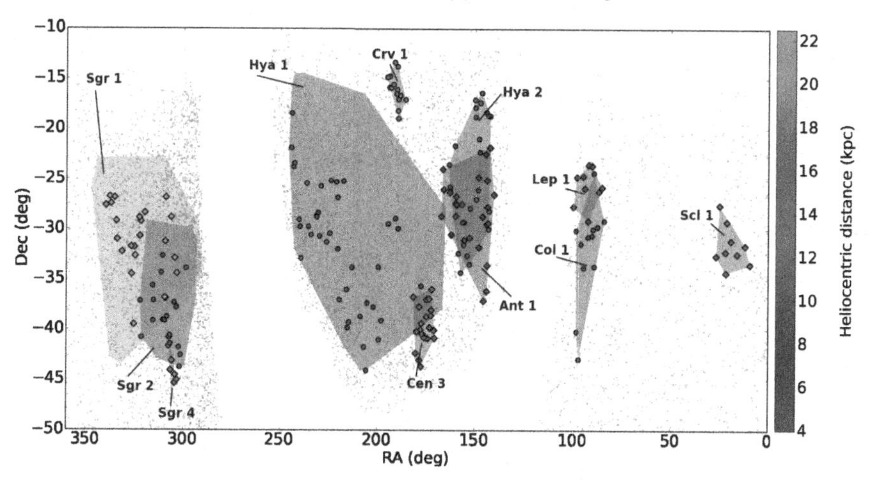

Figure 1. Spatial distribution of the targeted RRL overdensities (shaded polygons). Filled symbols correspond to RRLs with spectra while gray points are all the RRLs in the T15 catalog. Heliocentric distances are color-coded according to the scale on the right.

observed by us in each of the targeted overdensities. RVs have been measured for 123 stars, and corrected for their pulsation phase using the Sesar (2012) templates. [Fe/H] has been estimated for 99 stars using the method of Layden (1994). Typical errors are $\sim 20\,\mathrm{km\,s^{-1}}$ and 0.2 dex, respectively. Our preliminary analysis tentatively suggests a connection between Crv 1 and the Virgo Stellar Stream (Duffau *et al.* 2006), and reveals an interesting bimodal radial velocity distribution for Hya 2. Hya 1 is of special interest since Casetti-Dinescu *et al.* (2015) identified a moving group in this part of the sky. Our RVs together with proper motions will allow us to study in detail the nature of this object. Testing of the relation between Sgr 1 and the Sagittarius tidal stream using both data (e.g., Koposov *et al.* 2012) and disruption models (e.g., Law & Majewski 2010) will also be afforded by our data. Detailed comparisons with models of the smooth halo will be carried out to help confirm or reject these connections, and to assess their significance.

Acknowledgements

This project is supported by CONICYT through International Research Project DPI 20140066. Additional support is provided by the Ministry for the Economy, Development, and Tourism's Iniciativa Científica Milenio through grant IC 120009, awarded to the Millennium Institute of Astrophysics; by Proyecto Fondecyt Regular #1141141; and by Proyecto Basal PFB-06/2007. C.N. and G.H. gratefully acknowledge support from CONICYT-PCHA/Doctorado Nacional grants 2015-21151643 and 2014-63140099, respectively. C.N. also acknowledges additional support from a SOCHIAS grant through Gemini/CONICYT Project #32140015.

References

Casetti-Dinescu, D. I., Nusdeo, D. A., Girard, T. M., *et al.* 2015, *ApJ*, 810, 4
Duffau, S., Zinn, R., Vivas, A. K., *et al.* 2006, *ApJ*, 636, 97
Law, D. R. & Majewski, S. R. 2010, *ApJ*, 714, 229
Layden, A. C. 1994, *AJ*, 108, 1016
Koposov, S. E., Belokurov, V., Evans, N. W., *et al.* 2012, *ApJ*, 750, 80
Sesar, B. 2012, *AJ*, 144, 114
Torrealba, G., Catelan, M., Drake, A. J., *et al.* 2015, *MNRAS*, 446, 2251

The General Assembly of Galaxy Halos: Structure, Origin and Evolution
Proceedings IAU Symposium No. 317, 2015
A. Bragaglia, M. Arnaboldi, M. Rejkuba & D. Romano, eds.

doi:10.1017/S1743921315010443

Study of the Milky Way's hot coronal gas with its dwarf galaxies

Stefano Pasetto[1], Mark Cropper[1], Yutaka Fujita[2], Cesare Chiosi[3] and Eva K. Grebel[4]

[1] Mullard Space Science Laboratory, University College London, Holmbury St Mary, Dorking, Surrey RH5 6NT, United Kingdom
email: s.pasetto@ucl.ac.uk

[2] Department of Earth and Space Science, Graduate School of Science, Osaka University,1-1 Machikaneyama-cho, Toyonaka-shi, Osaka 560-0043, Japan [3]Dept. of Physics & Astronomy "Galileo Galilei", University of Padua, Vicolo dell'Osservatorio, 5, 35141 Padova PD, Italy, [4]Astronomisches Rechen-Institut, Zentrum für Astronomie der Universität Heidelberg, Mönchhofstr 12-14, 69120, Heidelberg, Germany

Abstract. A large amount (5×10^{10} $M_\odot$) of hot gas is thought to exist in an extended ($\approx 200kpc$) hot diffuse halo around the Milky Way. We investigate the competitive role of the different dissipative phenomena acting on the onset of star formation of this gravitationally bound systems in this external environment. Ram pressure, Kelvin-Helmholtz and Rayleigh- Taylor instabilities, and tidal forces are accounted for separately in an analytical framework and compared in their role in influencing the star forming regions. We present an analytical criterion to elucidate the dependence of star formation in a spherical stellar system on its surrounding environment, useful in observational applications as well as theoretical interpretations of numerical results. We consider the different signatures of these phenomena in synthetically realized colour-magnitude diagrams (CMDs) of the orbiting system, thus investigating the detectability limits and relevance of these different effects for future observational projects. The theoretical framework developed has direct applications to the cases of our MW system as well as dwarf galaxies in galaxy clusters or any primordial gas-rich star cluster of stars orbiting within its host galaxy.

Keywords. Galaxy: halo, Galaxy: fundamental parameters, galaxies: dwarf

---

1. Milky Way galactic halo model

We model a time-dependent Milky-Way-like galaxy considered as a gravitational environment in which to study the orbital evolution of a dwarf galaxy. A mass growth rate is applied only to the halo component of the Milky Way (MW) while all the other MW components are kept constant with a parameter distribution as in Pasetto *et al.* 2011, 2012a,b. We examine whether different profiles of hot intergalactic medium (HIGM) surrounding the MW halo can leave different observable traces on star formation of the orbiting dwarf galaxy and hence on its CMDs. For this purpose we simulate the evolution of a generic dwarf galaxy close to the MW with different HIGM distribution.

2. Results

The star formation history of the orbiting galaxy is reconstructed in Fig.1 by the methods presented in Pasetto *et al.* 2012a and Fujita (1998) for different HIGMs profiles. We found that the different HIGMs are able to influence the evolution of the molecular cloud distribution considered in the orbiting dwarf galaxy and leave different signatures on the CMDs (synthetically realized with the technique in Bertelli *et al.* 2009). This is

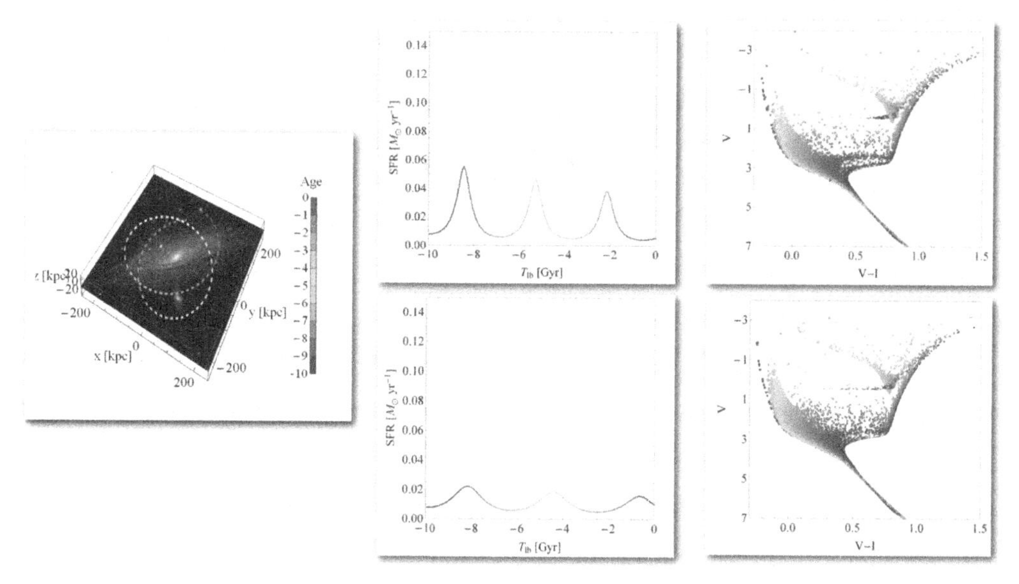

Figure 1. On the same orbit (left plot) the different electron density distributions are considered in calculating the star formation rate (central panels) and the synthetic CMDs (right panels).

more evident at the level of the sub-giant branch stars. We aim to realize a methodology able to compare automatically results with observations at least for the Local Group case and able to disentangle the different contributions to the star formation history of satellites in the Local Groups (e.g., Grebel, Gallagher, & Harbeck, 2003) thus placing a constraints on the orbits directly from the CMDs.

References

Bertelli, G., Nasi, E., Girardi, L., & Marigo, P., 2009, A&A, 508, 355
Fujita, Y., 1998, ApJ, 509, 587
Fukugita, M. & Peebles, P. J. E., 2006, ApJ, 639, 590
Grebel, E. K., Gallagher, J. S., III, & Harbeck, D., 2003, AJ, 125, 1926
Pasetto, S., Bertelli, G., Grebel, E. K., Chiosi, C., & Fujita, Y., 2012, A&A, 542, A17
Pasetto, S., *et al.*, 2012, A&A, 547, A70
Pasetto, S., Grebel, E. K., Berczik, P., Chiosi, C., & Spurzem, R., 2011, A&A, 525, A99

The General Assembly of Galaxy Halos: Structure, Origin and Evolution
Proceedings IAU Symposium No. 317, 2015
A. Bragaglia, M. Arnaboldi, M. Rejkuba & D. Romano, eds.

doi:10.1017/S1743921315006869

Identifying Remote Halo Giants in High-Latitude Fields with Kepler 2

Ruth C. Peterson

SETI Institute, 189 N Bernardo Ave, Mountain View, CA 94043

Abstract. This work sketches how SDSS *ugr* colors and Kepler 2 in halo fields can identify red giants 50 – 100 kpc distant with minimal metallicity bias. For these mildly-reddened, metal-poor giants, $(g-r)_0$ yields the effective temperature $T_{\rm eff}$ to 100 K. K2 can detect the p-mode oscillations of red giants and measure their frequency of maximum power $\nu_{\rm max}$. This sets the luminosity $L_{\rm bol}$ and thus the distance, plus an estimate of metallicity [Fe/H].

Keywords. Galaxy: halo, Galaxy: stellar content, stars: abundances, stars: atmospheres, stars: distances, stars: Population II, techniques: photometric

---

Red giants are the best tracers of the remote halo. At their high luminosities they remain visible to 100 kpc, and their presence at all metallicities and ages encourages an unbiased sample of the stellar population. Here we note how Kepler (K2) and SDSS can find them.

In K2 fields in the Galactic halo, reddening is low, so the effective temperature $T_{\rm eff}$ of a star known to be a metal-poor giant can be determined from its dereddened $(g-r)_0$ color alone. Figure 1 shows the minimal spread in $(g-r)_0$ at a given $T_{\rm eff}$ at various metallicities [Fe/H], for observed metal-poor giants and for the model atmospheres of Castelli & Kurucz (2003) (CK03).

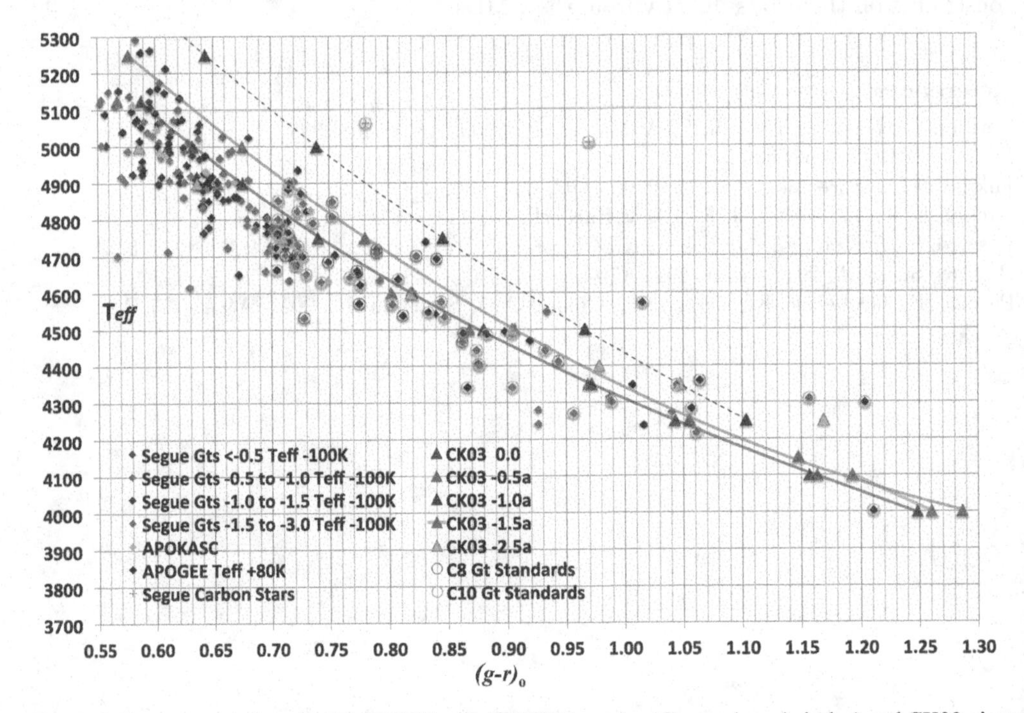

Figure 1. Temperatures from stellar SEGUE and APOGEE spectra (diamonds and circles) and CK03 giant models (triangles) are plotted vs. $(g-r)_0$. The model $T_{\rm eff}$ and log g values, from Yong *et al.* (2013) at [Fe/H] = −1.6, are 5125K/2.75, 4900K/2.30, 4750K/2.00, 4600K/1.60, 4500K/1.30, 4350K/1.00, and 4250K/0.75.

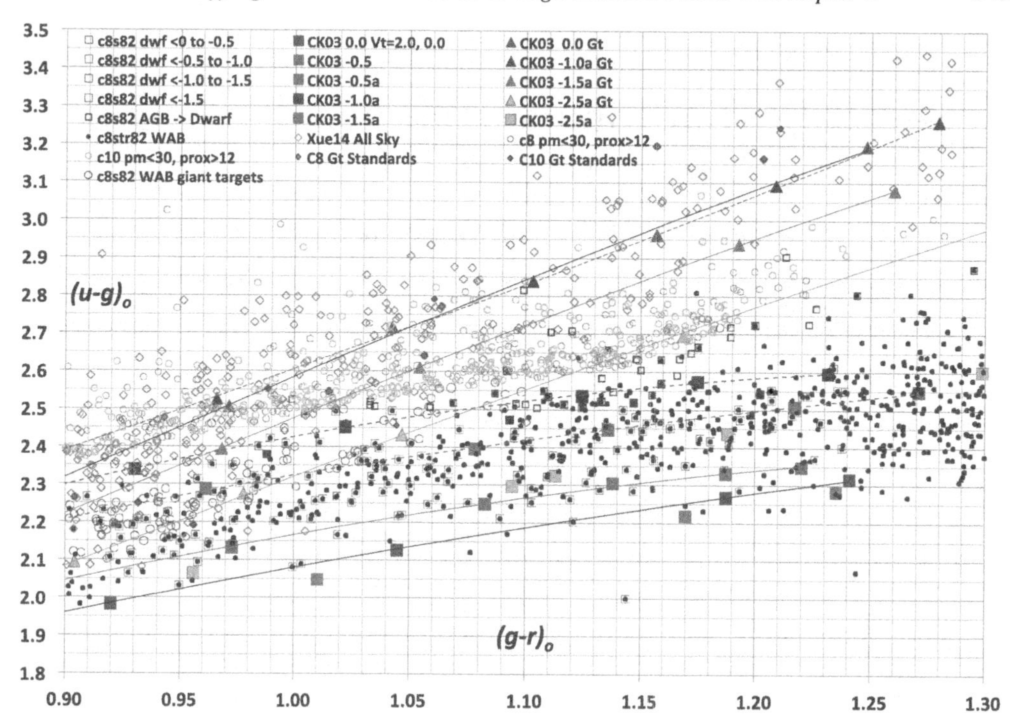

Figure 2. The dereddened colors of stars known to be giants (diamonds) or dwarfs (squares) are compared to the model colors of giants (triangles) and dwarfs (filled squares). Open circles are the dereddened colors for our proposed targets. Dark maroon circles are Bhatti *et al.* 2010 (WAB) stars, whose photometry is more precise. Note the paucity of cool giant WAB targets, hence their selection from the entire C8 and C10 fields.

K2 can identify remote giants asteroseismologically by detecting their p-mode oscillations. Those of dwarfs peak at frequencies too high to detect in long-cadence light curves, but those of cool, metal-poor giants fall within range, peaking at $\nu_{\rm max} \sim 200 - 2\,\mu$Hz. Despite roll corrections, noise trials suggest K2 light curves should detect p-mode oscillations to Kepler magnitudes $K_{\rm p} \sim 16.5$ for the most luminous metal-poor giants, reaching stars 100 kpc away.

In Figure 2 we show K2 targets we have proposed in the halo fields C8 and C10 (open circles). Their dereddened $(u-g)_{\rm o}$ and $(g-r)_{\rm o}$ colors are those of cool giants, as seen from model CK03 colors (with $u-g$ increased by 0.2) of giants (diamonds) and dwarfs (squares), and observed colors of C8 and C10 SEGUE/APOGEE giants (filled diamonds) and the all-sky Xue *et al.* (2014) giants (open diamonds). Giants are scarce: the high-quality Bhatti *et al.* (2010) (WAB) photometry (small filled circles) of Stripe 82 within C8 shows that foreground dwarfs dominate.

Applying this color selection to all SDSS + K2 halo fields should identify ~ 100 new giants. Set by stellar gravity, $\nu_{\rm max}$, plus $T_{\rm eff}$ from $(g-r)_{\rm o}$, fix the giant luminosity $L_{\rm bol}$. This sets its distance, and also yields an estimate of [Fe/H], since giant luminosity also depends on stellar metallicity (Xue *et al.* 2014). Thus SDSS + K2 alone can characterize the remote halo population.

References

Bhatti, W. A., Richmond, M. W., Ford, H. C., *et al.* 2010, *Ap. J. Supp.*, 186, 233

Castelli, F. & Kurucz, R. L. 2003, *IAU Symp. No. 210*, Modeling of Stellar Atmospheres, eds. N. Piskunov *et al.*, CD-ROM poster A20; also astro-ph 0405087

Xue, X.-X., Ma, Z., Rix, H.-W., *et al.* 2014, *Ap. J.*, 784, 170

Yong, D., Meléndez, J., Grundahl, F., *et al.* 2013, *MNRAS*, 434, 3542

The General Assembly of Galaxy Halos: Structure, Origin and Evolution
Proceedings IAU Symposium No. 317, 2015
A. Bragaglia, M. Arnaboldi, M. Rejkuba & D. Romano, eds.

doi:10.1017/S1743921315008832

Several evolutionary channels for bright planetary nebulae

Michael G. Richer[1] **and Marshall L. McCall**[2]

[1]Instituto de Astronomía, Universidad Nacional Autónoma de México, Apartado Postal 106, 22800 Ensenada, Baja California, México email: richer@astrosen.unam.mx
[2]Department of Physics and Astronomy, York University, Toronto, Ontario L3T 3R1, Canada email: mccall@yorku.ca

Abstract. The populations of bright planetary nebulae in the discs of spirals appear to differ in their spectral properties from those in ellipticals and the bulges of spirals. The bright planetary nebulae from the bulge of the Milky Way are entirely compatible with those observed in the discs of spiral galaxies. The similarity might be explained if the bulge of the Milky Way evolved secularly from the disc, in which case the bulge should be regarded as a pseudo-bulge.

Keywords. (ISM:) planetary nebulae: general, stars: evolution

1. Introduction

Planetary nebulae (PNe) are the penultimate evolutionary stage of stars in the 1-8 $M_\odot$ mass range. Bright PNe (large absolute luminosities in the [O III] $\lambda5007$ line) are useful for studying aspects of stellar and galactic evolution in the nearby universe. The bright end of the luminosity function of PNe in [O III] $\lambda5007$ (PNLF) has been used as a secondary distance indicator for galaxies of all morphological types (e.g., Ciardullo *et al.* 1989). While the physics behind the constancy of the PNLF peak luminosity is not clearly understood, empirical evidence indicates that the brightest PNe in all galaxies reach similar luminosities, have nebular shells whose kinematics are similar, and arise from stellar progenitors that have undergone similar nucleosynthetic processes (Ciardullo *et al.* 2002; Richer & McCall 2008; Richer *et al.* 2010a). The internal kinematics evolve with time, but not necessarily in the same way in all galaxies. It is tempting to suppose that the same stellar progenitors produce the brightest PNe in all galaxies, but it is unclear whether this is feasible since the dominant stellar populations vary from galaxy to galaxy, both in mass and metallicity.

PNe are dynamical systems. The state of both the central star and the nebular shell change substantially with time. Models predict that, initially, the luminosity in Hβ increases very rapidly, but that the [OIII] $\lambda5007$ luminosity overtakes it some time later (e.g., Schönberner *et al.* 2007). As a result, both the Hβ luminosity and 5007/Hβ ratio vary with time, probing to some extent the luminosity and temperature, respectively, of the central star. How the 5007/Hβ ratio varies as a function of the Hβ luminosity will depend upon the details of the central star and nebular shell and how they vary with time (e.g., the nebular mass and its distribution, the opacity of the nebular shell to ionizing photons, and the relative evolutionary rates of the central star and nebular shell).

We can use observations of the Hβ luminosities and 5007/Hβ ratios for bright PNe in different galaxies to determine whether the details of the evolution are the same in all galaxies. If bright PNe in all galaxies arise from stars with the same masses, metallicities, and ages, the distribution of Hβ luminosities and 5007/Hβ ratios should be the same. In Fig. 1, we see that they appear to differ, at least between the discs of spiral galaxies

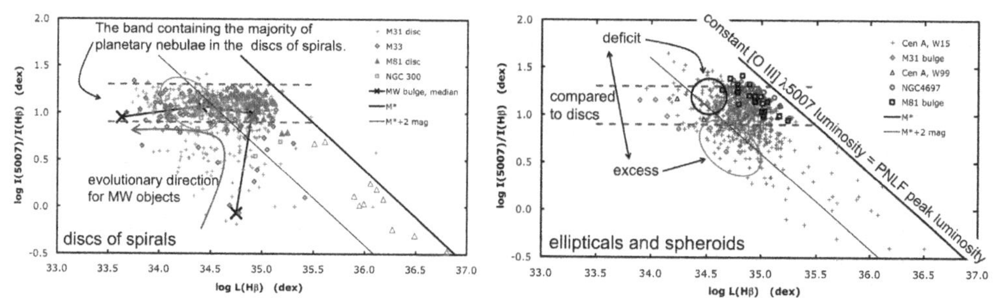

Figure 1. These figures present the distribution of the ratio of [O III] λ5007/Hβ as a function of the Hβ luminosity for bright PNe in nearby galaxies, separated into discs and spheroids. In both panels the bold diagonal line indicates the peak luminosity of the PNLF while the lighter diagonal line indicates an [O III] λ5007 luminosity 2 mag fainter. In both panels, the horizontal lines indicate the band occupied by PNe in the discs of spirals. In general, the disc systems (left panel) are nearer than the ellipticals and spheroids, explaining the much greater range of Hβ luminosities probed in disc systems. In neither case are the samples statistically complete. Even so, there are clues that the distributions of the brightest PNe differ between discs and spheroids. The ellipses indicate where the fainter PNe dominate in both samples. The circle indicates a lack of bright PNe with high 5007/Hβ in spheroids. Data sources: Balick *et al.* (2013, ApJ, 774, 3); Bresolin *et al.* (2010, *MNRAS*, 404, 1679); Corradi *et al.* (2015, ApJ, 807, 181); Fang *et al.* (2013, ApJ, 774, 138); Jacoby & Ciardullo (1999, ApJ, 515, 169); Jacoby & Ford (1986, ApJ, 304, 490); Kniazev *et al.* (2014, AJ, 147, 16); Kwitter *et al.* (2012, ApJ, 753, 12); Magrini *et al.* (2009, ApJ, 696, 729); McCall (2014, *MNRAS*, 440, 405); Méndez *et al.* (2005, ApJ, 627, 767); Richer *et al.* (1999, A&ASS, 135, 203); Richer *et al.* (2008, ApJ, 689, 203); Richer *et al.* (2010, ApJ, 716, 857); Richer & McCall (unpublished); Roth *et al.* (2004, ApJ, 603, 531); Sanders *et al.* (2012, ApJ, 758, 133); Stanghellini *et al.* (2010, A&A, 521, A3); Stasińska *et al.* (2013, A&A, 552, A12); Walsh *et al.* (1999, A&A, 346, 753); Walsh *et al.* (2015, A&A, 574, A109).

and ellipticals and the bulges of spirals, even for the PNe with the highest luminosities. First, the majority of objects more than 2 mag below the PNLF peak in disc systems (left panel) have larger 5007/Hβ ratios than those in ellipticals and spheroids (compare the ellipses in both panels). Though their [O III] λ5007 luminosities are similar, the Hβ luminosities differ (those of the PNe in spheroids are brighter), so there is no reason they should be missing from observations of the discs of spirals, if they exist. Second, compared to disc systems, the ellipticals and spheroids have an apparent deficit of PNe with luminosities 1-2 mag below the PNLF peak luminosity, but with high 5007/Hβ ratios (circle, right panel). More sensitive spectroscopy is needed to verify its reality.

These differences may be exploited to determine the origin of a PN population in some cases. In the left panel of Fig. 1, the median values are shown for the four evolutionary phases of bright PNe from the Milky Way bulge, ordered as indicated, from Richer *et al.* (2008, 2010b). Clearly, the locus defined by these median values are compatible with the the distribution of PNe in discs , perhaps evidence that the Milky Way's bulge evolved secularly from its disc.

References

Ciardullo, R., Jacoby, G. H., Ford, H. C., & Neill, J. D. 1989, *ApJ*, 344, 715)
Ciardullo, R., Feldmeier, J. J., Jacoby, G. H. *et al.* 2002, *ApJ*, 577, 31
Richer, M. G., López, J. A., Pereyra, M., *et al.* 2008, *ApJ*, 689, 203
Richer, M. G. & McCall, M. L. 2008, *ApJ*, 648, 1190
Richer, M. G., López, J. A., Díaz-Méndez, E. *et al.* 2010a, *RMxAA*, 46, 191
Richer, M. G., López, J. A., García-Díaz, Ma.-T. *et al.* 2010b, *ApJ*, 716, 857
Schönberner, D., Jacob, R., Steffen, M., & Sandin, C. 2007, *A&A*, 473, 467

The General Assembly of Galaxy Halos: Structure, Origin and Evolution
Proceedings IAU Symposium No. 317, 2015
A. Bragaglia, M. Arnaboldi, M. Rejkuba & D. Romano, eds.

doi:10.1017/S1743921315006961

The Milky Way's halo in 6D: *Gaia*'s Radial Velocity Spectrometer performance

George Seabroke[1], Mark Cropper[1], David Katz[2], Paola Sartoretti[2], Pasquale Panuzzo[2], Olivier Marchal[2], Alain Gueguen[2], Kevin Benson[1], Chris Dolding[1], Howard Huckle[1], Mike Smith[1] and Steve Baker[1]

[1]Mullard Space Science Laboratory, University College London, Dorking, Surrey, UK
[2]Observatoire Paris-Site de Meudon, GEPI, Paris, France

Abstract. *Gaia*'s Radial Velocity Spectrometer (RVS) has been operating in routine phase for over one year since initial commissioning. RVS continues to work well but the higher than expected levels of straylight reduce the limiting magnitude. The end-of-mission radial-velocity (RV) performance requirement for G2V stars was 15 km s^{-1} at $V = 16.5$ mag. Instead, 15 km s^{-1} precision is achieved at $15 < V < 16$ mag, consistent with simulations that predict a loss of 1.4 mag. Simulations also suggest that changes to *Gaia*'s onboard software could recover ~0.14 mag of this loss. Consequently *Gaia*'s onboard software was upgraded in April 2015. The status of this new commissioning period is presented, as well as the latest scientific performance of the on-ground processing of RVS spectra. We illustrate the implications of the RVS limiting magnitude on *Gaia*'s view of the Milky Way's halo in 6D using the *Gaia* Universe Model Snapshot (GUMS).

Keywords. techniques: radial velocities, surveys, stars: kinematics, Galaxy: halo

1. *Gaia*'s Radial Velocity Spectrometer performance

Gaia's Radial Velocity Spectrometer (Cropper & Katz 2011) data are processed on-ground by the Data Processing and Analysis Consortium (DPAC) Co-ordination Unit (CU) 6 Spectroscopic Processing pipeline (Katz *et al.* 2011). The pipeline formally runs at the CU6 Data Processing Centre. We present offline tests of the CU6 pipeline running at the Mullard Space Science Laboratory. As already presented in Cropper *et al.* (2014), Fig. 1 (left) presents tentative evidence that the CU6 pipeline is able to achieve the end-of-mission RV precision predicted by simulations that include the straylight. The majority of the stars in Fig. 1 (left) are G dwarfs so the measured $\sigma_{RV} \sim 15$ km s^{-1} at $15 < V < 16$ mag is consistent with simulations that predict a loss of 1.4 mag.†

2. The Milky Way's halo in 6D

The fixed limiting magnitude of $G_{RVS} = 16.2$ mag onboard RVS was changed to an adaptive one in September 2015. Now the limiting magnitude ranges from $15.5 < G_{RVS} < 16.2$ mag as a function of time, corresponding to the straylight pattern to save telemetry on spectra that contain more noise than signal. Selecting Milky Way halo stars from the GUMS simulation (Robin *et al.* 2012) with $G_{RVS} < 16.2$ mag suggests that RVS is collecting spectra for ~1.8 million stars in the Milky Way's halo, the majority of which will be metal-poor giants. Having approximately verified the RV performance model in Section 1 with preliminary measurements, we apply this model to each GUMS star. This predicts an *upper* limit (due to the adaptive limiting magnitude) on the number of halo stars in the final *Gaia* catalogue with $\sigma_{RV} < 15$ km s^{-1}: ~800,000, ~42% of the observed halo stars (red/dark grey dots in Fig. 1 right). Fig. 1 (left) presents tentative evidence that σ_{RV} can be measured up to 30 km s^{-1}: ~1.2 million stars, 67% of the observed halo

† `http://www.cosmos.esa.int/web/gaia/science-performance`

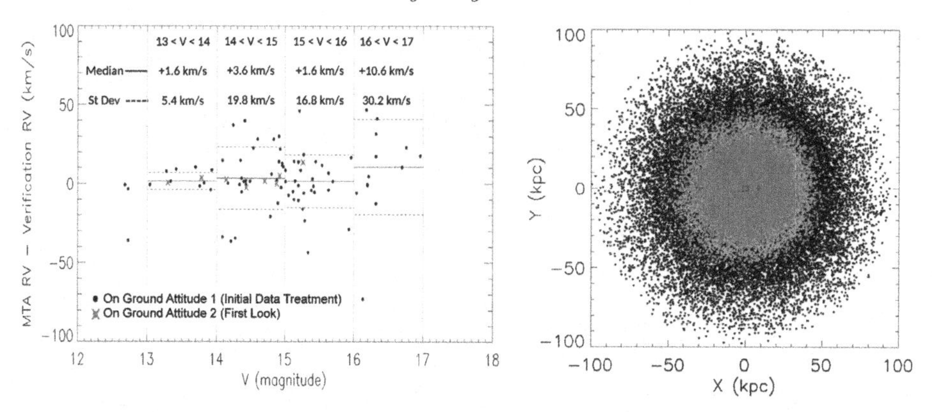

Figure 1. *Left*: Measured end-of-mission (40 transits) RV performance as a function of magnitude. *Right*: GUMS simulation of Milky Way halo stars with $G_{RVS} < 16.2$ mag looking face on: X and Y are in the Galactic plane with the Sun at the origin (square), where the cross is the Galactic centre. Each star is colour-coded according to its predicted end-of-mission σ_{RV}, assuming every star is a metal-poor K1 giant: $\sigma_{RV} < 15$ km s^{-1} (red/dark grey), $15 < \sigma_{RV} < 30$ km s^{-1} (green/light grey), $\sigma_{RV} > 30$ km s^{-1} (black).

stars (green/light grey dots in Fig. 1 right). Beyond this limit, it has not been tested that the data follow Poisson statistics and so whether the predictions are valid.

RVS spectra were obtained with a fixed ACross Scan (AC) window width of 10 pixels. The RVS S/N simulator suggests that adapting the AC width to the observing conditions could mitigate the impact of straylight, recovering ~0.14 mag. This would increase the aforementioned predicted upper limit to ~900,000, a 13% increase (all Galactic components increase by several million). *Gaia*'s onboard software was updated in April 2015 to observe with adaptive AC widths. This new functionality has been successfully commissioned but depends very much on RVS calibrations and so is still being optimised.

3. Summary

Gaia-RVS is already the largest ever spectroscopic survey (5.4 billion spectra observed in its first year). This means it will also be the largest ever survey of the Milky Way's halo. It will provide ~1 km s^{-1} precision radial velocities for $V < 12$ mag, which is the planned CU6 contribution to the second *Gaia* data release (2017, to be confirmed). With *Gaia* astrometry, this will provide ~10,000 *local field* halo stars (all Galactic components ~2 million) in 6D. RVS spectra will also be used by DPAC CU8 to derive abundances (Fe, Ca, Ti, Si), T_{eff}, $\log g$, [M/H] and A_0 for $V < 12$ mag, ready for later data releases. CU6-measured *v*sin*i* for these stars means that *Gaia* will measure these ~10,000 local field halo stars and the ~2 million all-Galactic-component stars in a total of 15 dimensions.

We present preliminary evidence that *Gaia*-RVS can provide ~15 km s^{-1} precision end-of-mission radial velocities at $15 < V < 16$ mag. At this precision and with *Gaia* astrometry, *Gaia* will provide ~800,000-900,000 halo stars (all Galactic components ~75-100 million) in 6D in the final *Gaia* catalogue (2022, to be decided). When analysed with CU8-derived [M/H] from *Gaia*'s BP/RP spectra of these stars, *Gaia*'s chemo-6D-kinematic mapping out to ~30-50 kpc from the Sun will revolutionise our understanding of the Milky Way halo's structure, origin and evolution.

References

Cropper, M. & Katz, D. 2011, *EAS Publications Series*, 45, 181

Cropper, M., Katz, D., Sartoretti, P., *et al.* 2014, *EAS Publications Series*, 67, 69

Katz, D., Cropper, M., Meynadier, F., *et al.* 2011, *EAS Publications Series*, 45, 189

Robin, A. C., Luri, X., Reylé, C., *et al.* 2012, *A&A*, 543, A100

The General Assembly of Galaxy Halos: Structure, Origin and Evolution
Proceedings IAU Symposium No. 317, 2015
A. Bragaglia, M. Arnaboldi, M. Rejkuba & D. Romano, eds.

doi:10.1017/S1743921315007036

Morphology and Structures of Nearby Dwarf Galaxies

Mira Seo and H. B. Ann

Dept. of Earth Science, Pusan National University
609-735 , Busan, Korea
email: mrseo@pusan.ac.kr, hbann@pusan.ac.kr

Abstract. We applied GALFIT and STARLIGHT to the r-band images and spectra, respectively, of $\sim 1,100$ dwarf galaxies to analyze the structural properties and stellar populations. In most cases, single component with $n = 1 \sim 1.5$ well describes the luminosity distribution of dwarf galaxies. However, a large fraction of dS0, dE_{bc}, and dE_{blue} galaxies show sub-structures such as spiral arms and rings. There is a bimodal distributions of stellar ages in dS0 galaxies. But other sub-types of dwarf galaxies show a single peak in the stellar distributions.

Keywords. galaxies, dwarf, structure, population

1. Introduction

Understanding dwarf galaxies is crucial in the mordern cosmology because they are most dominant populations in the universe. However, until recently, there is no detailed morphological studies of dwarf galaxies. Ann, Seo, & Ha (2015) determined morphological types of the local galaxies ($z < 0.01$) using color images of the Sloan Digital Sky Survey (SDSS) DR7 distinguishing sub-types of dwarf elliptical-like galaxies. The present study aims at understanding the structural properties of dwarf elliptical-like galaxies using r-band SDSS images and spectral data.

2. Data

We used ~1,100 dwarf elliptical-like galaxies in the catalog of Ann, Seo, & Ha (2015). They divided dwarf elliptical-like galaxies into 5 sub-types: dwarf lenticular galaxies (dS0), dwarf elliptical galaxies (dE), blue-cored dwarf elliptical galaxies (dE_{bc}), blue dwarf elliptical galaxies (dE_{blue}), and dwarf spheroid galaxies (dSph). Among the 5 sub-types, dS0, dE, and dSph galaxies show the presence of nucleation. Along with the morphological types, we used the r-band images and spectra of dwarf elliptical-like galaxies to analyze the luminosity distributions and stellar populations.

3. Methods

We performed 2D-photometric decompositions of dwarf elliptical-like galaxies using GALFIT (Peng, Ho, Impey & Rix 2002, 2010). Residual images are obtained by subtracting the single component model images from the observed galaxy images. We computed spectral synthesis models using STARLIGHT (Cid Fernandes *et al.* 2005) and obtained light-fraction population vectors (x_j) in 15 different age groups and 6 metallicity groups. We derived the local background density by the nth nearest neighbor technique. We used n = 5, ΔV=500 km/s, and M*=-15.24.

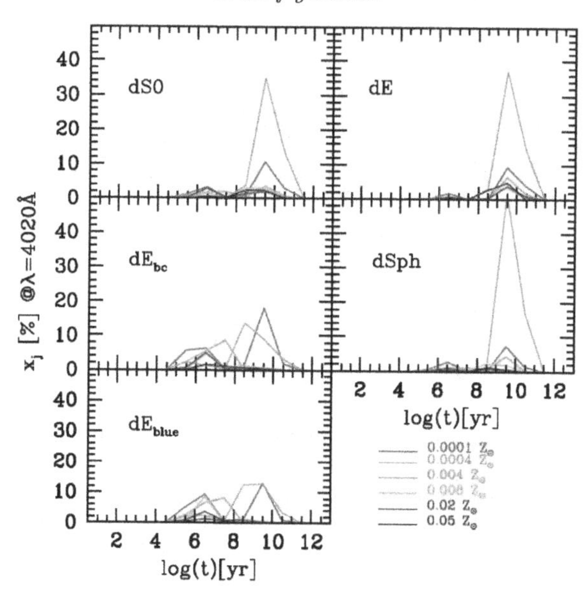

Figure 1. Age distributions in the population vector (x[%]) of dwarf elliptical-like galaxies, distinguished by their metallicity

4. Results and Discussion

There is no significant difference in the physical parameters, residual features and their environment between nucleated dwarfs (dE_n, $dSph_n$, $dS0_n$) and un-nucleated dwarfs (dE_{un}, $dSph_{un}$, $dS0_{un}$). There are a variety of features in the residual images such as lens, bar, and spiral arms. These features are more frequent in dS0, dE_{bc} and dE_{blue} galaxies. The residual images of dE_{blue} galaxies suggest that some dE_{blue} galaxies have nuclei which are difficult to be noticed due to their colors similar to their global colors. Half of dS0 galaxies show a variety of sub-structures in their residual images.

Fig. 1 shows the age distributions of dwarf elliptical-like galaxies sorted by the metallicities. The dE, and dSph galaxies have mostly old populations while dE_{bc} and dE_{blue} galaxies have a mixture of populations, characterized by a bimodal distribution. The dS0 galaxies show intermediate distributions. Combined with metallicity distribution, old stellar populations are mostly metal poor but young stellar populations have similar fractions of metal poor and metal rich stars. It is of interest to see that age distributions of stars with intermediate metallicity ($z = 0.004z_\odot$) display bimodal distributions in dS0 galaxies while there are single peaks at old stellar populations in dE and dSph and at young stellar populations at intermediate age populations in dE_{bc} and dE_{blue}.

References

Ann, Seo, Ha 2015, *ApJS*, 217, 27
Cid Fernandes, R., Mateus, A. , Sodre, L., Stasinska, G., & Gomes, J. M. 2005, *MNRAS*, 358,363
Peng, C. Y., Ho, L. C., Impey, C. D., & Rix, H.-W. 2002, *AJ*, 124, 266
Peng, C. Y., Ho, L. C., Impey, C. D., & Rix, H.-W. 2010, *AJ*, 139, 2097

The General Assembly of Galaxy Halos: Structure, Origin and Evolution
Proceedings IAU Symposium No. 317, 2015
A. Bragaglia, M. Arnaboldi, M. Rejkuba & D. Romano, eds.

doi:10.1017/S1743921315007115

Discovery of new dwarf galaxies around NGC4631 with Subaru/Hyper Suprime-Cam

Mikito Tanaka[1], Masashi Chiba[2] and Yutaka Komiyama[3]

[1]Frontier Research Institute for Interdisciplinary Sciences, Tohoku University, Sendai, Japan
email: mikito@astr.tohoku.ac.jp

[2]Astronomical Institute, Tohoku University, Sendai, Japan

[3]Subaru Telescope, National Astronomical Observatory of Japan

Abstract. We have observed on-going interacting galaxies (NGC4631 and NGC4656) using Subaru/Hyper Suprime-Cam and reduced the data using HSC pipeline and conducted photometry based on DAOphot. Then, we have detected 8 new dwarf galaxy candidates in the outer region of NGC4631 and confirmed the three candidates previously reported by Karachentsev *et al.* 2014. The 3 or 4 candidates detected in this study may be a star-forming dwarf irregular galaxy and the other 7 candidates may be an old dwarf spheroidal galaxy based on these stellar populations. It looks like that the effective radius - absolute magnitude relation of dwarf galaxies in NGC4631 group is similar to the relation of the Local Group and the other galaxy systems.

Keywords. galaxies: dwarf – galaxies: stellar content – galaxies: halo – galaxies: individual (NGC 4631, NGC 4656)

1. Introduction

Current large telescopes enable us to investigate resolved stellar populations of halos of more distant galaxies than the Milky Way (MW) and the Andromeda galaxy. Halos of several galaxies of different type, from ellipticals to late spirals, were studied. For example, Tanaka *et al.* (2011) carried out imaging observations of the halo of NGC55, which is less massive tham MW and M31, using Subaru/Suprime-Cam, and identified some overdense stellar structures and low-metallicity stellar populations in its outskirts. The presence of halo substructures and a heated thick disk with the abnormally large scale height suggests that NGC55 has hierarchically formed in the same way as M31 and MW. Such a recent discovery of the diversity of halo structures whose detailed physical parameters such as metallicity and luminosity of halos depend on morphology of host-galaxy suggests the possibility that formation scenario of galactic halos is also depending on morphology of host-galaxy.

2. This work

In order to investigate structure and state of an outer region of on-going interacting galaxy we have observed a nearby on-going interacting galaxy group mainly consisting of NGC4631 and NGC4656 using the Hyper Suprime-Cam (HSC) on the Subaru telescope. The exposure times were 48 min in i-band and 36 min in g-band. The seeing was over 1". In this study, we adopt the distance of 7.1 Mpc for this group from Tikhonov *et al.* (2006).

We have basically reduced our HSC image using HSC pipeline ver.3, and we have performed photometric calibration and astrometry based on SDSS DR7 data covering whole regions of our HSC field-of-view. We have newly discovered 8 dwarf galaxy candidates

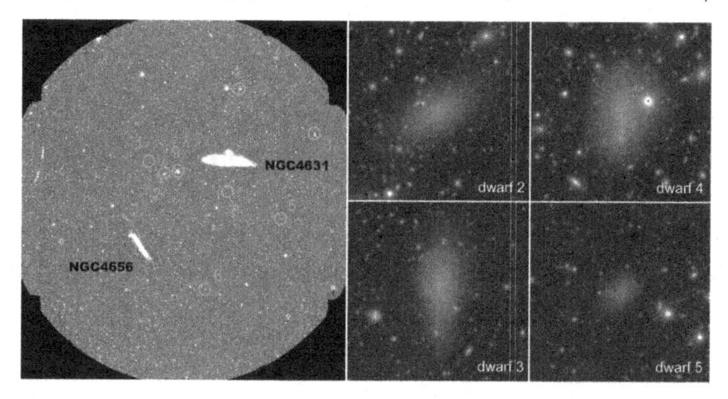

Figure 1. Left: the spatial distribution of dwarf galaxies in NGC4631 group. Right: 4 representative samples of all dwarf galaxy candidates we have photometrically detected.

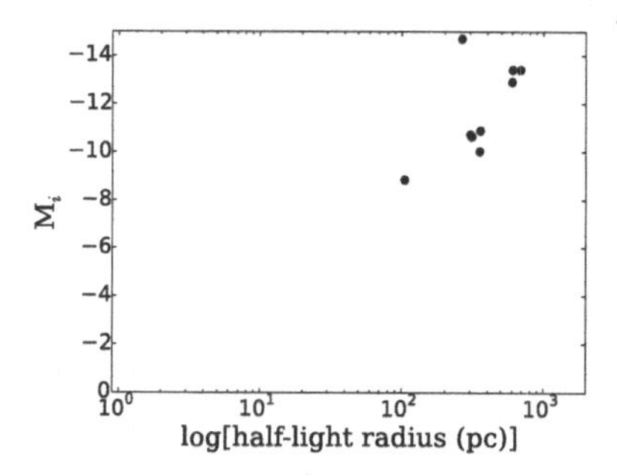

Figure 2. The effective radius vs. absolute magnitude relation of NGC4631's dwarf galaxies. Because of the low surface brightness of Dwarfs 10 and 11 close proximity to a bright star of Dwarf 7, we cannot determine physical parameters of them in this shallow observation.

(green circles in Fig. 1) by visual inspection. In addition, we have confirmed the 3 dwarf candidates which Karachentsev *et al.* (2014) found in their previous study (three red circles in Fig. 1). Apparently, these dwarf galaxies are isotropically distributed around NGC4631 (not NGC4656) in contrast to dwarf galaxies of MW and M31 distributed in a thin plane (e.g., Pawlowski *et al.* 2012 and Ibata *et al.* 2013).

On the other hand, from our HSC i-band image, we determine a position of dwarf candidates with SExtractor, and conduct surface photometry using PyRAF/ELLIPSE package. Furthermore, we measure physical properties such as position angle, ellipticity, Sersic indices and effective radius. Fig. 2 shows the effective radius vs. absolute magnitude relation for dwarf galaxies in NGC4631 group. The relation suggests that these dwarf galaxies with brighter total luminosity probably tend to be more spatially extended. Furthermore, the relation is similar to the one of the Local Group and the other galaxy systems (e.g., Gilmore *et al.* 2007), implying that the dwarf galaxy system of the NGC4631 group may have formed through similar processes as the Local Group.

References

Gilmore *et al.* 2007, *ApJ*, 663, 948
Ibata *et al.* 2013, *Nature*, 493, 62
Pawlowski *et al.* 2012, *MNRAS*, 423, 1109
Tanaka *et al.* 2011, *ApJ*, 738, 150
Tikhonov *et al.* 2006, *astro-ph/0603457*
Karachentsev *et al.* 2014, *astro-ph/1401.2719*

The General Assembly of Galaxy Halos: Structure, Origin and Evolution
Proceedings IAU Symposium No. 317, 2015
A. Bragaglia, M. Arnaboldi, M. Rejkuba & D. Romano, eds.

doi:10.1017/S1743921315007048

CNO abundances in giants of the peculiar globular cluster NGC 1851

G. Tautvaišienė,[1] A. Drazdauskas,[1] C. Lardo,[2] S. L. Martell,[3] E. Pancino,[4,5] E. Stonkutė,[1] and Gaia-ESO Consortium

[1]Institute of Theoretical Physics and Astronomy, Vilnius University, Gostauto 12, 01108, Vilnius, Lithuania, email: grazina.tautvaisiene@tfai.vu.lt

[2]Astrophysics Research Institute, Liverpool John Moores University, 146 Brownlow Hill, Liverpool L3 5RF, United Kingdom

[3]School of Physics, University of New South Wales, NSW 2052, Sydney, Australia

[4]INAF-Osservatorio Astronomico di Bologna, Via Ranzani 1, 40127, Bologna, Italy

[5]ASI Science Data Center, Via del Politecnico SNC, 00133, Roma, Italy

Abstract. We provide CNO and Fe abundance investigations for a sample of up to 45 NGC 1851 giants. High-resolution spectra were obtained with the VLT UVES spectrograph in the framework of the Gaia-ESO Survey. The stars in our sample can be separated into two groups with a difference of 0.1 dex in the mean metallicity, 0.3 dex in the mean C/N, and no significant difference in the mean values of C+N+O.

Keywords. Stars: abundances, stars: evolution, globular clusters: individual (NGC 1851)

---

1. Introduction

It has been long suspected that NGC 1851 is not chemically homogeneous (e.g., Walker 1992). By now, it is clear that NGC 1851 has two distinct subgiant branches, however explanations for their origin so far lack consensus. Some authors explain them by two generations of stars, the first being primordial, while the second one being born from the ejecta of a fraction of the stars of the first population (e.g., see the review by Gratton *et al.* 2012). There are suggestions that NGC 1851 originated by merging of two globular clusters (e.g. Campbell *et al.* 2012), or is a naked nucleus of a captured and disrupted dwarf galaxy (e.g. Bekki & Yong 2012; Marino *et al.* 2014). van den Bergh (1996) and later Carretta *et al.* (2010) joined two hyphoteses into one and suggested that NGC 1851 may have been formed by the merger between parental globulars that were once located within a dwarf spheroidal galaxy. The CNO content of NGC 1851 stars is a crucial observational constrain to test the proposed globular clusters formation scenarios. In the following, we present CNO abundances and metallicity determinations for a sample of up to 45 RGB stars in NGC 1851 from high-resolution spectral observations and archival data in the framework of the Gaia-ESO Survey.

2. Observations and method of analysis

Observations were conducted with the FLAMES multi-fiber facility on the ESO VLT telescope. Spectra of high-resolving power ($R \approx 47\,000$) were obtained in a wavelength interval of 4700–6840 Å with a gap of about 50 Å in the centre. Signal-to-noise ratios in the observed spectra range from 40 to 180 depending on the stellar brightness.

A differential model atmosphere technique which was applied for the determination of the atmospheric parameters is described by Smiljanic *et al.* (2014). Approximate uncertainties of the main atmospheric parameters are 55 K, 0.13 dex, and 0.07 dex for

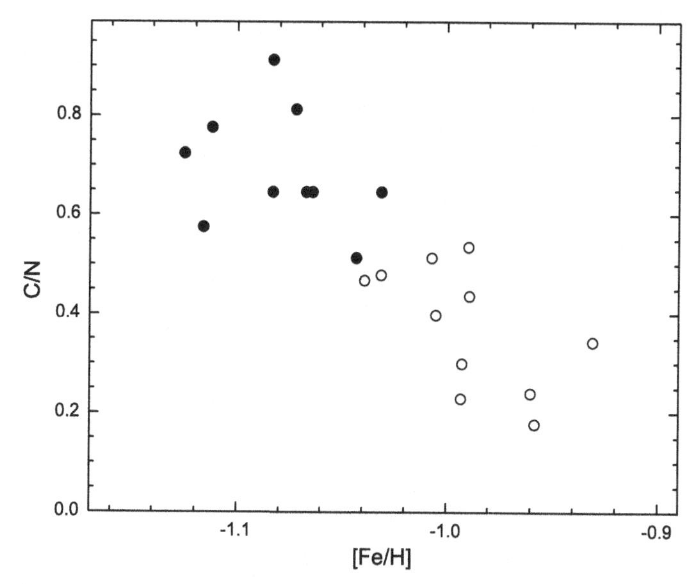

Figure 1. C/N ratios in respect to [Fe/H] for the NGC 1851 metal-rich (open circles) and metal-poor (filled circles) stars.

$T_{\rm eff}$, log g, and [Fe/H], respectively. The C_2 Swan (1,0) band head at 5135 Å and C_2 Swan (0,1) band head at 5635.5 Å were investigated in order to determine the carbon abundance. The interval 6470–6490 Å containing CN bands was used for the nitrogen abundance analysis. The oxygen abundance was determined from the forbidden [O I] line at 6300.31 Å. More details about the analysis can be found in Tautvaišienė *et al.* (2015).

3. Results

We were able to separate NGC 1851 stars between metal-rich and metal-poor ones according to their metallicity and CNO content. Metal-rich and metal-poor stars have averaged metallicities -0.99 ± 0.03 and -1.08 ± 0.03, respectively. Fig. 1 shows how the sampled stars are located in the C/N vs. [Fe/H] plane.

The metal-rich and metal-poor subsamples have distinct abundances of nitrogen and consequently different values of C/N ratios. The mean value of the C/N ratio in metal rich stars is 0.40 ± 0.14, and in metal-poor stars is 0.68 ± 0.12. We will present the full analysis of this interesting cluster in a forthcoming paper.

Acknowledgements. Based on data products from observations made with ESO Telescopes at the La Silla Paranal Observatory under programme ID 188.B-3002. This work was supported by the grant from the Research Council of Lithuania (MIP-082/2015).

References

Bekki, K. & Yong, D. 2012, *MNRAS*, 419, 2063
Campbell, S. W., Yong, D., Wylie-de Boer, E. C. *et al.* 2012, *ApJ* (Letters), 761, L2
Carretta, E., Gratton, R. G., Lucatello, S. *et al.* 2010, *ApJ* (Letters), 722, L1
Gratton, R., Carretta, E., & Bragaglia, A. 2012, *A&ARv*, 20, 50
Marino, A. F., Milone, A. P., & Yong, D. *et al.* 2014, *MNRAS*, 442, 3044
Smiljanic, R., Korn, A. J., Bergemann, M. *et al.* 2014, *A&A*, 570,122
Tautvaišienė, G., Drazdauskas, A., Mikolaitis, Š. *et al.* 2015, *A&A*, 573, 55
Walker, A. R. 1992, *PASP*, 104, 1063
van den Bergh, S. 1996, *ApJ*, 471, L31

The General Assembly of Galaxy Halos: Structure, Origin and Evolution
Proceedings IAU Symposium No. 317, 2015
A. Bragaglia, M. Arnaboldi, M. Rejkuba & D. Romano, eds.

doi:10.1017/S1743921315008601

The Stellar Age-$T_{\rm eff}$-Kinematical Asymmetry in the Solar Neighborhood from LAMOST

H. J. Tian[1,2], C. Liu[1], J. L. Carlin[3], Y. H. Zhao[1] and X. L. Chen[1]

[1]National Astronomical Observatories, Chinese Academy of Sciences, Beijing 100012, China
[2]China Three Gorges University, Yichang, 443002, China; hjtian@lamost.org
[3]Earlham College, 801 National Road West, Richmond, IN 47374, USA

Abstract. With the velocity de-projection technique, we derived the averaged 3 dimensional local velocity distribution using only the line-of-sight velocity for the 200,000 FGK type main-sequence stars from the LAMOST DR1 data. Taking the effective temperature as a proxy for age, we investigate the variation of the velocity distribution as a function of $T_{\rm eff}$ and disk height within $100 < |z| < 500$ pc. Using the mean velocities of the cool stars, we derive the solar motion of $(U_\odot, V_\odot, W_\odot)$=$(9.58 \pm 2.39, 10.52 \pm 1.96, 7.01 \pm 1.67)$ km s^{-1} with respect to the local standard of rest (LSR). Moreover, we find that the stars with $T_{\rm eff} > 6000$ K show a net asymmetric motion of $\langle U \rangle \sim 2$ km s^{-1} and $\langle W \rangle \sim 3$ km s^{-1} compared to the stars with $T_{\rm eff} < 6000$ K. And their azimuthal velocity increases when $|z|$ increases. The asymmetric motion in the warmer stars is likely because they are too young and not completely relaxed.

Keywords. Galaxy: disk – Galaxy: kinematics and dynamics – solar neighborhood

1. Introduction

The velocity distribution of the stars in the solar neighborhood plays a key role in understanding the global structure, dynamical features, and the evolution of the Milky Way. Although it is often approximated with a multi-dimensional Gaussian profile, the velocity distribution of the stars in the solar neighborhood is actually very complicated. Observations have found many substructures (Zhao *et al.* 2009; Xia *et al.* 2014), which may be associated with the perturbation of the Galactic bar and spiral arms, or belong to old tidal debris of disrupted clusters or dwarf galaxies (Dehnen 2000; Antoja *et al.* 2011). These substructures may shift the mean velocity slightly away from zero by a few km s^{-1}. Many observational evidences have proved the kinematical asymmetry (Carlin *et al.* 2013; Widrow *et al.* 2012; etc.).

The velocity distribution can be characterized by the velocity ellipsoid, which reflects the mass distribution and evolution of the Milky Way, assuming that most of the detected stars are in equilibrium. Many works have found that the age of stars is correlated with the velocity distribution. Specifically, older stars show larger velocity dispersion, and vice versa (Parenago 1950). This is usually thought to be because scattering of the disk stars increases over time. The age–velocity dispersion relation (AVR) reflects the evolution history of the Galactic disk.

2. Data and Results

LAMOST FGK-type Stars. The LAMOST Survey has delivered the first data release (DR1), which contains 1,085,404 stellar spectra with estimated stellar atmospheric parameters as well as line-of-sight velocities. The distances to stars are determined with uncertainty of ~20% from isochrone fitting by Carlin *et al.* (2015). With the criteria

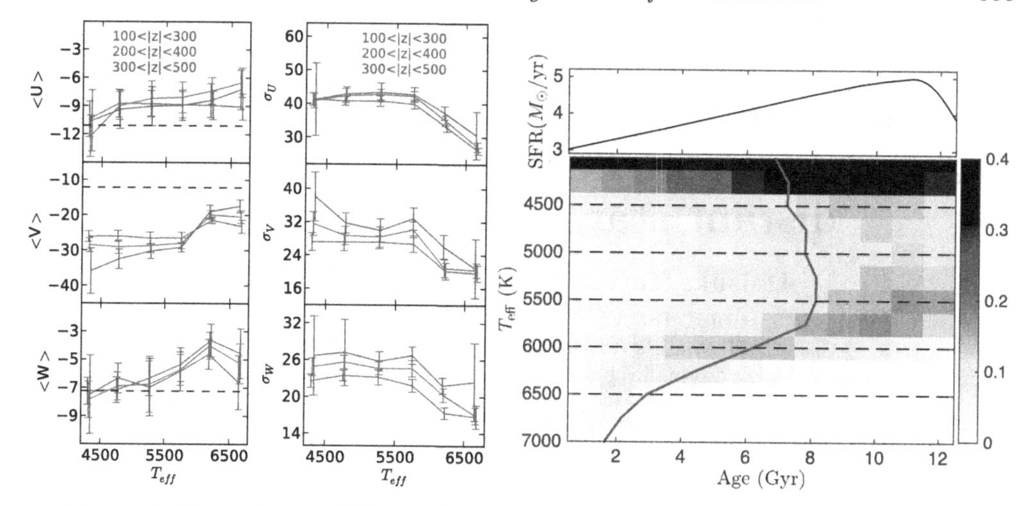

Figure 1. The left and middle panels present the variation of the mean velocities and ellipsoids as a function of $T_{\rm eff}$ in three disk heights, the right panel shows the age-$T_{\rm eff}$ relation.

provided by Tian *et al.* (2015, hereafter T15), we selected a total of 209,316 FGK type main-sequence stars to investigate the kinematics of the solar neighborhood.

The Mean Velocities and Ellipsoids. The LAMOST FGK sample spans a large area of the sky, which allows us to use the de-projection technique described in T15 to measure the 3 dimensional velocity and its ellipsoid from only the one dimensional line-of-sight velocities. The left and middle panels of Fig. 1 demonstrate the correlations of the three velocity and the three velocity ellipsoid components with $T_{\rm eff}$ in three slices of $|z|$. The most obvious feature is that the three mean velocities and ellipsoids are all correlated with $T_{\rm eff}$ for all $|z|$ bins. The warmer stars with $T_{\rm eff} > 6000$ K show a net asymmetric motion. The velocity dispersions show clear trends either along $T_{\rm eff}$ or $|z|$. All three velocity dispersions show an abrupt drop at around $T_{\rm eff} \sim 6000$ K in all $|z|$ bins. Using the cool and old stars (probably in equilibrium), we derived the solar motion of ($U_\odot$, $V_\odot$, $W_\odot$)=(9.58 ± 2.39, 10.52 ± 1.96, 7.01 ± 1.67) km s^{-1} with respect to the LSR.

The Age-$T_{\rm eff}$ Relation. Given a star formation history and a star formation rate (the right-top sub-plot in Fig. 1), we can set up the stellar distribution in age vs. $T_{\rm eff}$ plane using synthetic isochrones and an initial mass function, to reveal the reason why the kinematic features for the stars with $T_{\rm eff} > 6000$ K are significantly different from those for the cooler stars. The blue thick line shows the mean age at different $T_{\rm eff}$. It shows that for the stars with $6000 < T_{\rm eff} < 7000$ K, the mean age is only 4 Gyr. The $T_{\rm eff}$ of the abrupt change in age is perfectly consistent with that of the sudden change in the mean velocity and velocity dispersions. The stars with $T_{\rm eff} > 6000$ K are too young, it is the probable reason why those stars show significant bulk motion in all three orientations.

References

Antoja, T., Figueras, F., Romero-Gómez., M. *et al.* 2011, *MNRAS*, 418, 1423
Carlin, J. L., DeLaunay, J., Newberg, H. J. *et al.* 2013, *ApJL*, 777, 5
Carlin, J. L., Liu, C., Newberg, H. J. *et al.*, 2015, *AJ*, 150, 4
Dehnen, W. 2000, *AJ*, 119, 800
Tian, H. J., Liu, C. *et al.* 2015, *ApJ*, 809, 145
Widrow, L. M., Gardner, S., Yanny, B. *et al.* 2012, *ApJL*, 750, 41
Xia, Q. R, Liu, C., Mao, S. *et al.* 2014, *MNRAS*, 447, 2367
Zhao, J. K., Zhao, G., & Chen, Y. Q., 2009, *ApJL*, 692, 113

The General Assembly of Galaxy Halos: Structure, Origin and Evolution
Proceedings IAU Symposium No. 317, 2015
A. Bragaglia, M. Arnaboldi, M. Rejkuba & D. Romano, eds.

doi:10.1017/S1743921315008431

Baryonic inflow and outflow histories in disk galaxies as revealed from observations of distant star-forming galaxies

Daisuke Toyouchi and Masashi Chiba

Astronomical Institute, Tohoku University, Aoba-ku, Sendai 980-8578, Japan
email: toyouchi@astr.tohoku.ac.jp
email: chiba@astr.tohoku.ac.jp

Abstract. Gas inflow and outflow are the most important processes, which determine the structural and chemical evolution of a disk galaxy like the Milky Way. In order to get new insights into these baryonic processes in Milky Way like galaxies (MWLGs), we consider the data of distant star-forming galaxies and investigate the evolution of the radial density profile of their stellar components and the associated total amount of gaseous inflow and outflow. For this purpose, we analyze the redshift evolution of their stellar mass distribution, combined with the scaling relations between the mass of baryonic components, star formation rate and chemical abundance for both high- and low-z star-forming galaxies. As a result, we find the new relations between star formation rate and inflow/outflow rate as deduced from these distant galaxies, which will provide fundamental information for understanding the structural and chemical evolution of MWLGs.

Keywords. Disk galaxies, Gas inflow and outflow, Structural and chemical evolution

1. Our method

We investigate the time evolution of the surface density profiles of gas inflow and outflow rates in MWLGs, denoted as $\Sigma_{\rm in}$ and $\Sigma_{\rm out}$, respectively. For this purpose, we solve the mass conservation equations for gas and heavy elements at any time, t, and radius, R, along the disk. We then calculate $\Sigma_{\rm in}(t, R)$ and $\Sigma_{\rm out}(t, R)$ given the time evolutions of the radial profiles of surface mass densities of gas, stellar and heavy elements. For these later information in MWLGs, we adopt the recent observational results for distant star-forming galaxies, including the growth of the stellar mass distribution based on the abundance matching method (van Dokkum *et al.* 2013), their scaling relation between the mass of baryonic components, star formation rate and chemical abundance (Mannucci *et al.* 2010), as well as the supposed evolution of their radial metallicity gradients (Vila-Costas & Edmunds 1992; Stott *et al.* 2014). Thus, we derive the gas inflow and outflow histories in MWLGs as revealed from the observations of distant star-forming galaxies (see Toyouchi & Chiba 2015 for details).

2. New insights into galactic gas inflow and outflow processes

Based on this method, we obtain the radial profiles of $\Sigma_{\rm in}$ and $\Sigma_{\rm out}$, both of which are higher at smaller R at all epochs, and find that their central values monotonically decline with decreasing redshifts. We also investigate the dependences of $\Sigma_{\rm in}(t, R)$ and $\Sigma_{\rm out}(t, R)$ on a star formation rate density, $\Sigma_{\rm SFR}(t, R)$. This procedure allows us to inspect whether the assumption adopted in many previous studies, namely that inflow and outflow rates in galaxies are proportional to their SFR, is indeed valid or not. Figure

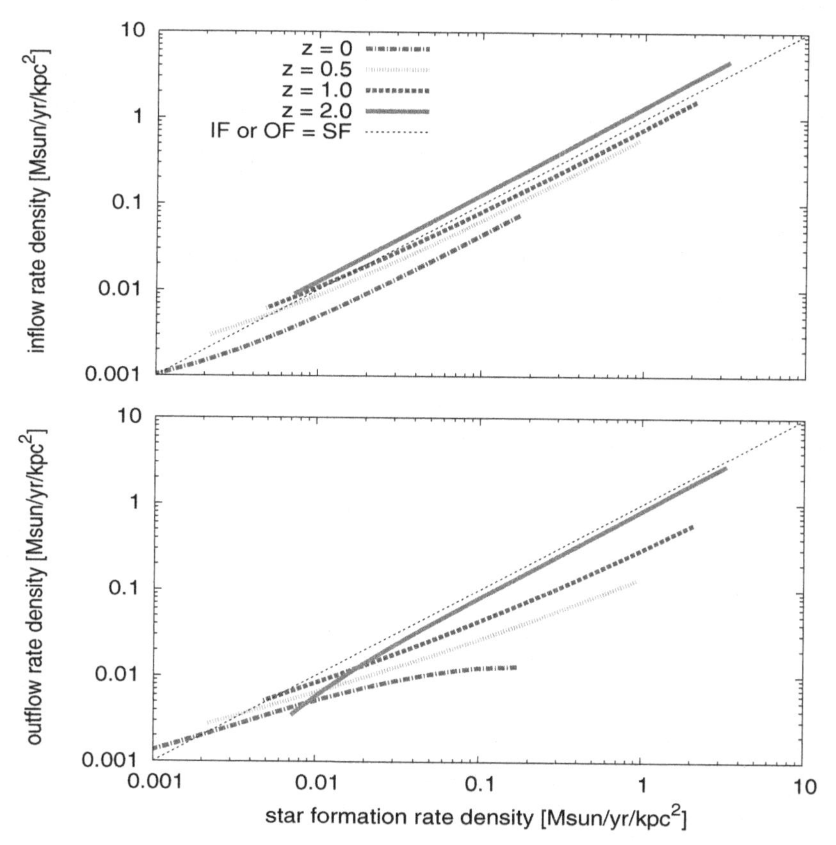

Figure 1. The inflow and outflow rate densities (top and bottom panels, respectively) as a function of the star formation rate density at redshifts of $z = 0$, 0.5, 1 and 2.

1 shows $\Sigma_{\rm in}$ and $\Sigma_{\rm out}$ as a function of $\Sigma_{\rm SFR}$ at redshifts of 0, 0.5, 1.0 and 2.0. Thin dotted lines show $\Sigma_{\rm in} = \Sigma_{\rm SFR}$ (top panel) and $\Sigma_{\rm out} = \Sigma_{\rm SFR}$ (bottom panel) for comparison. We find that although $\Sigma_{\rm in}$ is always approximately proportional to $\Sigma_{\rm SFR}$, such a proportional relation is not necessarily satisfied for $\Sigma_{\rm out}$. This result for outflow may imply that the assumption of proportionality of outflow rate to star formation rate is simplistic. We also find that these properties of outflow rate can be simply understood in the framework of momentum-driven or energy-driven wind mechanism. To conclude, this work allows us to distinguish the main driving mechanism for galactic outflows, which can significantly influence the structural and chemical evolution of star-forming galaxies.

References

Mannucci, F., Cresci, G., Maiolino, R., Marconi, A., & Gnerucci, A. 2010, *MNRAS*, 408, 2115
Stott, J. P., Sobral, D., Swinbank, A. M., *et al.* 2014, *MNRAS*, 443, 2695
van Dokkum, P. G., Leja, J., Nelson, E. J., *et al.* 2013, *ApJL*, 771, L35
Vila-Costas, M. B. & Edmunds, M. G. 1992, *MNRAS*, 259, 121

The General Assembly of Galaxy Halos: Structure, Origin and Evolution
Proceedings IAU Symposium No. 317, 2015
A. Bragaglia, M. Arnaboldi, M. Rejkuba & D. Romano, eds.

doi:10.1017/S1743921315008650

Stellar orbital properties as diagnostics of the origin of the stellar halo

Monica Valluri[1], Sarah R. Loebman[1,2], Jeremy Bailin[3], Adam Clarke[4], Victor P. Debattista[4], Greg Stinson[5]

[1]University of Michigan, USA email: mvalluri@umich.edu, [2]Michigan Society of Fellows, [3]University of Alabama, USA, [4]University of Central Lancashire, UK, [5]Max Planck Institute for Astronomie, Germany

Abstract. We examine metallicities, ages and orbital properties of halo stars in a Milky-Way like disk galaxy formed in the cosmological hydrodynamical MaGICC simulations. Halo stars were either accreted from satellites or they formed *in situ* in the disk or bulge of the galaxy and were then kicked up into the halo ("in situ/ kicked-up" stars). Regardless of where they formed *both types show surprisingly similar orbital properties*: the majority of both types are on short-axis tubes with the same sense of rotation as the disk – implying that a large fraction of satellites are accreted onto the halo with the same sense of angular momentum as the disk.

Keywords. Galaxy: halo, Galaxy: kinematics and dynamics, Galaxy: stellar content

---

1. Introduction

The orbital properties of halo stars from the MaGICC simulated galaxy g15784 – a realistic Milky Way sized disk galaxy at $z = 0$ (Stinson *et al.* 2012) – are used to assess whether it is possible to determine the birth site of halo stars based on individual orbital properties. The dark matter halo of this galaxy is mildly triaxial at all radii. The stars were picked to lie within 100 kpc of the center of the galaxy but not within the disk (i.e at $|z| > 3$ kpc and $R > 25$kpc). The formation sites of halo stars were determined by examining 100 snapshots of the simulation. "Accreted halo stars" form in satellites outside the virial radius and are subsequently added to the halo via tidal stripping. Stars that form "in-situ" in the main disk or bulge and are subsequently kicked into the halo are referred to as "in situ/ kicked-up" stars. The phase space coordinates of individual stars at $z = 0$ were used to numerically integrate individual orbits in the frozen potential corresponding to the full galaxy potential. Frequency analysis of the orbits was used to classify orbits (Valluri *et al.* 2010, 2012) into the major orbit families found in triaxial halos: box orbits, short axis tubes (SAT), long axis tubes (LAT) and chaotic orbits.

2. Orbits of accreted and "kicked-up" halo stars

Orbits of 14,000 accreted stars 14,000 "kicked-up" stars were classified. Table 1 shows the percentage of accreted and "kicked-up" stars belonging to each orbit family. The majority of orbits are on SATs with the same sense of rotation as the disk, regardless of where they formed. This is probably because satellies are often accreted together along the same large-scale filaments that contribute to the hierarchical growth of the galaxy (Helmi *et al.* 2011). More surprising is the fact that the fractions of box orbits and chaotic orbits are independent of formation site: however these orbit families are more centrally concentrated in the halo and are likely to have experienced more chaotic scattering (Valluri *et al.* 2013). We find that the distribution of orbital chaoticity in this high resolution simulation is identical to that in controlled simulations (Valluri *et al.* 2010).

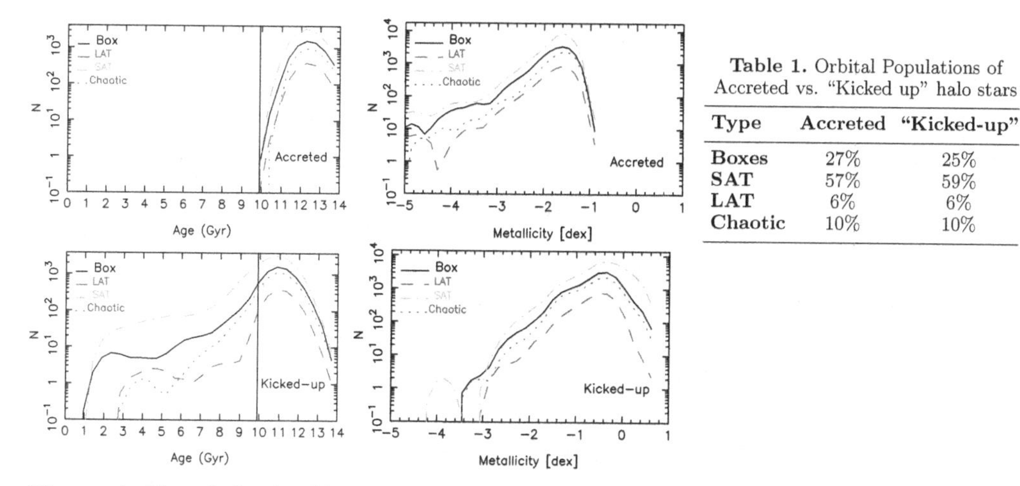

Table 1. Orbital Populations of Accreted vs. "Kicked up" halo stars

Type	Accreted	"Kicked-up"
Boxes	27%	25%
SAT	57%	59%
LAT	6%	6%
Chaotic	10%	10%

Figure 1. Kernel density histograms showing number of orbits of each family as a function of stellar age (left) and metallicity (right) for accreted stars (top row) and "in situ/ kicked up" stars (bottom row). Orbits of all families are similarly distributed in each panel. The thin vertical lines (at 9.9 Gyr) marks the youngest accreted stars.

Figure 1 shows that "accreted" and "in situ" stars have slightly different distributions with age and metallitiicy, but all orbit families in a given panel show similar distrbutions in these quantities. Accreted stars are older on average than "in situ" stars but the peak of the "in situ" stars is also old. Likewise there is significant overlap in their metallicity distributions. Snaith *et al.* (2015) show that while the accreted halo stars have higher α-abundances on average than "in situ" halo stars, there is overlap in this space too.

3. Conclusions

Orbital properties of halo stars are independent of whether they were accreted or whether they were formed in the disk/bulge and then "kicked-up" into the halo. Both types of stars are mainly on short-axis tubes ($\sim 60\%$), probably because the accretion of satellites occurs on a few large scale filaments that also feed the disk. Chaotic scattering by the central bulge put stars in the inner halo on box or chaotic orbits (Valluri *et al.* 2010). In this MaGICC galaxy the overlap in the ages, metallicities and orbital properties of halo stars makes it difficult to uniquely identify exactly where they formed. This analysis is being repeated for other disks to assess the dependence of this result on accretion history.

Acknowledgements

MV is supported by NASA-ATP grant NNX15AK79G. SRL is supported by the Michigan Society of Fellows, VPD is supported by STFC Consolidated grant # ST/M000877/1. JB acknowledges support from HST-AR-12837, provided by NASA through a grant from the Space Telescope Science Institute.

References

Helmi, A. *et al.* 2011, *ApJ* 733, L7
Snaith, O. N. *et al. MNRAS*, submitted
Stinson, G. S., *et al.* 2012, *MNRAS*, 3506
Valluri, M., Debattista, V. P. *et al.* 2010, *MNRAS*, 403, 525
Valluri, M., Debattista, V. P. *et al.* 2012, *MNRAS*, 419, 1951
Valluri, M., Debattista, V. P., Stinson, G. S., Bailin, J. *et al.* 2013, *Ap J.*, 767, 93

The General Assembly of Galaxy Halos: Structure, Origin and Evolution
Proceedings IAU Symposium No. 317, 2015
A. Bragaglia, M. Arnaboldi, M. Rejkuba & D. Romano, eds.

doi:10.1017/S1743921315007061

Building Blocks of the Milky Way's Stellar Halo

Pim van Oirschot[1], Else Starkenburg[2], Amina Helmi[3] and Gijs Nelemans[1,4]

[1]Department of Astrophysics/IMAPP, Radboud University Nijmegen, P.O. Box 9010, 6500 GL Nijmegen, The Netherlands, email: P.vanOirschot@astro.ru.nl

[2]Leibniz-Institut fur Astrophysik Potsdam, An der Sternwarte 16, D-14482 Potsdam, Germany

[3]Kapteyn Astronomical Institute, University of Groningen, P.O. Box 800, 9700 AV, Groningen, The Netherlands

[4]Institute for Astronomy, KU Leuven, Celestijnenlaan 200D, 3001 Leuven, Belgium

Abstract. We study the assembly history of the stellar halo of Milky Way-like galaxies using the six high-resolution Aquarius dark matter simulations combined with the Munich-Groningen semi-analytic galaxy formation model. Our goal is to understand the stellar population contents of the building blocks of the Milky Way halo, including their star formation histories and chemical evolution, as well as their internal dynamical properties. We are also interested in how they relate or are different from the surviving satellite population. Finally, we will use our models to compare to observations of halo stars in an attempt to reconstruct the assembly history of the Milky Way's stellar halo itself.

Keywords. Galaxy: halo, Galaxy: evolution

We post-process the Aquarius simulations (Springel *et al.* 2008) with the Munich-Groningen semi-analytic galaxy formation model (Starkenburg *et al.* 2013, and references therein) to model six Milky Way-like galaxies (A-F). The stellar masses of the spheroids (stellar halos including the inner few kpc bulge region) of these galaxies are ranging from $5 \cdot 10^9 - 2 \cdot 10^{10} M_\odot$, which are typically more massive than the analogous stellar halos in the GALFORM semi-analytic model (Cooper *et al.* 2010).

In total we find approximately 100 building block galaxies per spheroid, many of which however contribute only a small amount of halo stars. In agreement with Cooper *et al.* (2010), we find that the stellar halo is built up mainly by a few main progenitor galaxies. In Figure 1 we show the Age-Metallicity maps of eight building blocks of halo C compared with the Age-Metallicity maps of eight surviving satellites of that halo of similar stellar mass. The main progenitor building blocks are visualized in the upper four panels. Comparing these panels with the four in the second row, we see that the [Mg/H] values for these two classes of objects are very comparable (approximately solar) but the building blocks stop forming stars earlier on average. In the third and fourth row, we plot Age-Metallicity maps of $\sim 10^5$ $M_\odot$ building blocks and surviving satellites respectively. These smaller mass systems are clearly different from the more massive systems in that they stop their star formation at earlier times. Furthermore, these bottom two rows look more alike.

We plan to do a statistical analysis on the differences between the building blocks and the surviving satellites for all the halos, looking at the Mass-Metallicity relations and average star formation rates amongst others. We will compare our findings to results in

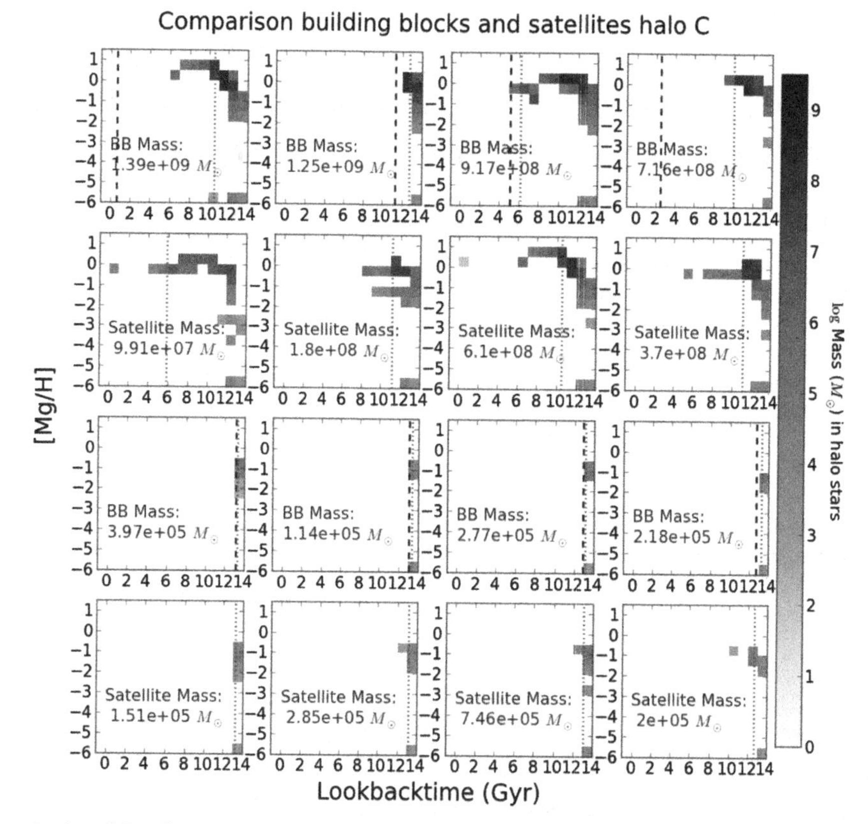

Figure 1. Age-Metallicity maps of building blocks of halo C (first and third row) compared with those of surviving satellites (second and fourth row) of that halo of similar stellar mass. The blue dotted line in each panel represents the time of infall (when the galaxy became a satellite), the black dashed lines in the building block panels indicate the time of merging with the main halo (see Starkenburg *et al.* 2013, how exactly this is defined).

the literature (eg. Font *et al.* 2006). Furthermore, we plan to couple a binary population synthesis code to this semi-analytic galaxy formation model in order to study binary evolution in a cosmological context. Doing this, one obtains inputs for the population synthesis modelling from the galaxy formation model such as a realistic star formation rate on the one hand, and inputs for the semi-analytic galaxy formation model from the population synthesis code such as yields from Supernovae (see also Yates *et al.* 2013) on the other hand.

References

Springel *et al.* 2008, *MNRAS*, 391, 1685
Starkenburg *et al.* 2013, *MNRAS*, 429, 725
Cooper *et al.* 2013, *MNRAS*, 406, 744
Font *et al.* 2006, *ApJ*, 638, 585
Yates *et al.* 2013, *MNRAS*, 435, 3500

The General Assembly of Galaxy Halos: Structure, Origin and Evolution
Proceedings IAU Symposium No. 317, 2015
A. Bragaglia, M. Arnaboldi, M. Rejkuba & D. Romano, eds.

doi:10.1017/S174392131500856X

How the first stars shaped the faintest gas-dominated dwarf galaxies

Robbert Verbeke, Bert Vandenbroucke and Sven De Rijcke

Sterrenkundig Observatorium, Ghent University
Krijgslaan 281, S9, 9000 Gent, Belgium
email: robbert.verbeke@ugent.be

Abstract. Cosmological simulations predict that dark matter halos with circular velocities lower than 30 km/s should have lost most of their neutral gas by heating of the ultra-violet background. This is in stark contrast with gas-rich galaxies such as e.g. Leo T, Leo P and Pisces A, which all have circular velocities of ~15 km/s (Ryan-Weber *et al.* 2008, Bernstein-Cooper *et al.* 2014, Tollerud *et al.* 2015). We show that when we include feedback from the first stars into our models, simulated dwarfs have very different properties at redshift 0 than when this form of feedback is not included. Including this Population-III feedback leads to galaxies that lie on the baryonic Tully-Fisher relation over the entire mass range of star forming dwarf galaxies, as well as reproducing a broad range of other observational properties.

Keywords. dwarf galaxies, first stars, galaxy formation, galaxy evolugtion.

---

1. Introduction

It is thought that the first generation of stars (Population-III or Pop-III stars), that formed out of pristine gas in the early universe, would have a very top-heavy initial mass function (Susa *et al.* 2014). For a given stellar population, such a Susa-IMF would result in ~4 times more feedback from supernovae type II and 40 times more in the form of ionizing UV radiation. Here, we use computer simulations to investigate the influence this Pop-III feedback has on the evolution of dwarf galaxies and whether it can alleviate tension between cosmological simulations and observed faint dwarf galaxies.

2. Baryonic Tully-Fisher relation

The baryonic Tully-Fisher relation links the total baryonic mass of a galaxy to its circular velocity (e.g. McGaugh 2012), or equivalently, its dynamical mass. Galaxy simulations generally have a hard time reproducing this relation down to the faintest galaxies. Their stellar mass is usually too high for their circular velocity, because these simulations form too many stars at high redshift. When we include feedback from the first generation of stars, the star formation rate at high redshift is significantly reduced. In Figure 1, we show that these simulations (stars) agree with the observed baryonic Tully-Fisher relation over the entire range of star forming dwarf galaxies. The simulations without Population-III feedback (triangles) on the other hands have stellar masses that are too high for their circular velocities.

3. Other properties

The process of galaxy formation is very challenging to model and the observable properties of simulated galaxies will strongly depend on the chosen set of parameters. While

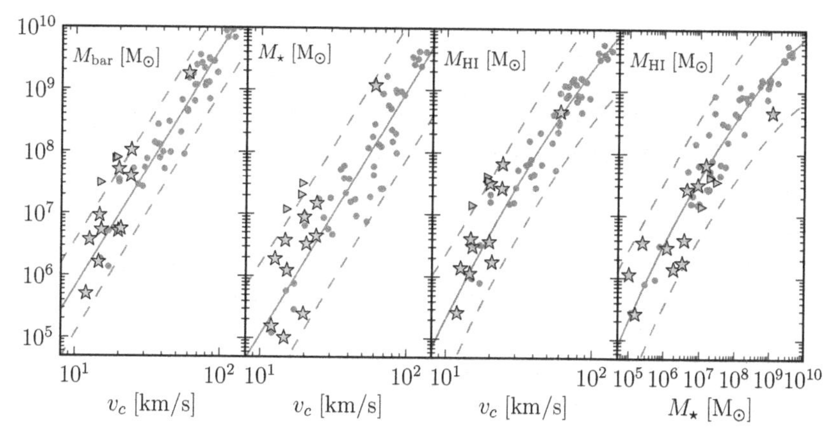

Figure 1. Baryonic Tully-Fischer relations. Baryonic mass (a), stellar mass (b), neutral gas mass (c) versus circular velocity and neutral gas versus stellar mass (d), compared to observations (grey circles; McGaugh 2012). Stars indicate simulations including Pop-III feedback, triangles are simulations without Pop-III feedback.

parameters may be tuned to get one or several galaxy properties in agreement with observations, reproducing a broad range of them is not a trivial task. However, failure to reproduce one or more observed galaxy properties may indicate that an important physical process is missing in the models. This missing ingredient might in turn strongly affect the previous results. We thus stress the importance of comparing a broad range of galaxy properties to observations. We have compared the star formation histories, metallicity, metallicity distribution functions, $M_\star - M_{\rm halo}$-relation, the HI-distribution, optical colours, the half-light radius, central surface brightness, stellar velocity dispersion, star formation rate, the Kennicutt-Schmidt relation and the total mass within the half-light radius and 300 pc, and found good agreement with observations. A more in depth discussion of these properties can be found in Verbeke *et al.* (submitted).

4. Conclusions

We have shown that when including feedback from the first stars in computer simulations, we obtain dwarf galaxies that agree very well with real dwarfs. Since this is generally not true for simulations without pop III stars, we conclude that the first stars that formed in the universe had an important impact on the formation of the faintest dwarf galaxies, which can still be seen today.

References

Bernstein-Cooper, E. Z., Cannon, J. M., Elson, E. C., *et al.* 2014, *AJ*, 148, 35
McGaugh, S. S. 2012, *AJ*, 143, 40
Ryan-Weber, E. V., Begum, A., Oosterloo, T., *et al.* 2008, *MNRAS*, 384, 535
Susa, H., Hasegawa, K., & Tominaga, N. 2014, *ApJ*, 792, 32
Tollerud, E. J., Geha, M. C., Grcevich, J., Putman, M. E., & Stern, D. 2015, *ApJL*, 798, L21
Verbeke, R., Vandenbroucke, B., & De Rijcke, S., *Submitted to ApJ*

The General Assembly of Galaxy Halos: Structure, Origin and Evolution
Proceedings IAU Symposium No. 317, 2015
A. Bragaglia, M. Arnaboldi, M. Rejkuba & D. Romano, eds.

doi:10.1017/S1743921315008765

The Dynamical Evolution of Galactic X-ray Coronae in Clusters

Rukmani Vijayaraghavan[1] and Paul Ricker[2]

[1]Dept. of Astronomy, University of Virginia, rukmani@virginia.edu
[2]Dept. of Astronomy, University of Illinois at Urbana-Champaign, pmricker@illinois.edu

Abstract. Galaxies in group and cluster environments are subject to ram pressure stripping by the hot intracluster medium, resulting in gas loss and the eventual suppression of star formation. Recent *Chandra* observations of galaxies in group and cluster environments show that 60 - 80% of these galaxies have compact (1-4 kpc), hot ($\sim$ 1 keV) X-ray coronae centered on their cores. These coronae have survived stripping and evaporation in the cluster, and their long-term survival poses a test of our understanding of the physical processes in the ICM. In this poster, I summarize results from Vijayaraghavan & Ricker (2015), where we simulated the evolution of populations of galaxies and their hot coronal gas in group and cluster environments, and evaluated their detectability with existing and future X-ray catalogs.

Keywords. hydrodynamics, methods: numerical, X-rays:galaxies:clusters, galaxies:clusters:general, galaxies:halos, galaxies:evolution

1. Background

The dominant baryonic component in clusters of galaxies is the hot (10^7 K) intracluster medium (ICM) plasma, composing about 15% of the total mass in these systems. Orbiting cluster galaxies are subject to ram pressure by the ICM, which can remove most of their hot and cold interstellar medium (ISM) gas. Evidence for ram pressure stripping of galactic gas has been observed in the form of stripped tails that trail cluster galaxies. Ram pressure stripping in addition to tidal stripping and thermal conduction can remove all the gas bound to galaxies. The eventual consequence of gas loss from cluster galaxies is that these galaxies have significantly lower gas fractions and star formation rates compared to field galaxies. *Chandra* observations of galaxies in cluster and group environments (e.g. Sun *et al.* 2007, Jeltema *et al.* 2008) show that $60 - 80\%$ of these galaxies have compact ($1 - 4$ kpc), hot (~ 1 keV) X-ray emitting circumgalactic coronae centered on their cores. These coronae have survived ram pressure and tidal stripping, harassment, and evaporation due to thermal conduction in the cluster for many dynamical times. Their survival for timescales comparable to the Hubble time therefore poses a test of our understanding of the physical processes in the ICM. In Vijayaraghavan & Ricker (2015), we attempted to understand the behavior of these coronae under gravitational and hydrodynamic physical processes alone. We simulated the evolution of populations of galaxies and their hot coronal gas in group and cluster environments and evaluated their detectability with existing and future X-ray catalogs.

2. Results & Conclusions

In Vijayaraghavan & Ricker (2015), we present the results of N-body + hydrodynamic simulations of 26 galaxies in an isolated group of mass 3.2×10^{13} $M_{\odot}$ and 152 galaxies in an isolated cluster of mass 1.2×10^{14} $M_{\odot}$. These simulations were performed with the FLASH code with minimum spatial resolution of up to 1.6 kpc and 10^6 $M_{\odot}$

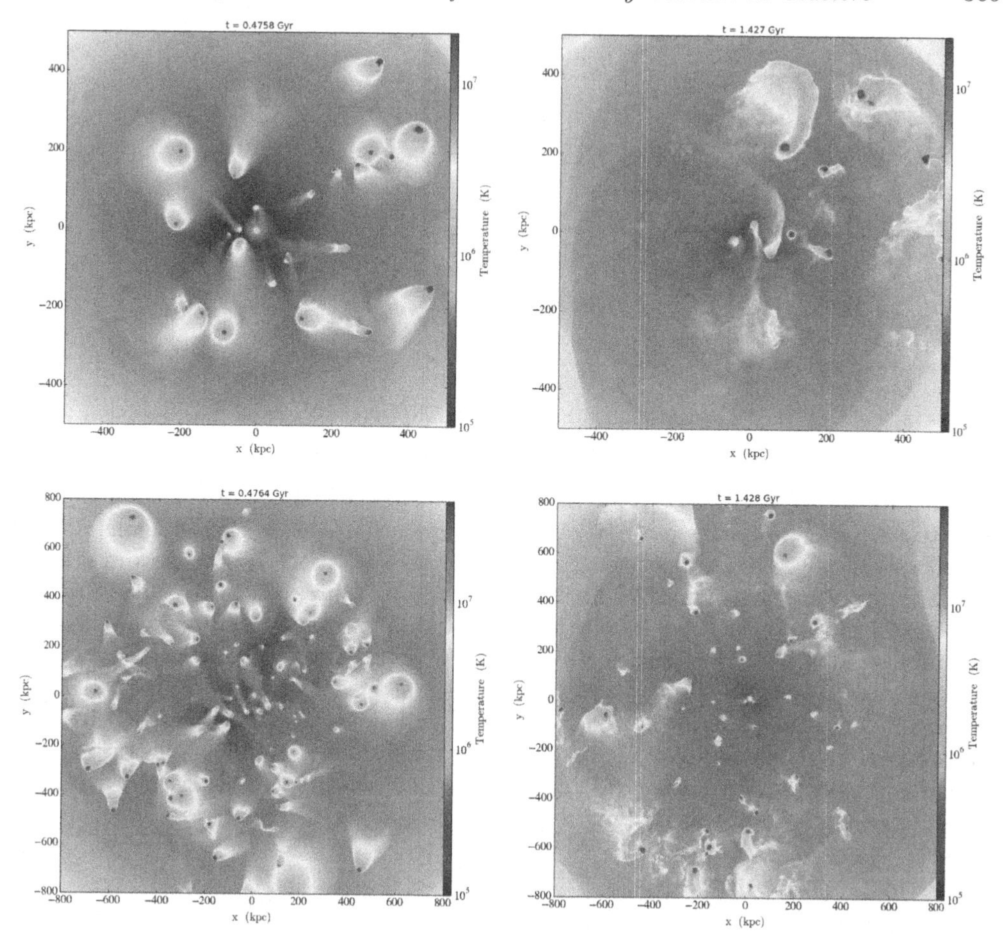

Figure 1. The evolution of gas in the group and cluster galaxies, as seen in maps of emission measure-weighted temperature. Galaxies are stripped of their gas by the ICM, and the stripped gas trails galaxies in their orbits in the form of wakes before mixing with the ICM.

particles. Galaxies in these simulations are NFW spheres with a collisionless dark matter component and hot gas component with 10% of the total mass of the galaxy.

Figure 1 shows ram pressure stripping in action for a range of group and cluster galaxies. Ram pressure is a drag force that removes gas when the local gravitational restoring force is not strong enough to overcome the opposing force of ram pressure. Stripped gas trails galaxies in their orbits, forming shear instabilities at the ISM-ICM interface, before eventually dissipating within the ICM. By ~ 3 Gyr, most galaxies have lost all their gas. The amount of gas removed depends on the mass of the galaxy and the host. Galaxies in the less massive group have smaller velocities and experience weaker ram pressure compared to galaxies in the massive, high velocity dispersion cluster. Group galaxies therefore lose gas at a slower rate than cluster galaxies. In a given environment, more massive galaxies are more resistant to ram pressure stripping due to their higher gravitational restoring forces.

We generated synthetic *Chandra* X-ray observations with 40 ks and 400 ks exposure times of the simulated group and cluster, including their galaxies. Galaxy wakes and tails are visible up to ~ 1 Gyr in the 40 ks image, and their surviving central coronae up to ~ 2 Gyr, albeit at low significance levels above the cluster background. Galac-

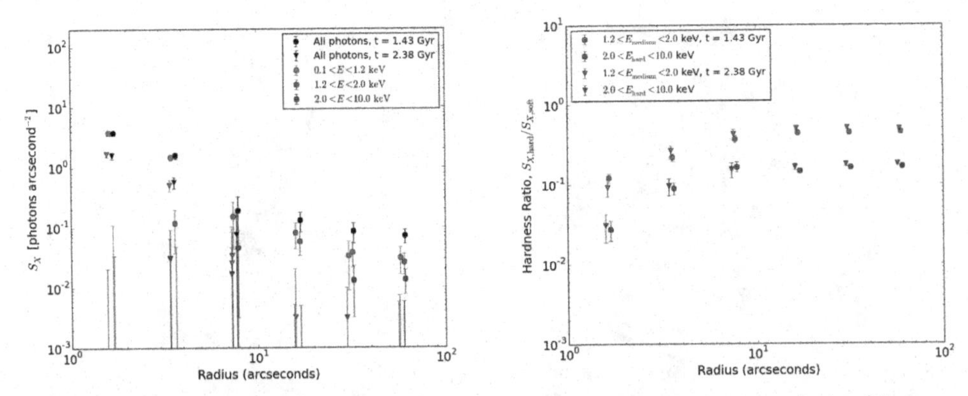

Figure 2. Left: Stacked , background subtracted radial profile for cluster galaxies (that are at least 200 kpc from the cluster center in projection). Right: Hardness radio profile.

tic tails are visible up to 2 Gyr in the 400 ks images. Galactic coronal emission can be detected observationally in short exposure observations by stacking regions around individual cluster galaxies identified in other wavebands. There is an excess in stacked galactic surface brightness profiles at $r \lesssim 10$ arcsec in group and cluster galaxies up to 2.38 Gyr in the low energy $0.1 < E < 1.2$ keV band. This excess persists on subtracting the correspondingly stacked emission centered on points diametrically opposite known galaxy centers. The X-ray emission from cluster galaxies declines faster than that of group galaxies, since galaxies in massive clusters experience stronger ram pressure. Additionally, the emission from galaxies at small galaxy-centric radii manifests itself in measurements of the hardness ratio ($E_{\rm hard}/E_{\rm soft}$), as a noticeable decrease in hardness ratio in the regions with significant galactic emission. These results are illustrated in Figure 2.

References

Jeltema, T. E., Binder, B., & Mulchaey, J. S. 2008, *ApJ*, 679, 1162
Sun, M., Jones, C., Forman, W., *et al.* 2007, *ApJ*, 657, 197
Vijayaraghavan, R. & Ricker, P. M. 2015, *MNRAS*, 449, 2312

The General Assembly of Galaxy Halos: Structure, Origin and Evolution
Proceedings IAU Symposium No. 317, 2015
A. Bragaglia, M. Arnaboldi, M. Rejkuba & D. Romano, eds.

doi:10.1017/S1743921315007152

Globular cluster clustering around ultra compact dwarf galaxies in the halo of NGC 1399

Karina Voggel[1], **Michael Hilker**[1] **and Tom Richtler**[2]

[1]European Southern Observatory, Garching, Germany, email: kvoggel@eso.org
[2] Universidad de Concepción, Concepción, Chile

Abstract. We tested the spatial distribution of UCDs and GCs in the halo of NGC 1399 in the Fornax cluster. In particular we tried to find out if globular clusters are more abundant in the vicinity of UCDs than what is expected from their global distribution. A local overabundance of globular clusters was found around UCDs on a scale of 1 kpc compared to what is expected from the large scale distribution of globulars in the host galaxy. This effect is stronger for the metal-poor blue GCs and weaker for the red GCs. An explanation for these clustered globulars is either that they are the remains of a GC system of an ancestor dwarf galaxy before it was stripped to its nucleus, which appears as UCD today. Alternatively these clustered GCs could have been originally part of a super star cluster complex.

Keywords. galaxies: clusters: individual: Fornax, galaxies: dwarf, galaxies: star clusters

---

1. Introduction and method

The first Ultra-Compact Dwarf Galaxies (UCDs) were detected in the Fornax cluster (Hilker *et al.* 1999). Their discovery filled the empty gap in size and magnitude between globular clusters (GCs) and galaxies that was not populated before. The formation and origin of these objects is still unclear. Three main ways of forming an UCD are discussed: 1.) They are the stripped remnant nuclei of a larger dwarf galaxy (Bekki al. 2003), 2.) they are from the high mass extension of the GC mass function (Murray *et al.* 2009) or 3.) they are the remnants of a super star cluster complexes that merged together and formed the UCD (Fellhauer & Kroupa 2002).

We are using the catalogue of globular clusters around NGC 1399 provided by Dirsch *et al.* (2003) to study their spatial distribution around UCDs. We derived the radial surface density distribution of GCs in the halo. The resulting power-law distribution is then assumed to be the expected value of GCs at each radial distance to NGC 1399. Then the local GC density around UCDs is computed and compared to the expected GC density from a global GC distribution.

2. Results and conclusion

The derived density ratio between expected and measured density, is shown in the left panel of figure 1. This clustering signal is the average for all 206 UCDs (green triangles). The same analysis was done for red and blue GCs only, which is shown in its respective colors. In the right panel the same figure is shown for the subset of 100 UCDs that have a GC candidate within 1 kpc to emphasize the difference in clustering between this population and the comparison GC population plotted in black. We find an increasing GC density around UCDs at scales smaller than 1 kpc for the full sample, which is distinguished from the GCs in their clustering properties. In the right panel of fig.1 this

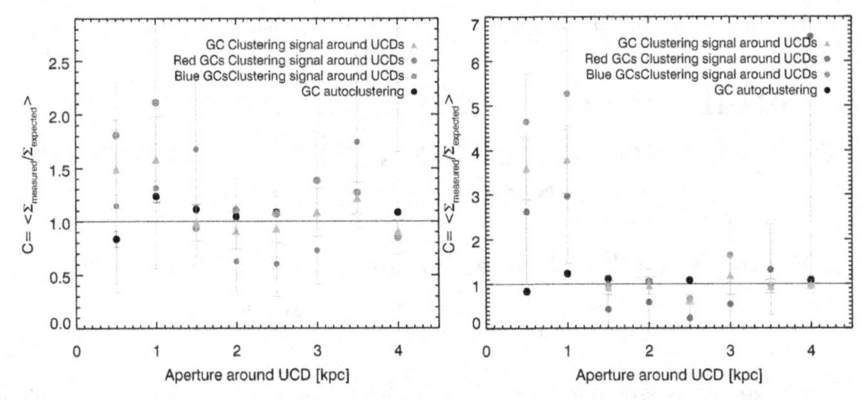

Figure 1. Left panel: clustering signal of GCs around UCDs at radial distances from the UCDs for the full GC population (green), and the red and blue populations. Black data points are the autoclustering of GCs. At each radial distance the clustering signal is the average for the 206 UCDs. Right panel: Same plot for 100 UCDs, which have GC candidate within $r < 1\,\mathrm{kpc}$.

difference is more pronounced, when selecting only a subset of UCDs. When testing the clustering of blue and red GCs separately we find that the clustering shows the same type of signal, but the association of metal-poor blue GCs with the UCDs is stronger than for the red ones.

One explanation for the clustering of GCs in the vicinity of UCDs could be that UCDs are the nuclei of an ancestor dwarf galaxy that possessed its own GC system. The detected clustering would then constitute the remains of the GC population of the initial galaxy before it was affected by stripping. This GC clustering is also expected in the case when globular clusters merge towards the center of their host galaxy via dynamical friction. As the observed clustering is strongest for the blue GC subpopulation, this supports the view that some UCDs with blue companions are the stripped nuclei of metal-poor (dwarf) galaxies.

A second explanation for the clustering of GCs around UCDs could be that these UCDs were formed in a large star cluster complex. This complex then merged to form the UCD (Fellhauer & Kroupa 2002). The observed overdensites of star clusters around UCDs could then be the remains of this process before a fully merged object is created. The simulations of Brüns *et al.* (2011) predict that these large star cluster complexes have typical merging timescales of 1 Gyr. In some cases though, GC substructure around the central object is still detectable after 5 Gyrs. Thus, if a UCD and its surrounding GCs are formed by super star clusters, we expect young ages and high metallicities.

We conclude that UCDs can be distinguished into two different populations: 1) UCDs that harbour a population of close-by satellite point sources, most probably low mass star clusters; 2) UCDs that have the same statistical clustering properties as 'normal' globular clusters. Further details on this work can be found in Voggel, Hilker & Richtler (2015).

References

Bekki, K., Couch, W. J., Drinkwater, M. J., & Shioya, Y., 2003, *MNRAS*, 344,399

Brüns, C., Kroupa, P., Fellhauer, M., Metz, M., & Assmann, P. 2011, *A&A*, 529, A138

Dirsch, B., Richtler, T., Geisler, D., *et al.*, 2003, *AJ*, 125, 1908

Fellhauer, M., & Kroupa, P., 2002, *MNRAS*, 330, 642

Hilker, M., Infante, L., Vieira, G., Kissler-Patig, M., & Richtler, T., 1999, *A&AS*, 134, 75

Murray, N., 2009, *ApJ*, 691, 946

Voggel, K., Hilker, M., & Richtler, T. 2015, *submitted to A&A*

The General Assembly of Galaxy Halos: Structure, Origin and Evolution
Proceedings IAU Symposium No. 317, 2015
A. Bragaglia, M. Arnaboldi, M. Rejkuba & D. Romano, eds.

doi:10.1017/S1743921315008522

Age-metallicity-velocity relation of stars as seen by RAVE

Jennifer Wojno[1], Georges Kordopatis[1], Matthias Steinmetz[1], Gal Matijevič[1], Paul J. McMillan[2], the RAVE Collaboration

[1]Leibniz-Institut für Astrophysik Potsdam (AIP),
An der Sternwarte 16, 14482 Potsdam, Germany

[2]Lund Observatory, Lund University, Department of Astronomy and Theoretical Physics,
Box 43, SE-22100 Lund, Sweden

Abstract. Throughout the past decade, significant advances have been made in the size and scope of large-scale spectroscopic surveys, allowing for the opportunity to study in-depth the formation history of the Milky Way. Using the fourth data release of the RAdial Velocity Experiment (RAVE), we study the age-metallicity-velocity space of $\sim$100,000 FGK stars in the extended solar neighborhood in order to explore evolutionary processes. Combining these three parameters, we better constrain our understanding of these interconnected, fundamental processes.

Keywords. stars: ages, stars: abundances, stars: kinematics and dynamics, Galaxy: formation

1. Introduction

Stellar ages are a crucial component of exploring the evolutionary history of the Milky Way. However, there are a number of obstacles facing reliable age estimations, as stellar ages are degenerate with a number of other factors in modern evolutionary models (e.g. metallicity, extinction, evolutionary tracks, see Soderblom 2010). Distances and age estimates for field stars were included in the fourth data release of RAVE (DR4) (Kordopatis *et al.* 2013). Binney *et al.* (2014) describe the method used to derive these parameters, and provide a comparison of the determined distances for a sample of stars with parallaxes measured by Hipparcos. We assess the validity of these ages, by generating a mock RAVE catalog using the stellar population synthesis code *Galaxia* (Sharma *et al.* 2011). This mock catalog is compared to a sample of RAVE stars which have been selected using a number of quality criteria (i.e., signal-to-noise ratio of the spectra, pipeline convergence, single stars only).

2. Trend Computations and Interpretation

Age-Metallicity Relation. First, we investigate the age-metallicity relation (AMR) for our RAVE sample. The distribution was modeled using a simple piecewise linear function, and fit to the data taking into account the uncertainties in both age and metallicity. The resulting fits are shown in Fig. 1. For the RAVE sample, the AMR for stars younger than $\sim$7 Gyr is flat, while there is a clear correlation for older stars. The 'knee' of the distribution occurs at $\sim$6.5 Gyr, however, it is important to note that the position of this knee is heavily influenced by biases introduced by the age determination method. These trends are also seen in the mock RAVE catalog, and agree with trends shown for similar samples of local solar neighborhood stars (d < 200 pc), such as in Haywood *et al.* (2013) and Bensby *et al.* (2014).

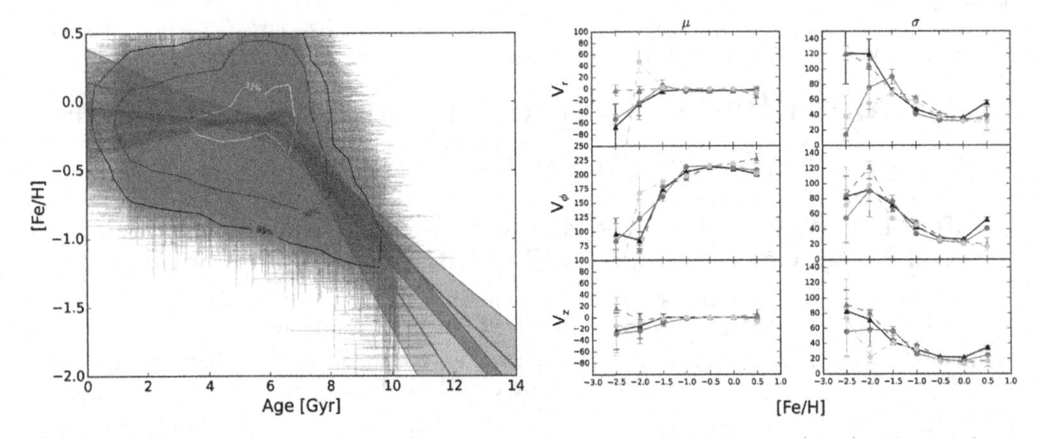

Figure 1. Left: Age v. [Fe/H] distribution for a sample of RAVE stars (grey). The resulting fit and uncertainties are shown in red. This same model was fit to the mock RAVE catalog generated by the *Galaxia* code, and the resulting fit is shown in green.
Right: Mean velocities (left) and velocity dispersion (right) for each velocity component as a function of metallicity. RAVE trends are indicated by the solid lines, and the mock RAVE catalog by dashed lines. In each panel, dwarfs are indicated by red circles, and giants by black triangles.

Velocity-Metallicity Relation. In order to ensure that the sample consists of only bound stars, we limited the selection to those stars with $V_{tot} < V_{esc}$, where we adopted the lower limit for the Galactic escape speed determined in Piffl *et al.* (2014) of 492 km s^{-1}. The resulting mean velocities and velocity dispersions are shown in Fig. 1. The velocity dispersions decrease with increasing metallicity, although the most metal-poor stars deviate greatly from the general trend. However, these metallicity bins are sparsely populated, and likely suffer from small number statistics. In each of the components, there is also a distinct increase in velocity dispersion for the most metal-rich stars.

3. Conclusions

From the comparisons shown in Fig. 1, it is clear that the age estimates in RAVE DR4 are useable for future analysis. We recover physical phenomena (such as asymmetric drift) present in the velocity-metallicity space of the RAVE data with the mock catalog, as well as match the trends in the velocity dispersions as a function of metallicity with those as a function of age, indicating a correlation. For future studies, the RAVE sample will be divided into age bins, in order to better explore the parameter space for a given mono-age population, where these age estimates may be better constrained by including photometric measurements in addition to the stellar parameters derived from spectra (Wojno *et al.*, in preparation). Furthermore, the procedure outlined in Binney *et al.* (2014) will be updated and improved; in particular, the priors used in determining ages, which may influence any biases in the currently observed trends, will be relaxed.

References

Bensby, T., Feltzing, S., & Oey, M. S. 2014, *A&A*, 562, 71
Binney, J., Burnett, B., Kordopatis, G., *et al.* 2014, *MNRAS*, 437, 351
Haywood, M., Di Matteo, P., Lehnert, M., Katz, D., & Gómez, A. 2013, *A&A*, 560, 109
Kordopatis, G., Gilmore, G., & Steinmetz, M. *et al.* 2013, *AJ*, 146, 134
Piffl, T., Scannapieco, C., Binney, J., *et al.* 2014, *A&A*, 562, 91
Sharma, S., Bland-Hawthorn, J., Johnston, K. V., & Binney, J. 2011, *ApJ*, 730, 3
Soderblom, D. R. 2010, *ARAA*, 48, 581

The General Assembly of Galaxy Halos: Structure, Origin and Evolution
Proceedings IAU Symposium No. 317, 2015
A. Bragaglia, M. Arnaboldi, M. Rejkuba & D. Romano, eds.

doi:10.1017/S1743921315006729

The LAMOST Complete Spectroscopic Survey of Pointing Area at Southern Galactic Cap

Hong Wu[1], Ming Yang[1,2], Man I Lam[1,3], Fan Yang[1], Chao-Jian Wu[1], Tian-Wen Cao[1] and LAMOST Collaboration

[1]Key Laboratory of Optical Astronomy, National Astronomical Observatories, Chinese Academy of Sciences, 20A Datun Road, Chaoyang District, Beijing, 100012, China
email: hwu@bao.ac.cn
[2]IAASARS, National Observatory of Athens, Vas. Pavlou & I. Metaxa, Penteli 15236, Greece
email: myang@noa.gr
[3]Shanghai Astronomical Observatory, Chinese Academy of Sciences, Shanghai 200030, China

Abstract. The LAMOST Complete Spectroscopic Survey of Pointing Area (LCSSPA) at Southern Galactic Cap (SGC), is one of the LAMOST Key Project, designed to complete the spectroscopic observations of all Galactic and extra-galactic sources in two selected fields of 20 degrees^2 at SGC, with the limiting magnitude of $r = 18.1$ mag. The main purposes of the project are focused on the completeness of the LAMOST ExtraGAlactic Surveys (LEGAS), testing the selection methods of galaxies and stars, and obtaining the basic performance parameters of the LAMOST telescope. Meanwhile the scientific studies include galaxies, clusters of galaxies, variable sources (quasars and variable stars), infrared excess stars and luminous infrared galaxies. The project has considerable scientific value since it is the most complete spectral data in LEGAS up to now. The project completed its observation in the early of 2014, and obtained at least 5000 spectra of galaxies and 25000 spectra of stars.

Keywords. catalog, surveys, galaxies: distances and redshifts, galaxies: clusters: general

1. Overviews of the LCSSPA Project

As an important part of LAMOST scientific survey, LAMOST ExtraGAlactic Surveys (LEGAS) aims to take hundreds of thousands of spectra for extra-galactic objects over 8000 degrees^2 of the Northern Galactic Cap (NGC) and 3500 degrees^2 of the Southern Galactic Cap (SGC) in five years. As a part of the LEGAS, the LCSSPA project focuses on the completeness of the galaxy survey. It also provides an opportunity to carry out different scientific studies from stars to galaxies, such as galaxies, clusters of galaxies, variable sources (quasars and variable stars), infrared excess stars and luminous infrared galaxies, etc.

Two selected fields of 20 degree^2 are located at $R.A. = 37.88°$, $Dec. = 3.44°$ (hereafter Field A) and $R.A. = 21.53°$, $Dec. = -2.20°$ (hereafter Field B) at SGC, and represent relatively lower and higher density regions of galaxies, based on the distribution of galaxy clusters of Abell rich galaxy clusters catalog (Abell *et al.* 1989). The fields are also located in the region of the South Galactic Cap u-band Sky Survey (SCUSS, Zou *et al.* 2015).

The targets are constituted by all the stars and galaxies from SDSS DR9 photometric data with r-band magnitude limit of 18.1, quasar candidates selected by the optical-infrared color-color criteria (Wu & Jia 2010, Wu *et al.* 2012), and u-band variable candidates selected by SDSS and SCUSS u-band catalogues.

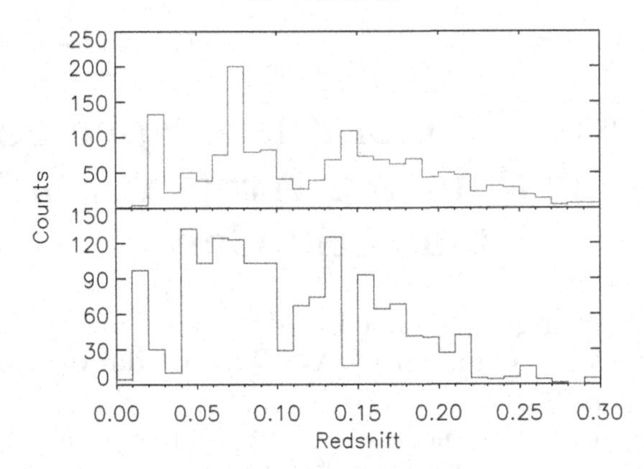

Figure 1. The redshift distributions of Field A (top) and B (bottom).

2. Observations and Data Reduction

The observations began in September 2012 and finished in January 2014. All the objects were divided into bright (B, $r = 14.0 \sim 16.0$ mag) and faint (F, $r = 16.0 \sim 18.1$ mag) plates. The typical exposure times for B and F plates are 3×600 s and 3×1800 s, and vary according to the weather condition and telescope performance. The galaxies with low S/N ratio, quasars and u-band variable sources are observed repeatedly, from twice to eleven times. We observed 2447 (out of 2519) galaxies in Field A and 2995 (out of of 3104) galaxies in Field B.

The raw data have been reduced with LAMOST 2D and 1D pipelines (Luo *et al.* 2004, 2012, 2014). To get reliable redshifts, an additional process is applied to 1D spectra. All the emission and absorption lines are checked and identified visually. The final uncertainties in redshifts are less than 0.001.

3. Preliminary Results

In total, for 1528 of 2447 observed galaxies ($\sim$62%) in Field A and 1573 of 2995 observed galaxies in Field B ($\sim$53%), we obtain reliable redshifts, ranging from 0.001 to 3.148. The distributions of redshifts (less than 0.3) of the two fields are shown in Figure 1.

Based on these data, we have obtained some preliminary results. First, we discovered 64 new LIRGs in LCSSPA Field A (Lam *et al.* 2015) and analyzed their infrared properties. Secondly, using the Gaussian Mixture Model, we determine the members of galaxy cluster Zwicky 721 and present its abnormal structure (Yang *et al.*, in progress). Thirdly, we identified spectra of u-band variable objects at LCSSPA (Cao *et al.*, in progress).

References

Abell, G. O., Corwin, H. G., Jr., & Olowin, R. P. 1989, *ApJS*, 70, 1
Lam, M. I., Wu, H., Yang, M., *et al.* 2015, *RAA*, 15, 1424
Luo, A.-L., Zhang, Y.-X., & Zhao, Y.-H. 2004, *SPIE*, 5496, 756
Luo, A.-L., Zhang, H.-T., & Zhao, Y.-H., *et al.* 2012, *RAA*, 12, 1243
Luo, A., Zhang, J., Chen, J., *et al.* 2014, *IAUS*, 298, 428
Wu, X.-B. & Jia, Z. 2010, *MNRAS*, 406, 1583
Wu, X.-B., Hao, G., Jia, Z., Zhang, Y., & Peng, N. 2012, *AJ*, 144, 49
Zou, H., Wu, X. B., & Zhou, X. *et al.* 2015, *PASP*, 127, 94

The General Assembly of Galaxy Halos: Structure, Origin and Evolution
Proceedings IAU Symposium No. 317, 2015
A. Bragaglia, M. Arnaboldi, M. Rejkuba & D. Romano

doi:10.1017/S1743921315006584

A catalog of M-type star candidates in the LAMOST data release 1

Jing Zhong[1], Sébastien Lépine[2], Jing Li[1], Li Chen[1], Jinliang Hou[1]

[1]Key Laboratory for Research in Galaxies and Cosmology, Shanghai Astronomical Observatory, Chinese Academy of Sciences, 80 Nandan Road, Shanghai, China
email: jzhong@shao.ac.cn

[2]Department of Physics & Astronomy, Georgia State University, 25 Park Place, Atlanta, GA 30303, USA

Abstract. In this work, we present a set of M-type star candidates selected from the LAMOST DR1. A discrimination method with the spectral index diagram is used to separate M giants and M dwarfs. Then, we have successfully assembled a set of M giants templates from M0 to M6, using the spectra identified from the LAMOST spectral database. After combining the M dwarf templates in Zhong *et al.* (2015a) and the new created M giant templates, we use the M-type spectral library to perform the template-fit method to classify and identify M-type stars in the LAMOST DR1. A catalog of M-type star candidates including 8639 M giants and 101690 M dwarfs/subdwarfs is provided. As an additional results, we also present other fundamental parameters like proper motion, photometry, radial velocity and spectroscopic distance.

Keywords. stars: fundamental parameters — stars: late-type — catalogs — surveys

1. Overview of the M-type spectral templates

In our previous work (Zhong *et al.* 2015a), a set of M dwarf templates has been developed as references for automatically identifying and classifying the M dwarfs in the LAMOST spectroscopic data. With these well defined M dwarf templates, the template-fit method was used to determine the spectral type of LAMOST stars. As we described in Zhong *et al.* (2015a), although our M dwarf templates provide a reliable estimate of spectral classification, a fraction of M giants are mis-classified because we lack M giant templates in our spectral library. To solve this problem, we created a new M-type spectral templates library by combining the M dwarf/subdwarf templates and the M giants templates.

Since the surface gravity is totally different for giants and dwarfs, one can use the spectral features as gravitational indicators to determine the luminosity classes. First, we used the template-fit method to select M-type spectra which positively present the characteristic molecular features. Then the spectral indices of TiO5, CaH2 and CaH3, as defined by Reid *et al.* (1995) and Lépine *et al.* (2007), were calculated. Figure 1 shows the spectral indices diagram for all M type stars we identified in the LAMOST DR1. Two populations are clearly distinguishable in this spectral indices diagram. Giants with weaker CaH molecular bands are located on the upper branch, which is consistent with the giant/dwarf discrimination by Mann *et al.* (2012).

As shown in Covey *et al.* (2007), the SDSS $r - i$ color for late-type stars has shown good relationship with the Morgan-Keenan (MK) spectral subtypes, which spans about 2 mag from M0 to M10. To provide spectral subtypes along the temperature sequence for M giants, we choose the SDSS $r - i$ color as an indicator to classify M giant subtypes. Then, a set of criteria were adopted to select high quality LAMOST spectra as training spectra for each spectral subtype grid. Finally, approximately 200 high quality giant

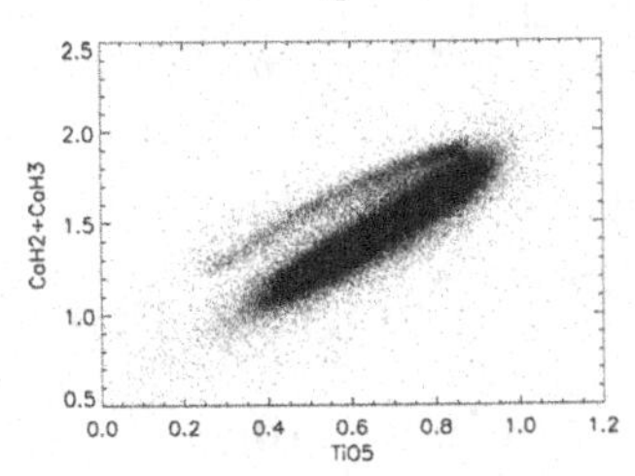

Figure 1. The M type stars distribution in the CaH2+CaH3 against TiO5 diagram. Two branches in this diagram clearly indicate the two populations. Because of the weaker CaH molecular bands, about 10000 M giants are located in the upper branch. Comparison with previous results, Lépine *et al.* (2007) shows that the stars distributed in the lower branch are mainly M dwarfs/subdwarfs. The clear separation of different populations in this diagram indicate the great potential of using spectral indices to distinguish M giants and M dwarfs.

spectra with good photometry in SDSS were left as the training spectra to assemble a grid of temperature sequence. Since in our sample there is no giant candidate with $r - i$ color greater than 2.0 mag, the synthetic M giant templates span the spectral subtypes from M0 to M6.

To correct the radial velocity for each training spectrum, we manually used the IRAF package to measure the wavelength correction to the zero-velocity rest-frame. For each spectral subtype bin, at least five training spectra were combined to create the synthetic template spectra.

In the whole M-type templates, there are M dwarf templates with temperature from K7.0 to M8.5 and metallicity from dMr to usdMp, and the M giant templates from M0 to M6. The total number of M type templates is 223.

2. Results

Based on the M-type templates, we re-run our spectral classification pipeline (Zhong *et al.* 2015a) to automatically identify and classify M-type stars with spectra from the LAMOST DR1 data. In the classification pipeline, the template-fit method is used by calculating the chi-square values between the LAMOST spectrum and each of the template spectra. Then, the template spectrum which has the minimum chi-square value is considered as the best-fit, and its spectral subtype is used to mark the corresponding LAMOST spectrum.

After passing through the 2,204,696 LAMOST DR1 spectra to our spectral classification pipeline, we identified 8639 M giants and 101,690 M dwarfs/subdwarfs. The excluded spectra were marked as non-M type spectra of which most are earlier type objects like AFGK stars, and a small fraction of spectra were too noisy to be classified.

Finally, we have successfully assembled a set of M giant templates from M0 to M6 by using the LAMOST DR1 spectra. The M-type star candidates are cataloged in Zhong *et al.* (2015b). We present the information of celestial coordinates, JHK$_s$ infrared magnitudes in 2MASS, spectral subtypes, radial velocity and derived spectroscopic distance.

References

Covey, K. R., Ivezić, Ž., Schlegel, D., *et al.* 2007, *AJ*, 134, 2398
Lépine, S., Rich, R. M., & Shara, M. M. 2007, *ApJ*, 669, 1235
Mann, A. W., Gaidos, E., Lépine, S., & Hilton, E. J. 2012, *ApJ*, 753, 90
Reid, I. N., Hawley, S. L., & Gizis, J. E. 1995, *AJ*, 110, 1838
Zhong, J., Lépine, S., Hou, J., *et al.* 2015, *AJ*, 150, 42
Zhong, J., Lépine, S., Li, J., *et al.* 2015, *RAA*, 15, 1154

Author index

Printed in the United States
by Baker & Taylor Publisher Services